Vibration Control Systems

Scrivener Publishing
100 Cummings Center, Suite 541J
Beverly, MA 01915-6106

Publishers at Scrivener
Martin Scrivener (martin@scrivenerpublishing.com)
Phillip Carmical (pcarmical@scrivenerpublishing.com)

Vibration Control Systems

Kundan Meshram

and

Prashant Lahre

Scrivener Publishing

WILEY

This edition first published 2026 by John Wiley & Sons, Inc., 111 River Street, Hoboken, NJ 07030, USA and Scrivener Publishing LLC, 100 Cummings Center, Suite 541J, Beverly, MA 01915, USA

For more information about Scrivener publications please visit www.scrivenerpublishing.com.

Wiley Global Headquarters
111 River Street, Hoboken, NJ 07030, USA

For details of our global editorial offices, customer services, and more information about Wiley products visit us at www.wiley.com.

The manufacturer's authorized representative according to the EU General Product Safety Regulation is Wiley-VCH GmbH, Boschstr. 12, 69469 Weinheim, Germany, e-mail: Product_Safety@wiley.com.

AI disclosure statement: The author acknowledges the use of AI-based tools for minor language editing or grammatical correction. However, all intellectual contributions, analyses, and conclusions presented in this chapter are entirely the author's own.

Library of Congress Cataloging-in-Publication Data

ISBN 9781394358250

Cover image: Generated with AI using Adobe Firefly
Cover design by Russell Richardson

Set in size of 11pt and Minion Pro by Manila Typesetting Company, Makati, Philippines

Printed in the USA

10 9 8 7 6 5 4 3 2 1

Contents

Preface

This book, *Vibration Control Systems,* explores structural vibration control by combining important theories with new technology and integration methods. The material begins with teaching about Mass-Spring Dampers, Hydraulic Dampers, Tuned Mass Dampers, and Liquid Circular Angular Acceleration Systems, helping readers see their role in vibration control.

The book then covers techniques for combining damping systems, which are important for stabilizing complex civil structures. PSO and ANN tools are used to design optimized systems and better implement predictive control. Using PSO, ANN, and Transient Flow Models together is presented as a new way to improve seismic resilience.

Other chapters describe methods that use Remote Sensing Devices for instant monitoring, allowing control and repair as soon as they are needed. Tying together traditional approaches and AI tools, this book becomes useful for students, researchers, and engineers willing to improve structural safety and performance.

Kundan Meshram
Prashant Lahre
Bilaspur, India

1

Introduction to Vibration Engineering

Abstract

Engineering design together with structural and machinery maintenance and safety functions heavily depend on vibration engineering expertise across multiple engineering areas. The article examines vibration engineering essentials by categorizing various vibrations as free and forced and deterministic with random components. Structures and safety standards can be seriously impaired by vibrations without proper control. Different damping systems including both viscous and frictional elements function as essential components for vibration reduction. The research area now examines two superior damping approaches: Mass Spring Damper (MSD) systems and Liquid Circular Angular Accelerometers (LCAA).

***Keywords*:** Damper system (DS), vibration, mass spring damper (MSD), liquid circular angular accelerometers (LCAA)

1.1 Vibration Engineering

Engineering sectors frequently encounter vibration which describes the periodic motion of mechanical frameworks around their stable positions. The phenomenon presents itself through several expressions which include suspension bridge sway and intricate machinery vibrations [1]. The basic principles behind these vibrations hold major importance for both engineering design procedures and maintenance activities. Engineering systems exhibit four different vibration types that include independent free vibrations as well as forced vibrations with external triggering and predictable patterns and unpredictable random vibrations which stem from stochastic forces or natural environmental influences. The classification system forms the starting point to evaluate and control vibrations within engineering structures [2]. Engineering structures along with their mechanisms experience significant adverse effects when vibrations remain unmanaged. Without vibration control, the structural components face serious damage due to fatigue along with diminished operational performance and

Kundan Meshram and Prashant Lahre. Vibration Control Systems, (1–14) © 2026 Scrivener Publishing LLC

enhanced maintenance expenses and lower safety conditions. Aerospace industry components demonstrate premature destruction when vibrations remain unmanaged which results in potential catastrophic events. Civil engineering projects experience deteriorating infrastructure because of bridge and building vibrations that occur chronologically [3]. Knowledge about vibration together with its effects demonstrates why adequate vibration control systems must be implemented.

Engineering systems require damping systems to properly manage and reduce vibrations as their central control method. This technology operates by converting both kinetic and potential energy from vibrations into dissipated forms which decrease oscillation amplitude levels. The field of engineering makes use of multiple damping system designs that serve different purposes [4]. Viscous damping operates as an energy dissipation mechanism that transforms kinetic movement energy into thermal vitality inside a viscoelastic substance. The second method of damping utilizes frictional forces to produce resistance that dissipates energy. Structural damping involves the internal damping within a material due to its inherent properties. These mechanisms serve as the foundation upon which more advanced damping systems are developed and integrated into diverse engineering applications.

The selection and implementation of appropriate damping systems are integral in the design and maintenance of engineering systems as shown in Figure 1.1. By strategically incorporating damping systems into the design phase, engineers enhance the performance, longevity, and safety of

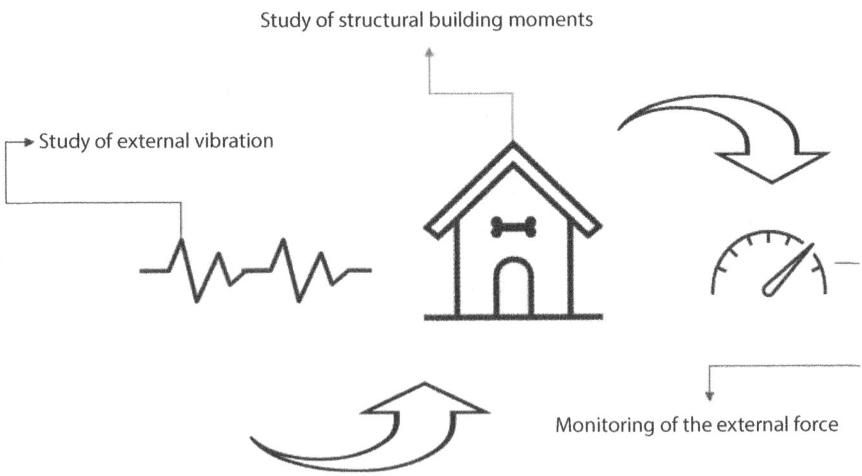

Figure 1.1 Study of vibration assessment.

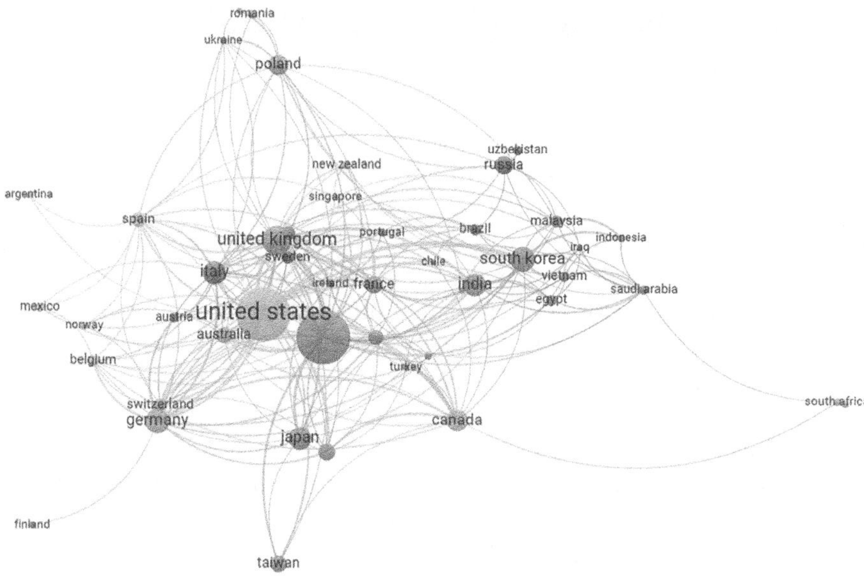

Figure 1.2 Bibliography shows the study about vibration-damping systems country-wise.

structures and machinery [5]. When addressing vibrations during design, it becomes possible to develop specialized solutions that minimize potential problems effectively. Optimal system performance depends heavily on damping systems because engineering systems continue to grow more complex and demanding. The core concepts and system designs with integration approaches of particular damping control technologies which encompass Mass Spring Damper systems and Liquid Circular Angular Accelerometers and Hydraulic Dampers and Tuned Mass Dampers as well as additional superior damping methods. The main objective is to provide extensive knowledge regarding engineering vibration control alongside an understanding of multiple damping systems involved in this process. The country-wise vibration-damping sytems is shown in Figure 1.2.

Engineering professionals specializing in vibration study how to analyze and manage mechanical and structural system vibrations. This field regulates the analysis of vibrations in both mechanical equipment and large civil engineering projects ranging from vehicles to buildings and bridges to dams [6]. The application of vibration engineering within civil engineering structures becomes essential since it directly affects the structural integrity as well as safety together with operational capabilities.

1.2 How Vibration Engineering is Useful in Civil Engineering

Engineering success rests on structural integrity and safety principles which guarantee dependability and operational quality of different structures alongside systems. Structures need precise engineering plans plus complete testing before they can stay safe through professional ongoing monitoring. Structural integrity describes an all-encompassing evaluation of how structures perform when loaded during their operational lifespan to withstand failure and stress as well as deformation [7]. Safety engineering enables the elimination of hazards while simultaneously preventing risks and potential damage to human beings as well as properties and environmental elements. Safety protocols exist to minimize potential risks and protect against accidental occurrences that can result in either injuries to people or deaths or environmental damage.

Safety measures merge directly with structural integrity requirements in industries like civil engineering along with aerospace and automotive and manufacturing since structural failures would result in serious risks. The civil engineering sector needs to maintain bridge and dam and building structural integrity because failure would endanger people and disrupt neighborhoods. Aerospace operations depend on strict safety regulations together with component integrity examinations for preventing aircraft component failures which endanger passenger safety [8]. The preservation of structural integrity and safety happens through the strong role played by regulatory bodies with industry standards supporting them. Regulatory bodies together with industry standards set mandatory guidelines and performance standards while industry-specific codes that engineers and their sectors must follow to achieve basic safety thresholds. Through vibration engineering,

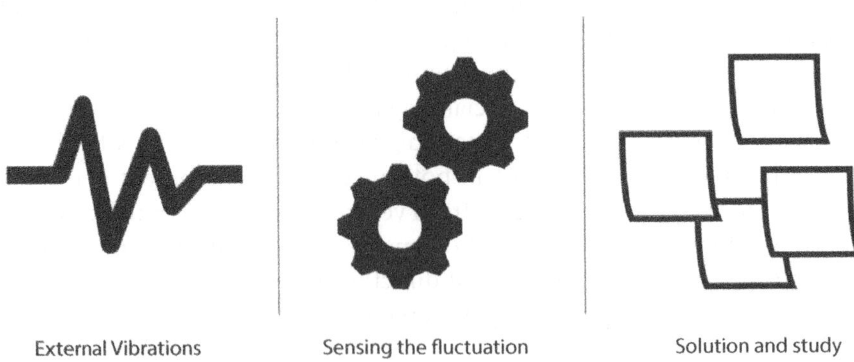

Figure 1.3 External vibrations sensing in civil engineering.

one can determine how various dynamic loads like wind, earthquakes, and machinery vibrations impact the state of structures as shown in Figure 1.3. Engineering analysis helps predict the natural vibrational frequencies to identify where resonances that could lead to structural damage may occur.

1.3 Design Optimization and Performance Enhancement

Applied design optimization and performance enhancement present a fundamental engineering innovation that focuses on systematic improvement of functionality along with efficiency for multiple systems and products and structures. The comprehensive process requires specific combinations of innovative thinking and scientific methods and technological improvements to achieve maximum results in different fields of application. Design optimization operates through a well-organized method to enhance and refine products and systems alongside their operation methods. The main goal of this process involves achieving the highest possible level of desired features accompanied by reduced inefficiencies along with minimized destructive elements. Engineers conduct an iterative process that includes computational modeling and simulation teaming up with advanced algorithms to analyze extensive design spaces before they can locate the best suitable configuration that meets their predefined requirements [9]. The augmentation of different metrics and attributes which boost system functioning and efficiency defines performance enhancement.

a. The use of sophisticated programming software and algorithmic systems allows the assessment of system and structural behaviors under different operating conditions through computational modeling. The method expedites fast evaluation of numerous design alternatives for determining superior configurations.
b. Multidisciplinary Optimization facilitates different engineering professions (mechanical and electrical materials) to create a whole design optimization by handling multiple performance benchmarks and limitations together.
c. Research on materials through Material Science and Engineering leads to either the creation of new materials or improved versions with enhanced features for weight-strength properties and resistance properties and heat conduction and flexibility to advance system efficiency.

Several industries actively work to achieve optimizations of their designs while enhancing their operational performances. The automotive industry needs engineers to create stronger materials which also weigh less in order to achieve better fuel efficiency together with safer vehicles. System optimization studies done on renewable energy devices work to boost both performance capture and operational efficiency for solar panels and wind turbines. The optimization process for structural design that incorporates natural frequencies together with damping ratios and dynamic force response uses principles from vibration engineering [10]. Such design methods produce stabilized structures which ensure strength during contact with external forces. Excessive vibrations that occur in constructed buildings and bridges produce negative effects on the comfort levels of those who occupy or utilize them. Design procedures that employ vibration engineering produce structures that reduce perceptible vibrations to improve human comfort and usability. The optimization process of structural design requires implementation of revolutionary strategies combined with modern technologies to reduce vibration effects which improves structure's operational efficiency and resilience.

1.4 Structural Design Tailored for Seismic Vibration Control

The method provides complete insight into the propagation of vibrations across the system which helps locate crucial zones for assessment. Vibration control depends heavily on installing damping systems inside structural buildings. Engineers strategically combine designed damping mechanisms such as tuned mass dampers, viscoelastic materials, and active vibration control systems for purposes of vibration dissipation and attenuation to minimize their harmful effects. Engineers use particular polymers and composite materials as an intrinsic method to reduce vibrations because these materials possess built-in damping properties [11]. Modification of material properties enables engineers to lower natural frequencies while improving damping properties which leads to vibration control systems. Engineers make structural modifications or redesigns as a method to modify natural frequencies and strengthen critical points and implement vibration-dampening elements. The addition of braces or modification of the shape changes how the structure vibrates. Linking sensitive structures and systems to vibration sources by installing isolation mounts with flexible connectors or vibration isolation pads decreases vibration movement to protect vital built-in components. These systems modify the structural reaction by making automatic adjustments while vibrations take place. Transformation of these solutions needs deep knowledge about structural dynamics together with precise vibration patterns found in individual environments. Civil engineering projects that include tall buildings or bridges

Table 1.1 The differences between buildings designed to withstand external vibrations and non-vibration structural.

Buildings designed for external vibrations	Buildings for non-vibration structural purposes
Emphasizes resilience against external forces like seismic activity, wind loads, or machinery-induced vibrations.	Primarily focuses on stability, safety, and functionality without special considerations for external vibrations.
Often utilizes damping systems (Tuned Mass Dampers, Hydraulic Dampers) or flexible materials to mitigate vibrations.	Typically uses conventional building materials without specific damping or vibration control mechanisms.
Incorporates dynamic analysis, modal testing, and dampening strategies during design to minimize resonance and oscillations.	Design focuses on static load-bearing capabilities and architectural aesthetics without specific vibration-damping tactics.
Includes Tuned Mass Dampers, Hydraulic Dampers, Vibration Isolators, and other vibration control systems as integral parts.	Lacks specialized dampening systems or vibration isolators within the structural design.
Ensures occupant comfort by reducing perceptible vibrations due to external sources, optimizing building usability.	Prioritizes stability and functionality without specific emphasis on minimizing perceptible vibrations.
Often integrates structural health monitoring systems to assess vibrations and structural integrity continually.	May incorporate basic monitoring but not specialized for vibration-induced assessments.
Commonly in high-rise buildings, aerospace, seismic-prone regions, and structures near heavy machinery or transportation hubs.	Typical in residential, commercial, or industrial buildings not exposed to significant external vibration sources.

require specific design approaches to combat wind-induced vibrations because this phenomenon threatens stability and safety according to [12]. The prevention of vibrations stands vital for industrial machinery operations because it helps to delay equipment degradation which keeps systems running efficiently Table 1.1 shows the Vibration and Non-Vibration structural building to study.

1.5 Risk Assessment and Monitoring

Through risk assessment and monitoring engineering along with management practices have adopted a fundamental conceptual change which develops structured techniques for determining and assessing along with minimizing risks during various operation cycles. Stakeholders use this method beforehand to detect and control unpredictable factors which threaten their targets or produce negative results. Proactive risk management relies on risk assessment as its essential foundation through which stakeholders conduct sequential evaluations of predicted risks including predicted outcomes. Proactive risk management requires a complex assessment process which includes hazard identification together with risk assessment and development of strategies for risk management or mitigation.

Risk assessment contains two main methodological approaches which use qualitative risk identification through expert assessment of data and quantitative systems which perform probabilistic model calculations and statistical analyses using historical information. Risk assessment tools help organizations establish priority rankings of their risks as well as enable them to identify critical risks which guide resource distribution for risk minimization strategies [13]. Risk assessment receives its complementary nature from monitoring due to its continuous surveillance and parameter measurement and analysis method which detects emerging risks and potential issues. Systems incorporating sensors along with data collection technology and analytical methods observe and track essential risk-related variables which affect assessed risks through their likelihood and impact indicators. Valid monitoring methods operate as real-time monitoring or periodic assessments according to the system requirements and the type of existing risk. Civil engineering utilizes structural health monitoring technology for non-stop observation of structural integrity factors to identify impending weaknesses so prevent catastrophic failures.

1.6 Integration in Various Domains

The process involves evaluating financial threats together with market instabilities and operational uncertainties for decision-making purposes that protect business interests. Performing risk assessments in health care

Table 1.2 Advantages and disadvantages of studying vibrations within engineering.

Advantages	**Disadvantages**
Enhanced Safety: Identifies potential structural weaknesses and helps prevent failures due to resonance or fatigue.	Complexity: Understanding and mitigating vibrations require advanced analysis and specialized knowledge.
Optimized Designs: Allows for the optimization of structural designs to withstand dynamic loads and improve performance.	Costly Solutions: Implementing vibration control systems or damping mechanisms is expensive.
Improved Durability: Mitigates structural damage caused by dynamic loads, extending the lifespan of engineering systems.	Increased Complexity in Design: Incorporating damping systems may complicate the architectural or engineering design.
Enhanced Efficiency: Enables the creation of more efficient mechanical systems and machinery.	Maintenance Requirements: Specialized systems might require regular maintenance, adding to operational costs.
Occupant Comfort: Minimizes perceptible vibrations, enhancing occupant comfort in buildings and vehicles.	Increased Design Time: Analyzing and designing for vibrations lengthen the design phase.
Predictive Maintenance: Allows for predictive maintenance by monitoring and detecting potential failures in advance.	Limited Applicability: Not all structures or systems require extensive vibration analysis.
Research and Innovation: Drives innovation in materials, control systems, and design techniques.	Limited Accessibility: Advanced analysis tools might not be accessible to all engineers or industries.

facilities protects both patient safety and workplace operational security. Risk managers evaluate how climate-related aspects and pollution along with environmental transformations should be managed to avoid ecological damage. Engineering projects together with structural design and infrastructure development require risk identification to guarantee structure and system reliability and safety and extend their useful life. Risk assessment integration with monitoring creates organizations that proactively manage threats and successfully distribute funds while implementing preventive methods to suppress risks before they materialize [14]. The approach enables stakeholders to make well-informed choices that strengthen operational resilience and project/system and operational longevity within constantly changing uncertain environments, Table 1.2 shows the advantages and disadvantages of the vibration study.

1.7 Environmental Impact and Regulations: Vibration Control Measures

Engineering projects must properly evaluate vibration control measures on the environment while following all seismic vibration control regulations specifically for sensitive structures and systems [15]. Users must comprehend the environmental impacts of vibration control methods and meet regulations meant to protect the environment when implementing these techniques as shown in Figure 1.4.

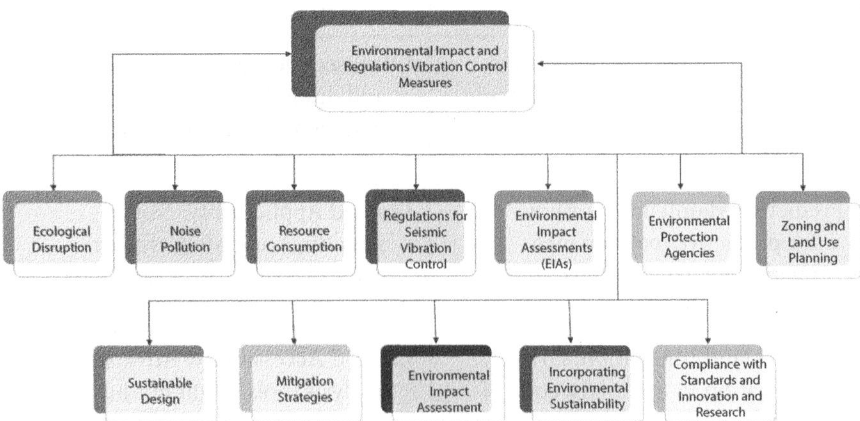

Figure 1.4 Environmental impact and regulations vibration control measures.

a. Ecological Disruption arises from vibration control mechanisms which include the installation of dampers and structural modifications because they affect local ecosystems. Installation of vibration control systems through heavy machinery disturbance during construction could result in habitat destruction which causes soil erosion and alters wildlife environments and natural habitats.
b. The implementation of certain vibration control techniques that use machinery or perform construction work can lead to substantial noise pollution the environment. The prolonged exposure to extreme noise pollutes wildlife ecosystems and threatens nearby communities' health and disrupts biological habitats.
c. Resources become depleted during the entire lifespan of vibration control systems because they need energy along with materials and substantial amounts of water to produce and maintain these systems. It is essential to evaluate environmental impacts during every phase of these systems to determine their total sustainability levels.
d. Seismic Vibration Control receives guidance through building codes and standards that exist in most countries. Such regulatory requirements outline necessary structural design aspects with specified materials together with building procedures meant to protect inhabitants and avoid significant property damage during seismic activity.
e. Organizations perform Environmental Impact Assessments (EIAs) before starting vibration control implementation to determine environmental consequences. Environmental inspectors assess damage to both natural ecosystems and the severity of acoustic pollution alongside resource management status.
f. Governments through zoning and land use planning authorities regulate the land utilization within seismic areas prone to earthquakes. The authorities set rules during zoning that specify construction types and methods alongside earthquake reinforcement standards to reduce seismic risks.
g. Environmental Protection Agencies serve as authority bodies to enforce compliance with all environmental rules and laws. Environmental agencies in the locality check construction sites to verify that vibration control systems protect the environment from further damage.

h. Engineering practices now focus more on sustainability as they implement environmental sustainability while achieving seismic regulations.
i. Project sustainability takes priority through eco-friendly vibration control systems which maintain energy efficiency throughout entire project duration.
j. Building projects use mitigation strategies that lower both ecological disturbances and wildlife interruptions through scheduling activities during low-activity periods and implementing noise reduction methods.
k. Environmental agencies in the locality check construction sites to verify that vibration control systems protect the environment from further damage.
l. Engineering practices now focus more on sustainability as they implement environmental sustainability while achieving seismic regulations.
m. Project sustainability takes priority through eco-friendly vibration control systems which maintain energy efficiency throughout the entire project duration.
n. Building projects use mitigation strategies that lower both ecological disturbances and wildlife interruptions through scheduling activities during low-activity periods and implementing noise reduction methods.

References

1. Trower, S., *Senses of vibration: A history of the pleasure and pain of sound*, A&C Black, New York, NY, USA, 2012.
2. Bachmann, H., *Vibration problems in structures: practical guidelines*, Springer Science & Business Media, Basel, Switzerland, 1995.
3. Korkmaz, S., A review of active structural control: challenges for engineering informatics. *Comput. Struct.*, 89, 23–24, 2113–2132, 2011.
4. Cao, M.S., Sha, G.G., Gao, Y.F., Ostachowicz, W., Structural damage identification using damping: a compendium of uses and features. *Smart Mater. Struct.*, 26, 4, 043001, 2017.
5. Taylor, D.P., Buildings: design for damping, in: *Proceedings of the Boston Society of Civil Engineers, BSCES, Lecture Series, "Dynamics of Structures" USA*, 1999.

6. Liu, P., Chen, A.Y., Huang, Y.-N., Han, J.-Y., Lai, J.-S., Kang, S.-C., Wu, T.-H., Wen, M.-C., Tsai, M.-H., A review of rotorcraft unmanned aerial vehicle (UAV) developments and applications in civil engineering. *Smart Struct. Syst.*, 13, 6, 1065–1094, 2014.
7. Plumbridge, W.J., Matela, R.J., Westwater, A., *Structural integrity and reliability in electronics: enhancing performance in a lead-free environment*, Springer Science & Business Media, Dordrecht, Netherlands, 2007.
8. Bahr, N.J., *System safety engineering and risk assessment: a practical approach*, CRC Press, Boca Raton, FL, USA, 2014.
9. Martins, J.R.R.A. and Ning, A., *Engineering design optimization*, Cambridge University Press, Cambridge, UK, 2021.
10. Genta, G., *Vibration of structures and machines: practical aspects*, Springer Science & Business Media, Dordrecht, Netherlands, 2012.
11. Chakraborty, B.C. and Ratna, D., *Polymers for vibration damping applications*, Elsevier, Amsterdam, Netherlands, 2020.
12. Xu, Y.-L., *Wind effects on cable-supported bridges*, John Wiley & Sons, USA, 2013.
13. Cooper, D., Grey, S., Raymond, G., Walker, P., *Project Risk management guidelines managing risk in large projects and complex procurements*, John Wiley & Sons, Hoboken, NJ, USA, 2005.
14. Heng, Y.-K., *War as risk management: strategy and conflict in an age of globalised risks*, Routledge, London, UK, 2006.
15. Wenzel, H. and Pichler, D., *Ambient vibration monitoring*, John Wiley & Sons, Hoboken, NJ, USA, 2005.

2

Fundamentals of Mass Spring Damper System and Liquid Circular Angular Accelerometer

Abstract

The subject explains basic principles of Mass Spring Damper (MSD) systems along with Liquid Circular Angular Accelerometers (LCAA) pertaining to vibration control and engineering dynamics. MSD systems create the basis for understanding mechanical vibrations by serving as essential blocks for vibration expertise. The following section explains the detailed operational principles of LCAAs through an analysis of their applications for high-precision angular acceleration measurements. A historical analysis of MSD systems combined with their application growth becomes part of the text together with explorations of MSD and LCAA implementation in civil engineering vibration control. This discussion addresses all aspects of system benefits and drawbacks together with simulation characteristics to provide total understanding about how these systems function in modern engineering designs for effective vibration control and motion stability.

***Keywords*:** Integration system, mass spring damper (MSD), mechanical vibration, system vibration

2.1 Introduction of MSD and LCAA

MSD systems create the basis for understanding mechanical vibrations by serving as essential blocks for vibration expertise. The following section explains the detailed operational principles of LCAAs through an analysis of their applications for high-precision angular acceleration measurements. A historical analysis of MSD systems combined with their application growth becomes part of the text together with explorations of MSD and LCAA implementation in civil engineering vibration control [1]. This discussion addresses all aspects of system benefits and drawbacks together with simulation characteristics to provide total understanding of

Kundan Meshram and Prashant Lahre. Vibration Control Systems, (15–26) © 2026 Scrivener Publishing LLC

how these systems function in modern engineering designs for effective vibration control and motion stability as shown in Figure 2.1.

Analysis of simple harmonic motion in MSD systems enables the understanding of advanced vibrational dynamics which appear in practical applications [2]. The authorization of differential equations together with mathematical models allows readers to understand how MSD systems respond under different forces and disturbances. Research focuses on both unforced free vibrations that occur without external forces together with forced vibrations that develop under external excitations. The text explains why resonance appears as a central phenomenon in MSD systems. The analysis of resonance plays a vital role in engineering applications because it relates system's natural frequency with external excitation frequency to avoid damaging structural vibrations. Proper damping ratios together with specific frequencies play an essential role in reducing resonance so systems operate without instability. Engineering practitioners use MSD systems in multiple domains where they implement them in mechanical suspension systems and shock absorbers as well as civil building and bridge structures. The text introduces readers to fundamental concepts about the mass-spring-damper principles. The study presents complete knowledge about elementary harmonic oscillation along with resonance effects and practical MSD system deployment in engineering practice. The obtained knowledge establishes foundations for investigating the advanced damping systems together with their design applications for successful vibration control.

The LCAA exists as a sophisticated instrument which measures angular accelerations in a highly accurate manner. This chapter delves into the intricate workings, applications, and significance of the LCAA in various engineering fields. The basis of LCAA operations depends on how liquids inside circular chambers react to detect angular accelerations. When the

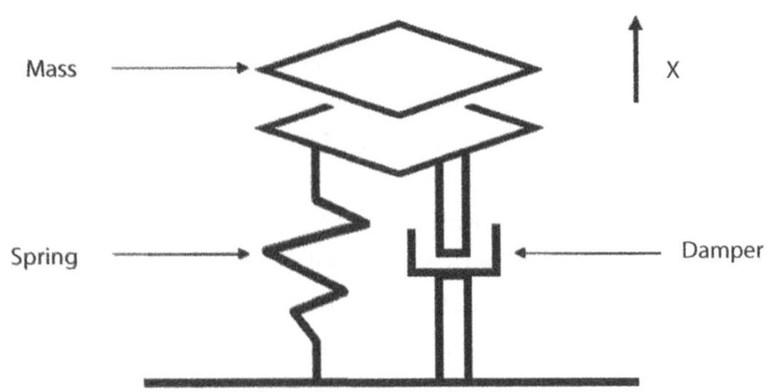

Figure 2.1 Mass spring damper system (MSD).

device undergoes angular acceleration the liquid inside redistributes itself because of the Coriolis Effect thus creating a detectable change in position. Knowing this principle enables complete understanding of how the LCAA operates. Aspects regarding the LCAA's individual components together with their design factors help understand the device's operational characteristics [3]. A carefully designed device features three main elements that include the circular liquid storage chamber and sensor systems for liquid displacement detection with subsequent displacement interpretation as angular acceleration data. Detailed analyses of LCA construction together with its materials and sensor technology demonstrate how they function during operation.

The LCAA functions by applying Coriolis Effect principles in a circular chamber that contains liquid as shown in Figure 2.2. Liquid within the chamber displaces throughout angular acceleration when the device experiences angular motion and the Coriolis force takes effect. The sensor measures detected displacement to calculate the angular acceleration of the device. The detailed discussion explains the foundation physics which investigates both the Coriolis Effect together with fluid dynamic behaviors of liquid molecules in the LCAA. A detailed analysis of LCAs shows how sophisticated the devices are designed to operate [4, 6]. The circular chamber houses specially designed liquid that enables precise measurement capability through its precise engineering specifications. The MEMS technology-based sensors measure the small movements that result from Coriolis force detection.

The essential role that LCAs play in aerospace engineering includes their function to support aircraft and spacecraft navigation systems and inertial

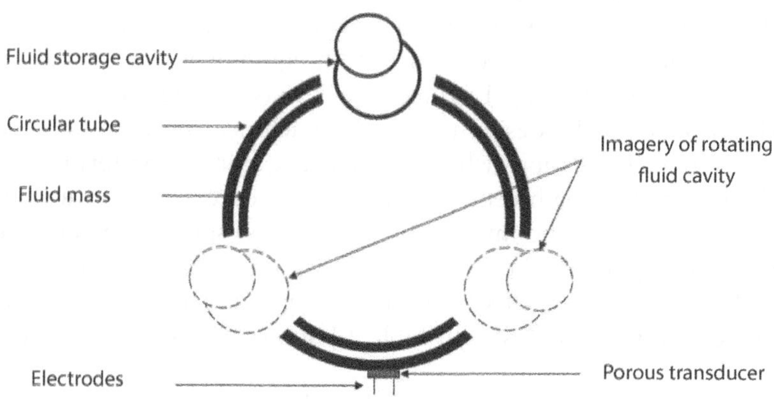

Figure 2.2 Liquid circular angular accelerometer (LCAA).

navigation stability control systems. The automotive sector uses them for stability control features in vehicles while robotic systems depend on them to sense motions precisely to maintain stability. LCAs demonstrate their ability to serve multiple technological applications through these practices. Furthermore, the chapter delves into the calibration, accuracy, and limitations of LCAs. The calibration method enables precise measurement and evaluations of accuracy levels provide data reliability information for these devices. Users need to understand and address the limitations of LCAs when using them because these sensors have restricted performance at various temperatures along with difficulties from external sources.

2.2 History of MSD and LCAA

The engineering community has been using mass-spring-damper systems to control vibrations in different structures since scientists and engineers began studying them centuries ago [5]. Initial principles of the mass-spring-damper system were discovered their scientific origins through early researchers and mathematicians. Harmonic motion theory and oscillations theory originated from Galileo Galilei and Christiaan Huygens and Isaac Newton through their respective discoveries. Studies on motion laws and pendulum behavior by these scientists enabled scientists to understand both oscillations and vibrations. During the 19th century scientists made groundbreaking discoveries about vibrations as well as damping procedures. Engineers together with scientists studied spring mechanisms and damping during this time frame for multiple purposes including vibration reduction within machinery and railroad car suspension development.

When the Industrial Revolution started the priority for machinery and building vibration control became essential. Engineers conducted research against oscillations which threatened both equipment operational safety and engineering efficiency. Engineers used mass spring damper systems widely throughout the design process of machinery because these components reduced vibrations while improving equipment performance and durability. The twentieth century witnessed an exceptional advancement in both development and implementation of mass spring damper systems throughout engineering practices. Modern engineering together with technological advancements allowed engineers to achieve exact designs for implementing these systems; Table 2.1 shows the aspects of MSD and LCAA.

Sensor designers have incorporated liquid-based systems as a method to improve angular acceleration measurement accuracy as well as sensitivity and reliability across multiple applications. From its historical beginnings, the LCAA followed the general patterns of inertial sensor evolution for specialized measurements within aerospace fields and navigation systems and structural engineering which requires accurate angular acceleration

Table 2.1 MSD and LCAA.

Aspect	Mass-spring-damper (MSD)	Liquid circular angular accelerometer (LCAA)
Function	Dampens structural vibrations by dissipating kinetic energy.	Measures angular acceleration in systems experiencing rotational motion.
Purpose	Reduces structural oscillations, enhancing stability.	Measures angular acceleration, providing insights into rotational motion.
Components	Mass (Inertia), Spring (Restoring Force), Damper (Energy Dissipation).	Sensors, Gyroscopes, Electronic Components for measuring angular motion.
Application	Commonly used in buildings, bridges, machinery for vibration control.	Utilized in navigational systems, vehicles to measure rotational dynamics.
Impact on Systems	Controls vibrations, improves structural stability, reduces resonance.	Provides data on rotational motion, aiding stability assessment.
Engineering Emphasis	Focuses on structural dynamics and minimizing vibrations.	Concentrates on understanding rotational dynamics and stability.
Mechanical Composition	Physical mass, elastic element, and energy dissipating mechanism.	Utilizes sensors and electronic components to measure angular motion.
Control Over System Vibrations	Actively or passively controls structural oscillations.	Observes and records angular motion for stability assessment.

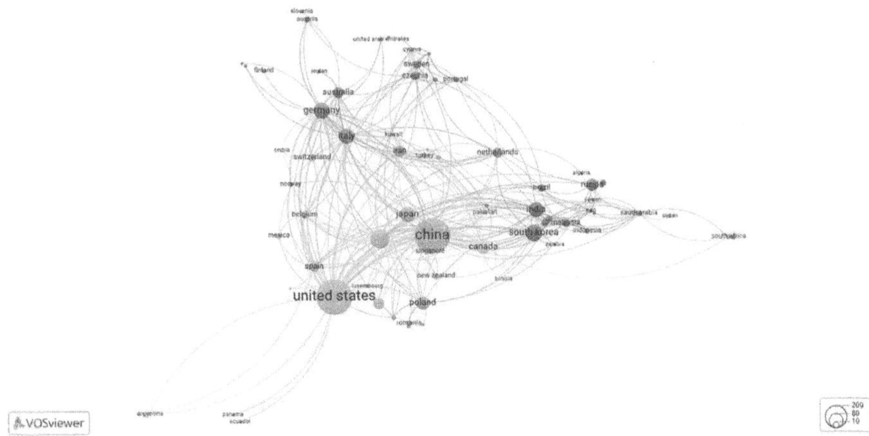

Figure 2.3 Bibliographic reviews.

measurement. The evolution of the LCAA as a sensor represents the wider historical development of sensor technology which culminated in accurate angular acceleration measurement for various industrial and scientific uses but lacks detailed documentation about its origins and creator. The country wise bibliographic reviews shown in Figure 2.3.

2.3 Uses of MSD and LCAA in Civil Engineering for Vibration Control

Mass Spring Damper systems serve as fundamental engineering mechanisms which reduce vibrations by finding civil engineering applications for structural stability control, as well as earthquake protection purposes and comfort maximization. MSD systems installed in earthquake-prone buildings serve to absorb seismic energy that minimized ground motion effects. These systems manage building oscillations and help lower earthquake-related structural damage for both safety purposes and structural integrity needs [7]. Bridges face vibrations from three main sources which include traffic loads, wind and seismic events. Building structures that reach great heights experience wind-induced vibrations that pose dangers to user comfort and create stress on the framework. These systems operate in MSD structures to manage and minimize oscillations which both protect the occupants along with guaranteeing structural stability. Mass spring dampers along with other damping systems get incorporated into different structures to minimize vibrations from both internal and external forces. The installation of these systems decreases both structural instability and excessive movements and oscillations.

The Mass Spring Damper exists as one of the basic engineering mechanisms that utilize civil applications to stabilize structures and protect buildings from earthquakes and enhance occupant comfort. Buildings that integrate MSD systems use the systems to consume seismic energy resulting in decreased ground motion effects. Building oscillations are managed through these systems that reduce earthquake-generated structural damage for ensured safety and to preserve structural integrity [7]. The three primary sources of action create bridge vibrations during traffic movements and when exposed to wind and seismic activity. The integration of MSD systems within bridge designs cuts down vibrations thus providing structures with longer durability and cheaper upkeep. Tall structures facing wind-generated vibrations both endanger human safety and put strain on their underlying framework. Structural oscillations in MSD systems operate seemingly to protect people while assuring stability in buildings. Mass spring dampers serve together with multiple damping systems in different structures to minimize vibrations that result from internal and external forces. These systems help lower both structural instability and overreactions and unstable movement when installed.

Civil engineering vibration control relies on the precise measurements provided by the Liquid Circular Angular Accelerometer (LCAA) which functions as a critical monitoring device. This tool finds important applications throughout multiple essential fields of civil engineering domain. LCAAs' operation in structural health monitoring and vibration analysis depends on their ability to measure accurate angular acceleration data from vibrations [8]. The advanced monitoring system through LACAs provides complete analysis of vibrations which helps detect structural irregularities early as a preventive mechanism for structural safety maintenance. The foremost advantage of LCAAs rests in their optimization function for damping systems. These devices use measured angular accelerations to improve the performance quality of damping systems particularly tuned mass dampers which enhance structural stability against dynamic forces. Seismic response engineering together with earthquake research depends heavily on LCAAs to record and analyze angular acceleration measurements which occur in seismic events. The collected information helps researchers understand how structures respond to earthquakes therefore they use this knowledge to create resilient earthquake-resistant designs. Towers and skyscrapers must have LCAAs as mandatory components to track wind-affecting angular movements. The monitoring system enables predictions and managerial decisions about sway that help protect building occupants and safeguard high-rise buildings' structures.

Structural assessments demand constant monitoring and live data acquisition methods which perform the primary role. The LCAAs allow real-time acceleration monitoring because of their ability to perform continuous measurement of angular acceleration values. The real-time monitoring enables experts to determine the vibration effects on stability as well as performance measurements. The implementation of LCAAs helps engineers conduct dynamic load assessments by providing knowledge regarding how dynamic surcharges affect structures. Structural designs receive deeper optimization because this data enhances rotational force resistance capacity in built structures. LCAAs function as construction tools for angular movement control since they protect structural integrity by both improving stationary positioning and reducing dangerous displacements. The continuous assessment of bridge and dam safety relies on these devices that supply technical information about structural health along with torsional data and stability evaluation results, Table 2.2 shows the importance of the system.

Table 2.2 Advantages and disadvantages of the liquid circular angular accelerometer (LCAA) sensor.

Advantages	**Disadvantages**
High accuracy in measuring angular motion	Costlier compared to some other sensors
Suitable for dynamic angular acceleration	Complex calibration and setup procedures
Excellent sensitivity to angular movements	Susceptible to environmental interference
Resistant to external vibrations and shocks	Limited accuracy under extreme conditions
Capable of measuring small angular changes	Requires specialized handling and maintenance
Wide operating temperature range	Higher power consumption in some applications

2.4 Simulation of MSD and Applications of LCAA

Through computational modeling of Mass Spring Damper (MSD) systems engineers can use predictions to evaluate how structures comply with vibrations during the design phase [9]. The simulation outcomes give essential information about vibration responses and the damping efficiency of the system. The main simulation outcomes for MSD systems include measuring structural natural frequencies because they guide understanding of vibration responses. The calculated results include Mode shapes which indicate the frequency-specific patterns of structural or mechanical deformation or motion. The obtained results facilitate the identification of resonance points along with parameter optimization for better resonance vibration mitigation performance. Tests demonstrate that MSD systems use simulations to reduce structural vibrations by demonstrating effective vibration reduction. The calculated value of the damping ratio shows how effectively the system will reduce energy while controlling oscillations. Engineers use the obtained results to evaluate how well the damping system works before making adjustments to maximize its performance effectiveness.

The MSD system and its integrated structure undergo stress and strain distribution analysis by means of structural simulations. Structured analysis of an MSD system provides vital information to validate that material components resist external vibrations better than their failure limits. Simulations demonstrate that a structure with MSD technology shows proper reactions to different outside pressure patterns including seismic movements and wind forces and equipment-caused movements. System stability evaluation along with safety assessment becomes possible through complete understanding of this response. The assessment of various damping mechanisms operating within an MSD system is performed by engineers through results analysis of simulations. The evaluation process leads toward choosing and perfecting the right damping solution or multicomponent damping solution based on specific use requirements. The optimization process for MSD system parameters including mass and stiffness and damping coefficients becomes possible through simulations which lead to desired vibration control results. Simulation validation allows designers to check whether the system fulfills safety standards and achieves its performance requirements.

Applications of LCAs show broad industrial use throughout numerous sectors of operation. LCAs manifest as essential components of vehicle stability systems in automotive engineering because they improve safety and control functions. The precise motion sensing along with control capabilities of robotics comes from using LCA systems which enable robots to move and handle objects efficiently. This part discusses various applications demonstrating the broad utility of LCAs in modern technology development [10]. The precision and dependability of LCAs need careful calibration steps for their achievement. The provided chapter reveals specific calibration methods that lead to exact measurement results. A thorough understanding of LCA performance challenges can be gained through investigations about accuracy metrics and performance-limiting factors including temperature sensitivity and sensitivity to external interferences. This chapter defines core operational knowledge of complex devices alongside their industrial functions while explaining necessary elements required to obtain exact angular acceleration measurements. The obtained information establishes knowledge that allows people to better appreciate how LCAs work within modern vibration control systems and their application in engineering designs for stability control, Table 2.3 shows the importance of the sensors of the system.

Table 2.3 Advantages and disadvantages of sensors used in the mass spring damper (MSD) system.

Advantages	Disadvantages
Accurate measurement of structural movement	Costly procurement and installation
Real-time data acquisition for analysis	Requires regular calibration and maintenance
Enables precise feedback for control systems	Susceptible to environmental influences
Facilitates early detection of anomalies	Data processing complexities
Aids in predictive maintenance strategies	Potential for data inaccuracies
Enhances system monitoring capabilities	Compatibility issues with certain structures

2.5 The Analytical Approach Modeling - MSD for Vibration Control

Data analysis through the system requires researchers to develop mathematical expressions based on classical mechanics principles which describe how the system behaves. Through this method we achieve precise mathematical equations for system position along with velocity data together with other measurement values which do not require numerical simulations.

Newton's second law enables the writing of the equation of motion for undamped linear MSD systems.

$$m\frac{d^2x}{dt^2} + kx = F(t) \tag{2.1}$$

Where,

- m stands for mass in kg
- k stands for the constant of applied spring
- x denotes the displacement of mass
- F(t) external vibration
- $\frac{d^2x}{dt^2}$ acceleration - second derivation w.r.t. time

To include damping in the system, we introduce a damping term using the velocity of the mass:

$$m\frac{d^2x}{dt^2} + c\frac{dx}{dt} + kx = F(t) \tag{2.2}$$

Where:

The variable c represents the damping coefficient that measures value in units of N s/m.

The solution of this second-order ordinary differential equation becomes possible by applying two approaches either through a specific F(t) assumption or through homogeneous equation resolution followed by unknown coefficient analysis or different techniques for forcing term handling.

The steady-state solution for a simple harmonic force $F(t) = f_0 \, sin(wt)$ *can be determined through a solution of* $x(t) = A \, sin(wt + \varnothing)$, *while A represents the amplitude and w and ∅ denote the forcing frequency and phase angle, respectively.*

Applying the differential equation to an assumed solution allows one to determine A and ∅ in order to get the complete system response to harmonic forcing.

The solution of complex forcing functions together with nonlinear system problems demands specialized analytical methods like the Laplace transform alongside Fourier series and additional specialized approaches according to the problem characteristics. Analytical methods create an advanced understanding of system behavior which provides exact response predictions for various operational conditions through mathematical methods instead of numerical simulation.

References

1. Castaldo, P., *Integrated seismic design of structure and control systems*, Springer International Publishing, New York, 2014.
2. Cirstea, M., Dinu, A., McCormick, M., Khor, J.G., *Neural and fuzzy logic control of drives and power systems*, Elsevier, Newnes, Oxford, UK, 2002.
3. Boehm, B., Anchoring the software process. *IEEE Software*, 13, 4, 73–82, 1996.
4. Cheng, S., Fu, M., Wang, M., Li, X., Xiao, M., Wang, T., Modeling for fluid transients in liquid-circular angular accelerometer. *IEEE Sens. J.*, 17, 2, 267–273, 2016.
5. Genta, G., *Vibration dynamics and control*, vol. 616, Springer, New York, 2009.
6. Nakayama, Y., *Introduction to fluid mechanics*, Butterworth-Heinemann, Oxford, UK, 2018.
7. Housner, G.W.E.A., Bergman, L.A., Caughey, T.K.F., Chassiakos, A.G., Claus, R.O., Masri, S.F., Skelton, R.E., Soong, T.T., Spencer, B.F., Yao, J.T.P., Structural control: past, present, and future. *J. Eng. Mech.*, 123, 9, 897–971, 1997.
8. Glisic, B. and Inaudi, D., *Fibre optic methods for structural health monitoring*, John Wiley & Sons, Hoboken, NJ, USA, 2007.
9. Sandberg, G. and Ohayon, R. (Eds.), *Computational aspects of structural acoustics and vibration*, vol. 505, Springer Science & Business Media, Vienna, Austria, 2009.
10. Kemppi, P., *Next generation satellite navigation systems*, VTT Technical Research Centre of Finland, Espoo, Finland, 2007.

3

Hydraulic Dampers and Tuned Mass Dampers

Abstract

This paper delivers an extensive investigation of hydraulic dampers (HD) and tuned mass dampers (TMD) through a historical review followed by an evaluation of their vital functions in vibration control for civil engineering structures. The research traces the technological development of HD and TMD from their inception and their advanced influence on engineering structures. Civil engineering implements Mass-Spring-Damper (MSD) systems and Liquid Circular Angular Acceleration (LCAA) in several ways which effectively reduce vibrations. These technologies require absolute integration for solving dynamic load and seismic event problems. The paper performs MSD system simulation work and demonstrates various LCAA applications in vibration control through research-based analysis and real-world showcase examples of its effective and adaptable nature. The research draws its central attention to analytical methods used for modeling MSD in vibration control to expose theoretical concepts of these systems. The research integrates analytical modeling with practical applications alongside historical methods to build complete knowledge about building vibration control systems using hydraulic dampers along with tuned mass dampers.

Keywords: Hydraulic damper (HD), tuned mass damper (TMD), integrated system, practical analysis

3.1 Introduction

Hydraulic components function as essential mechanical systems for vibration control because they provide multiple ways to combat vibrations within various engineering systems as shown in Figure 3.1. The chapter performs an in-depth investigation of hydraulic dampers by explaining their core concepts together with design details as well as their use in engineering systems. Such devices work by dispersing vibrational energy because fluids move inside controlled areas [1]. The section elaborates on

Kundan Meshram and Prashant Lahre. Vibration Control Systems, (27–46) © 2026 Scrivener Publishing LLC

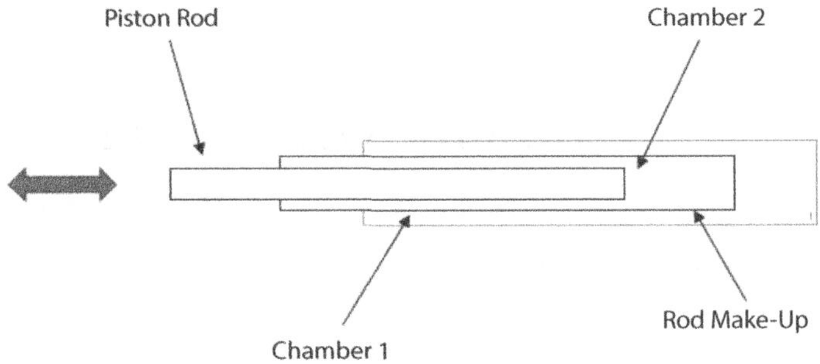

Figure 3.1 Hydraulic damper.

the operational basis which shows how restrictive passageways let hydraulic fluids transform kinetic energy into thermal energy. Readers who thoroughly learn these principles attain knowledge about the physical forces that guide hydraulic damping systems.

This section studies both design elements and operational components of hydraulic dampers with great depth. The text examines dampers from a structural perspective to explain their key elements which include pistons as well as cylinders and valves and hydraulic fluids. Explicating details about fluid selection alongside material choices and valve engineering demonstrates the complex nature of making efficient hydraulic damping systems. The following section describes the comprehensive set of conditions where hydraulic dampers bring value to different fields. Suspension systems of automobiles incorporate these devices for smooth driving by absorbing shockwaves and vibrations. Civil engineering structures including buildings and bridges implement hydraulic dampers to fight against seismic forces and wind-induced vibrations for reducing structural damage [2]. Hydraulic dampers enhance operational stability while managing equipment oscillations in different types of machinery and industrial equipment. The different application ranges demonstrate the broad usefulness of hydraulic dampers in numerous engineering practices.

The chapter provides in-depth sections on hydraulic damper calibration as well as maintenance procedures and performance boundaries. A clear description outlines the process needed to calibrate hydraulic systems in order to achieve long-term optimal performance. Processing methods and practical constraints regarding fluid leaks and temperature sensitivity and cavitation provide essential information about handling hydraulic dampers properly. The forward-thinking approach gives readers insight into possible enhancements that lead to understanding the future growth and enhanced usage of hydraulic dampers for advanced vibration control systems throughout engineering applications.

Tuned mass dampers (TMD) offer complex engineering applications which actively control vibrations found in different systems. The investigation in this chapter explores TMDs by revealing their operational rules as well as their engineering applications and system design concepts for vibration control in engineered structures. The basic operation of TMDs functions by implementing secondary mass-spring-damper systems to primary structures where these systems reduce and dampen out-of-control vibrations [3]. A detailed analysis of TMD principles investigates the secondary mass component which performs vibration frequency resonance and achieves efficient vibrational energy absorption. The ability to understand the effective control of vibrations by TMDs depends directly on mastering these fundamental principles.

This chapter provides detailed investigation of both TMD design requirements alongside their vital components. The chapter explores the detailed approach for designing secondary mass and spring and damping elements which must exactly replicate the vibration characteristics of the primary structure. Engineers need to understand different aspects of TMD

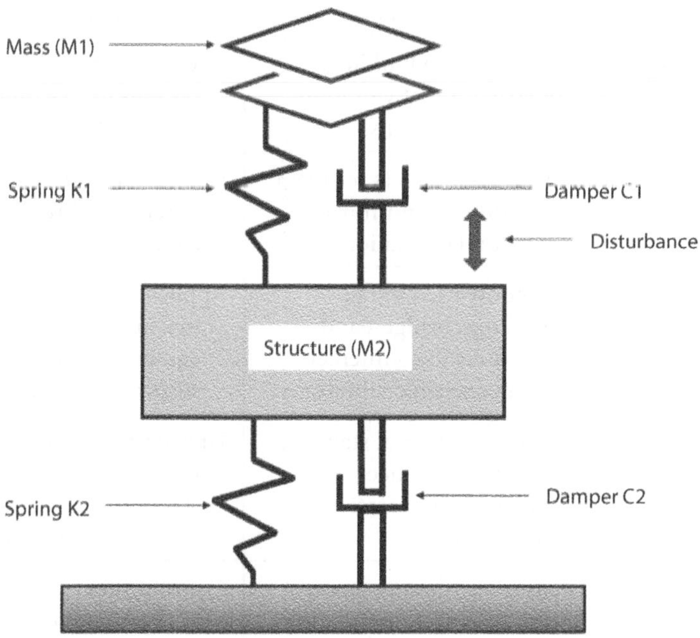

Figure 3.2 Tuned mass damper.

implementation through a discussion of natural frequency adjustment and choice of appropriate ratios and proper materials for effective energy dissipation mechanisms. EMDs traverse multiple applications which strongly benefit from TMD implementation. The utilization of TMD technology in civil engineering infrastructure includes skyscrapers and bridges and other constructions where the devices counteract vibrations from wind

Table 3.1 The differences between a tuned mass damper (TMD) and a hydraulic damper (HD).

	MSD	**HD**
Function	Reduces structural vibrations by counteracting resonance.	Mitigates vibrations by dissipating kinetic energy as heat.
Mechanism	Employs a mass-spring system tuned to the structure's frequency to dampen vibrations.	Uses fluid or hydraulic resistance to dissipate energy and reduce vibrations.
Components	Mass, spring, damper elements tuned to the structure's natural frequency.	Hydraulic fluid, chambers, valves for energy dissipation.
Installation	Installed within buildings or structures at vibration-sensitive locations.	Integrated into building systems or machinery susceptible to vibrations.
Control of Vibrations	Actively counteracts resonant vibrations to minimize structural oscillations.	Passively absorbs energy to reduce vibrations and enhance stability.
Applicability	Commonly used in tall buildings, bridges, and structures to counteract wind or seismic vibrations.	Applied in heavy machinery, industrial equipment, and vehicle suspension systems.
Impact on Systems	Enhances structural stability, reduces resonance, and controls excessive vibrations.	Diminishes vibrations, improves operational stability, and reduces wear in machinery.
Engineering Emphasis	Focuses on vibration control strategies and minimizing structural oscillations.	Concentrates on damping systems and mitigating vibrations in machinery.

forces and earthquakes to boost stability. TMDs serve multiple purposes in mechanical systems because they eliminate vibrations, which protects both equipment efficiency and operational lifespan as shown in Figure 3.2.

Researchers have intensively studied the calibration methods along with optimization processes and operational boundaries of TMDs. There is a clear description in [4] of the calibration protocols for TMDs which focus on finding their best resonance and vibration absorption state. The evaluation of TMD parameters and configurations through discussion brings clarity to their performance enhancement in various structural conditions. The paper reviews real-world TMD application barriers which include space availability and financial matters as well as installation complexities for existing infrastructure while evaluating future possibilities for TMD technology development. Improved TMD systems would result from developments in adaptive mechanisms as well as new materials for enhanced efficiency together with predictive control systems integration, Table 3.1 shows the system of damper.

3.2 History of HD and TMD

Researchers have intensively studied the calibration methods along with optimization processes and operational boundaries of TMDs. There is a clear description in [4] of the calibration protocols for TMDs which focus on finding their best resonance and vibration absorption state. Analyses regarding the optimization of TMD parameter settings as well as configurations identify methods to enhance their performance across multiple structural buildings. The paper reviews real-world TMD application barriers which include space availability and financial matters as well as installation complexities for existing infrastructure while evaluating future possibilities for TMD technology development. Improved TMD systems would result from developments in adaptive mechanisms as well as new materials for enhanced efficiency together with predictive control systems integration.

The development of engineering technology led hydraulic dampers to gain multiple uses in civil infrastructure applications. Hydraulic dampers began to protect not only machines but also bridges and buildings and tall structures that required protection against dynamic forces. Designs integrating hydraulic dampers served to lower the impact of wind-induced oscillations along with seismic events as well as external forces that threatened structural safety. Designers used advanced fluid dynamics theories to create complex hydraulic systems which achieved optimal damping performance in their hydraulic systems. The development of specialized hydraulic dampers featuring precise settings enabled designers to use them effectively in various structural applications. Hydraulic dampers were adopted as fundamental components for damping systems where they work alongside tuned mass dampers and semi-active or active damping systems. Their historical development has been driven by intentions to perfect technologies which achieve better vibration suppression as well as structural stability and occupant comfort within civil engineering work. Civil engineering relies on hydraulic dampers as necessary foundation components today because they provide multiple solutions to handle structural vibrations and bolster both seismic stability and damping methods and preserve diverse infrastructure safety. The historical development of engineering systems manifests as an ongoing effort to boost stability during constructive activities while fighting against natural forces.

The development of engineering technology led hydraulic dampers to gain multiple uses in civil infrastructure applications. Hydraulic dampers began to protect not only machines but also bridges and buildings and tall structures that required protection against dynamic forces. Designs integrating hydraulic dampers served to lower the impact of wind-induced oscillations along with seismic events as well as external forces that threatened structural safety. Improved hydraulic damper designs developed hand in hand with technological advances in materials production and calculation tools as well as manufacturing methods. Designers used advanced fluid dynamics theories to create complex hydraulic systems which achieved optimal damping performance in their hydraulic systems. The development of specialized hydraulic dampers featuring precise settings enabled designers to use them effectively in various structural applications. Hydraulic dampers were adopted as fundamental components for damping systems where they work alongside tuned mass dampers and semi-active or active damping systems. The design history indicates an ongoing effort between engineers to develop better vibration control solutions and structural stability systems and occupant comfort applications in civil engineering projects. Civil engineering relies on hydraulic dampers as necessary foundation components today because they provide multiple solutions to handle structural vibrations and bolster both seismic stability and damping methods and preserve diverse infrastructure safety [2]. TMD designs reached prominence at the end of the 1960s and beginning of the 1970s due to technological advancements. Following his involvement in designing Tokyo Tower Takashi Fujita made key contributions to TMD technology development. The seismic events prompted Fujita to design innovative features which linked a tuned mass with hydraulic dampers to minimize tower oscillations. After the late 1960s, numerous primary buildings incorporated TMDs as part of their construction projects worldwide. TMDs found their application in Taipei 101 alongside Petronas Towers and other high-rising buildings as part of their structural design to decrease swaying motions from wind along with seismic events. Engineers used TMDs to enhance the comfort of occupants while improving the stability of these structures successfully. Engineering specialists have created superior

Table 3.2 Advantages and disadvantages of TMD in structural building.

Advantages	Disadvantages
Effective at mitigating specific resonant vibrations	Increased complexity in design and installation
Improved structural stability during vibrations	Space requirements may impact architectural design
Reduction in structural fatigue and damage	Initial cost and maintenance considerations
Versatility in applications across various structures	Tuning requirements for different frequencies
Potential reduction in operational and maintenance costs	Limited effectiveness for wide frequency ranges

Table 3.3 Advantages and disadvantages of hydraulic damper (HD) in structural building.

Advantages	Disadvantages
Effective in dissipating energy over a wide range of frequencies	Complex design and integration
Improved structural stability during dynamic events	Space requirements may impact architectural design
Absorption of high-energy vibrations	Initial cost and maintenance considerations
Potential reduction in structural damage	Hydraulic fluid leaks pose a risk
Adaptable to various structural configurations	Regular maintenance and monitoring required

algorithms and simulation programs to enhance TMD design precision which allows their use across various architectural forms. Engineers implement tuned mass dampers as fundamental elements for structural engineering to provide specific control systems that both stabilize tall buildings and bridges and keep structures secure. Barriers have evolved historically toward the establishment of innovative mechanisms to manage structural oscillations and maximize structural resistance against dynamic forces within civil engineering practice. Tables 3.2 and 3.3 show the importance and role of the system.

3.3 Integration of HD and MSD

The implementation of hydraulic dampers takes place during structural design stages to achieve effective vibration control and improved stability in different civil infrastructure systems.

The process includes several key aspects:

a. The structural requirements guide engineers to find vulnerable areas for vibration implementation so engineers can position hydraulic dampers effectively. Engineers strategically deploy hydraulic dampers during design while evaluating building height together with materials and predicted dynamic loads as well as vibration sources including wind speed and seismic forces and equipment movement.
b. Hydraulic dampers receive installation at particular locations throughout the structural framework. The placement choice between vertical or horizontal positions depends upon how vibrations are expected to move through the system. The mounting technique provides strong connections and enables adequate system movement to process energy which effectively manages vibrations.
c. Main modifications to the structural framework may become essential for building hydraulic dampers. Engineering designers create a solution by adding reinforcement to special locations for supporting the damper weight while maintaining structural compatibility and seamless integration of damping elements into building design.
d. The hydraulic system needs proper design of fluid dynamics and control systems to function properly. Engineers review fluid pathways together with pressure control mechanisms

and valve operations in order to design customized damping solutions for hydraulic dampers. Control systems function with integrated dampers to enable the modification of dynamic responses through changes in present environmental conditions.

e. Before operation the hydraulic damper system requires multiple testing phases to prove performance satisfaction. The vibration reduction capability of the system undergoes testing through simulation and physical evaluation under multiple loading scenarios together with natural environment parameters.
f. After installation both monitoring activities and system parameter adjustments are carried out by post-installation engineers on the hydraulic damper system. The vibrational control performance of the system advances by modifying fluid viscosity together with adjustments of damping coefficients while simultaneously observing system dynamic behavior.
g. The proper functioning of Hydraulic damper systems depends on ongoing maintenance checks and continuous monitoring because without this management they cannot work correctly. Hydraulic dampers benefit from regular inspections and fluid assessment tests and performance tests which ensure their operation stays optimal from beginning to end thus extending the longevity of structural stability.

Control engineers need to implement whole infrastructure scanning for analyzing structural frameworks alongside fluid dynamics study and feedback system design to enhance performance leading to better vibration regulation and civil engineering structure safety improvements.

Addition of tuned mass dampers to civil engineering structures follows a complete procedure to both stabilize the system and regulate vibrations according to civil engineers.

a. Engineers use thorough assessments to discover structural areas in buildings that exhibit vibrations caused by wind pressure and seismic movements and equipment operation. Engineering evaluations lead to the recognition of specific building positions where TMD benefits would produce maximal results.

b. TMD implementations occur at crucial points which correspond to the structural plan. Engineers identify the best locations and dimensions of TMD arrangements based on how these components will counteract the natural frequencies of building structures to minimize all forms of movement. Engineers integrate the TMD system with architects and structural designers *via* an integration procedure which maintains both design style and structural integrity.
c. The best conditions for oscillation can be found at upper structural areas so TMDs get mounted in stable positions to attach to these elements. The equipment attachment mechanisms first establish stability at their points then activate movements that cut down vibrations.
d. The TMD's mass components and its damping elements along with spring stiffness receive full technical calibration to find frequencies which counterbalance natural structural frequencies. When properly adjusted the damping system plus the vibrating energy absorption leaves the system capable of minimizing structural vibrations.
e. Dynamic analysis paired with testing helps secondary team members to identify how the TMD operates when subjected to varying load conditions. The effectiveness of the damper as a vibration suppressor can be tested physically by controlled excitation tests and shake table evaluations.
f. The TMD unit requires precise installation steps which determine proper placement alignment since its insertion into the structure. The standard maintenance requirements feature implemented methods for checking the operational state of damper systems. The TMD system requires periodic checks and exhibits needed recalibrations and demands maintenance work addressing all structural components linked to it.
g. Changes to structure design along with modifying environmental factors require TMD systems to undergo modification programs along with structural adaptations. Engineers conduct continuous parameter reviews of TMD systems by creating processes that monitor changes in system variables which include diverse loading conditions and structural designs and damping improvement capabilities.

h. Real-time Control Systems equipped with particular advanced setups can function as integrated components of TMD systems.

3.4 Simulation of HD and TMD

Determining hydraulic dampers through detailed simulation methods enables researchers to discover ways to improve their performance in structural vibration control as shown in Figure 3.3 and Figure 3.4.

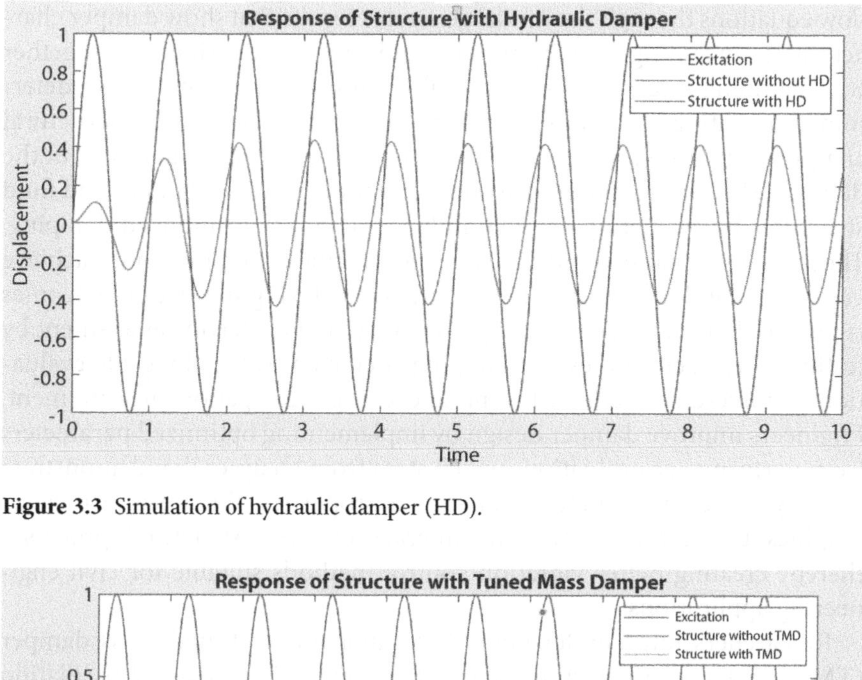

Figure 3.3 Simulation of hydraulic damper (HD).

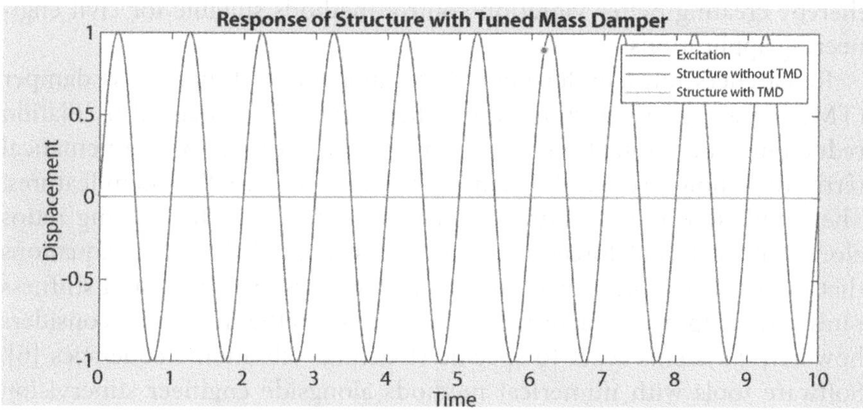

Figure 3.4 Simulation of TMD.

The simulation process starts with a mathematical model framework according to [5]. The mathematical equations analyze fluid flow behavior within the damper to evaluate its impact on structural dynamics by including factors related to fluid viscosity and structural mechanic properties and pressure gradient regulation.

The core component of CFD simulation exists throughout the entire procedure. Engineers provide the hydraulic damper with complete 3D modeling to create computational elements suitable for analysis purposes. The meshed system provides users with tools to define system boundaries and define both inflow and outflow constraints and required fluid specifications for generating accurate simulation data. Computational solving of flow equations through the model generates results that show damper characteristics including velocity profiles and pressure distributions together with damping forces. Simulation of fluid flow operates after its predetermined operating boundaries. The simulation moves forward to structural analysis by uniting fluid dynamics results with the structural model of the damper. The simulation models the interaction between damper and fluid structures by evaluating external influences with fluid-induced damping. The precise simulation model needs experimental data validation to achieve verification and refinement of its accuracy. Analyzing simulation outcomes is pivotal. Education specialists perform dampers behavior assessment by testing which determines the damping coefficient value alongside evaluation of force-displacement testing and energy dissipation measurement. Engineers improve damper design by implementing optimized parameters from performance evaluations which stem from analyses that demonstrate fluid dynamic and structural functional properties. Engineers use visual graphics to view fluid-dynamic interactions with structural processes thereby creating better vibration control methods suitable for civil engineering applications.

Researchers must understand all components of tuned mass damper (TMD) systems to determine their effectiveness in structural oscillation reduction. The simulation launch needs the creation of mathematical terms that show structural dynamic properties and TMD system features. The simulation unites natural frequencies together with damping ratios along with external forces which define structural dynamics equations that merge with TMD equations describing mass properties and stiffness rates and damping characteristics. The analysis in the simulation considers how TMD systems react to specific structural vibration frequencies [6]. Software tools with numerical methods alongside engineer supervision enable engineers to conduct dynamic analysis of TMD efficiency in vibration suppression. Engineers must inspect resonance features in addition to understanding energy transfers when they study TMD functionality and discover the proper requirements that make vibration reduction effective. Engineers use computational simulations to inspect multiple scenarios which they achieve through changes in TMD parameters such as mass size and stiffness and damping ratio adjustments.

The iterative process works to optimize TMD performance through matching of its resonance frequency with structural excitation frequency which enables maximum energy dissipation while reducing structural vibrations as shown in Figure 3.4. Hence simulation models need experimental data validation together with theoretical models analysis to verify their accuracy. Visual presentation methods such as graphics or animated displays assist engineers to understand how TMD structures influence vibrating dynamics. Such observations reveal how much energy the system dissipates alongside decreases in displacement and how successful the TMD remains in resisting dynamic forces.

3.5 Uses of HD and MSD

Hydraulic dampers make extensive civil engineering applications possible because their fundamental vibration control function helps manage structural dynamics between different systems. The hydraulic dampers use fluid mechanics principles to deliver adaptable solutions among different fields of application.

a. Hydraulic dampers serve as primary elements for structurally reducing vibrations that come from natural forces such as wind pressure and seismic events and mechanical equipment actions. The installed dampers inside buildings and bridges together with towers work to distribute vibrational energy while regulating building movements so they support better structural stability rates. The mechanical elements absorb kinetic energy to make them conductive at reducing unwanted motions as they protect both buildings from damage and personnel from discomfort.
b. The main protective mechanism of hydraulic dampers in Seismic Response Mitigation helps decrease earthquake damage for structures built in high seismic zones. The structure's built-in resilience increases when these dampers combine energy-dissipation functions with damping of natural structural vibrations. Seismic design systems that integrate hydraulic dampers turn buildings into less vulnerable

systems and ensure safety for building occupants during earthquakes.

c. Hydraulic dampers serve bridge and infrastructure protection purposes by controlling movements in exposed structures triggered by traffic flow and high winds. The devices function through bridge bearings and suspension systems to restrict movements and reduce vibration and boost structural operation duration. The application of these devices for bridge oscillation management helps to conserve both structural safety and durability.
d. Industrial machinery stability functions are supported by hydraulic dampers through their capability to provide equipment support systems within industrial operations. The ability of hydraulic dampers to accept kinetic energy allows them to decrease the vibrations of moving parts thus prolonging the operational lifetime of mechanical components. This method keeps the smooth operation of industrial facilities while equipment extends its life span.
e. The installation of hydraulic dampers in tall structures functions to restrict harmful movements in high-rise buildings thus safeguarding the comfort of building occupants. Such devices reduce both movement sway and angular motions in structures to create better comfort experiences particularly for tall buildings that undergo perceptible wind-inflicted vibrations.
f. The functioning of these systems requires hydraulic dampers to serve as part of their damping systems comprising tuned mass dampers and active damping systems. The damping system generates heat through temperature changes to turn kinetic energy into heat thereby providing optimal damping methods for structural stability.

TMD Stability Strategy to Mitigate

a. TMDs operate as structural vibration control units when connected to buildings and bridges and towers to minimize movements of structures under external interferers. The installation of secondary mass-spring-damping systems designed for matching structural frequencies enables the reduction of resonance amplitudes resulting in decreased structural vibrations from wind forces and seismic movement or machine activity.

b. Earthquake-prone areas require TMD implementations as fundamental components of seismic design programs. The energy exchange capabilities of TMDs create a successful method for reducing damaging forces from seismic waves that protects the structure from harm. Structures gain advantages from TMD systems because they make seismic structures stronger and diminish the natural vibrations which occur in seismic events.
c. Tall buildings require TMDs for stability because these devices successfully control building sway effects that occur in high-risk wind-prone structures. The building structure or elevated areas house these dampers that control natural building oscillations to protect residents from harm. These devices restrict buildings from moving horizontally so they simultaneously enhance structural stability and lower noticeable movements that occur in tall structures.
d. TMDs employ bridge and infrastructure control methods through oscillation control systems that result from wind riding forces and traffic-related movement.
e. These structures provide bridge designers with a solution to minimize structural vibrations which aids in maintaining bridge infrastructure durability for long periods. Controlling bridge oscillations through their implementation provides structures with better stability and extends their operational lifespan.
f. Within the industrial sector TMDs serve to stabilize machinery because they absorb vibrations which occur on equipment. The dampers consume operational kinetic energy to decrease mechanical vibrations that protect equipment components from degradation. The durability of machines along with smooth operational integrity improves when TMDs are used in industrial facilities.
g. The controlled systems which utilize Damping Systems gain their performance from TMDs as essential elements. The calibrated functionality of mass-spring-damping in these systems supports the successful implementation of semi-active and active damping methods. The structural stability and vibration control improve because of TMDs which perform energy dissipation.

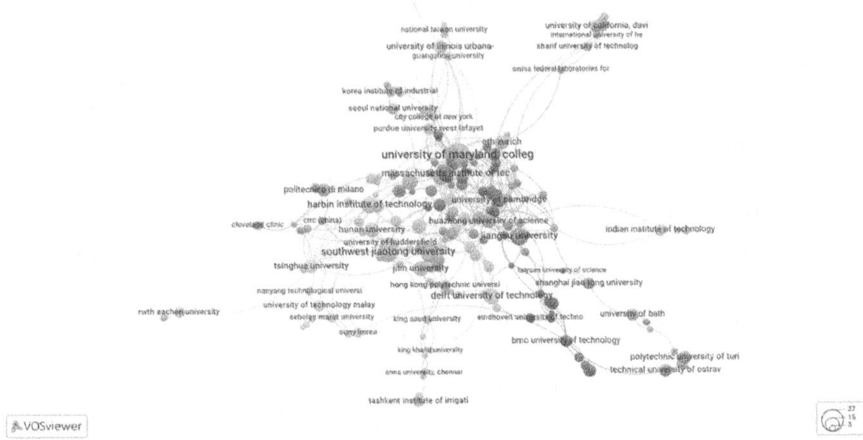

Figure 3.5 Bibliographic reviews university-wise.

h. The use of tuned mass dampers represents a customized approach in civil engineering for vibration control by external forces. The bibliographic reviews of university-wise as shown in Figure 3.5.

3.6 Analytical Approaches for HD and TMD

Hydraulic dampers (HDs) represent essential structural components adopted by civil engineering practitioners to regulate vibrations and boost structural support systems. Fluid dynamics modeling represents the initial step in analytical methods which explore and maximize such dampers along with other analytical approaches. The Navier-Stokes equations with their fundamental status serve as fundamental tools for fluid flow analysis within HDs. The equations together with Reynolds number analysis help determine whether the damper flow is laminar or turbulent which affects the damping mechanics [7]. Analysis of pressure-flow characteristics allows scientists to study fluid dynamics which generate damping forces inside the damper. Mathematical modeling functions as an essential element for achieving two goals that include coefficient estimation and transfer function generation. Engineers conduct damping coefficient estimation by developing analytical models that rely on fluid properties with additional usage of geometry and flow characteristics. The damping models calculate estimates regarding the capabilities of dampers. Engineers derive improved understanding of how dampers perform at various frequencies with the help of transfer functions.

The analytical methods must include both structural dynamics and fluid-structure interaction. The Finite Element Analysis (FEA) software enables modeling of integrated fluid dynamics with structural elements through simulation capabilities. The integrated insights between fluid flows and structural responses through this approach support optimal performance optimization of the damper system in structural applications. A basic equation for hydraulic damping analysis is the damping force calculation:

$$F_{damping} = c.v \tag{3.1}$$

Where:

$F_{damping}$ refers to the damping force and the damping coefficient is represented by c and the fluid velocity is denoted by v.

The force of damping equals c x velocity v while its coefficient stands as c.

The term v represents the velocity which fluid moves through the damping device.

The mathematical expression demonstrates how fluid velocity along with the damping coefficient determines the amount of generated damping force inside the damper unit. Civil engineering structures benefit from improved damping efficiency through optimal analysis of the relation between damping force and fluid velocity.

Multiple approaches studying TMDs exist to model the damping behavior of these devices in structural systems. The initial analysis phase begins by identifying vulnerable areas of vibration and establishing the best locations for TMD installation [8]. Mathematical modeling serves as the foundation of TMD examination because investigators establish governing equations for how dampers operate. The natural frequency of the structure represents one fundamental TMD analysis mathematical equation.

$$f_{natural} = \frac{1}{2\pi}\sqrt{\frac{k_{eff}}{m_{eff}}} \tag{3.2}$$

Where:

The system's natural frequency is represented by $f_{natural}$ and the effective stiffness of structure-TMD system is denoted by k_{eff} and effective mass is shown by m_{eff}.

The calculation incorporates k_{eff} to represent the overall stiffness of structure-TMD system.

The effective system mass is represented by m_{eff}.

The formula explains how the effective stiffness along with effective mass influences the natural frequency of structure-TMD systems which remains essential for TMD design optimization.

The analytical approach handles structure-specific dynamic assessment and performs numerical examinations of TMD movement behavior. Engineers use simulation tools to examine how the TMD operates with the structure while changing loads for property tuning that target specific frequencies of structural vibrations. Engineers employ two optimization techniques namely parameter tuning and sensitivity analysis to achieve ideal parameter settings of the TMD. Engineers reach their goal of improved structural stability by adjusting mass, stiffness, and damping characteristics to match the damper frequencies with structural natural frequencies which results in vibration reduction. Civil engineering scientists use mathematical modeling together with structural dynamics analysis and simulation techniques as analytical approaches to study tuned mass dampers.

3.7 Other Advanced Damping Systems

The text provides a broad analysis of multiple progressive damping systems which outpace previous standard systems discussed. The report investigates multiple complex damping systems as solutions for difficult vibration control situations in engineering fields. The text explains the complex nature of sophisticated damping systems through an examination of viscoelastic dampers and electromagnetic dampers and active damping systems. Viscoelastic materials show elastic properties along with viscosity which allows them to deliver adjustable damping capabilities for different applications [9]. This section details material science and engineering principles of creative damping mechanisms through comprehensive examination of these operational methods.

The control of vibrations in electromagnetic dampers occurs through electromagnetic forces which yield superior accuracy alongside adaptable performance. The detailed explanations about electromagnetic induction and coil systems and control methods offer understanding of complex engineering within these systems. Active damping systems represent contemporary technology because their damping characteristics automatically adjust using feedback information from the present moment. The article includes details about the control algorithms along with sensors and actuators used in active damping systems which demonstrate their suitability for adaptive vibration control [10]. These advanced damping systems support diverse applications because of their utility. Each damping mechanism serves different engineering needs through its specific features which make it suitable for high-rise buildings as well as aerospace structures and automotive systems and precision machinery. The extensive review demonstrates how these advanced damping systems are crucial and useful in various technical areas of engineering.

The document examines the calibration methods and optimization techniques as well as practical limitations of sophisticated damping mechanisms. The paper discusses calibration processes which lead to optimal performance together with optimal damping characteristic strategies but addresses implementation challenges and cost limitations. Knowledge about these factors reveals the full spectrum of implementation realities and advantages which engineers encounter while incorporating advanced damping systems into their designs. The insights obtained from these documentation methods display future tendencies in advanced damping technology developments.

References

1. Tokhi, M.O. and Veres, S.M. (Eds.), *Active sound and vibration control: Theory and applications*, vol. 62, Iet, London, United Kingdom, 2002.
2. Korkmaz, S., A review of active structural control: challenges for engineering informatics. *Comput. Struct.*, 89, 23–24, 2113–2132, 2011.
3. Elias, S. and Matsagar, V., Research developments in vibration control of structures using passive tuned mass dampers. *Annu. Rev. Control*, 44, 129–156, 2017.
4. Taghipour, J., Dardel, M., Pashaei, M.H., Vibration mitigation of a nonlinear rotor system with linear and nonlinear vibration absorbers. *Mech. Mach. Theory*, 128, 586–615, 2018.
5. Jugulkar, L.M., Singh, S., Sawant, S.M., Fluid flow modeling and experimental investigation on automobile damper. *Constr. Build. Mater.*, 121, 760–772, 2016.
6. Lu, P., Chen, S., Zheng, Y., Artificial intelligence in civil engineering. *Math. Probl. Eng.*, 2012, 2012.
7. Oberkampf, W.L. and Trucano, T.G., Verification and validation in computational fluid dynamics. *Prog. Aerosp. Sci.*, 38, 3, 209–272, 2002.
8. da Costa, M.M.A., Castello, D.A., Magluta, C., Roitman, N., On the optimal design and robustness of spatially distributed tuned mass dampers. *Mech. Syst. Signal Process.*, 150, 107289, 2021.

9. Zhang, W., *Dynamics of coupled systems in high-speed railways: Theory and practice*, Academic Press, Beijing, China, 2019.
10. Harne, R.L. and Wang, K.-W., *Harnessing bistable structural dynamics: For vibration control, energy harvesting and sensing*, John Wiley & Sons, Chichester, United Kingdom, 2017.

4

Neural Network-Driven Evaluation of Tuned Mass Damper Systems for Enhanced Structural Resilience

Abstract

The natural disaster of earthquakes inflicts deadly harm to buildings and ruins their construction while cutting down the period structures that remain functional. The combination of foundation slides alongside weak soil together with ground-shaking vibrations leads to structural damage. This investigation noted that Tuned Mass Dampers operate at maximum capability through MATLAB simulation. The research examines both TMD system applications and different system configurations alongside the range of device distribution and amounts. Virtual results for the parameters frequency, acceleration, displacement, and noise vs. time were calculated to show comparative data in this research. As the dynamic amplitude of the system grew stronger the performance output of TMDs became more efficient. The study demonstrates that TMD systems and their ANN predictions using regression coefficients (R) show high validation rates at 0.9998 for all measurements thus enabling buildings to reduce structural reactions against earthquake damage.

Keywords: Tuned mass damper (TMD), damping ratio, ANN, R^2

4.1 Introduction

Passive systems function as the most common vibration-control technology mainly because these systems use simple mechanical elements. The energy source from excitations produces input energy that passive systems absorb in order to reduce both structural response and the resulting damage from the excitations. Part of their benefit comes from not requiring any outside power source or any assessment process for determining magnitude measurement. Multiple researchers display various passive system designs in their works. The collection includes friction control devices and

Kundan Meshram and Prashant Lahre. Vibration Control Systems, (47–64) © 2026 Scrivener Publishing LLC

fluid flow viscous dampers together with seismic base isolation systems and tuned liquid (fluid) dampers and tuned mass dampers. Electrohydraulic actuators generate control forces as active systems require a dependable power connection in order to function [1, 2]. The deployment of active wind and earthquake response control needs large-scale outside power supply because engineers must utilize small stiffness modifications for better structural damping. Wind and earthquake forces cause the system to function through its regulation mechanism.

The structural responses receive substantial influence from this particular part. Control systems include Sensors as well as Actuators which work together with a Controller through a specified control algorithm. The system contains three essential components which form its structure (control law). The implementation of adaptive system features within semi-active systems enables the overall system performance to improve according to [3]. Semi-active systems lose valuable energy because they distribute it without control to the surrounding environment [4]. These devices use gathered data to adjust performance settings of the damper system for optimal functionality in real-time situations. Semi-active dampers include a passive damping device as one of their essential components. Cable dampers have four main components including Sensors, a control computer, a control actuator, and a control device according to [5]. Various control methods unite passive internal damping with active external damping along with active control following passive methods in developing conceptual control procedures. Each system plays a specific role in generating innovative hybrid control systems [6] while cooperating with the others. Such multi-faceted systems combine components from hybridized approaches to enable the removal of single-procedure limitations through the incorporation of positive features from the integrated techniques [7]. Engineers prefer to use passive devices as their first choice because these systems exhibit straightforward understanding and implementation. There are numerous system types available which leads to this conclusion.

The working efficiency of a Tuned Mass Damper (TMD) focuses on a particular frequency which corresponds to the building it supports as shown in Figure 4.1. The natural frequencies that affect high-rise and slender buildings should be considered due to factors including wind speed and ambient temperatures and humidity changes over their lifespan [8]. A reliable design for the TMD needs to be established because it must function effectively during the entire building period. A basic TMD design requires three main elements which include mass (m) together with spring (k) and damping device (c). A mass unit in the system accepts building movement energy while the attached spring creates stability and the attached damping

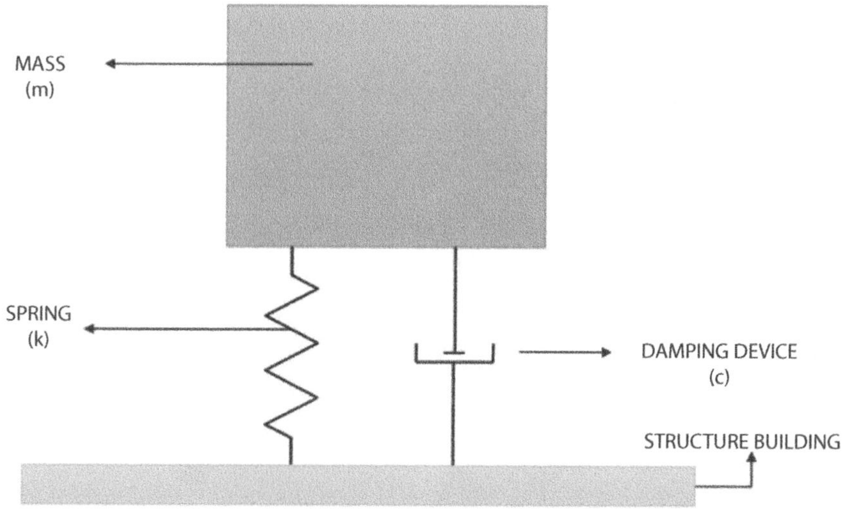

Figure 4.1 Tuned mass damper (TMD).

element releases this energy. The TMD system gets implemented on the structure (denoted M in Figure 4.1) thus providing vibration compensation to decrease building reactions. The performance of TMD systems becomes optimized through parameter adjustments comprising mass dimension and stiffness strength and damping function which leads to successful vibration reduction from external forces like wind and seismic events.

The TMD's performance stays effective for vibration control by having regular checks and adjustments to maintain its parameters which adapt to the building's changing natural frequency.

4.2 Literature Review

The investigation of TMD has been conducted by multiple researchers since its inception as shown in Figure 4.2 and Figure 4.3. Systems applied the principle of TMD to reduce ship motions and hull vibrations for the first time back in 1909 according to Frahm. In 1940 Hartog created an analytical model that explained possible TMD applications for vibration damping. Researchers later modified TMD parameters to achieve their highest output during sinusoidal_excitations. Several experts studied mass-to-frequency and damping-to-mass ratios as well as different single-DOF and multi-DOF structures with TMD subjected to various seismic stimulations [9, 10]. The optimized parameters might increase the system response

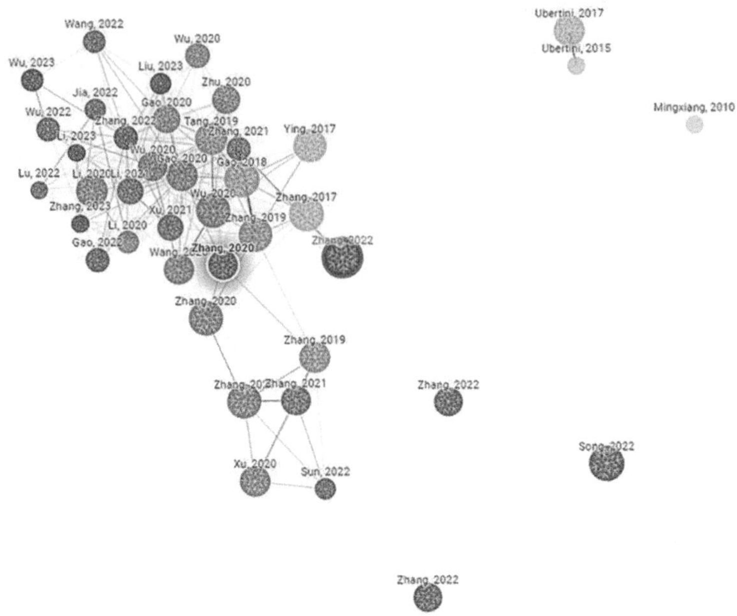

Figure 4.2 Connected papers according to authors.

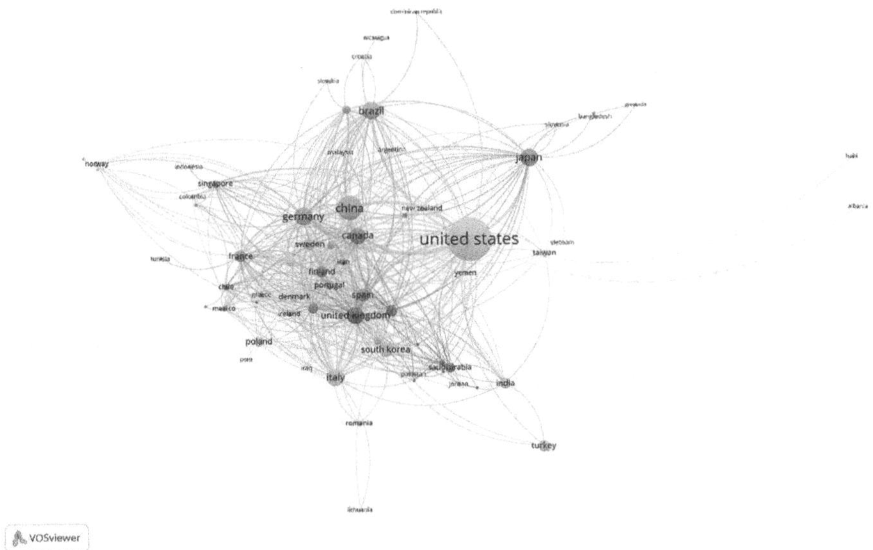

Figure 4.3 Literature reviews according to the country.

time when it encounters changing acceleration and displacement values. A tuned mass damper operates on a flexible foundation while experiencing seismic effects [11]. This research uses a frequency-independent model which supports examination of diverse soil and structural characteristics. The model-building evaluation of TMD effectiveness relies on its response to stationary stimulation tests. Research findings demonstrated that strong Soil-structure interaction creates substantial reduction in TMD system performance during seismic events. The structure's maximum response continues to diminish when the velocity of soil shear simulation reduces. The application of TMD structures proves challenging when used for protecting structures built over soft soil because of the significant damping properties exhibited by the soil-structure system. The study includes the North-South (NS) component of the El Centro, California hesitation to accurately evaluate the performance of the Tuned Mass Damper. The investigators installed a 36-story building with Tokyo-based bi-axial eccentricity as the test subject for their hybrid mass damper (HMD). With its gear-type pendulum component the HMD system extends the natural period of its auxiliary mass along with its linear actuator mechanism that ensures quiet operation which collectively improves the system's operational effectiveness. Two methods of HMD control transverse torsion-coupled vibration while experts have developed numerous feedback techniques to optimize HMD system responses to external inputs. A comprehensive evaluation of the control system took place through independent free vibration assessments alongside wind vibration tests that occurred as a result of Typhoon 9810 during 1998. A 1998 vibration test on the structure showed controlled accelerations reached only 63% and 47% of uncontrolled measurements of acceleration [12]. The design of active multiple-tuned mass damper (AMTMD) serves to reduce building responses to ground accelerations. The team accomplished this achievement through independent stiffness and mass modifications before implementing the mode-reduced order technique that reduced structural natural vibrations. A numerical searching technique examined the drug efficacy changes from dynamic parameters of AMTMD operating with feedback acceleration gain coefficient and frequency ratio and damping ratio included. When compared to MTMD technology AMTMD systems demonstrated better capability to reduce structural vibrations that result from subsurface acceleration motion. The author executed a quantitative analysis between a Multiple Tuned Mass Damper (MTMD) and a tuned mass damper (TMD) through a dedicated research [13, 14]. Researchers applied white noise treatment to a construction frame of three stories dimension while performing shaking table prototyping. Experimental tests demonstrated that MTMD system achieved better floor acceleration reduction performance compared to TMD systems. The research validated the dynamic behavioral properties of the structure through extensive analysis of experimental and numerical information. The vibration response of single-degree-of-freedom buildings under uniform ground acceleration was analyzed through quantitative research of multiple active-passive tuned mass dampers (MAPTMD). The MAPTMD regulates acceleration response whereas it preserves displacement and velocity response to ensure force control. Through experimental research the author studied the ability of multiple-tuned mass dampers to decrease seismic reactions in three-story structures. Study of both structural dynamics and damping features enabled researchers to evaluate their separate properties for determination. The authors ran their research from computer simulations until the completion of shaking table testing that used white noise stimulation. Research looked into the relationship between soil foundation stiffness because the building foundation could move relative to its surrounding environment [15]. To decrease soil-structure interaction TMDs need proper tuning which synchronizes with the basic structural foundation resonance. The structure needs TMD damping that exceeds critical damping levels to achieve smooth response transition. A study focused on how tuned mass dampers used energy dissipation for seismic loads in inelastic construction elements [16]. The research proved that the TMD system effectively seized energy from buildings before converting it into damping energy that reduced the waste of plastic and damping energy. The study evaluated plastic energy spectra with different structural yielding levels to determine how TMD could decrease the energy response of such systems. The author proved that wind-induced loads and seismic events cause strong vibrations in both tall buildings and long-span bridges with adverse results. The researchers employed STMD (structurally tuned mass damper) as a system to handle vibration control. The researchers utilized a multi-objective optimization technique to discover the Pareto front consisting of all Pareto efficient solutions [17]. Technologists employ this concept on regular occasions. The optimization technique determined the best combination of damping and frequency ratios in the STMD system to minimize peak displacement along with peak acceleration.

4.3 Methodology

Such static analysis applies the simple linear equation [A]x=B because static research results do not provide time-based adjustments. The most demanding equation used for dynamic analysis requires time-dependent variables. The motion equation of a system completely evades analytical solutions when subject to ground accelerations that change randomly over time. Only one degree of freedom restricts such outcomes in the system. Differential equation integration through numerical temporal steeping serves as an appropriate technique to resolve these types of computational problems. The achievement of this goal requires adoption of two separate procedures. Explicit methods should replace the implicit integration techniques. An implicit solution exists when present-time evaluations need preceding-time data for calculations. Technical professionals refer to this integration technique in time domain as Euler Time Integration Scheme. The solution stability remains constant regardless of the time increment lengths used in this approach.

A system is unconditionally stable when such characterization becomes possible. This methodology requires computing the stiffness matrix inverse because the method uses this input for directly solving the U (gamma) vector. Computing power becomes necessary due to the requirement to calculate inverse in this step. Beta method uses the computed information according to Newmark. The main objective of explicit analysis is to determine "U" through its solution process. An explicit analysis will always select the diagonal matrix as its mass matrix because this matrix is additionally known as a lumped matrix. The necessary operation for the mass matrix inversion allows us to eliminate complex stiffness matrix inversion methodologies. This matrix inversion requires only one step to reverse each entry on the diagonal portion of the matrix. Carrying out this operation proves straightforward because it bypasses Euler Time integration leaving the process thus unstable. The used integration method does not employ Euler Time. The main drawback of using this strategy presents itself in this way. This system progresses only through small time steps without exception. Amplitudes, frequencies, and phases: The performance level which the TMD system achieves in structure response reduction depends on fundamental variables such as mass ratio, damping ratio, frequency ratio and soil-structure interaction strength. Three different methods inspect the performance outcomes of TMD systems: earthquake simulation testing, wind-induced vibration testing, and free vibration testing. News parameters for the TMD system can reach desired performance levels through optimizations performed through numerical methods and multi-objective optimization techniques. The TMD system proves its effectiveness in vibration control for structures that experience seismic loads as well as wind-induced loads and other types of dynamic loads as shown in Figure 4.4 and Figure 4.5.

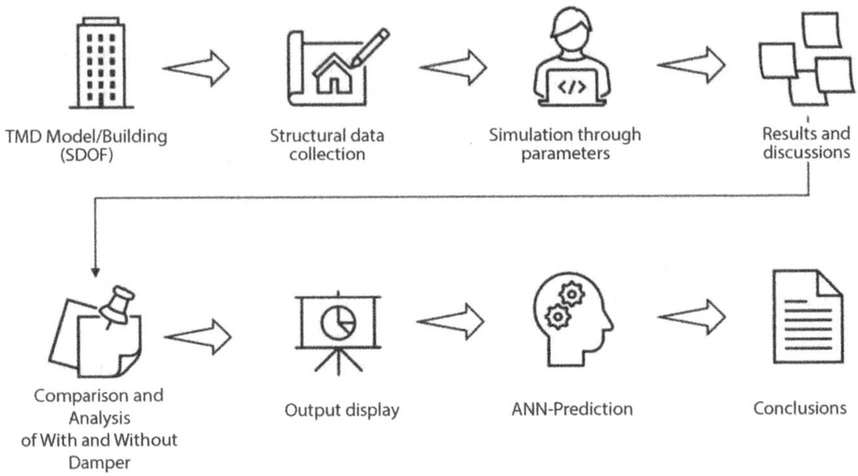

Figure 4.4 Methodology.

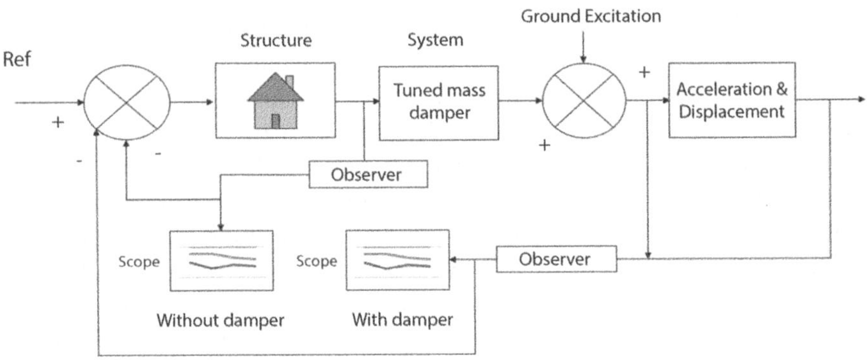

Figure 4.5 Simulated methodology.

4.4 Results and Discussion

The examination of structural vibration control through Tuned mass dampers depends on the evaluation of both system parameters and external force parameters during the investigation. The performance of a TMD to reduce vibrations depends on its mass, damping and stiffness parameters (m, b, k) together with the external force (w_force) frequency and amplitude. The response of the TMD depends on building storey numbers alongside the frequencies within its operational range. Standardizing analysis across structures of different dimensions along with soil conditions becomes possible through the non-dimensional formulation which is linked to the TMD frequency and its structural relationship measured by the F0 and F1 parameters. The response of the system dramatically depends on the soil-structure-TMD interaction because this relationship affects the transmission of forces and energy between the structure foundations and TMD. All measures for evaluating the performance of a Tuned mass damper include acceleration together with stiffness and damping ratio and displacement measurements used to assess how well the system reduces vibrations while preserving structure stability and safety. Parameters as follows:

1. The TMD system's mass (m) should equal a portion of the main structure's weight and is indicated in kilograms (kg).
2. The damping coefficient of TMD systems is expressed through damping (b). The damping function enables the system to release energy which diminishes the intensity of vibrations. (N s/m)
3. The TMD system's stiffness value is represented by this parameter. The system stiffness delivers a force which opposes the movement of the system. (N/m)
4. The applied frequency matches the natural frequency of the structure. (Hz)
5. Number of storeys serves as an item to define.
6. Non-dimensional formulation uses selected parameters including mass ratio alongside damping ratio and frequency ratio to produce better TMD system designs.
7. The behavior of the system depends on the precise analysis of how the TMD interacts with both the structure and the soil.
8. A structure's acceleration response needs specific attention during the design of TMD systems because this parameter directly impacts the system's functionality. The TMD system needs to successfully decrease the structural acceleration response value to (m/s2).
9. The displacement response of the structure should be considered as a main parameter when designing TMD systems. The displacement response reduction capability of the TMD system must be effective at m.

4.5 Relationship between Displacement, Acceleration with Time, and Noise, the Frequency with Time

The generated MATLAB simulation visualizes the relationship that exists between displacement acceleration and time through programming simulation methods as given in Figure 4.6. The earthquake activity produces acceleration as a function of time metric along with vibration and building shaking.

MATLAB programming simulation produces a graph that displays the time-based difference between noise and frequency as shown in Figure 4.7.

Both the building structure and tuned mass damper experienced their worst catastrophe during the earthquake event. Our damper operated

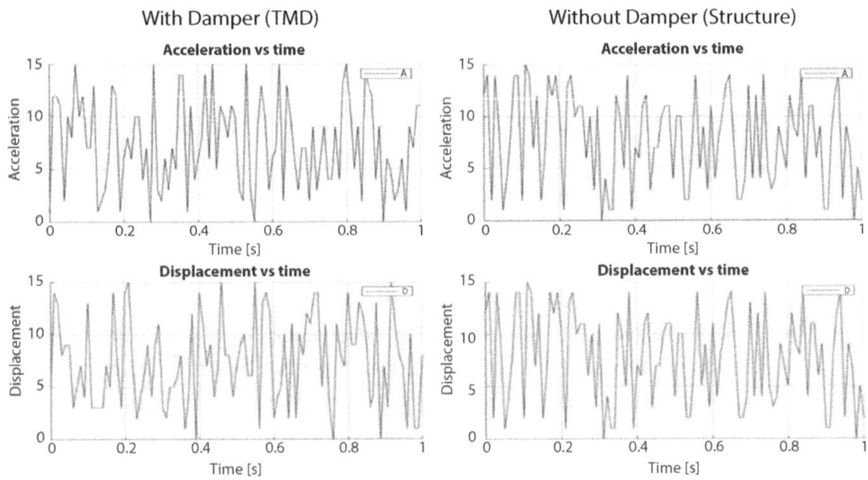

Figure 4.6 Acceleration, displacement vs. time(s).

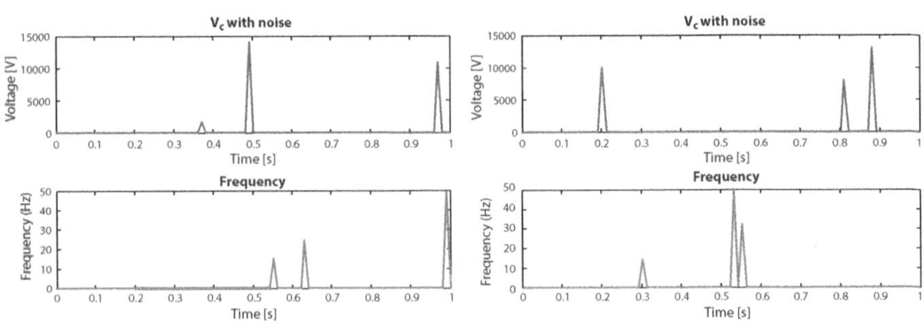

Figure 4.7 Noise, frequency vs. time(s).

initially yet frequency fluctuations resulted in hesitation according to points 8, 9, and 10 on the simulation results. The tuned mass damper achieves stabilization of the situation by completely absorbing earthquake energy as the structure reaches movement stability. The simulation demonstrates a comparison between the period with and without the damper installed as Damper moment and Structure moment as shown in Figure 4.8, Figure 4.9, and Figure 4.10.

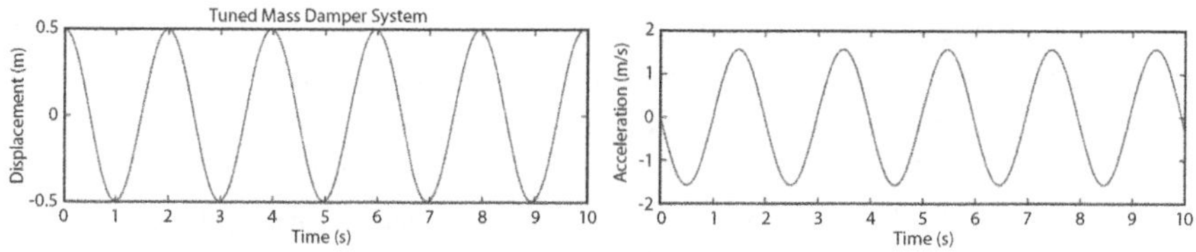

Figure 4.8 Without tuned mass damper moment of structure.

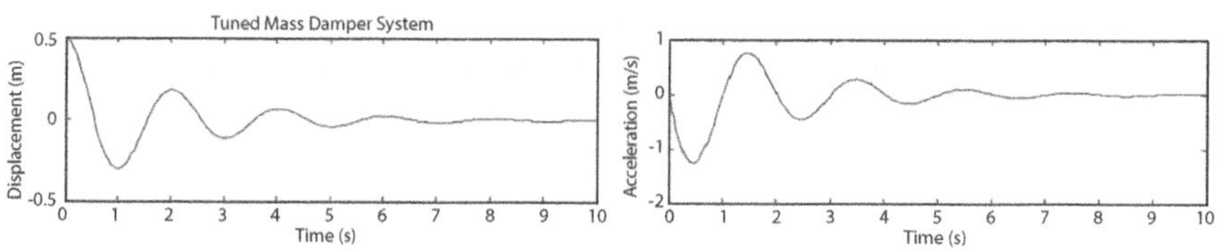

Figure 4.9 With tuned mass damper moment of system.

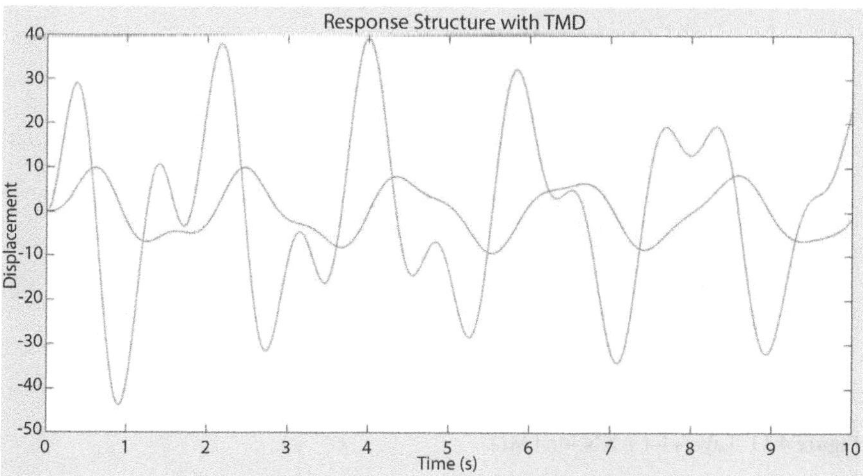

Figure 4.10 Response of structure with TMD.

4.6 Artificial Neural Network (ANN)

The implementation of the model together with forecasted Tuned Mass Damper parameters led to productive outcomes. The trained network demonstrated outstanding approximation abilities of structural characteristics by completing its training through a thorough process that utilized numerous TMD parameter sets and optimized outcome pairs. The ANN mastered complicated relationships that linked input features of mass alongside stiffness and damping with their optimized TMD parameters because it processed and recognized fundamental data patterns effectively as given in Figure 4.11. ANN-based predictions for TMD parameters displayed both high accuracy levels along with reliable consistency during performance evaluation. The network predictions matched the optimized values precisely while maintaining strong regression properties based on the obtained results. Trained and optimized network parameters allowed for performance enhancement making ANN an effective tool for TMD parameter estimation. The ANN showed outstanding generalization capabilities by giving predictions that moved towards the predicted optimized TMD parameters even when assessing different datasets. The property of

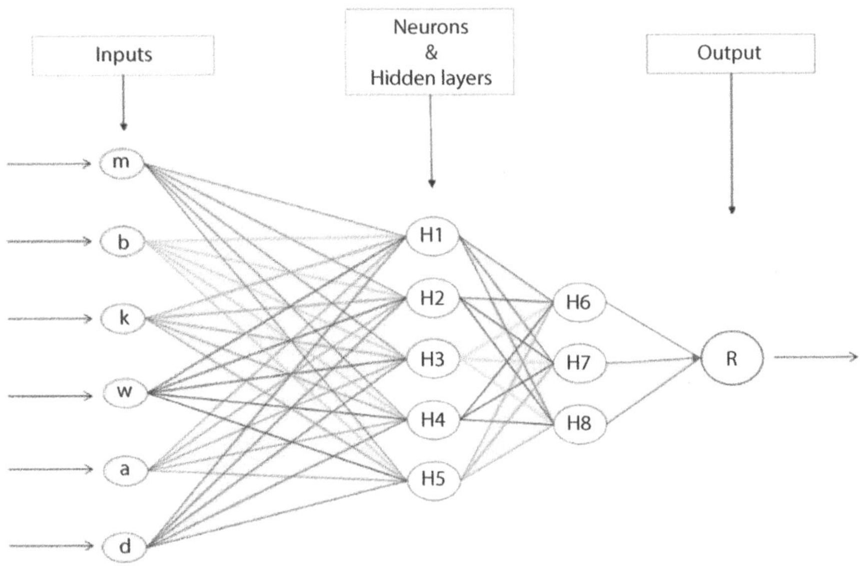

Figure 4.11 Layers for ANN for TMD.

generalization strengthens the network as a suitable choice for real-world application deployment across multiple scenarios outside its training database. A combination of ANN capabilities to analyze complex connections and deliver precise TMD parameter forecasts proves its effectiveness as it suggests potential uses in developing optimally robust structural systems when exposed to vibrational forces especially in areas vulnerable to earthquakes.

The analysis of predictions through regression Figures 4.12 and 4.13 provided essential predictive capabilities for the trained Artificial Neural Network (ANN). Visual analysis of the regression plots confirmed high precision in the model's estimation process based on the match between predicted and optimized TMD parameters. The ANN successfully tracked the linear relationship that existed between its predicted TMD outputs and the actual TMD data from the dataset. The regression line produced predictions that differed from the identity line thus demonstrating the model's ability to accurately predict essential structural features. The reliability of the model was strengthened by this analysis because it demonstrated stable and accurate performance in TMD parameter estimation to enhance its use in structural vibrational force optimization.

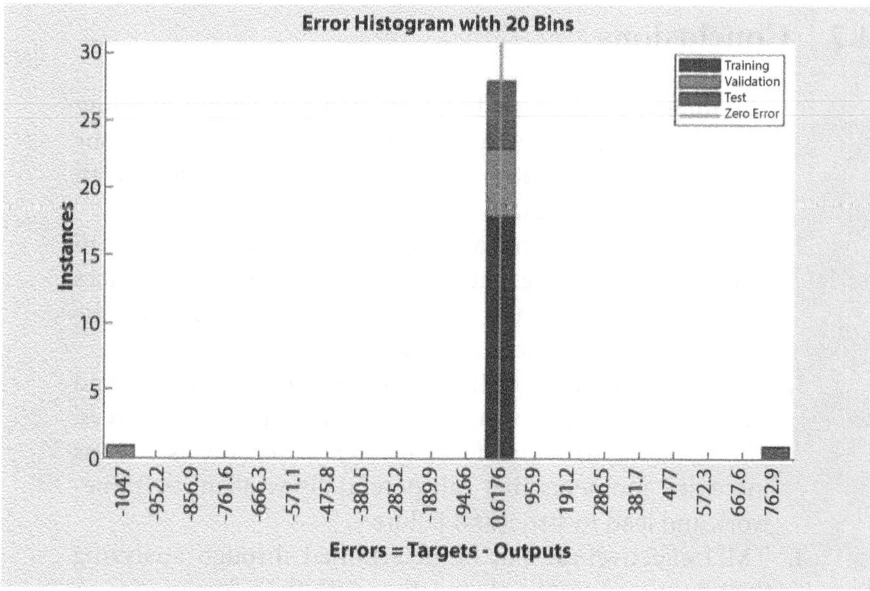

Figure 4.12 ANN – histogram error.

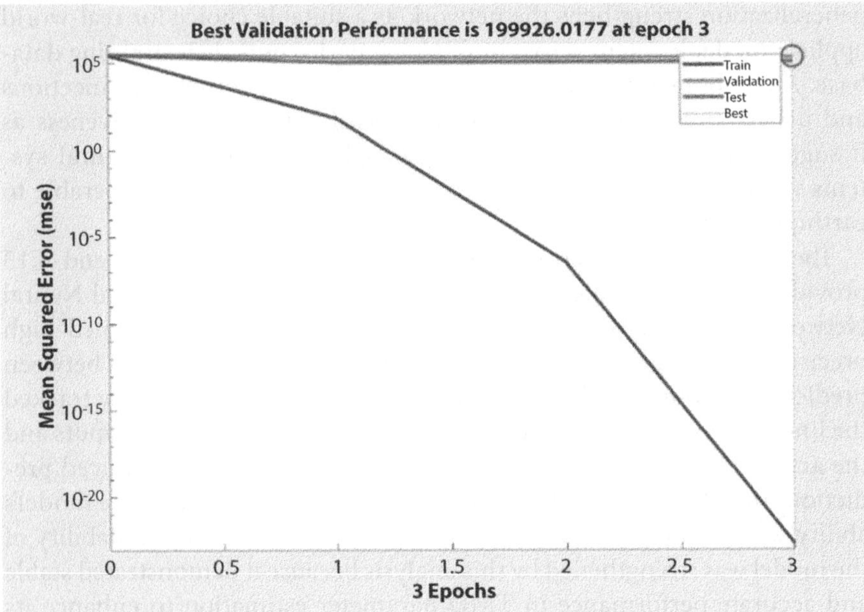

Figure 4.13 ANN – mean square error.

4.7 Conclusions

1. The implementation of TMD mechanisms within building structures lowers the structural response to seismic actions through decreased movements and acceleration along with lower deflection rates.
2. A direct link between TMD and the structure enables it to absorb seismic force energy which subsequently reduces building vibrations and deflections. The response of the structure becomes more controlled and stable as a result.
3. A building without TMD experiences notably harsh and extended structural vibrations because of seismic forces. The building structure faces greater deflection as well as increased moments that can potentially weaken its framework and lead to structural failure.
4. TMD effectiveness can be determined through analyzing the vibrations' amplitude together with their frequencies. A correctly installed TMD reduces vibration magnitude which

results in better control over building movement. Properties of moment vary negatively as TMD deployment takes effect in structures.

5. Building designers should consider TMD implementation during construction since doing so enhances seismic performance which prevents harm to occupants. Using TMDs in structures leads to remarkable reduction of structural displacement combined with reduced acceleration rates during earthquake events.
6. At a high level of accuracy the ANN computed parameters of the Tuned Mass Damper by processing input features which yielded results that matched the predicted optimal values.
7. The ANN predicted TMD parameter values with accuracy because regression analysis showed a direct linear relationship between the predicted results and actual measurements.
8. The ANN system maintained performance reliability for new untested TMD parameter data through established validation tests including cross-validation along with out-of-sample evaluation.
9. The ANN demonstrates its value to structural engineering practice through precise TMD parameter estimation that helps structural optimization for fighting vibrational forces notably during seismic activities.
10. ANN technology demonstrates outstanding capability to understand complex TMD parameter information which could lead to better designs for resilient structures exposed to vibrational forces.
11. The best ANN model fit for TMD produced validation R value of 0.9998 while generating all R values at 0.99987.
12. The approach of training an ANN through MATLAB provides engineers with a flexible method to predict TMD parameters at high accuracy levels which benefits multiple structural engineering fields.

References

1. Elias, S. and Matsagar, V., Research developments in vibration control of structures using passive tuned mass dampers. *Annu. Rev. Control*, 44, 129–156, 2017, https://doi.org/10.1016/j.arcontrol.2017.09.015.
2. Hrovat, D., Barak, F., Rabins, M., Semi-Active Versus Passive Or Active Tuned Mass Dampers For Structural Control. *ACSE*, 109, 3, 691–705, 1992, https://doi.org/10.1061/(ASCE)0733-9399(1983)109:3(691).
3. Pinkaew, T. and Fujino, Y., Effectiveness of semi-active tuned mass dampers under harmonic excitation. *Eng. Struct.*, 23, 7, 850–856, 2000, https://doi.org/10.1016/S0141-0296(00)00091-2.
4. Ada, M. and Ayvaza, Y., The Structure-Soil-Structure Interaction Effects on the Response of the Neighbouring Frame Structures. *Lat. Am. J. Solids Struct.*, 1, 19, 2019, https://doi.org/10.1590/1679-78255762.
5. Rana, R. and Soong, T.T., Parametric study and simplified design of tuned mass dampers. *Eng. Struct.*, 20, 3, 193–204, 1998, https://doi.org/10.1016/S0141-0296(97)00078-3.
6. Hoang, N. and Warnitchai, P., Design of multiple tuned mass dampers by using a numerical optimizer. *Earthquake Eng. Struct. Dyn.*, 34, 2, 125–144, 2004. DOI: 10.1002/eqe.413.
7. Garrido, H., Curadelli, O., Ambrosini, D., Improvement of tuned mass damper by using rotational inertia through tuned viscous mass damper. *Eng. Struct.*, 56, 2149–2153, 2013, http://dx.doi.org/10.1016/j.engstruct.2013.08.044.
8. McNamara, R.J., Tuned Mass Dampers. *Arch. Comput. Methods Eng.*, 20, 4, 419–431, 2013, DOI 10.1007/s11831-013-9091-7.
9. Sadek, F., Mohraz, B., Taylor, A., Chung, R., A method of estimating the parameters of tuned mass dampers for seismic applications. *Earthquake Eng. Struct. Dyn.*, 26, 6, 617–635, 1997, https://doi.org/10.1002/(SICI)1096-9845(199706)26:6%3C617::AID-EQE664%3E3.0.CO;2-Z.
10. Gorini, D.N. and Chisari, C., Impact of soil-structure interaction on the effectiveness of Tuned Mass Dampers. *Wiley*, 51, 6, 1501–1521, 2020. DOI:10.1002/eqe.3625.
11. Wu, J., Chen, G., Lou, M., Seismic effectiveness of tuned mass dampers considering soil–structure interaction. *Gen. Int. Assoc. Earthquake*, 28, 11, 1427–1447, 1999, https://doi.org/10.1002/(SICI)1096-9845(199911)28:11<1219::AID-EQE861>3.0.CO;2-G.
12. Nagashima, I., Maseki, R., Asami, Y., Hirai, J., Abiru, H., Performance of hybrid mass damper system applied to a 36-storey high-rise building. *Gen. Int. Assoc. Earthquake*, 30, 11, 1615–1637, 2001, https://doi.org/10.1002/eqe.84.
13. Chen, G. and Wu, J., Active multiple tuned mass dampers for structures under the ground acceleration. *Gen. Int. Assoc. Earthquake*, 32, 5, 793–810, 2002, https://doi.org/10.1002/eqe.136.

14. Chen, G. and Wu, J., Experimental study on multiple tuned mass dampers to reduce seismic responses of a three-storey building structure. *Gen. Int. Assoc. Earthquake*, 32, 5, 793–810, 2003, https://doi.org/10.1002/eqe.249.
15. Chen, J.W., Performance enhancement of bridge infrastructure systems: Long-span bridge, moving trucks and wind with tuned mass dampers. *Eng. Struct.*, 30, 11, 3316–3324, 2004, https://doi.org/10.1016/j.engstruct.2008.04.035.
16. Wong, K.K., Seismic Energy Dissipation of Inelastic Structures with Tuned Mass Dampers. *ASCE*, 134, 2, 163–172, 2008, https://doi.org/10.1061/(ASCE)0733-9399(2008)134:2(163).
17. Abed, A. and Moustachi, O.E.K., Pareto Optimal Design of the Tuned Mass Damper. *Civ. Eng. Archit.*, 10, 1, 236–245, 2022, DOI: 10.13189/cea.2022.100120.

5

Integration of Damping Systems for Vibration Control

Abstract
The research provides an extensive examination of damping system integration methods in engineering applications which require vibration control measures. Researchers focus on combined damping units together with control mechanisms and algorithms as well as progressive damping techniques while also exploring prediction models, feedback methods, builder-controlled frameworks, maintenance routines, and economic assessments. The research study first presents vibration control as critical before it moves sequentially through hybrid damping units and their mathematical equations then PID-based control systems with an overview of multi-stage approaches. The analysis includes performance-enhancing predictive simulation methods as well as sensor implementations for feedback control loops and adaptability assessment of complex structures. The later sections in this work address maintenance requirements and reliability measurements as well as practical financial aspects. The research develops a unified approach that boosts the operational excellence and adaptability as well as reliability of damping technologies for researchers and industry professionals alike.

***Keywords*:** Damper, integrated system, vibration, PID, hybrid system

5.1 Introduction

Multiple strategies using diverse damping systems need to be integrated for successful vibration control of specific engineering problems. The content expands vibration control fundamentals to explain how various damping elements work together for total effective vibration mitigation [1]. The chapter analyzes the benefits of hybrid damping systems that utilize various damping mechanisms to produce an enhanced result. This discussion examines hybrid systems by focusing on their strengths which unite precise performance from specific damping technologies with general usability from other components [2]. Experienced engineers discover hybrid

Kundan Meshram and Prashant Lahre. Vibration Control Systems, (65–88) © 2026 Scrivener Publishing LLC

system applications through observation of their implementation results across various engineering applications.

The text discusses multi-stage damping systems that incorporate arranged sequential damping methods which deliver cumulative damping through each stage. Researchers study multi-stage damping arrangements as they provide detailed knowledge about damping methods used sequentially to control vibrations affecting complex engineering structures [3]. The chapter demonstrates how predictive modeling and simulation tools function in the integration process of damping systems. Detailed discussions about advanced simulation techniques and modeling software demonstrate that engineers use predictions to analyze and optimize integrated damping systems' performance when implementing them [4]. The early design assessment through this approach enables the optimization of integrated damping solutions and a proof of their effectiveness while control strategy improvements focus on adaptive and learning-based control algorithms as essential elements of integrated damping systems. Real-time adjustments through these algorithms make use of sensor feedback and system variations or environmental changes to improve the dynamic external force response of integrated damping systems [5].

Researchers conduct detailed analysis of practical matters involving systems' costs alongside scalability and maintenance requirements to demonstrate the implementation limitations of multiple damping solutions [6]. Multiple engineering sectors such as aerospace craft and automotive industry and civil engineering use integrated damping solutions as demonstrated through practical case studies [7]. Successful implementation of diverse damping systems for vibration control requires evaluation of multiple components along with appropriate methodologies.

Let's discuss some integrated components:

5.2 Component 1: Hybrid Damping Units

5.2.1 Introduction

The integration of damping technologies through pioneering approaches has now become a leading strategy to achieve structural stability and mitigate vibrations in various engineering systems according to [8]. Hybrid Damping Units combine different damping capabilities to develop a flexible and effective solution which controls vibrations. The integration of viscoelastic dampers and tuned mass dampers through hybrid Damping Units makes them stand as a unified innovation that brings multiple damping technologies together [9]. Viscoelastic dampers have a wide frequency response by distorting during stress events and tuned mass dampers specialize in counteracting resonant frequencies that occur within structures. Engineers seek to use distinct methods of damping as an integrated system to benefit from their individual strengths in comprehensive vibration control approaches [10]. The combination of damping systems generates a complete solution that goes beyond the capabilities of standalone damping systems. A combined damping system offers great potential to handle vibrations across multiple frequency bands along with diverse amplitude ranges. Viscoelastic and tuned mass dampers unite in a single system which excels at vibrational control of environmental forces and machinery operations as well as seismic activities [11]. Engineering professionals aim to transform vibration control using Hybrid Damping Units because these sophisticated adaptable systems provide an effective solution for ensuring structural stability during a wide range of engineering applications. The engineering field presently adopts Hybrid damping units as an advanced method to build systems that efficiently handle structural vibrations. Engineers integrate viscoelastic dampers (energy-absorbing devices) with tuned mass dampers (vibration-reducing oscillating masses) in this component of the design.

Viscoelastic dampers gain reputation through their stress-deformation mechanism which enables energy absorption. They show efficient performance at all known vibration frequencies while effectively eliminating energy generated from different vibrational sources [12]. Tuned mass dampers operate through mechanical oscillation that fights against the natural vibrations of the structure at target frequencies. Engineers combine different damping systems through hybridization to access their individual advantages. Engineers develop this system because they want to build a vibration control system which is more flexible and effective. The combination approach delivers increased applicability because it extends operational flexibility for various vibration frequencies and magnitudes [13]. The unified system uses viscoelastic dampers which dissipate energy during different frequency ranges together with tuned mass dampers that specifically target resonant frequencies. The system provides advanced vibration control of structures and machinery and any system which experiences vibrations because of its combined functionality [14]. This system provides observable advantages because it responds well to different vibration conditions as shown in Table 5.1. The system provides structural integrity alongside safety through its ability to handle vibrations which come from diverse origins including natural forces and machine operations as well as earthquake episodes.

Table 5.1 Advantages of (viscoelastic dampers – tuned mass damper) for hybrid damping units.

Damping technology	Characteristics	Advantages
Viscoelastic Dampers	Absorbs and dissipates energy through deformation	Effective across a wide range of vibration frequencies Efficient energy dissipation
Tuned Mass Dampers	Counters specific resonant frequencies	Precision in targeting resonant frequencies. Effective at mitigating specific vibrations
Hybrid Damping Units	Integration of both technologies	Broader range of vibration control Versatility in handling diverse frequency ranges and amplitudes

5.2.2 Methodology

Structural vibration analysis serves to determine all essential vibration elements including natural frequencies and mode shapes as well as force response behavior. Selection of critical dampening zones occurs after performing this evaluation because it determines the locations of maximum vibration and resonance. The established findings call for selecting viscoelastic or tuned mass dampers as appropriate damping systems that conform to structural requirements and frequency goals [15–17]. The integration of damping technologies needs to follow a properly designed plan which keeps structural functionality intact throughout the process. A set of mathematical algorithms needs to be created to demonstrate how hybrid damping systems interact with structural dynamics and their components. Performance predictions of the system under various vibrational conditions become possible with simulation tools while design improvements receive support. The laboratory should produce prototypes that encompass chosen damping elements to test functionality alongside damping effectiveness. Optimal performance happens when validated designs finish an iterative development process by modifying properties of mass and stiffness and material characteristics [18, 19]. Testing of damping units

Table 5.2 Advantages and disadvantages of hybrid damping units.

Advantages	**Disadvantages**
Effective control across a wide frequency range	Complex design and implementation
Enhanced adaptability to specific frequencies	Higher initial cost
Optimized energy dissipation	Space requirements
Improved structural stability	Maintenance and tuning requirements
Synergy of diverse damping mechanisms	Specificity in application

in real-life structures consisting of buildings, bridges, or machinery validates operational effectiveness. A system performance monitoring process which continues from design through testing until field application is completed ensures reliability while providing future improvement direction.

Different engineering considerations and analyses determine the integration factors when combining damping technologies within Hybrid Damping Units as given in Table 5.2. Structural analysis of vibrations helps engineers determine which frequency bands and vibration amplitudes control the behavior of the system before choosing damping technologies that focus on these vital frequencies. The assessment process identifies proper working relationships among different damping systems with structure and inter-damping configurations.

5.2.3 Analytical System Equations for Hybrid Damping Unit

Single-Degree-of-Freedom (SDOF) Systems:

Equations of motion for hybrid damping units are represented similarly to traditional damping systems:

$$m\ddot{x} + c_{passive}\dot{x} + k_{passive}x + c_{active}\dot{x} + k_{active}x = F(t) \qquad (5.1)$$

The system incorporates two coefficients representing damping with c while k stands for stiffness to describe passive and active damping elements.

5.3 Component 2: Control Systems and Algorithms

5.3.1 Introduction

Engineers rely on Control Systems and Algorithms as a fundamental field which delivers several managerial approaches for governing system responses to vibrations and external forces. The components established structural stability and support machinery performance and operational efficiency within different industries [20]. Control systems implement different methods which include both passive structures using built-in properties and active systems that perform real-time vibration opposition. Several algorithms including PID and adaptive and fuzzy logic and optimal control provide advanced decision-making capabilities to these systems because they base their responses on system dynamics and environmental inputs. These engineering systems and automatic algorithms have widespread use in structural engineering as well as mechanical systems and aerospace and automotive industries and robotics together with biomedical devices and energy networks and environmental control systems [21]. Multiple capabilities of these systems enable technological progress toward better safety and efficiency and higher precision throughout various engineering and technological fields.

5.3.2 Control Systems and Algorithms

The control system represents an important solution for structural vibration management that enables system stability benefits along with security improvements and operational efficiency across engineering systems. A complete vibration control system integrates sensors with controllers as well as actuators to detect vibrations before implementing algorithm-based corrective procedures. The field of vibration control includes three primary concept categories that consist of passive control units and active control units and semi-active control units. Dampers and isolators make up passive systems which absorb vibrations naturally without power input yet remain unadoptable. The combination of smart materials in semi-active systems allows for damping property adjustment which grants better flexibility than passive systems without the need for as much power consumption as active systems [22].

Control Systems in Vibration Management:

a. Structural responses to vibration receive control through systems that result in changes to stability measures together with performance levels and safety protocols.
b. The system contains three essential modules.
c. The sensors function to discover as well as measure vibrations which originate either from environmental conditions or mechanical systems while delivering instantaneous data.
d. The structure receives commands from controllers that result in changes through actuators which work to fight or minimize vibrations.
e. The controllers take input from sensors through algorithms which help find ideal system reactions and synchronize actuator functions.
f. These systems operate through passive as well as active and adaptive strategies while using predictive strategies with feedback loops.
g. These control systems find extensive implementation across structural and mechanical fields and aerospace industries and automotive sectors for improving system operational performance and product lifespans and reliability.

Role of Algorithms in Control Systems:

a. The fundamental decision process of control systems functions through algorithms that determine the operational relationship between sensors and controllers and actuators.
b. The system turns measured sensor data into valuable information as it detects vibrations in combination with their frequency data, alongside their intensity levels.
c. Real-time controller analysis and response generation requires implementation of mathematical models and rules or optimization approaches through controllers.
d. The guidance system determines how actuators modify their performance metrics and force applications for vibration management.
e. The system must include extensive collection of methodologies.
f. Traditional algorithms: Such as PID (Proportional-Integral-Derivative) control.

g. The implementation of adaptive predictive and model-based approaches which focus on particular system dynamics specifies the advanced algorithms.
h. An efficient intelligent control system requires the synchronization of data interpretation with actuator responses for maintaining dynamic system stability.

5.4 PID (Proportional-Integral-Derivative) Control

The classical feedback control algorithm known as PID control serves industries around the world. The system operates through an automatic output adjustment system which depends on three essential terms.

The control procedure increases its output dimension proportionally with the existing measurement of error that emerges from actual versus desired system results.

The integral term sums past errors together while operating during a specific time period to develop compensating actions for handling enduring errors.

Audio mistakes into future errors through examining the speed at which errors change.

The process of combining proportional at the current error input with integral total error accumulation from the past and derivative error prediction for future usage enables PID controllers to minimize reach and actual response differences for enhanced system stability. PID controllers maintain widespread use because they provide easy control and broad functional capability with simple parameter adjustment processes, which take much time to implement effectively.

Adaptive control systems use input feedback to modify their operational strategies automatically, therefore making them effective for different situations. Fuzzy logic control functions like a human brain through rules-based logic and linguistic control variables to handle situations where an accurate model would be hard to obtain. Theoretical approaches for controlling systems minimize operational expenses through advanced computational tools yet need perfect system models along with powerful computing systems. System complexity alongside data requirements and precision needs and operation constraints determine which control algorithm among PID and adaptive and fuzzy and optimal will be the most appropriate choice.

Adaptive Control Systems: The system adjusts its behavior depending on environmental conditions and variations within the system.

The control parameters should be updated continuously through learning processes, suitable for systems with variable dynamics.

Fuzzy Logic Control: The system implements rule-based logic by using linguistic terms which include "very hot" and similar expressions.

The control system uses human-like logical approaches to process vague or imprecise information.

This system proves useful when there is specialized expertise present alongside difficulties in developing mathematical models.

Optimal Control Systems: Best control actions should be sought through performance criteria-based assessments.

The optimization tools calculus of variations and dynamic programming should be used to achieve desired results.

The system must use highly precise models together with powerful computing capabilities.

Algorithm Selection: The determination of which control system to use depends on how complex the system is as well as the availability of data and accuracy requirements and available computing power.

Each control type (PID, adaptive, fuzzy, optimal) serves specific applications.

5.5 Applications of Control Systems and Algorithms in Various Fields

Structural Engineering:

Strategic control systems together with algorithmic protocols become necessary tools for inhibiting vibrations within buildings as well as bridges and other types of infrastructure. Structural systems equipped with these systems improve safety along with system stability and enhance their lifespan by absorbing:

a. Seismic events
b. Wind-induced forces
c. Operational and environmental loads
d. Vibrational energy is absorbed or redirected through integrations of dampers along with isolators and active control mechanisms.
e. These preventive measures boost structural resistance and disaster readiness thus engineers can achieve extended operation time for built infrastructure.

Mechanical Engineering:
Vibrational control depends on control systems and algorithms for their operation in rotating machines and equipment.

a. Vibrational control prevents:
b. Premature wear and tear
c. Reduced mechanical efficiency
d. Potential mechanical failure
e. Professional engineers frequently implement active vibration isolation techniques together with suppression methods for their applications.
f. The reliability of systems and their operational performance need special attention in tools that require precision and in manufacturing units as well as turbines and engines.

Aerospace and Automotive Engineering:
Modern control technology helps aerospace as well as automotive companies fight vibrations while protecting their aircraft and space missions and vehicles.

a. Passenger comfort
b. Structural integrity
c. System efficiency and performance
d. Control systems operate to neutralize the vibration effects which stem from engine procedures and flight mechanics as well as street surfaces and elements from the external environment.
e. Transportation systems experience smoother operation and lower fatigue loading, as a result of which their safety and operational durability increase.

Vibrations within these applications require control systems and algorithms to control the vibrations while simultaneously managing their effects. These systems and algorithms substantially increase safety standards while improving operational performance as well as extending the life of both structures and machinery and vehicles and dynamic systems across engineering sectors.

1. Control systems implemented in robotics guarantee robots and automated machines maintain accurate positioning while providing stability during movement operations including drones and autonomous vehicles and robots.

2. Control systems within biomedical engineering space manage medical equipment together with prosthetic devices as well as physiological systems thus delivering precise medications while operating therapeutic technologies and implantable medical technology.
3. The control systems implement optimization methods for energy management through power grids and renewable resources along with smart grids to guarantee secure stable and efficient delivery of power.
4. Environmental engineering depends on control systems to regulate the treatment of waste while maintaining water quality standards together with monitoring air protection measures that allow proper regulation enforcement and preservation of natural resources.
5. Process Control and Manufacturing implements control systems to optimize industrial manufacturing processes and chemical processes and machine operations for precise results and waste management reduction.

5.5.1 Methodology

Structural dynamic research that combines control systems with algorithms requires a well-defined process starting from dynamic structural analysis with mathematical modeling at its foundation. The control objectives become defined after understanding the system behavior at which point appropriate algorithms are chosen from available options according to system specifications. The following development phase requires constitution of algorithms through simulation assessment to confirm their capacity in managing system outcomes effectively [23]. A series of integration procedures happens after development to ensure the control system connects flawlessly with its associated algorithms. The optimization process together with parameter refinement optimizes system performance while sensitivity testing determines system stability levels. The method continues by testing under different environmental circumstances before field implementation to acquire real-world findings. System operation improves through continuous feedback together with proper documentation and consistent maintenance procedures which maintain operational efficiency in both the short- and long-term [24]. Structural dynamic research that combines control systems with algorithms requires a well-defined process starting from dynamic structural analysis with mathematical modeling at its foundation. The control objectives become defined after understanding the system behavior at which point appropriate algorithms are chosen from available options according to system specifications. The following development phase requires constitution of algorithms through simulation assessment to confirm their capacity in managing system outcomes effectively. A series of integration procedures happens after development to ensure the control system connects flawlessly with its associated algorithms. The performance maximum is achieved by optimization and tuning processes alongside sensitivity analyses that check system robustness. The method continues by testing under different environmental circumstances before field implementation to acquire real-world findings.

5.5.2 Output

a. Improved Stability: Maintains structural integrity under dynamic conditions.
b. The system operates with better functionality because it suppresses oscillations while delivering enhanced operational performance.
c. Vibration Control: Suppresses harmful vibrations and resonance.
d. The system can adapt its operation to handle different loads together with changing environmental conditions.
e. Damage Reduction happens because structures experience minimal structural wear and tear with this method.
f. Real-time Response: Enables instant monitoring and adjustments.
g. The optimized design system results in safer structures that become cost-efficient.
h. The continuous learning mechanism continuously adjusts through feedback mechanisms.
i. Greater Resilience: Increases resistance to seismic and wind forces.

5.6 Component 3: Multi-Stage Damping Strategies

5.6.1 Introduction

The goal of innovation in structural dynamics involves vibration management to enhance safety together with durability and performance outcomes. The multi-stage damping method helps operation teams achieve specific vibration control by using multiple structured layers. Mechanism: Combines passive, active, and semi-active damping methods for broad-spectrum vibration mitigation. Engineers need to have exact engineering skills together with vibration behavior comprehension to successfully implement damping elements through precise strategic positions [25–27].

5.6.2 Applications and Advantages

Structural durability and stability improve through multi-stage damping while it lowers the risks connected to material fatigue [28]. The applications of passive devices include buildings and bridges as well as machinery systems and aerospace operations.

Adaptability: Features real-time optimization and continuous improvement through feedback loops. Engineers employ this precise method because it ensures both structural durability along with reliability according to [29].

Main Components:

- Tuned mass dampers together with passive components operate through the implementation of viscoelastic materials as well as friction and tuned mass dampers.
- These systems perform corrective operations through sensor-based applications of algorithmically controlled forces through actuators.
- Semi-active elements join passive systems with active ones in order to deliver adjustable damping control capabilities.

Damping Layer Strategy:

- The structure contains layers which distribute themselves sequentially from top to bottom.
- Each system within this approach operates to control vibrations which fall within its assigned frequency ranges.

Monitoring Systems:

- Sensors identify what types of vibrations exist in addition to the intensity of those vibrations.
- Feedback Loops: Allow real-time control and adjustment.

Control and Actuation:

- The algorithms through sensors determine how much damping force should be applied to structures.
- The application of forces through actuators enables vibration control in the system.

Design Tools:

- Frequency Analysis: Identifies critical vibration modes.
- FEA works as a simulation method to identify optimal locations where damping systems should be placed.

Integration Measures:

- Knowledge in structural dynamics and control systems represents the necessary field experience for this work.
- The fundamental requirement for excellent vibration control is both precise structural placement of damping devices and tuning accomplishments.

Optimization:

- The system allows real-time data adjustments since it supports feedback-based refinements.
- The system demonstrates flexibility to handle modifications in its environment.

Maintenance:

- The system stores documentation useful for tracking performance as well as tracking system upgrades.

Routine Checks: Ensures long-term reliability.

5.6.3 Methodology

A systematic approach must be deployed for Multi-Stage Damping Strategies implementation in structural dynamics since they manage vibrations within different frequency bands [30]. The method starts with dynamic structural analysis to find vital frequencies with their origins.

Moving forward the research group develops particular design targets. Damping mechanisms of passive, active, and semi-active types are selected before getting arranged in multiple stages for specific vibration control. The system parameters receive mathematical evaluation through modeling and simulation before final adjustment. The structure receives precisely installed damping systems following their design and optimization stages. Under real conditions performance of the system is achieved through strict testing and validation procedures. The system obtains continuous feedback which enables the execution of adaptive adjustments. The application of proper documentation together with maintenance protocols ensures efficiency maintains itself over the long run. The advanced method demonstrates its capability through application of precise engineering with adaptable characteristics for achieving strong structural responses as shown in Table 5.3.

5.6.4 Analytical Approaches

Multiple formulations serve to analyze Multi-stage Damping Strategies through mathematical equations which describe the operational behavior of damping devices throughout the stages. The exact formulations between Multi-Stage Damping Strategies differ extensively depending on what damping mechanisms are used along with their interrelationships in the structural system as given in Table 5.4.

Single-DOF (Degree of Freedom) Systems

The equations of motion for a single-degree-of-freedom system which includes damping mechanisms become:

$$m\ddot{x} + c_1\dot{x} + k_1x + c_2\dot{x} + k_2x + \cdots + c_n\dot{x} + k_nx = F(t) \quad (5.2)$$

m is the mass of the system.

The system displacement represents the value of x.

The system incorporates both damping coefficients c_I and stiffness constants k_I at each step of its operation.

The system reacts to its external force through the variable F(t).

Multi-DOF Systems

The equations of motion evolve into complicated matrices for systems that possess multiple degrees of freedom in structural and building applications.

$$M\ddot{X} + C_1\dot{X} + K_1 + C_2\dot{X} + K_2X + \cdots + C_N\dot{X} + K_nX = F(t) \quad (5.3)$$

Table 5.3 Methodology multi-stage damping strategies.

Step	Description
1. Structural Analysis	Conduct a thorough analysis of the structural dynamics, identifying critical frequencies and modes of vibrations.
2. Vibration Assessment	Assess the sources and characteristics of vibrations affecting the structure, considering external and internal factors.
3. Design Objectives	Define specific objectives for damping, considering targeted frequency ranges and types of vibrations to be mitigated.
4. Damping Mechanisms	Select appropriate damping mechanisms (passive, active, or semi-active) suitable for the identified vibration types.
5. Sequential Arrangement	Design a sequential arrangement of damping stages, each targeting specific frequencies or types of vibrations.
6. Mathematical Modeling	Develop analytical models and equations to represent the behavior of the damping mechanisms within the structural system.
7. Simulation and Analysis	Utilize simulations and analysis tools (e.g., finite element analysis) to evaluate the effectiveness of the damping strategies.
8. Optimization and Tuning	Optimize the damping system parameters based on simulation results and fine-tune the arrangement for optimal performance.
9. Integration and Implementation	Integrate the damping mechanisms into the structural design, ensuring compatibility and precise placement.
10. Testing and Validation	Conduct comprehensive testing to validate the performance of the damping strategies under various conditions.
11. Monitoring and Feedback	Implement monitoring systems to gather real-time data for continuous feedback and adaptive adjustments.
12. Documentation and Maintenance	Maintain detailed records, manuals, and protocols for ongoing maintenance and future enhancements.

Table 5.4 Multi-stage damping strategies in structural buildings.

Application	Description
Skyscrapers	Mitigation of wind-induced vibrations, reducing discomfort for occupants, and preventing structural fatigue.
Bridges	Controlling oscillations caused by traffic loads or wind, improving ride comfort, and ensuring structural integrity.
Seismic Protection	Absorbing and dissipating seismic energy to reduce structural damage and enhance resilience during earthquakes.
Tall Structures	Addressing sway and vibrations due to natural forces, enhancing stability and occupant comfort.
Heritage Structures	Preserving historical buildings by minimizing vibrations and structural degradation, ensuring long-term conservation.
Industrial Facilities	Managing vibrations caused by machinery operation, preventing structural wear and optimizing operational safety.
Hospitals & Laboratories	Minimizing vibrations to maintain stability for sensitive equipment and critical medical operations.
Sports Arenas	Reducing vibrations during events to ensure spectator comfort and structural safety.
Residential Buildings	Enhancing overall structural performance, reducing vibrations due to occupant activities.
Mixed-Use Complexes	Managing diverse sources of vibrations from various functions within a single structure.

M is the mass matrix.
X represents the vector of displacements for all degrees of freedom.
C_i and K_i are the damping and stiffness matrices at each stage.

5.7 Component 4: Predictive Modeling and Simulation

5.7.1 Introduction

Engineers need Predictive Modeling and Simulation to analyze how damping systems perform before their deployment in real operations. Advanced simulation tools enable engineers to create system behavior models under various conditions including earthquake and wind loads which follow refinements through testing sequences and allow them to validate outcomes by matching computer simulations to theoretical findings and experimental measurements to minimize the need for real prototypes [31,–33].

5.7.2 Methodology

The response behaviors from simulations direct engineers to optimize their designs of damping systems through different load scenarios. The simulation outcomes enable engineers to both find the best possible configuration while assessing possible dangers and to validate accuracy levels against modeled data or experimental information and support decisions based on the collected data. The system's ability to resist failures improves through this sequential method that produces documented data for future development.

5.8 Component 5: Feedback Mechanisms and Sensors

5.8.1 Introduction

Integrated damping systems use core components of sensors and feedback mechanisms to bring adaptability through their capabilities to acquire real-time data and perform dynamic structural analysis and automatic response adjustments as shown in Figure 5.1 and Table 5.5. Environmental conditions monitoring is achieved through sensors that report data while feedback systems activate fast damping controls for structural stability

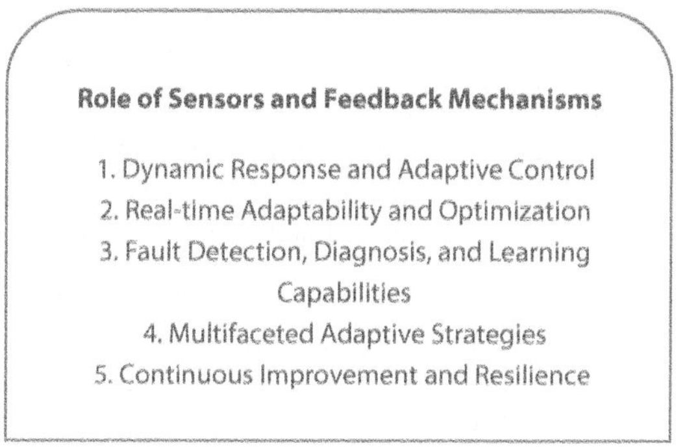

Figure 5.1 Role of sensors and feedback mechanisms.

maintenance. The damping system enables error detection and automated response and feature learning and shows resistance to seismic activity, wind effects, and operational conditions. The system produces real-time data together with adaptive control protocols as well as performance metrics and anomaly diagnostics which are documented alongside methodologies to achieve optimized damping systems in various conditions.

5.9 Component 6: Structural Adaptability and Scalability

5.9.1 Introduction

This component ensures damping systems can evolve and adjust to structural changes, environmental conditions, and new technologies without losing effectiveness as given in Table 5.6. It emphasizes flexibility, resilience to factors like temperature or seismic forces, and scalability to integrate new damping elements. Robustness and redundancy ensure reliability during component failures. Compatibility with advanced tech—such as IoT and AI—enables smarter control and monitoring. Adaptive control strategies allow real-time adjustment and learning from historical data. Key outcomes include verified adaptability, environmental resilience, scalable integration, robust performance, technological compatibility, and compliance with standards through comprehensive documentation and validation as shown in Figure 5.2.

Component 7: Maintenance and Reliability Protocols

Component 8: Cost-Efficiency and Practical Implementation

Table 5.5 Methodology of feedback mechanisms and sensors.

Step	Description
1. Requirement Analysis	Define system requirements, including types of sensors needed, data acquisition frequency, and required feedback mechanisms.
2. Sensor Selection and Integration	Research and select appropriate sensors (accelerometers, strain gauges, etc.) compatible with the structural dynamics. Integrate sensors into the system.
3. Calibration and Testing	Calibrate sensors to ensure accuracy and reliability. Conduct testing to validate sensor functionality under different conditions.
4. Data Acquisition and Processing	Implement data acquisition systems to collect real-time structural response data. Develop processing techniques for analyzing raw sensor data.
5. Feedback Loop Design	Design feedback mechanisms and control strategies to interpret sensor data and trigger system adjustments.
6. Integration with Control Systems	Integrate sensor data and feedback mechanisms into the overall control system architecture for dynamic adjustments.
7. Dynamic Response Simulation	Simulate real-time responses and adjustments based on sensor data to evaluate system behavior under different scenarios.
8. Performance Evaluation	Evaluate system performance metrics such as response time, accuracy of adjustments, and system stability under varying conditions.
9. Fault Detection and Diagnostics	Implement algorithms for fault detection based on sensor data anomalies. Develop diagnostic protocols for system analysis.
10. Continual Improvement	Establish protocols for ongoing system learning and improvement based on historical sensor data and machine learning algorithms.
11. Documentation and Reporting	Document sensor integration methodologies, system architectures, and feedback loop designs, and generate comprehensive reports for reference.

Table 5.6 Methodology steps for structural adaptability and scalability.

Step	Description
1. Requirement Analysis for Adaptation	Define the system's adaptability requirements, including potential structural changes and scalability needs.
2. Environmental Resilience Assessment	Assess the system's capability to withstand diverse environmental factors and external forces.
3. Scalability Evaluation	Evaluate the system's capacity for expansion, integration of new damping mechanisms, and scalability.
4. Redundancy and Robustness Strategies	Design redundancy mechanisms and ensure robustness to maintain functionality and reliability.
5. Compatibility with Advanced Tech	Analyze and ensure compatibility with advanced control systems, IoT devices, and AI integration.
6. Development of Adaptive Control	Design adaptive control algorithms for real-time adjustment based on sensor data and system learning.
7. Documentation and Records Management	Maintain comprehensive documentation detailing adaptations, scalability plans, and system records.
8. Validation and Testing Protocols	Develop testing protocols to validate system adaptability, scalability measures, and compliance.
9. Adherence to Regulatory Standards	Ensure compliance with industry standards and regulations despite system modifications or expansions.

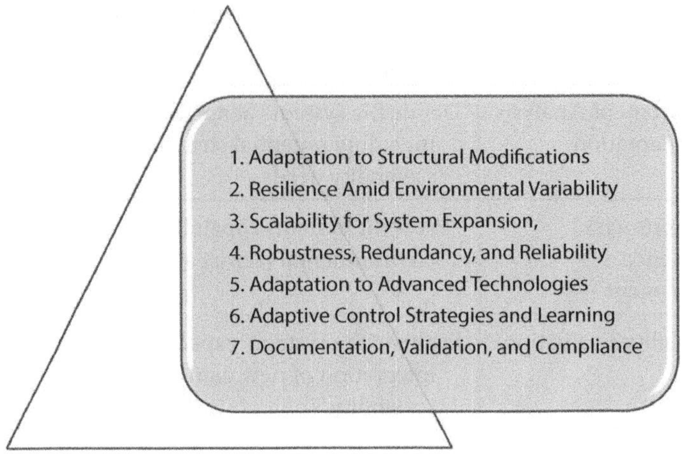

Figure 5.2 Results and outputs from implementation.

References

1. Kavyashree, B.G., Patil, S., Rao, V.S., Review on vibration control in tall buildings: from the perspective of devices and applications. *Int. J. Dyn. Control*, 9, 1316–1331, 2021.
2. Van Der Schaft, A.J. and Schumacher, H., *An introduction to hybrid dynamical systems*, vol. 251, springer, London, UK, 2007.
3. Chen, B., Zhao, S.-L., Li, P.-Y., Application of Hilbert-Huang transform in structural health monitoring: a state-of-the-art review. *Math. Probl. Eng.*, 2014, 2014.
4. Murray-Smith, D.J., *Modelling and simulation of integrated systems in engineering: issues of methodology, quality, testing and application*, Elsevier, Cambridge, UK, 2012.
5. Feng, D. and Feng, M.Q., Computer vision for SHM of civil infrastructure: From dynamic response measurement to damage detection–A review. *Eng. Struct.*, 156, 105–117, 2018.
6. De Groot, R.S., Alkemade, R., Braat, L., Hein, L., Willemen, L., Challenges in integrating the concept of ecosystem services and values in landscape planning, management and decision making. *Ecol. Complex.*, 7, 3, 260–272, 2010.
7. Brownjohn, J.M.W., De Stefano, A., Xu, Y.-L., Wenzel, H., Aktan, A.E., Vibration-based monitoring of civil infrastructure: challenges and successes. *J. Civ. Struct. Health Monit.*, 1, 79–95, 2011.
8. Casciati, F., Magonette, G., Marazzi, F., *Technology of semiactive devices and applications in vibration mitigation*, John Wiley & Sons, Chichester, UK, 2006.

9. Vasanth, X.A., Paul, P.S., Lawrance, G., Varadarajan, A.S., Vibration control techniques during turning process: a review. *Aust. J. Mech. Eng.*, 2019.
10. Nakra, B.C. and Dig., S.V., Vibration control with viscoelastic materials-III. *Shock Vib. Dig.*, 1984.
11. Worden, K., Bullough, W.A., Haywood, J., *Smart technologies*, World Scientific, Singapore, 2003.
12. Ibrahim, R.A., Friction-induced vibration, chatter, squeal, and chaos—part II: dynamics and modelling. *Appl. Mech. Rev.*, 227–253, 1994.
13. Kaveh, A., *Advances in metaheuristic algorithms for optimal design of structures*, Springer International Publishing, Switzerland, 2014.
14. Preumont, A., *Vibration control of active structures: an introduction*, vol. 246, Springer, NY, USA, 2018.
15. Doebling, S.W., Farrar, C.R., Prime, M.B., A summary review of vibration-based damage identification methods. *Shock Vib. Dig.*, 30, 2, 91–105, 1998.
16. Dai, J., Xu, Z.-D., Gai, P.-P., Li, H.-W., Effect of frequency dependence of large mass ratio viscoelastic tuned mass damper on seismic performance of structures. *Soil Dyn. Earthquake Eng.*, 130, 105998, 2020.
17. Beards, C., *Structural vibration: analysis and damping*, Elsevier, Cham, Switzerland, 1996.
18. Pandey, D.K., Mishra, S.K., Chakraborty, S., A tuned liquid mass damper implemented in a deep liquid storage tank for seismic vibration control of short period structures. *Struct. Des. Tall Special Build.*, 31, 8, e1928, 2022.
19. Di, F., Sun, L., Chen, L., Optimization of hybrid cable networks with dampers and cross-ties for vibration control *via* multi-objective genetic algorithm. *Mech. Syst. Signal Process.*, 166, 108454, 2022.
20. Chattopadhyay, D.K. and Raju, K.V.S.N., Structural engineering of polyurethane coatings for high performance applications. *Prog. Polym. Sci.*, 32, 3, 352–418, 2007.
21. Serpanos, D. and Wolf, M., *Internet-of-things (IoT) systems: Architectures, algorithms, methodologies*, Springer, Oxford, UK, 2017.
22. Dwivedi, Y.K., Hughes, L., Ismagilova, E., Aarts, G., Coombs, C., Crick, T., Duan, Y., *et al.*, Artificial Intelligence (AI): Multidisciplinary perspectives on emerging challenges, opportunities, and agenda for research, practice and policy. *Int. J. Inf. Manage.*, 57, 101994, 2021.
23. Wilson, B., *Systems: concepts, methodologies, and applications*, John Wiley & Sons, NY, USA, 1991.
24. Hammond, W.E., *Design methodologies for space transportation systems*, Aiaa, Reston, VA, USA, 2001.
25. Harne, R.L. and Wang, K.-W., *Harnessing bistable structural dynamics: for vibration control, energy harvesting and sensing*, John Wiley & Sons, Hoboken, NJ, USA, 2017.
26. Mitchell, D., Blanche, J., Zaki, O., Roe, J., Kong, L., Harper, S., Robu, V., Lim, T., Flynn, D., Symbiotic system of systems design for safe and resilient autonomous robotics in offshore wind farms. *IEEE Access*, 9, 141421–141452, 2021.

27. Suresh, B.N. and Sivan, K., *Integrated design for space transportation system*, Springer India, New Delhi, 2015.
28. Aloisio, A., Pasca, D.P., De Santis, Y., Hillberger, T., Giordano, P.F., Rosso, M.M., Tomasi, R., Limongelli, M.P., Bedon, C., Vibration issues in timber structures: A state-of-the-art review. *J. Build. Eng.*, 107098, 2023.
29. KC, S. and Gautam, D., Progress in sustainable structural engineering: a review. *Innovative Infrastruct. Solutions*, 6, 2, 68, 2021.
30. Avci, O., Abdeljaber, O., Kiranyaz, S., Hussein, M., Gabbouj, M., Inman, D.J., A review of vibration-based damage detection in civil structures: From traditional methods to Machine Learning and Deep Learning applications. *Mech. Syst. Signal Process.*, 147, 107077, 2021.
31. Harris, H.G. and Sabnis, G., *Structural modeling and experimental techniques*, CRC press, Boca Raton, FL, USA, 1999.
32. Kwok, K.C.S. and Samali, B., Performance of tuned mass dampers under wind loads. *Eng. Struct.*, 17, 9, 655–667, 1995.
33. Takewaki, I., *Building control with passive dampers: optimal performance-based design for earthquakes*, John Wiley & Sons, Chichester, UK, 2009.

6

Applications and Studies of Damping Systems

Abstract
A brief yet descriptive evaluation of damping systems in their performance with different engineering applications is available in this section. The engineers in use now integrate LCAAs and TMDs and use FVDs and hydraulic as well as active damping systems to integrate aerospace industry and civil infrastructure and automotive transportation - industrial system. The unavailability of damping systems in the existing engineering environment gives a possibility to make important improvements regarding system safety, efficiency, and stability performance.

Keywords: Hydraulic damper (HD), active damper system (ADS), tuned mass damper (TMD)

6.1 Introduction

6.1.1 Case Study: Aerospace Applications

Stability maintenance and passenger safety in aerospace architecture are dependent on damping systems since these systems are critical in safety aspects of such a dynamic industry. The Liquid Circular Angular Accelerometers (LCAAs) are key elements of precision navigation systems in this environment [1]. The inherent ability of LCAAs to reduce natural vibration hence protect their indispensable worth by making aircraft sensor granting the TMD implementation the significance of a key goal attainment in aerospace engineering development [2]. Lawrence Anticipatory Control Accelerometers (LCAAs) particularly counteract aerodynamic-induced vibrations since they improve aircraft structural integrity, and flight security to passengers [3]. In this story, the aerospace sector demonstrates how LCAAs and TMDs demonstrate their importance in producing precision in the airspace environment with structural resistance supported with enhanced safety measures in the airspace environment [4].

Kundan Meshram and Prashant Lahre. Vibration Control Systems, (89–102) © 2026 Scrivener Publishing LLC

6.1.2 Liquid Circular Angular Accelerometers (LCAAs) in Precision Navigation

Aerospace engineering requires damping systems for aircraft stability and passenger safety as part of an important safety element of this dynamic area. In this environment, Liquid Circular Angular Accelerometers (LCAAs) are used to carry out key tasks by Precision navigation systems [1]. The inherent capacity of LCAAs to reduce natural vibrations guarantees their key role in enhancing aircraft sensors, whereas the process of implementation of TMD is a vital engineering success in aerospace engineering development [2]. Special protective devices called Lawrence's Anticipatory Control Accelerometers (LCAAs) have been created to counteract aerodynamic-induced vibrations, hence reinforcing the framework of the aircraft and safety of passengers during flights [3].

6.1.3 Methodology

Research on the aerospace damping systems involves the following sequential steps:

Requirement Identification and Scope Definition

The identification of exact damping requirements of aerospace damping needs, that include vibration control and precision guidance as well as structural support, is vital for exploration of aerospace damping needs. Research Scope Definition involves establishing parameters that center on the analysis of both Liquid Circular Angular Accelerometers (LCAAs) and Tuned Mass Dampers (TMDs) as essential damping technology systems.

Literature Review and Case Study Selection

- A study of current research must be performed regarding the utilization of damping systems within aerospace applications.
- The research should include detailed analysis of LCAAs and TMDs through specific case studies which represent their aerospace implementations.

Data Collection and Analysis

- Technical specifications together with performance data and operational details regarding LCAAs and TMDs must be collected from aerospace manufacturers or research papers.
- Comparative Analysis: Analyze the effectiveness, limitations, and benefits of LCAAs and TMDs in their respective applications.

Simulation and Modeling (If Applicable)

- Laboratory experiments should simulate LCAAs to validate their improvements for aircraft navigation precision and conduct tests to showcase TMDs' ability to reduce aircraft structure vibrations.
- Modeling and simulation techniques should be used to display the effects of these damping systems on aircraft stability together with passenger experience.

Performance Evaluation and Validation

- The evaluation method for aerospace requirements performance must be developed for both LCAAs and TMDs.
- Laboratory tests should validate how precisely LCAAs support navigation systems as well as how well TMDs attenuate airplane vibrations.

Documentation and Report Preparation

- A structured format will gather all obtained data combined with analysis results as well as simulation findings and validation evidence.
- A structured report contains detailed information about aerospace damping applications methodology together with findings and conclusions as well as recommendations.

Peer Review and Validation

- The research findings with professionals from aerospace engineering or damping technology disciplines are subject to expert verification of the validated conclusions and results.
- The study will implement feedback through expert input to enhance study quality by refining its content.

Presentation and Dissemination

- The team should create presentation materials for the dispersion of case study findings and research methodology.
- The study findings will be publicly presented at aerospace community events such as conferences as well as professional publications and industry seminars.
- A standardized methodology assists in assessing the significance of aerospace damping systems where LCAAs and TMDs are reviewed for their effectiveness. Researchers perform detailed investigations followed by analysis and verification as well as proper documentation to develop extensive understanding of aircraft safety performance and comfort aspects.

6.1.4 Simulation for the System

MATLAB simulation development of aerospace applications to Liquid Circular Angular Accelerometers and Tuned Mass Dampers necessitates mathematical modeling of the system dynamics as well as simulation of their effects concerning aircraft stability and precision navigation and vibration reduction as shown in Table 6.1. The nature of the complexity of the system normally means that a detailed aerospace model will inevitably contain several interconnected subsystems with intricate algorithmic variances.

6.2 Case Study: Civil Engineering and Infrastructure

HD applications are advantageous for tall buildings and bridges because their important role in combating wind-induced vibrations and extending

Table 6.1 Methodology case study 1 – aerospace applications.

Step	Description
1. Requirement Identification	Identify specific damping needs in aerospace, focusing on vibration reduction and structural integrity.
2. Literature Review and Case Selection	Review existing research and select case studies highlighting LCAAs and TMDs in aerospace applications.
3. Data Collection and Analysis	Gather technical specifications, operational details, and performance data of LCAAs and TMDs.
4. Simulation and Modeling (if applicable)	Conduct simulations to visualize the impact of LCAAs on navigation accuracy and TMDs on vibration reduction.
5. Performance Evaluation	Develop metrics to evaluate the effectiveness of LCAAs and TMDs in meeting aerospace requirements.
6. Validation Studies	Validate efficiency through real-world or simulated tests to ensure accuracy and efficacy.
7. Documentation and Report Preparation	Compile data, analysis, simulation results, and validation findings into a comprehensive report.
8. Peer Review and Validation	Subject the study to peer review by aerospace experts and incorporate feedback for refinement.
9. Presentation and Dissemination	Prepare presentations summarizing findings and disseminate them through conferences or publications.

service life in turbulent winds [6]. TMD applications in high-rises are considered as creative solutions to the sway problem because such applications have the ability to enhance safety standards especially in seismic places [7]. Whatever there is of civil infrastructure today must have damping systems to strengthen its structure and reduce movement-related challenges to produce secure buildings under dynamically altering conditions as shown in Table 6.2.

Table 6.2 Simulations and description.

Simulation component	Description
Liquid circular angular accelerometers (LCAA)	
1. Modeling Angular Acceleration	MATLAB model simulating angular acceleration sensors and their behavior in capturing aircraft movements.
2. Integration with the Navigation System	Simulate integration of LCAAs with the aircraft's navigation system, showcasing their impact on precision navigation.
3. Vibration Reduction Simulation	Develop simulations illustrating how LCAAs contribute to reducing vibrations during critical flight phases.
Tuned mass dampers (TMDs)	
1. Structural Modeling	Create MATLAB model representing aircraft elements and where TMDs would be integrated.
2. TMD Integration	Model the behavior of TMDs and simulate their effects on reducing aerodynamic-induced vibrations.
3. Impact on Structural Integrity	Simulations demonstrating TMDs contribute to maintaining structural integrity and passenger comfort.

6.2.1 Methodology

A tabulated systematic structure describes the systematic method for studying HD and TMD use in civil engineering projects and infrastructure. The given steps also provide a comprehensive analysis of applications for damping systems in tall structures and bridges.

The study approach allows the organized research and analysis of both HD and TMD use in civil engineering practice for the purpose of determining the effectiveness of the two systems during damping of vibrations and enhancing structural integrity and bridge stability.

6.2.2 Outcomes

The focus pertains to damping techniques used in civil engineering infrastrucre is shown in Table 6.3.

Table 6.3 Methodology case study 2 – civil engineering and infrastructure.

Methodology steps	Description
1. Requirement Analysis and Scope Definition	Assess specific structural vulnerabilities and define the study's scope, focusing on wind-induced vibrations and seismic activities.
2. Literature Review and Case Selection	Review existing research and select relevant case studies highlighting Hydraulic Dampers and TMD applications in tall structures and bridges.
3. Data Collection and Analysis	Gather technical specifications, performance data, and structural dynamics related to damping systems.
4. Simulation and Modeling	Develop structural simulations to visualize wind or seismic forces and model the effects of damping systems.
5. Performance Evaluation and Validation	Establish metrics to evaluate damping system efficiency and validate their impact through simulations or real-world data analysis.
6. Documentation and Report Preparation	Compile collected data, simulation outcomes, and analysis into a comprehensive report detailing findings and recommendations.
7. Peer Review and Validation	Subject the case study to expert review, validating findings and conclusions, and incorporating feedback for report enhancement.

High-Damping Devices (HD):

- High-Damping Devices show documented capability for wind vibration reduction.
- High-rise buildings together with bridges benefit from structural stability improvement through the implementation of damping systems which reduce structural fatigue.

Tuned Mass Dampers (TMDs):

- Seismic events produce effective results for minimizing the way structures sway structurally.

- Damping systems enable safer conditions for building inhabitants by keeping structures structurally stable in regions prone to earthquakes [5].

Enhanced Structural Resilience:

- The implementation of damping systems minimizes structural fatigue as well as wear down in the long term.
- Damping systems promote longer operational periods for infrastructure through their use.

Improved Occupant Comfort:

- The improvement of occupant comfort occurs because vibration reduction leads to reduced motion-induced discomfort.

Simulation Validation:

- An analysis through simulation proves the research findings.
- Show proof of how damping systems function in actual deployments.

Future-Oriented Recommendations:

- The document offers recommendations about improving damping system configurations.
- The investigation offers engineered recommendations for prospective civil engineering designs and execution.

Overall Significance:

- Civil infrastructure stability together with safety and extended service life requires effective implementation of HD and TMD.
- Research funding for damping systems should continue because their integration enables better structural resistance capabilities.

6.3 Case Study 3: Automotive Industry

Research shows that damping systems have become standard in the automotive industry because they create better performance and passenger comfort alongside each other [8]. Many academic projects show that Hydraulic Dampers serve as critical components in shock absorbers to reduce road vibrations as well as maintain ride smoothness and stability. The combination of these technologies actively contributes to vehicle dynamics refinement and substantial enhancement of driving experiences according to [9–11].

6.3.1 Methodology

A systematic method guides the examination of how damping systems connect with the automotive industry as part of this study. The research starts with defining particular automotive requirements which involves examining vehicle performance behavior and road understanding and passenger comfort characteristics during the base study establishment phase. Researchers review the development of damping systems through an assessment of Hydraulic Dampers, Shock Absorbers with an additional focus on advanced active systems for performance evaluation. Technical personnel gather performance-related data from actual models of vehicles. The research proceeds to analyze real-world systems during driving situations using virtual simulations which replicate the behavior of these damping techniques under diverse environmental circumstances. The gathered data goes into an extensive report where experts confirm findings before establishing how suspension systems make vehicles perform better and create enhanced passenger experiences.

6.4 Case Study 4: Industrial Machinery and Robotics

Equal measure priority belongs to damping systems in the industrial landscape of robotics and machinery since they provide stability during operations and high precision standards as shown in Table 6.4. The paper explores this technical field by examining how diverse damping systems affect operations such as viscoelastic dampers and active damping systems [12]. Practical examples show that heavy machinery companies apply these technologies particularly in dampening vibrations of machinery operations, while robotic systems use them to control the precise movement.

Table 6.4 Methodology case study 3 – automotive industry.

Methodology steps	Description
1. Requirement Identification and Scope Definition	Analyze specific vehicle dynamics and define the study's scope, emphasizing the role of damping systems in addressing automotive needs.
2. Literature Review and Technology Assessment	Review existing research and assess advancements in Hydraulic Dampers, Shock Absorbers, and active damping systems in the automotive sector.
3. Data Collection and Analysis	Gather technical specifications, performance data, and operational details of damping systems utilized in various vehicle models.
4. Case Studies and Scenario Evaluation	Explore real-world applications through case studies and evaluate the impact of damping systems in diverse driving scenarios.
5. Simulation and Modeling (if applicable)	Develop simulations and models to visualize and analyze damping system behavior under varied road conditions and driving scenarios.
6. Performance Evaluation and Validation	Establish metrics for quantitative evaluation and validate the efficiency of damping systems through simulations and real-world tests.
7. Documentation and Report Preparation	Compile collected data, simulation outcomes, and analysis into a comprehensive report detailing damping system impacts.
8. Expert Validation and Conclusion	Subject the case study to expert review for validation and refinement, concluding with key findings and implications.

This research explains how the damping systems operate and proves the significant components that enhance the operational performance by reducing the degradation of the equipment and increasing the reliability of the system in industrial robots and equipment [13]. The introduction is a basis for an extensive analysis of damping systems, which is realized through their use for enhancing operational stability as well as accuracy and effectiveness in industrial machinery and robotics as shown in Table 6.5.

Table 6.5 Methodology case study 4 – industrial machinery and robotics.

Simulation and modeling phase	Description
1. Machinery Operations Simulation	Develop simulations to replicate heavy machinery activities, illustrating the impact of viscoelastic dampers on vibrations and operational stability.
2. Robotic Motion Control Simulation	Create simulations depicting various robotic actions and movements, showcasing the role of active damping systems in precise motion control.
3. Mathematical Models for Damping Systems	Construct mathematical models representing the behavior and characteristics of viscoelastic dampers and active damping systems.
4. Dynamic Models for Active Damping	Build dynamic models simulating real-time adjustments by active damping systems in response to input signals or commands.
Purpose and Benefits	Predictive analysis, optimization insights, and cost-efficient testing for anticipating and enhancing damping system performance.

A table gives key information about the Simulation and Modeling phase by giving an account of different simulations along with mathematical models that assess the performance of damping systems in industrial machinery and robotics.

6.4.1 Simulation and Modeling Phase

In simulation and modeling, professionals create digital images, which are used to calculate how the products of damping systems respond when used with viscoelastic dampers and active systems on industrial machinery equipment.

- Developed simulations show the way these viscoelastic dampers influence the working of machinery equipment. Simulations operate heavy machinery operations by incorporating damping system effects that reduce vibration (shock and vibration) which enhance operational stability.

The virtual environment allows users to view how the damping system operates under different operational configurations such as different loads and speeds.

- Simulations govern robotic movements using robotic control replicas. Experimentation and virtual displays related to the impact of active damping systems on robotic movement control are presented through virtual platforms.

Modeling:

- The mathematical models build equations of viscoelastic dampers and active damping systems' behavior patterns. Using these mathematical models, the physical characteristics of damping systems can be observed whereas the output behavior of the damping systems can also be predicted in different operating situations.
- Real-time simulation models under conditions of dynamicity are used in order to show the behavior of active damping systems when using dynamic models. Visualizing active damping systems has models with vibrations and commands of movement as inputs (which reveal the system makes adjustments of rate of damping or stiffness parameters), resulting in reductions in oscillations and increased control.

Purpose and Benefits:

- Prediction of how damping systems perform in actual world situations can be achieved with the help of simulation and models made use of in predictive analysis.
- System optimization is achievable due to Optimization Insights that analyze key parameters and settings that influence performance.
- Virtual simulations as well as models support economical testing procedures for analysis until the damping systems are ready to be applied to industrial machinery and robotic systems.

References

1. Berthoz, A., *The brain's sense of movement*, Harvard University Press, Cambridge, MA, USA, 2002.
2. Cheng, S., Fu, M., Wang, M., Li, X., Xiao, M., Wang, T., Modeling for fluid transients in liquid-circular angular accelerometer. *IEEE Sens. J.*, 17, 2, 267–273, 2016.
3. Groenesteijn, J., Droogendijk, H., de Boer, M.J., Sanders, R.G.P., Wiegerink, R.J., Krijnen, G.J.M., An angular acceleration sensor inspired by the vestibular system with a fully circular fluid-channel and thermal read-out, in: *2014 IEEE 27th International Conference on Micro Electro Mechanical Systems (MEMS)*, pp. 696–699, IEEE, San Francisco, CA, USA, 2014.
4. Yu, X. and Zhang, Y., Sense and avoid technologies with applications to unmanned aircraft systems: Review and prospects. *Prog. Aerosp. Sci.*, 74, 152–166, 2015.
5. Alvarez, P., McElwain, B., Thesing, L., Edalath, S., Kukreti, A., Cohen, K., PD and fuzzy logic control for earthquake resilient structures. *Comput. Appl. Eng. Educ.*, 22, 1, 142–152, 2014.
6. Madhekar, S. and Matsagar, V., *Passive Vibration Control of Structures*, CRC Press, Boca Raton, FL, United States, 2022.
7. Titurus, B., Du Bois, J., Lieven, N., Hansford, R., A method for the identification of hydraulic damper characteristics from steady velocity inputs. *Mech. Syst. Signal Process.*, 24, 8, 2868–2887, 2010.
8. Gohardani, O., Chapartegui Elola, M., Elizetxea, C., Potential and prospective implementation of carbon nanotubes on next generation aircraft and space vehicles: A review of current and expected applications in aerospace sciences. *Prog. Aerosp. Sci.*, 70, 42–68, 2014.
9. Villasenor, A.G., *Design, modelling and control of inter-vehicle damping mechanisms for convoy transportation systems*, The University of Manchester (United Kingdom), ProQuest, Manchester, UK, 2006.
10. Ata, W.G. and Oyadiji, S.O., An investigation into the effect of suspension configurations on the performance of tracked vehicles traversing bump terrains. *Veh. Syst. Dyn.*, 52, 7, 969–991, 2014.
11. Jalili, N., A comparative study and analysis of semi-active vibration-control systems. *J. Vib. Acoust.*, 124, 4, 593–605, 2002.
12. Yu, Y., Li, F., Yao, G., Vibration response and isolation of X-shaped two-stage vibration isolators: Analysis of multiple parameters. *Nonlinear Dyn.*, 111, 17, 15891–15910, 2023.
13. Crowder, R., *Electric drives and electromechanical systems: applications and control*, Butterworth-Heinemann, Oxford, UK, 2019.

7

Transient and Transit Flow Models for Dynamic Vibration Analysis

Abstract

The research investigates vibration control applications in structural engineering through merger of transient modeling with transit flow modeling to improve structural stability and minimize unnecessary vibrations. It also examines transient structural response patterns to external forces through transient modeling that allows evaluation of sudden movement patterns and transient events disturbing structural stability and includes both transient modeling for structural system control strategy optimization and transit modeling for vibration control strategies to achieve structural safety excellence. The research employs transient modeling to investigate short-term changes and transit flow modeling for the optimization of mechanisms to generate productive vibration control solutions. The study produces data that supports the creation of predictive strategies and infrastructure planning on the part of proactive maintenance teams to result in optimum structural performance and reduced environmental impacts, and a more resilient urban infrastructure.

Keywords: Transit flow model (TFM), flow model, structure system, transient, vibration modeling

7.1 Introduction

Transient Flow Models are an important analytical tool for interpretation and control of dynamic vibrations in civil engineering structures [1, 16]. It is concerned with the use of transient flow analysis as a method of predicting, analyzing, and controlling the vibrations caused by fluid dynamics in structures. It discusses differential equations that control transient flow, and the importance of fluid-structure interaction at producing vibrations. Discussions on fluid-induced vibrations affect structures, caused by wind, water flow, or internal fluid dynamics vibrate structures, models help understand such phenomena and can be used for predictions [2].

Kundan Meshram and Prashant Lahre. Vibration Control Systems, (103–122) © 2026 Scrivener Publishing LLC

Here, the role of transient flow analysis in predicting and controlling vibrations induced in fluid flow is highlighted. It shows computational fluid dynamics (CFD) combined with structural analysis aids in assessing the effect of fluid-induced forces on structures. In-depth discussions are provided for the fluid-structure interaction (FSI) investigations, which display transient flow models representing fluid dynamics interacting with structural responses [3]. Case studies showcase FSI applications with regard to predicting and controlling vibrations of bridges, offshore constructs and buildings, and show transient flow models combined with optimization techniques and damping strategies. It shows that such models help in optimizing the structural designs and damping systems that are required to optimize fluid-induced vibrations efficiently.

The Transit Flow Model forms a basic model used to analyze and simulate traffic of vehicle passengers or cargo in transit systems [4]. This model includes a range of factors that affect transit flow, including travel demand, routing patterns, operations of the vehicle and infrastructure design. Through the application of mathematical algorithms and simulation techniques, the transit flow model helps in optimizing transit networks, increasing operational efficiency, as well as increasing the overall efficiency of the transportation systems.

7.1.1 Key Components of Transit Flow Modeling

Travel Demand Analysis: Realizing the demand for transit service, such as number of passengers, trip origins, destinations, and travel patterns.
Network Design: Designing a transit system, transit routes, stations, and infrastructure in response to the travel demand.
Service Scheduling: Setting frequencies, schedules, and requirements of vehicles to fulfill demand and to maximize the efficiency of operations.
Vehicle Operations: The arrangement of the transit vehicles' movement schedule in order to provide timely and effective transit service.

7.1.2 Elements Considered in Transit Flow Modeling

Passenger Behavior: Analyzing commuting preferences, travel behaviors, and modal options for enhanced service of transit capacities.
Infrastructure Capacity: The evaluation of the ability of transit infrastructure, like stations or terminals, to manage the amount of passenger load satisfactorily.
Congestion Management: Reducing congestion and maximizing value of time from the use of effective route planning and traffic management strategies.

7.1.3 Applications of Transit Flow Models

System Optimization: Omitting transit operations through the selection of the best routes, frequencies, and vehicle deployments to adjust to travel demand efficiently.
Service Planning: Checkout transit services designed to serve the varied patterns of travel of commuter needs in urban areas.
Infrastructure Development: Shaping solution of transit infrastructure design and expansion to meet increasing demand and enhance system efficiency.
Passenger Experience: Improvements to commuters' experience of travel by reducing waiting times, decreasing congestion, and improving reliable convenient transit services.

7.1.4 Importance in Urban Mobility

Transit flow modeling is critical in the development of urban mobility strategies, facilitating effective supply of public transportation systems, minimizing the use of individual vehicles, and contributing to sustainable urban development. It assists policymakers, transportation planners, as well as the transit authorities in reaching intelligent decisions that will maximize the transit services and increase the quality of transportation services to the commuters generally [5–7].

7.2 Transient and Transit Flow Model

It can be said that Transient Modeling refers to the study of short- or transient phenomena in different systems. In transport or traffic engineering, the transient models address short-lived events or changes in traffic patterns [8]. It wishes to show dynamic changes that happen in a very short period, for instance, sudden spikes of traffic caused by event (accidents, road closures, or special events) or short-term variations in traffic patterns. Transient modeling facilitates the simulation of such short-term variation for determining its effect on traffic flow, travel time, congestion level, or efficiency of transport system [4]. It helps in resolution of short-term disruptions; optimization of traffic signal timing, and planning for special events etc.

7.2.1 Transit Flow Modeling

Transit Flow Modeling is concerned with the movement, behavior, and operations of transit systems (buses, trains, subways, etc.) [9]. It is directed to

Table 7.1 Aspects of transient modeling and transit flow modeling.

Aspect	Transient modeling	Transit flow modeling
Temporal Scope	Deals with short-term fluctuations or sudden changes.	Focuses on continuous operation and planning of transit systems.
Time Scale	Typically minutes to hours.	Considers longer-term operations and system efficiency.
Focus	Analyses temporary disruptions or changes affecting traffic flow.	Optimizes transit networks and public transportation systems.
Objectives	Assess the immediate impacts of transient events on traffic.	Enhance efficiency and manage commuter experiences in transit.
Impact Analysis	Evaluates temporary changes or fluctuations in traffic.	Focuses on sustained performance and efficiency of transit.

examining the dynamics of transit operations such as scheduling, routing, capacity utilization and overall system efficiency as given in Table 7.1. Transit flow modeling has the potential to optimize transit networks by identifying optimal routes, frequency, and vehicle deployments that will provide efficient transit service required by travel demand [10]. It helps improve commuter experience and decreases congestion and increases use of public transport.

7.2.2 Differences

Temporal Scope: Analogous to transient modeling which pertains to short time-scale variation or abrupt changes typically on the order of minutes – hours, transit flow modeling refers to transit systems' operation in the continuous mode and on its functioning on various time scales.
Focus and Objectives: Transient modeling seeks to capture and provide a solution to temporary disruptions or changes affecting traffic flow, while transit flow modeling herein focuses on planning, flow management, and optimization of public transport systems.
Impact Analysis: Transient modeling evaluates the direct effect of transitory events on traffic and transit flow modeling evaluates the long term performance and effectiveness of transit systems.

This table offers a very compact comparison emphasizing the differences between transient modeling that corresponds to short-time changes in the traffic volumes and transit flow modeling that represents continuous operation and optimization of the public transport systems during longer periods.

7.3 Vibration Control Strategies of Fluid Vibrations in Civil Engineering

The discussions emphasize the importance of fluid-induced vibrations in civil engineering structures. Motion of wind, fluid flow in water and motion of internal fluids result in forces acting on structures resulting in oscillations. Transient flow models help understand these phenomena and underline their important role in predicting and minimizing adverse effects [11]. This part focuses on the critical role played by transient flow analysis in predicting and controlling vibrations due to fluid flow. It describes the combined use of computational fluid dynamics (CFD) integrated into structural analysis, demonstrating such models to predict and analyze the effects of fluid-induced forces on structures [2]. There is a specific keenness on the part of FSI studies manifesting the complex nature of interaction between fluid dynamics and structural response. Case studies and simulations demonstrate transient flow models that depict these interactions that help in predicting and controlling vibrations in a multitude of civil engineering structures.

The models were combined with optimization techniques and damping strategies. It describes their key role of optimizing structural designs and damping systems to adequately reduce fluid-induced vibrations [12]. These models enable improvement of and optimization of designs as well as damping mechanisms. Practical implications of transient flow models in vibration control are exhibited in real-world applications & case studies. Examples include works of bridges, offshore works, buildings, and other infrastructures. These demonstrations demonstrate that transient flow analysis aids in vibration prediction and reduction caused by fluid dynamic variations. Prisons with the previous chapters by showing the additive interaction between transient flow models and predictive modeling, control strategies, and optimization methodologies presented in the book [13]. It shows that these models improve the understanding of fluid-induced vibrations and lead to holistic vibration control strategies. Builds through the synthesis of transient flow models, predictive modeling, control strategies, and optimization strategies discussed earlier. It demonstrates a transient flow analysis which supports ANN-based predictive models and hybrid controller systems for integrated vibration control strategy.

Validating Transient Flow Models for vibration control requires a number of steps in order to confirm their accuracy and consistent predictions in dealing with fluid-induced vibration in civil engineering construction. Collect experimental data using vibration sensors placed on structures under the influence of fluid-induced forces, e.g., wind tunnels, water-flow tests or in real-life situations. Make use of fluid dynamics models in the form of a transient flow to model fluid dynamics affecting the structure. Such simulations should mirror the same conditions of the collected experimental data; compare the results from the transient flow models with those of the actual experimental data [14]; superimpose the simulated vibration responses to the observed vibration to compare how accurate the transient flow attempts are; vary different parameters such as flow velocity, structural attributes, and fluid density in the transient flow models to do sensitivity analysis; assess the effects of changes in these parameters on the anticipated vibration responses; and analyze how the model responds rationally and reasonably to parameter modifications and how the sensitivity of the model corresponds to physical phenomena.

7.4 Model Validation with Benchmark Cases

Transients flow models are applied to these benchmark scenarios and the model predictions are compared with well-documented and validated experimental or numerical results [15]. Confirm that the transient flow models produce the same results as established benchmarks, proving the validity of the models in simulating fluid-induced vibrations accurately. Use transient flow models to actual world examples of various civil engineering structures subject to fluid-induced vibrations. Test the model prediction against the observed vibrations in these case studies. Verify the transient flow models predict the ratio due to the structural response and behavior under fluid-induced forces.

Employ mismatches between the model directions and observed data directly as feedback to fine-tune and calibrate the transient flow models consistently [16]. Simplify the models with new data and insights to ensure continuous validation and calibration in increasing their accuracy and applicability for the prediction and control of fluid-induced vibrations. Through systematic validation of transitory flow models using these methods, engineers are able to guarantee the believability and viability of these models in predicting and controlling fluid-induced vibrations of civil engineering structures. Analytic approach applicable to vibration control strategy based on transient flow includes a step-wise process to comprehend, forecast, and reduce fluid- induced vibrations in civil engineering structures. Initiate formulation and solution of the basic governing equations of transient fluid flow including Navier-Stokes equations, which provide for time-varying fluid behavior. Couple these fluid flow equations with structural dynamic equations, taking into the account the interaction between the fluid forces and the structural responses. This combination underpins the understanding of fluid structure interactions (FSI) [17]. Perform simulations depending on time (time-dependent) using computational fluid dynamics (CFD) methods. Simulate transient flow patterns over immersed structures subjected to forces from fluids (such as wind flow or flow of water) to obtain the time-dependent behavior of fluid-induced vibrations. The dynamic responses of the structure due to transient fluid forces should be predicted, as these forces provoke vibrations and influence the structural integrity.

7.5 Parametric Study

Sensitivity analyses are used where the fluid velocity, structure's shape, material characteristics, and environmental conditions, (i.e., wind speed, and water flow rates) are varied; analyze the changes in these parameters that influence fluid-induced vibrations; analyze the sensitivity of the structural responses to changes in fluid flow characteristics and change of structural properties; determine primary mode shapes of fluid-induced structures; analyze these mode shapes and engage with transient flow form; identify resonant frequencies that have fluid-induced forces colliding with structural natural frequencies; use optimization algorithms, e.g., genetic algorithms or particle swarm optimization, to optimize structural design and the damping system from transient flow simulations; and, based on the transient flow prediction [18], develop adaptive control strategy.

7.6 Simulating the Transient Flow Model

It is associated with a complex procedure that is crucial in determining the fluid behavior in systems under dynamic change. These equations, which are often complex partial differential equations, form the basis of developing modeling of transients including pressure pulses, flow oscillations, and changes in velocities. Numerical methods are essential to this simulation process.

Engineers discretize the governing equations either by finite difference, finite volume, or finite element methods. This subdivision of the system into smaller parts provides for the approximation and solution of complex differential equations, where fluid behaviors can be forecast for sequential time-step intervals. Exact boundaries and initial conditions will ensure that exact real-world examples are modeled accordingly. Engineers carefully define these conditions in order to reflect real system parameters and to make the simulation process accurate. Then advanced solvers and algorithms are used to iteratively solve the discretized equations over successive time levels in simulating the transient behavior of fluids in the system.

The validity is elicited through some strict validation and calibration routines, when simulated results are compared to known data or analytical solution. Engineers proceed by analyzing the results, examining pressure, velocity, temperature distributions, and other fluid properties at time. Visualization tools can help understand and uncover insights about these complex behaviors and how systems react to changing conditions. Such simulations allow for scenario analysis, which lets engineers better predict how the system would react to different dynamic conditions (towards the maze, guest flow changes, environmental perturbations, etc.). Lessons learned from these simulations are of tremendous value in optimizing system architectures, improving components, and developing appropriate control strategies. Ultimately, this simulation-based approach empowers engineers to implement enhanced efficiency, reliability, and performance in engineering systems whose engineering systems operate under transient flow.

7.7 Validation, Calibration, and Uses in Civil Engineering

Verify analytical predictions by comparing the outcome of simulation with the available experimental data or corresponding benchmark cases to confirm accuracy and reliability of prediction of fluid-induced vibrations [17]. The system continuously adjusts and refines the analytical model using validation results and feedback to improve the precision and predictive capability. By undertaking this analytical approach, the engineers analyze broadly the transient flow influence on engineering structures in civil engineering, estimate the fluid-induced vibrations, and develop useful measures to control and suppress fluid vibrations for structural integrity and safety benefits. The transient flow model is an essential tool applied in the field of civil, mechanical, or hydraulic engineering. Its main utility resides in its ability to analyze the fluid behavior in systems or structures with time, especially when flow conditions are dynamic.

7.7.1 Civil Engineering Applications

Water Distribution Systems: Transient flow models are used to ascertain pressure surges, water hammer effects and flow changes in pipelines and networks.
Hydraulic Structures: Knowledge of transient flows through dams, spillways, and reservoirs is of critical importance for forecasting water levels, pressures, and waves as a variable environment changes, such as in flood events or the rapid closing of valves.
Sewer Networks: Modeling transient flows helps examine surges, pressures, and velocities in sewers and avert active damage caused by abrupt changes or blockages in flow rates.

7.7.2 Mechanical Engineering Applications

Piping Systems: Transient flow models anticipate pressure transients and variations in flow in piping systems caused by alterations to pump operation, valve closure, or system start-up, thus preventing structural damage or leaks.
Heat Transfer Systems: The analysis of the transient flows is helpful to optimize the heat transfer systems by predicting temperature fluctuations and fluid activities in start-ups or alterations of operating conditions.

7.7.3 Computational Fluid Dynamics (CFD)

Dynamic Simulation: CFD-based transient flow simulations allow engineers to describe and analyze complex flow behaviors occurring in systems for the cases of different boundary conditions (BCs) or external forces.
Optimization and Design: Through analysis of transient flows, engineers will optimize system designs, or fine-tune components or strategies of system control to improve the performance and steady state of their system.

7.7.4 Environmental Engineering

Natural Disaster Prediction: Transient flow models help with forecasting the amount of water flowing in rivers and streams during water floods/storms, hence supporting the management of floodplains and preparation for disasters.
Wastewater Treatment: Knowledge of transient flows in treatment plants enables optimization of operation, surge capture, and avoiding hydraulic over of the treatment plant, thus improving the efficiency of the wastewater treatment process.

In general, transient flow models are indispensable instruments across the engineering disciplines, and they will help engineers simulate, predict, and optimize fluid behavior under changing conditions. Their application guarantees the design, operation, and management of robust and efficient systems and structures sensitive to dynamic flow scenarios.

7.8 Furthermore, the Analytical Formulation for Transient Flow Models

In the case of vibration control includes complex mathematical equations, which present fluid dynamics and their interaction with response of structures. It is difficult to present a short analytic formula that embraces all sides and direct correlation with vibration control of transient flow because of fluid – structure interactions. Still, it can be viewed that a simplified formulation of the Navier-Stokes equations with structural dynamic equations can serve as an underlying analytical model for investigation of transient flow-induced vibrations. The Navier-Stokes equations for fluid flow coupled with the structural equations to which the standard of proof is stated as:

For Fluid Flow

$$\rho\left(\frac{\partial u}{\partial t}+u\,.\,\nabla u\right)=-\nabla p+\mu\nabla^2 u+f_{external} \tag{7.1}$$

For Structural Dynamics

$$M\frac{\partial^2 x}{\partial t^2}+C\frac{\partial x}{\partial t}+Kx=f_{fluid} \tag{7.2}$$

Where,

ρ represents fluid density,

u is the vector field of fluid velocity,

p signifies pressure,

μ represents dynamic mechanical viscosity,

M, K, and C, respectively, are mass, damping and stiffness matrices for the structural system.

x is the structural displacement vector.

f_fluid denotes fluid-induced forces on structure.

The coupling of these equations represents the action of fluid forces on structural responses that permits the investigation of the transient flow effects on structural vibration. Numerical solutions to solve these coupled equations, under time-dependent boundary conditions and properties of structures, allow for the prediction of transient flow-induced vibrations. Analytical formulations generalized to include sufficiently advanced optimization tricks, or a control strategy, so structural parameters, or damping behaviors are back-traceable with respect to the analytical predictions to suppress vibrations. Though these equations are a base, actual models of transient flow for vibration control have more complex and detailed formulations depending upon the geometry, fluid properties, boundary conditions, and desired control strategies for the structural geometry. Computational fluid dynamics (CFD) combined with finite element analysis (FEA) is frequently used to model and analyze transient flow effects for the aim of vibration control in realistic engineering circumstances.

7.9 Study of TFM

Full transit flow model, through use of analytical approach, is a broad framework applied to comprehend and determine the dynamics of transit systems. It includes diverse analytical tools and mathematical models that simulate the flow of passengers or vehicles in a transit network [4]. This model is useful for analyzing the system performance, comparing various scenarios, and optimizing transit operations.

7.9.1 Components of a Full Transit Flow Model

7.9.1.1 Network Topology

First, the model defines the topological structure of the transit network (routes, stops, stations, connections, and frequencies of services). This provides the structure of the system of transit.

The first step in the development of a transit flow model entails developing a network topology, i.e. the architectural structure of the transit system. This basic stage represents the arrangement and interconnection of the transit network in space and includes routes, stops, stations, interchanges, and frequencies of the service. Roads and lines describe the route taken by some vehicles (trains), linked with stops and stations where people board or disembark from there. The network's connectivity designs the traffic routes that vehicles take specifying accessibility and effectiveness. Fruitfully, service frequencies at each route are a measure of the availability and dependability of transit services throughout a day. This basic element provides a detailed picture of the transit infrastructure hence shaping the subsequent stages in modeling, analysis, and optimization in the transit flow paradigm.

7.9.1.2 Passenger Demand Estimation

Analytical methods are applied to calculate the rate of the passenger demand, taking into account variables, such as population density, trip patterns, time of the day and purposes of trips. Rules such as gravity models or trip distribution models are used to forecast the number of passengers travelling from one origin and destination to another.

7.9.1.3 Traffic Assignment

Traffic assignment models distribute estimated passenger demand on transit routes and modes. This step decides how passengers select individual routes and modes having parameters such as travel time, fare, and service frequency, which is an important step in learning about the traffic volume and patterns enjoyed by transit within a region. This factor applies analytical approaches to assess passenger travel behaviors, including such parameters as population distribution, type of travel, purpose of trips, and temporal values. Flow of passengers between various origins and destinations through different destinations of transit network is predicted using various models like gravity models or trip distribution models. Utilizing statistical and mathematical methods, this component intends to predict the expected demand for the transit services, presenting the spatial distribution of the passenger needs and travel characteristics. This estimate is a building block for the coming steps of the model of transit flow, allowing for improved planning of the transit, allocation of resources, and optimization of transit operations to serve a forecasted demand and operate efficiently and effectively.

7.9.1.4 Operations Modeling

Analytical methods are employed in modeling the processes of transit operations based on considered schedules, headway, and dwell time as well as vehicle capacities. This was a modeling phase that facilitates transit movements in transit network in terms of operational constraints and service characteristics, focusing on distributing estimated passenger demand to specific transit routes and transit modes. This phase explores all the factors that passengers consider when selecting routes and modes (modes – travel time, cost, convenience, and frequency of service). Analytical approaches such as mathematical models and algorithms allocate the calculated demand of the passengers to various transit options and provide passenger distribution in the transit network. These are models of passengers' decision-making process explaining how they are likely to choose routes and modes given number of variables. By knowing passengers assign themselves into various transit services, this aspect helps in tuning transit operations, increasing transit usage, and increasing overall transit efficiency system.

7.9.1.5 Performance Evaluation

The model considers various performance metrics include travel time, waiting time, system capacity; congestion levels and reliability. This assessment is useful to pinpoint the potential bottlenecks, areas for improvement, and overall efficiency of the transit system where the analytical techniques are used to model and analyze transit operations. Analytical tooling is employed to model transit service dynamics while incorporating operational restrictions and service features. This module, by modeling the movement of vehicles through the network, enables a detailed analysis of the performance of transit systems, detection of possible bottlenecks, and estimation of the effects of operational parameters on level of service and passenger experience. This modeling period is essential for optimization of schedules, operational efficiency improvement, and guarantee of smooth functioning of transit services.

7.9.2 Analytical Approach in Transit Flow Modeling

7.9.2.1 Mathematical Modeling

Equation-based Models: Use mathematical equations to describe the characteristics of passenger flow, vehicle movement, and system dynamics.
Optimization Models: Come up with the optimization problems to optimize transit operations with limitations such as reducing time or increasing system capacity.

7.9.2.2 Simulation and Analysis

Simulation Tools: Use simulation software, model transit operations, and a flow of passengers in different scenarios. Analytical tools are used in this stage to measure a number of indicators of performance, including travel time and time waiting at stops or stations, capacity of the system, levels of congestion and reliability of services. With the analysis of these metrics the model identifies inefficiencies, bottleneck hotspots, or areas of suboptimal

quality of service in the transit network. In addition, it aids in the comprehension of the system's advantage over the passenger's needs and expectations. It acts as an important feedback mechanism for rationalizing transit operations and the formulation of transit infrastructures for improved service delivery and customer satisfaction.
Sensitivity Analysis: Test the reaction of the system to alterations in parameters, service frequency, or network structure.

7.9.2.3 Predictive Capability

Forecasting: Estimate future transit demand and system performance by using the historical data and trend analysis as given in Table 7.2.
Scenario Analysis: Assess the effect of proposed changes/interventions to system performance thus supporting decision-making to enhance system improvements.

Table 7.2 Transit flow models with artificial neural networks (ANNs).

Aspect	Explanation
Demand Prediction	ANNs analyze historical transit data to predict future travel demands, assisting in optimizing services and schedules based on anticipated passenger requirements.
Route Planning and Optimization	Using pattern recognition, ANNs optimize transit routes to minimize travel times and enhance service frequency, improving operational efficiency and commuter experiences.
Passenger Experience Enhancement	ANNs aid in reducing waiting times and enhancing reliability by predicting peak travel times and optimizing services for a more comfortable commuter experience.
Adaptive Decision-Making	Real-time adjustments based on continuous learning enable transit systems to adapt swiftly to changing conditions, optimizing operations and response mechanisms.
Infrastructure Planning and Investment	ANN-derived predictions guide long-term planning and investment decisions, aligning infrastructure developments more accurately with anticipated future demands.

7.10 Benefits of Analytical Approach in Transit Flow Modeling

Optimization: Determining best routes and schedules and service frequencies, to improve system efficacy.
Decision Support: Valuable information for planners, transit authorities; in planning, managing, and enhancing transit systems.
Resource Allocation: Facilitates an efficient allocation of facilities based on allocation of areas in need of infrastructure improvements/services improvements.

Essentially, a complete transit flow model with an analytical approach uses mathematical models and simulation methods to altogether analyze and optimize transit systems to assist in decisions and planning of a reliable and effective public transportation network.

7.11 Transit Flow Model with ANN Prediction

The doping of Transit Flow Models with Artificial Neural Networks (ANNs) is one of the most powerful tools to predict, simulate, and optimize transit systems. ANNs are good at learning patterns from data, when ANNs are coupled with transit flow modeling, they provide several advantages for prediction and optimization.

7.11.1 Demand Prediction

Travel Patterns Analysis: The massive data sets that ANNs analyze are full of past transit data, such as number of passengers, trip time, and route usage. Using the detection of recurring patterns and trends, ANNs are able to forecast future travel needs on various routes and periods.
Service Optimization: With respect to predictive capabilities of ANN, transit services can be optimized better. Changes in schedules, routes, or vehicle placements can be made depending on expected demand so that closer alignment is established with passenger requirements.

7.11.2 Route Planning and Optimization

Pattern Recognition: ANNs work on complicated transit data, analyzing the movement patterns of passengers, congestion level and traffic flow. This knowledge helps toward the establishment of optimal transit routes, in terms of traffic, travel preferences, and the peak hours.

Efficiency Improvement: Optimal routes based on ANN forecasts attempt to decrease travel times while increasing the frequency of services.

7.11.3 Passenger Experience Enhancement

Wait Time Reduction: ANNS provides predictions on peak travelling hours and congestion times, which work to improve the schedule and resource distribution. Decreased wait times and prevented overcrowding in stations will greatly enhance passenger experiences.
Reliability Improvement: Things like assured and repeatable transit services are necessary for commuter satisfaction. ANNs support predictions of service reliability on the part of transfer systems for the purpose of working effectively and consistently with the commuters' expectations.

7.11.4 Adaptive Decision-Making

Real-time Adjustments: ANN constantly learns from incoming data, and transit systems are therefore responsive to changing environment. Such flexibility makes it possible to make direct adjustments in scheduling or routings depending on the actual demand patterns.
Adaptive Control Strategies: Through the use of ANNs, transit systems are able to engineer adaptive control procedures that make a dynamic optimization of operation possible. This will guarantee that the system is responsive to changes in transit demand as well as remain efficient.

7.11.5 Infrastructure Planning and Investment

Future Projections: The predictions revealed using ANNs help in long-term planning and investment decisions. Such observations influence choices over infrastructure improvement, expansion, or new transit ventures, making these choices more representative of what is expected in the future.

The combination of Transit Flow Models and ANNs provides a holistic understanding of transit behavior and thus improved planning, optimization, and decision-making. It enables transit authorities to establish more effective, robust, and transit-friendly transit systems that can respond dynamically to dynamic urban surroundings.

7.12 Conclusion

The transit flow model is the mathematical model used to model and analyze the movement of vehicles or passengers in transit systems (buses, trains, subways, etc.). It includes the understanding of transit flow dynamics, it envisions travel demand, vehicle scheduling, planning routes and even comes up with a mechanism to manage congestion. The model helps optimize the transit systems, enhance efficiency, and promote much-improved urban mobility.

1. The simulation of transit flow is useful in optimizing such aspects as vehicle schedules, frequency, and routes. It helps in finding out the most effective way through which resources to accommodate travel demand should be used.
2. Knowledge of transit flow dynamics encourages better planning of the service, reduction of waiting hours, minimization of congestion and improved experience for commuters.
3. Transit flow models help in reducing traffic congestion by encouraging people use, reducing usage of personal vehicles, and enhancing traffic flow.
4. Insights gained from transit-flow models help in planning the infrastructure including designing the transit network, siting of station and capacity expansions giving sustainable and efficient transit systems.
5. Analysis of transit flow is useful in determining the environmental impact of transportation systems. Promoting public transit use has the effect of lowering carbon emissions and developing environmentally friendly city transportation.
6. Transit flow models are useful for policymakers and urban planners, so that they can make decisions on what to invest in, the fare structure or service expansion.
7. In general, transit flow modeling underpins the provision of efficient, accessible, and reliable techniques of transportation thereby facilitating better urban mobility and connectivity.
8. The combination of transient modeling and the interaction between Transit Flow Models and ANNs provides a complete picture of knowledge and a toolbox for the planning and optimization of transport.

9. Though transient modeling is useful in resolving the short term disruptions and fluctuations in the traffic systems, the combination of ANNs with the transit flow allows for proactive optimization and the adaptive control in the transit systems.
10. Combined, these approaches support more efficient flow control of traffic, reactive transit services, and enhanced urban mobility, casting a foundation for resilient and adaptive transportation systems in fast-changing urban settings.

Traffic flow model systems and, as a useful tool in developing transit policies, way of building efficient transit systems, as well as promoting sustainable urban development through optimization of transit operations and bettering the general transit experience of transit users.

References

1. Lu, Z., Wang, Z., Zhou, Y., Lu, X., Nonlinear dissipative devices in structural vibration control: A review. *J. Sound Vib.*, 423, 18–49, 2018.
2. Xing, J.T., *Fluid-Solid interaction dynamics: Theory, variational principles, numerical methods, and applications*, Academic Press, Cambridge, MA, USA, 2019.
3. Hirschhorn, M., Tchantchaleishvili, V., Stevens, R., Rossano, J., Throckmorton, A., Fluid–structure interaction modeling in cardiovascular medicine–A systematic review 2017–2019. *Med. Eng. Phys.*, 78, 1–13, 2020.
4. Cascetta, E., *Transportation systems analysis: models and applications*, vol. 29, Springer Science & Business Media, New York, NY, USA, 2009.
5. Lawrence, X.Y., Crison, J.R., Amidon, G.L., Compartmental transit and dispersion model analysis of small intestinal transit flow in humans. *Int. J. Pharm.*, 140, 1, 111–118, 1996.
6. Florian, M., A traffic equilibrium model of travel by car and public transit modes. *Transp. Sci.*, 11, 2, 166–179, 1977.
7. McGuire, K.J. and McDonnell, J.J., A review and evaluation of catchment transit time modeling. *J. Hydrol.*, 330, 3–4, 543–563, 2006.
8. Pavella, M., Ernst, D., Ruiz-Vega, D., *Transient stability of power systems: a unified approach to assessment and control*, vol. 581, Springer Science & Business Media, Boston, MA, USA, 2000.
9. Tao, S., Corcoran, J., Mateo-Babiano, I., Rohde, D., Exploring Bus Rapid Transit passenger travel behaviour using big data. *Appl. Geogr.*, 53, 90–104, 2014.

10. Tong, L.C., Zhou, L., Liu, J., Zhou, X., Customized bus service design for jointly optimizing passenger-to-vehicle assignment and vehicle routing. *Transp. Res. Part C Emerg. Technol.*, 85, 451–475, 2017.
11. Civan, F., *Reservoir formation damage: fundamentals, modeling, assessment, and mitigation*, Gulf Professional Publishing, Oxford, UK, 2023.
12. Tani, J., Takagi, T., Qiu, J., Intelligent material systems: application of functional materials, *Appl. Mech. Rev.*, 505–521, 1998.
13. Bennett, A.F., *Inverse modeling of the ocean and atmosphere*, Cambridge University Press, Cambridge, MA, USA, 2005.
14. Gorji, S., Seddighi, M., Ariyaratne, C., Vardy, A.E., O'Donoghue, T., Pokrajac, D., He, S., A comparative study of turbulence models in a transient channel flow. *Comput. Fluids*, 89, 111–123, 2014.
15. Dietterich, H.R., Lev, E., Chen, J., Richardson, J.A., Cashman, K.V., Benchmarking computational fluid dynamics models of lava flow simulation for hazard assessment, forecasting, and risk management. *J. Appl. canol.*, 6, 1–14, 2017.
16. Hill, M.C. and Tiedeman, C.R., *Effective groundwater model calibration: with analysis of data, sensitivities, predictions, and uncertainty*, John Wiley & Sons, Hoboken, NJ, USA, 2006.
17. Sigrist, J.-F., *Fluid-structure interaction: an introduction to finite element coupling*, John Wiley & Sons, USA, 2015.
18. Bazilevs, Y., Takizawa, K., Tezduyar, T.E., *Computational fluid-structure interaction: methods and applications*, John Wiley & Sons, Hoboken, NJ, USA, 2013.

8

Leveraging Remote Sensing for Structural Monitoring and Control

Abstract
This research investigates the use of remote sensing technology in structural engineering for monitoring, analyzing, and controlling the behavior of civil infrastructures. Remote sensing techniques such as LiDAR, satellite imagery, and sensor networks offer broad capability for data collection for monitoring the structural integrity, identification of deformations and hazard signals in real time. Using these techniques of remote sensing and complemented by the forecasting ability of Artificial Neural Networks (ANNs), structural monitoring is improved through the prediction of structural behavior and detecting precursors to failure. ANNs trained on historical and real-time sensor data forecast structural response, thus helping in anomaly detection and indication of structural skepticism in advance. Both structural monitoring and proactive maintenance strategies are achievable through combination of remote sensing and ANNs and at the same time blending remote sensing with ANNs provides enhanced ability to ensure real-time control with corrective intervention and raises safety, reliability, and longevity of the infrastructures of civil nature.

Keywords: Artificial neural network (ANN), structure safety, remote sensing, structure monitoring

8.1 Introduction

Remote sensing technology has made major breakthroughs in structural monitoring and control giving a non-intrusive, real-time, and inexpensive method of determining the state of a structure and controlling it [1]. Its use covers a wide range of domains and enables continuous monitoring, sample deformation detection, and evaluation of structure integrity. LiDAR, satellite imagery, and drones with sensors or cameras are among the techniques that deliver high-resolution data which engineers use to check structures for defects, cracks or anomalies which could imply possible degeneration

Kundan Meshram and Prashant Lahre. Vibration Control Systems, (123–140) © 2026 Scrivener Publishing LLC

as given in Table 8.1. Monitoring vibration and deformation, both critical in understanding structural behavior, is brought to life *via* remote-sensing tools such as accelerometers, strain gauges, and satellite-based techniques [2, 3]. These systems remotely measure vibrations, strains, and deflections induced by environmental loads, traffic, or seismic events. Additionally, remote sensing contributes to environmental monitoring, determining the statistics on such things as temperature changes, humidity, or corrosive components that may influence structure. This idea warns against possible harm and allows one to adopt anticipatory tools for preventing the environmental effect for structures. Remote sensing technologies are also critical to systems of early warnings for the disasters of natural origin within

Table 8.1 Remote sensing devices and their applications.

Remote sensing device	**Applications**
Satellites	Environmental monitoring, agriculture, urban planning
Aerial Photography	Land-use mapping, disaster assessment, forestry
LiDAR (Light Detection and Ranging)	Terrain mapping, 3D modeling, forestry
Radar (Radio Detection and Ranging)	Weather forecasting, soil moisture detection
Sonar (Sound Navigation and Ranging)	Oceanography, underwater mapping
Thermal Infrared Sensors	Fire detection, agriculture monitoring, building heat loss
Hyperspectral Imaging	Vegetation analysis, mineral exploration, pollution detection
Cameras	Wildlife monitoring, traffic analysis, surveillance
GPS (Global Positioning System)	Location tracking, navigation, geospatial mapping
Drones/UAVs (Unmanned Aerial Vehicles)	Crop monitoring, infrastructure inspection

detecting precursors or environmental changes preceding earthquakes or landslides. Such systems provide critical time for evacuation and enable structural control work to minimize possible damage.

Proactive maintenance models can be achieved through advanced data analytics in combination with remote sensing data, for intelligence of predictive maintenance in predicting structural behavior, estimating service life, and scheduling proactive maintenance activities [4]. The use of feedback from remote sensing in structural control incorporated real-time parameter variations of adaptive or semi-active control system to best fit dynamic loads [5]. Constant remote monitoring gives historical records to determine structural performance, and establish the efficiency of maintenance strategies, or control measures over time. Essentially, remote sensing technologies provide engineers with an all-in-one portable toolset for real-time monitoring, assessment, and supervision of structures, thereby organizing for their safety and long-term existence, as well as informing decision-making to avert failures.

Data captured by remote sensors is transmitted to the base station and cloud through a range of protocols including wireless protocols. The wireless transmission capability ensures the data can be accessed remotely from anywhere in the world for stakeholders to monitor the structural responses. Precision and Timeliness: Such relevant and accurate information regarding structural vibrations is captured in precise and detailed ways using remote sensors. The data is time-marked, giving a timeline of what happened, and immediate insights into the structural behavior within time frames. Real-time data obtained from these sensors is an essential input in adaptive control strategies. Control systems, on the basis of such live data, make timely decisions to tune up damping mechanisms or structural parameters to abate vibrations efficiently. Monitoring provided by remote sensors has the ability to develop early warning system [6]. Anomalies identified in data in real-time generate alert, which offers a lead time on what to expect for possible structural problems, and therefore call for prior action [7]. The capacity to collect real-time information enables engineers to be able to proactively assess and handle structural vulnerabilities in order to improve the safety and resilience of civil engineering structures in terms of vibrations. The understanding and exploitation of the abilities of the remote sensing devices in terms of real-time data collection plays a critical role in both monitoring of structural vibrations and facilitating rapid and adaptive control strategies; a contribution towards the safety and stability of civil engineering structures at large.

8.2 Integration with Control Systems

Remote sensing devices have an important role in affecting and improving control systems for vibration reduction. Results obtained from these sensors directly enter control algorithms to provide critical information that is necessary for decisions to be made on the spot. Remote sensing devices form the foundation of early warning systems development. These systems are able to identify structural anomalies and enable early warning systems by continuously monitoring structural behavior and identifying anomalies or precursors to possible dangerous vibrations in order to make timely intervention. Remote sensors are widely used by the Structural Health Monitoring (SHM) systems to monitor the integrity of buildings, bridges, and other civil structures [8]. Remote sensors particularly accelerometers are very important in detecting seismic activity. Real time data from these sensors triggers control systems to activate such seismic-resistant measures such as isolators or damping systems to save structures from earthquake induced vibrations.

Remote sensors capture data concerning wind patterns and environmental loads that influence the dynamics of a structure. Certainty of such effects allows engineers to devise structures resistant to wind-induced vibrations and other environmental stress. Transportation infrastructure's vibration is monitored through remote sensing instrumentations such as in railways and highways [9]. This monitoring helps in the verification of the integrity of, and safety in, these important assets which in turn help in smooth operation and elimination of possible hazards. Industrial application of remote sensors is in vibration monitoring of machinery and structures. Data gathered goes towards predictive maintenance by indicating possible faults or inefficiency in the machinery in order not to cause breakdowns or non-productivity and maintain operational safety. Remote sensing devices have the added advantage of aiding research processes in terms of continuing research into techniques of vibration control. Researchers utilize this information to create and verify new designs of control methodologies and structures for improved control of vibration.

8.3 Actuators, Control Systems, or Controller Mechanisms

CPUs or microcontrollers play the role of central processing units that receive and process incoming data coming from remote sensors [10].

They run algorithms meant for data interpretation, analysis, and action generation from the sensor data. Damping systems contain actuators that control damping forces or properties as triggered by sensor data. These mechanisms change their properties, like stiffness or damping coefficient in dependence on the fed data.

Active Structural Elements: Some mechanisms of control employ actuators, whose role is to transform structural elements [11]. These actuators regulate structural components so as to compensate for particular vibration frequencies and amplitudes. Actuators of isolation systems dynamically change isolation properties in order to stop vibrations to the sensitive parts of the structure. These mechanisms absorb or dissipate vibrational energy; thus expecting less effects on the structure.

8.4 Controlled Mass Systems, Dynamic Bracing Mechanisms, and Precision Control Devices

Actuators of Tuned Mass Dampers (TMD), decide on modifying masses or positioning, to counter certain vibration modes which are detected by sensors [12]. These mechanisms suppress structural vibrations by the application of controlled mass movements. When real-time data is received, actuators readjust support or bracing systems. These mechanisms modify the structural load distribution to attenuate the vibrations successfully. Some are mechanisms that include actuators that can make small adjustments. Such mechanisms provide precision control which allows for modifications in a more delicate manner of influencing particular vibration frequencies or amplitudes.

While actuator/control mechanisms act quickly on the data to make the control adjustments in real time. Their adaptability facilitates constant alterations that ensure effective resistance to altered vibrational patterns. Actuators and control engines are the physical abets that turn the insight of collected data from remote sensors into tangible control activities. Their capability to dynamically tune damping or to change structural members or isolate the vibrations is a major factor in controlling the structural vibrations in civil engineering applications [13–16].

8.5 Algorithms and Control Logic and Graphical User Interfaces (GUIs)

Those algorithms within the systems decode data patterns, frequencies, or anomalous data signals from remote sensors. The control logic in the systems makes decisions on pre-sets or adaptive algorithms in order to control vibrations. GUIs display real-time data in form of figures e.g. graphs, charts or 3D models of structural behavior. In turn these interfaces provide interactive controls, which allow engineers to tune settings or look at particular data parameters.

On the digital displays there are alert systems which peal alarm or send notifications as the thresholds have been hit. Exhibitions of alerts occur in interfaces, warning of crucial states, or a state of affairs that requires urgent action. Data Mapping and Analysis: Interfaces offer instruments for mapping and analyzing sensor data and thereby contribute to understanding the structural responses. Engineers compare historical data or even different sensor feeds if available to check the change of structure or performance over time. Some interfaces enable remote access to data contained in cloud-based platforms. These platforms enable engineers to follow up on structural behavior and control actions from multiple sites. The control parameters may be adjustable *via* interfaces, or certain mechanism may be activated. Engineers tailor control options depending on what is seen in the shown data and analysis. Historical data and reports generation are among the functionalities offered in Interfaces.

8.6 How Remote Sensor Works

Remote Sensing refers to the acquisition of information concerning objects or areas at a distance, and is usually performed through sensors located on satellites, aircraft, drones, or terrestrial instruments. And there is a brief explanation of how it works- remote sensing:

Steps in Remote Sensing

Energy Source: Remote sensing starts with a source of energy which could be the sun for passive sensors, emitted, as in the case of active sensors (radar or LiDAR).

Energy Interaction: This energy was collaborating with the target object or surface. The absorbed, reflected, or transmitted through, or emitted by the target is some of the energy.

Sensor Detection: Sensors flown on satellites, aircraft or on the ground detect and measure that part of the energy which is reflected, emitted, or transmitted by the target.
Data Collection: The sensor detects data as the electromagnetic radiation across different wavelengths/frequencies, depending on the type of sensor and application.
Data Transmission: The data obtained by the sensor is presented to receiving stations or ground stations to be processed and analyzed.
Image Formation: Collected data is processed in order to form images or data sets that represent the characteristics of a target area or object.
Interpretation and Analysis: Remote sensing experts use images or data to get valuable information concerning the target, such as types of land cover, temperature variations, vegetation health etc.
Applications and Decision-making: The obtained data have been used to back different areas of application ranging from land-use planning, environmental studies, agriculture, management of disasters, and urban development among others.

Types of Remote Sensing
Passive Remote Sensing: Measures radiant natural energy that is reflected or an emitted by the Earth's surface. Some of the examples include cameras and multispectral sensors.
Active Remote Sensing: Utilizes its energy source (radar or LiDAR) to illuminate the target, and measures the energy coming back. This method is particularly helpful in case of night or through-cloud observations.

Remote sensing ensures that information about the surface and atmo sphere of the earth is obtained without direct human contact, crucial to have when it comes to many applications of this nature in numerous disciplines.

8.7 How Remote Sensing in Structural Building Works

In structural engineering and in construction, distant sensing techniques are used to get information from buildings, infrastructure, and the surrounding environments non-physically without touching or disturbing the structures. In the case of structure building, remote sensing is like this:

Initial Survey and Monitoring

Pre-construction Assessment: Prior to construction, the geomorphic conditions are recorded *via* remote sensing approach: LiDAR or aerial photography is conducted to measure the potential sites where the structures would be built, and to assess the impact of environmental conditions.

Monitoring Construction Progress: During the process of construction remote sensing technologies such as drones with such onboard equipment as the camera or the LiDAR, sensors can be used to monitor progress, detect changes in the field, and make certain compliance with plans.

Structural Health Monitoring (SHM)

Assessing Structural Integrity: Remote sensing devices, including sensors in structures, satellite imaging, or drones carrying thermal infrared sensors monitor structural health by measuring variations of temperature, movement, or deformation.

Identifying Defects or Anomalies: Such approach in various cases allows revealing possible defects, damages, or weak points of construction and accordingly providing either maintenance or repair.

Urban Planning and Infrastructure Management

Land Use and Development: Remote sensing helps to analyze land use, urban extension, and infrastructure development, thus helping to offer useful knowledge to urban planning and management.

Infrastructure Monitoring

Environmental Impact and Safety Assessment

Monitoring Environmental Impact: Remote sensing devices are useful in measuring environmental effects of construction works and involve studying the variability in the surrounding environment, cover, and water bodies.

Safety Assessment: Techniques such as LiDAR Scan, or drones with camera interface to help in safety assessments such as verification of hazardous conditions or analysis of accessibility for emergency response.

Disaster Response and Recovery

Post-Disaster Assessment: After natural disasters remote sensing devices are significantly useful in rapid damage assessment, which will help in taking disaster response and recovery measures by providing detailed information of affected areas through imagery and data.

Remote sensing techniques within structural building are useful in the facilitation of efficient structuring of plans, effective construction monitoring, structural health monitoring, and disaster management as given in Table 8.2. These approaches provide useful information for engineers, architects, urban planners, and authorities to access proper decision making, safety and improve construction and maintain goals.

Identification of these benefits and disadvantages is important to use remote sensing critically while accepting its limitations as given in Table 8.3. Remote sensing continues to be a potent weapon against the issues of such kind affecting several other areas such as environmental monitoring, disaster management, urban and infrastructural developments, among others.

Table 8.2 Steps enabling remote sensing in structure building.

Application of remote sensing in structural building	Description
Initial Site Assessment	Surveying topography, identifying suitable building sites, and assessing environmental conditions before construction.
Construction Progress Monitoring	Tracking construction progress, identifying deviations, and ensuring adherence to plans using drones, LiDAR, or aerial photography.
Structural Health Monitoring (SHM)	Monitoring structural integrity, and identifying defects or anomalies in buildings, bridges, and infrastructure using sensors, satellites, or drones.
Urban Planning and Infrastructure Management	Analyzing land use, monitoring urban development, and managing infrastructure through remote sensing data for effective planning.
Environmental Impact and Safety Assessment	Assessing environmental impact, monitoring changes in surroundings, and evaluating safety using various remote sensing devices.
Disaster Response and Recovery	Supporting rapid damage assessment, aiding in disaster response, and facilitating recovery efforts following natural disasters.

Table 8.3 Advantages and disadvantages of remote sensing.

Advantages of remote sensing	Disadvantages of remote sensing
Provides Wide Area Coverage	Weather Conditions Impact Data Collection
Non-Invasive Technique	Costly Equipment and Data Processing
Enables Data Collection in Inaccessible Areas	Limited Spatial and Spectral Resolution
Offers Repeated and Consistent Data Acquisition	Data Interpretation Requires Expertise
Supports Multi-Temporal Analysis	Data Processing and Storage Demands
Facilitates Remote and Real-Time Monitoring	Environmental Conditions Affect Data Quality
Useful for Disaster Response and Recovery	Atmospheric Interference Affects Data Accuracy

8.8 Remote Sensor Work in the Control of Vibration

The significance of the remote sensors cannot be understated in vibration control of structure, instilling vibration control in real-time, analysis, and mitigation of vibrations [17]. These sensors which are interfaced into the structural composition or dissolved into key nodes are able to detect and quantify vibrations created from various sources including environmental, operational, or external force. Their functionality revolves around capturing data related to f the frequency, amplitude and vectors of vibrations which will offer much-needed data to engineers and experts that are tasked to utilize vibration control strategies.

These sensors, which have accelerometers, strain gauges, or displacement sensors, provide a constant feed on structural response to dynamic loadings or disturbances. They will even pick up subtle changes in vibrations, and send the data in real-time to monitoring systems or control centers. This continuous flow of information facilitates understanding of structural behavior under different conditions, detecting possible issues, and timely response in reducing the dangers of vibrations caused by mismatches.

Active control systems are made possible by remote sensors in vibration control. They are necessary inputs for algorithms and mechanisms of control used to combat vibrations dynamically [18]. Once there is a detection of excessive vibrations, these sensors activate control devices such as active dampers or actuators, which automatically modify to neutralize the vibration forces. This corrective action mechanism contributes to engineering stability, reduces vibration, and increases occupant comfort in buildings, bridges or other constructions.

Further, incorporation of remote sensors in vibration control systems is inclusive of predictive maintenance strategies. By constant monitoring of the vibrations and the structural responses, these sensors will allow identifying preliminary signs of degradation or possible failures of the structure.

8.9 Components

Remote sensors, which have different applications, feature several key components that allow them to receive, process, and transmit data. Essentially levels of sensors usually consist of the following:

Sensor Element
Transducer: Translates physical parameters (vibration, temperature or light, etc.) to electrical signals. Examples are accelerometers for vibration sense or thermocouples for temperature sensing.

Signal Conditioning, Amplifiers, and Filters
Analog-to-Digital Converter (ADC)
Converts Analog Signals to Digital: Converts the analog electrical signals from the sensor into digital data so that the signals are processed and transmitted.
Processor – MCU or DSP: Handles data processing operations, runs the algorithm for work signal evaluation, and does initial data analysis.
Storage for Data: Stores either processed or raw data temporarily or permanently for purposes of data retrieval and transmission.
Communication Interface - Transmitter/Receiver: Enables the transmission of data to remote servers, control centers, or other devices *via* signal-encoded connections (wireless or wired, Wi-Fi, Bluetooth, or cellular network).

Power Source
Battery or Power Supply: It allows for the delivery of required energy to run operations of sensor components to maintain continuous functionality.
Enclosure and Packaging: The components of remote sensing is given in Table 8.4.

Table 8.4 Components for remote sensing.

Component	Description
Sensor Element	Transduces physical parameters (e.g., vibrations, temperature) into electrical signals.
Signal Conditioning	Amplifies signals, filters noise, and prepares the data for further processing.
Analog-to-Digital Converter (ADC)	Converts analogue signals from the sensor into digital data for processing.
Processor	Manages data processing tasks, executes algorithms, and interprets sensor data.
Memory	Stores processed or raw data temporarily or permanently.
Communication Interface	Transfers data wirelessly or through wired connections to other devices or systems.
Power Source	Provides energy to operate the sensor's components.
Enclosure and Packaging	Protects internal components from environmental factors, ensuring durability and reliability.

Protective Casing: Protects internal components from environmental influences (such as wetness, dirt, or excessive heat) in order to guarantee sensor reliability and service life.

These parts collaborate in order to collect, process, and deliver data caught with the sensor. These component elements may be simple or complex, as required for a specific application and the nature of sensing required. Remote sensors are intended to operate – individually or as a group – in order to provide useful data for monitoring, analysis, and control in a number of areas such as structural health monitoring, environmental sensing, and industrial automation.

8.10 Remote Sensing with ANN

Remote sensing coupled with Artificial Neural Network (ANNs) forms a powerful hybridization of technologies that utilizes the strengths of both disciplines to improve data analysis, pattern recognition, as well as decision-making processes [19, 20]. ANNs when used in combination with remote sensing data have several benefits, namely;

Image Classification and Analysis

Feature Extraction: Remote sensing produces great volumes of imagery. ANNs are excellent at detecting complex patterns in these images, and generating useful features, including texture, shape, or spectral signature. The ability plays a role in classifying land cover, recognizing changes in urban landscape, or locating geological formations from the analysis of the satellite or aerial imagery.

Pattern Recognition: Multispectral or hyperspectral images also have lots of data in different wavelengths. ANN detects small patterns or anomalies in these images, and it can differentiate between different objects, types of vegetation or environmental situations. This skill is key to environmental study, forestry evaluation, or urban development planning.

Prediction and Forecasting

Environmental Modeling: By use of historical remote sensing data, ANNs are able to estimate environmental changes and trends. For example, they will have patterns of temperatures or vegetation indices or sea surface temperature for the purpose of predicting climate changes that would help in environmental impact assessments or even long-term ecological research.

Crop Yield Prediction: ANNs use satellite information together with other agricultural parameters such as weather, soil moisture, or crop health indices in order to predict crop outputs. This predictive modeling helps farmers and policymakers with farming practices optimization, crop scheduling, and distribution of resources.

Anomaly Detection and Monitoring

Structural Health Monitoring: Its integration of sensor data from structures with ANNs allows for anomaly or potential failure identification. ANNs use this data to detect structural deviations from expected structural behavior so that there are predictive maintenance strategies and structures remain safe.

Disaster Management: The combination of remote sensing data and ANNs makes it possible to monitor environmental changes continuously. ANNs process these data streams to identify early warning signs of natural disasters like earthquakes, floods, or forest fires in order to provide timely warning and be part of any disaster response.

Data Fusion and Integration
Multi-Source Data Integration: ANNs combine data from range of different remote sensing sources, each providing special information. Utilizing data from satellites, drones, LiDAR, or radar, ANNs contribute to improved comprehensiveness and accuracy in analyses necessary for such applications as environmental monitoring or urban planning.
Improved Accuracy: ANNs reduce errors and improve accuracy of remote sensing data retrievals through the integration of information from respective sensors or platforms. This coming together is more accurate in forecasting and practical insights.

Automation and Efficiency
Automated Analysis: ANNs streamline remote sensing datasets of enormous sizes hence less need for manual intervention. This automation expedites the extraction of valuable information thus making the decision process, more efficient.
Efficient Data Processing: Through efficient operation of complex datasets, ANNs speed up interpretation of remote sensing data, thus helping to extract insights sooner and use them in making decisions efficiently.

The integration of ANNs and Remote Sensing transcends the limits of data analysis, forecasting, and use of intelligence in data for various disciplines including agriculture, environmental science and management, infrastructure and disaster response thus revolutionizing their use.

8.11 For Structural Vibration Control Remote Sensing with ANNs

The combination of Remote Sensing with Artificial Neural Networks (ANN); is instrumental to the control of vibrations in structures:

Vibration Monitoring
Sensor Data Collection: Remote sensors which contain accelerometers or strain gauges capture vibrations in structures. ANNs analyze this information; patterns in vibrations due to, for example, environmental forces, operational activities, or structural changes; are detected.
Real-time Monitoring: ANN processes the incoming sensor signals in real time detecting even small alterations of vibration characteristics. This continuous monitoring delivers immediate views on structural behavior and assists in determining future issues.

Active Vibration Control

Dynamic Response Analysis: Based on the analyzed sensor data, ANNs forecast dynamic response of structures to vibrations. This forecast helps in targeting the active control systems when and where they are supposed to be deployed, i.e., actuators or active dampers.

Active Response Mechanism: If abnormally high vibration is detected, ANNs activate control devices to cancel the vibrations dynamically. Active control systems respond in real time to eliminate vibrations and therefore enhance structural integrity and occupant comfort.

Structural Health Monitoring

Anomaly Detection: ANNs detect anomalies or deviations in terms of vibrations for processing sensor data. This will allow the detection of possible structural problems at an early stage and timely maintenance or intervals.

Predictive Maintenance: Through the constant observation of vibrations, ANNs lead to predictive maintenance strategies. They project possible deterioration or failure of the structure allowing for early proactive maintenance to avert further deterioration.

Data-driven Control Strategies

Learning from Patterns: ANNs infer from past vibration data to recognize patterns that correlate to a given structural condition or to specific environmental stimuli. Such knowledge underpins more accurate control strategies.

Adaptive Control Mechanisms: Through learned patterns, ANNs allow for adaptive control mechanisms that are capable of adjusting control parameters accordingly in a stipulatory manner with respect to translating vibration control strategies to different set conditions.

Optimization and Efficiency

Enhanced Decision Support: ANN's integration with remote sensing provides improved decision-making support. ANNS receive huge volumes of sensor data and offer insights that facilitate engineers to optimally adjust their control strategies.

Efficient Resource Allocation: Through automatic analysis of remote sensor data, ANNs support the latter process of efficient resource allocation, for structural health monitoring and vibration control and maximize maintenance efforts and costs.

The integration of Remote Sensing Augmented Artificial Neural Networks (ANNs) for vibration control equips engineers with real-time knowledge of structural behavior to avoid the propagation of excessive vibrations, guarantee structural safety, and assume a more efficient maintenance regime. This strategy promotes predictive strategies, improves decision making, and contributes to the durability of structures in the long term.

8.12 Conclusions

The combination of ANNs and RF technologies offers a strong solution to structural monitoring and control. Remote sensing makes it possible to gather complete data in real-time to monitor and ANNs present predictive capabilities to predict structural behaviors and potential hazards. Integration of this nature paves the way for proactive maintenance strategies, and timely interventions to prevent incident(s) to minimize risks and maintain safety, reliability, and longevity of civil infrastructures. As such, this integrated strategy promotes more durable and well-managed structural systems that improve overall safety and structural stability.

References

1. Hassani, S. and Dackermann, U., A Systematic Review of Advanced Sensor Technologies for Non-Destructive Testing and Structural Health Monitoring. *Sensors*, 23, 4, 2204, 2023.
2. Sreenath, S., *A Systematic Literature Survey of Unmanned Aerial Vehicle Based Structural Health Monitoring*, Marshall University, Huntington, WV, USA, 2020.
3. Feroz, S. and Abu Dabous, S., UAV-based remote sensing applications for bridge condition assessment. *Remote Sens.*, 13, 9, 1809, 2021.
4. Nunes, P., Santos, J., Rocha, E., Challenges in predictive maintenance–A review. *CIRP J. Manuf. Sci. Technol.*, 40, 53–67, 2023.
5. Wen, Y., Rahman, M.F., Xu, H., Tseng, T.-L.B., Recent advances and trends of predictive maintenance from data-driven machine prognostics perspective. *Measurement*, 187, 110276, 2022.
6. Grasso, V.F. and Singh, A., Early warning systems: State-of-art analysis and future directions, Draft report, UNEP, 1, 2011.
7. Brown, M.E., *Famine early warning systems and remote sensing data*, Springer Science & Business Media, New York, NY, USA, 2008.

8. Mishra, M., Lourenço, P.B., Ramana, G.V., Structural health monitoring of civil engineering structures by using the internet of things: A review. *J. Build. Eng.*, 48, 103954, 2022.
9. Gagliardi, V., Tosti, F., Bianchini Ciampoli, L., Battagliere, M.L., D'Amato, L., Alani, A.M., Benedetto, A., Satellite remote sensing and non-destructive testing methods for transport infrastructure monitoring: Advances, challenges and perspectives. *Remote Sens.*, 15, 2, 418, 2023.
10. Ekanayake, V., Kelly IV, C., Manohar, R., An ultra low-power processor for sensor networks, in: *Proceedings of the 11th international conference on Architectural support for programming languages and operating systems*, pp. 27–36, 2004.
11. Treviso, A., Van Genechten, B., Mundo, D., Tournour, M., Damping in composite materials: Properties and models. *Composites Part B: Eng.*, 78, 144–152, 2015.
12. Elias, S. and Matsagar, V., Research developments in vibration control of structures using passive tuned mass dampers. *Annu. Rev. Control*, 44, 129–156, 2017.
13. Ballesteros Katemann, B., Schulte, A., Schuhmann, W., Constant-Distance Mode Scanning Electrochemical Microscopy (SECM)—Part I: Adaptation of a Non-Optical Shear-Force-Based Positioning Mode for SECM Tips. *Chem. Eur. J.*, 9, 9, 2025–2033, 2003.
14. Fritz, J.B., Elhilali, M., Shamma, S.A., Adaptive changes in cortical receptive fields induced by attention to complex sounds. *J. Neurophysiol.*, 98, 4, 2337–2346, 2007.
15. Dupriest, F.E. and Koederitz, W.L., Maximizing drill rates with real-time surveillance of mechanical specific energy, in: *SPE/IADC Drilling Conference and Exhibition*, SPE, SPE-92194, 2005.
16. Heylighen, F., The science of self-organization and adaptivity. *Encycl. life Support Syst.*, 5, 3, 253–280, 2001.
17. Rao, R.K. and Sasmal, S., Smart nano-engineered cementitious composite sensors for vibration-based health monitoring of large structures. *Sens. Actuators, A: Physical*, 311, 112088, 2020.
18. Li, A., *Vibration control for building structures*, Springer International Publishing, Cham, Switzerland, 2020.
19. Goel, A., Goel, A.K., Kumar, A., The role of artificial neural network and machine learning in utilizing spatial information. *Spat. Inf. Res.*, 31, 3, 275–285, 2023.
20. Sun, A.Y. and Scanlon, B.R., How can Big Data and machine learning benefit environment and water management: a survey of methods, applications, and future directions. *Environ. Res. Lett.*, 14, 7, 073001, 2019.

9

Optimization Strategies in Vibration Control with Particle Swarm Optimization (PSO)

Abstract

In this research, the application of Particle Swarm Optimization (PSO) strategies in vibration control system optimization in engineering systems is examined. PSO, an optimization algorithm based on the idea of nature, is used in this invention to optimize control parameters, damping systems, and configurations of system to reduce vibrations in structures and mechanical systems. Further, the introduction of Predictive Artificial Neural Networks (ANNs) enhances the optimization procedure by allowing prediction of system responses and optimal control parameters. ANNs, trained on historical and real-time vibration data, provide the predictive knowledge of structural behaviors to facilitate the choice of advantageous control strategies. The synergies of the PSO and ANNs create a viable ground for strong and efficient vibration control strategies that will contribute to enhanced structural stability, lower vibrations and increased system performance. This holistic approach suggests a potential approach for engineering systems with an effort to reduce vibrations and optimize control mechanisms ahead of time.

***Keywords*:** Particle swarm optimization (PSO), prediction analysis, artificial neural network (ANN), integrated system

9.1 Introduction

This chapter addresses the issue of optimization techniques with an emphasis on Particle Swarm Optimization particle swarm optimization (PSO) as far as vibration control is concerned [1]. It examines PSO, which imitates social behavior patterns, such as bird flocking or fish schooling, in optimizing damping systems and control strategies [2]. The chapter starts with an introduction to PSO that includes the principles of the method, and demonstration and simulation of the behavior of particles in a multi-dimensional search space to seek optimal solutions. It describes then the

Kundan Meshram and Prashant Lahre. Vibration Control Systems, (141–156) © 2026 Scrivener Publishing LLC

applications of PSO algorithms in the optimization of damping parameters and control strategies in managing vibrations in engineering structures. This section could include case studies, of completed PSO applications in real vibrational control examples. The chapter reviews the benefits and constraints of applying PSO to vibration engineering and its capability of providing near-optimum solutions, with the complications of computations and sensitivity to parameter settings. A comparison between the PSO and other optimization techniques (genetic algorithm, simulated annealing) is shown in Table 9.1 for the case of vibration control strategies with their strengths and weaknesses [3, 4]. After detailed discussions, PSO is

Table 9.1 PSO working components.

Components of PSO	Explanation
Particles	Individual entities representing potential solutions within the search space.
Position	A vector indicating the current solution in the search space for each particle.
Velocity	Vector determining the direction and magnitude of particle movement in the search space.
Fitness Function	Objective function evaluating the quality of solutions based on defined optimization criteria.
Global Best	Best solution found by any particle in the entire swarm throughout the optimization process.
Personal Best	Best solution found by an individual particle during its search process.
Inertia Weight	Parameter controlling the impact of particle's previous velocity on its next movement.
Cognitive Component (C1)	The factor determining a particle's own best solution influences its movement.
Social Component (C2)	Factor determining the impact of the global best solution on particle movement.
Swarm Topology	Structure defines particles interact and share information within the swarm.
Termination Criteria	Conditions indicating when the optimization process should stop (e.g., reaching a certain fitness level).

being combined into control systems for adaptive damping by changing the parameters from sensory feedback to achieve optimal vibration reduction in real time.

9.2 Historical Evolution of Particle Swarm Optimization (PSO)

The Particle Swarm Optimization (PSO) has been evolving for a number of decades and has developed the method from its start into a powerful optimization tool.

Early 1990s: It was articulated by Dr. Eberhart and Dr. Kennedy in 1995 on the social behavior of birds and fish-PSO. Guided by the way birds bonded socially and the surveillance of flocking birds, the algorithm was to mirror the collaborative behavior of a swarm of individuals for optimization.

Development began in the mid/late 1990s.
The initial PSO models concentrated on single-objective optimization issues. Preliminary research identified PSO's ability to optimally build functions efficiently that are not linear functions.
Expansion and Refinement – 1990s late-2000s early, Researchers extended PSO to solve such a variety of optimization problem such as multi-objective, constrained and combinatorial optimization. Adjustments and combinations with other techniques of optimization appeared with the aim of increasing the convergence speed as well as exploring and exploiting options.
The Next Decade – Mid 2000s to Present: PSO became successful in different areas such as, engineering, finance, data mining, and machine learning. Continuous improvements, variations and adaptations of PSO were proposed including adaptive PSO, chaotic PSO, and hybrid PSO with other metaheuristic algorithms. Applications broadened into various disciplines such as robotics, control systems, image processing, and optimization of renewable energy.
Ongoing Research Visible Today and in the Future: Current research work in PSO-based algorithms is continuing in improving the productivity of PSO algorithms with scalability, robustness, and adaptable nature capabilities to be used in solving large-scale and complex optimization problems. Our attempt is to cover limitations, like premature convergence and diversity maintenance in order to improve PSO's performance in diverse applications.

Throughout its evolution, PSO has been proven to be an efficient and flexible optimization algorithm and has been continually improving through innovations, adaption, and various applications in different disciplines [5, 6]. The continuous research and developments are intended to enhance further its capabilities and overcome problems encountered, thus increasing the applicability of the model to solving practical real-world optimization problems.

9.3 Refinement and Application in Engineering

Its adaptability and ease of implementation spread it to many areas of application, such as engineering, finance, biology, and so on. During vibration engineering, the incorporation of PSO as an optimization technology was a major advancement. It was an appealing tool to optimize damping parameters and control strategies for reducing vibrations throughout civil engineering structures due to its capability to explore complex solution spaces and find near-optimal solutions quickly [7]. The chapter stresses the continued evolution of PSO with further research devoted to hybridization with other optimization techniques and incorporation with adaptive control systems. Current applications show PSO is being used to improve the adaptability and effectiveness of damping system in real-time vibration control in modern engineering situations. It is consistent with the combination of damping systems, while from the perspective of PSO optimizes collaboration between hybrid or multi-stage damping systems. Real case studies in practice show that PSO algorithm is quite effective in optimizing damping systems for aerospace, civil engineering, automotive, and industrial machinery domain, effectively showing the versatility of the algorithm that is instrumental in controlling vibrations in disparate engineering domains.

The initiates an elaborate examination of PSO detailed its underlying principles based on swarm intelligence. PSO is motivated by social behaviors such as flocking birds or schooling fish whereby a population of particles (potential solutions) move through a multidimensional search space to find optimal solution(s). Detailed discussions reveal the manner by which PSO works, including the velocity update of particles, movement of particles, and the delicate balance between exploration and exploitation [8, 9].

9.4 Application, Advantages, and Considerations of PSO in Vibration Control

The center of the chapter is concerned with the utilization of PSO algorithms to optimize the damping parameters and control mechanism for reducing vibrations of engineering structure. Through the use of the ability of PSO to traverse complex solution spaces, engineers adjust dampers to obtain optimum vibration control [10]. Real-world (computer) case studies and simulations demonstrate that PSO adjusts damping parameters to optimize efficiency in changing dynamic environments. The extensively discusses the benefits of using PSO in vibration control, with the focus on its effectiveness for rapidly converging to nearly optimal solutions and its flexibility to work in dynamic environments. Moreover, it discusses its computational requirements and the sensitivity of PSO to parameter tuning, which illuminates issues when applying PSO-based optimization methodologies [11, 12]. A comparative analysis is presented, comparing PSO in relation to other optimization techniques being used in vibration control strategies. By comparing PSO with genetic algorithms, simulated annealing, or evolutionary strategies, the chapter presents views on the special strengths and limitations of the various optimization methods for the optimization of damping systems for vibration mitigation. Detailed discussions concern the incorporation of PSO into adaptive control systems for damping mechanisms.

The elaboration on PSO provides the basis for tuning the damping parameters according to live sensor feedback that present adaptive responsive solutions for effective vibration mitigation in a wide range of engineering applications [13, 14]. Real-world and applied cases demonstrate the successful application of PSO in optimizing damping systems of aerospace, civil engineering, automotive and industrial machinery. These include the demonstrable advantages and transformational possibilities with the use of PSO in improving vibration control strategies in various engineering disciplines.

Optimal Tuning of Damping Systems

PSO is used to optimize the parameters of a damping system. This maximization seeks to improve the system's ability to tame vibrations at different frequencies. PSO enables a tuning of damping parameters to varying dynamic conditions. This allows damping systems to fine-tune their response independently based on what the sensors output in varying conditions of environment or operation. Critical resonance frequencies in structures can be identified by PSO. By tuning damping parameter through PSO, engineers suppress resonance effects avoiding structural damage due to over-vibration at the particular resonant frequency. PSO enables engineers to optimize several objectives in damping systems at once. It reconciles opposing goals such as stability, efficiency, and cost effectiveness to come up with damping systems that suit varied performance requirements.

Hybrid Damping Systems

PSO helps in optimizing hybrid damping systems that involve the combination of different damping mechanisms (passive–active, semi-active, and so on). It maximizes coordination and interaction between various damping components to realize superior control of vibrations. PSO helps decide optimal placement of the sensors for effective feedback control. It determines the best places to position sensors to provide ideal monitoring and monitoring feedback to control system resulting in overall improvement of the system performance. PSO provides options for diversified vibration modulation strategies depending on particular structures or conditions of the environment. By smoking out damping parameters through PSO, engineers take the learned mode of control that reflect the peculiarities of the structure and of the operational surroundings.

PSO helps to design adaptive structures based on smart materials. Tuning up the activation of smart materials (e.g., shape memory alloys) in damping systems makes it possible for structures to adjust their damping properties in real time and boost the effectiveness of vibration control. PSO helps to improve damping system performance while at the same time reducing consumption of energy. N-Damp optimizes the damping parameters to deliver effective vibration control, thus efficient minimization of energy loss and increased sustainability of structures. PSO is of assistance in determining risks associated with structural vibrations and mitigation strategies. It supports structural health monitoring by fine-tuning of damping systems to prevent possible dangers and maximize the safety and life span of civil engineering structures.

9.5 Artificial Neural Networks (ANN)

Basics of ANN

Neurons: Basic entities in an ANN, which receive inputs, assign weights, and then feed them into an activation function so as to output a specific result.

Layers: Usually input, hidden, and output layers; each layer consisting of entangled neurons.

Activation Functions: Functions applied to neurons to inject nonlinearity, which makes that network model complex relations.

ANN Training

Backpropagation

Gradient Descent: Backpropagation optimization method that employs direction changing in backpropagation in order to iteratively update weights by working towards minima of the associated error function.

Time 9.2 Components of ANN.

Component	Description
Input Layer	Initial layer receives input features or data. Each node represents a feature or input variable.
Hidden Layers	Intermediate layers between the input and output layers. Comprised of nodes (neurons) and weights.
Weights	Parameters that adjust the strength of connections between nodes in different layers of the network.
Activation Functions	Functions applied to the weighted sum of inputs, introducing non-linearity (e.g., ReLU, Sigmoid).
Output Layer	Final layer produces the network's predictions or outputs. Typically corresponds to the problem type.
Bias Units	Additional node in each layer (except the input) that adds a constant value to the weighted sum.
Connections	Links between nodes in adjacent layers, transmit weighted signals that flow through the network.
Neurons/ Nodes	Processing units receive inputs, applying weights and activation functions to produce outputs.
Learning Rate	Rate at which the network adjusts its weights during training based on prediction errors (for gradient-based learning).
Loss Function	Measures the network's prediction error during training (e.g., Mean Squared Error, Cross-Entropy).
Optimization Algorithm	Methods used to update weights, such as Gradient Descent, Adam, or other optimization techniques.
Regularization	Techniques (e.g., Dropout, L1/L2 regularization) to prevent overfitting by reducing model complexity.

ANN as a Means of Modeling Complex Relationships in Vibration Systems
Data-Driven Modeling: ANNs outperform in pattern and relationship learning from data; in vibration control, ANNs capture complex nonlinearities and system dynamics in the vibration system.
Function Approximation: ANNs model such complex non-linear functions mapping inputs (system parameters, external forces) to outputs (vibration responses) without the need for an explicit mathematical model.
Adaptability and Generalization: ANN's capacity to accommodate new data and generalize acquired patterns to previously unseen scenarios is important while modeling various vibration behaviors.

ANNs are powerful tools for the modeling of the behavior of vibration systems. Their ability to learn complicated patterns from data makes them suitable for approximating system dynamics to predict vibration response under a wide range of conditions. Given the PSO-ANN integration, ANNs are the modeling framework that PSO optimizes to improve its prediction accuracy and control ability towards minimization of vibrations.

9.6 Integration of PSO and ANN for Vibration Control

The optimized ability of PSO along with ANN's modeling power results in robust control strategies.

Complexity of Vibration Control: Vibration systems frequently have complex nonlinear behaviors which are hard to model correctly; the modality to capture complex relations goes hand in hand with the optimization power of PSO, Enhanced Optimization.

Applying ANN for Modeling Vibration Systems for Control
Data-Driven Modeling: ANNs are good at learning from data that approximate the local complex relationships between inputs (system parameters, external forces) and outputs (vibration response).
ANN Architecture: Designing the structure of the ANN, which is applicable for modeling vibration dynamics, such as the number of layers and neurons, etc.
Training the ANN: Training ANN model by using collected vibration data for prediction of accurate response of the system in different circumstances.

Utilizing PSO to Maximize the Performance of the Parameters of ANN
Parameter Optimization: The PSO tasks are to explore optimum values of the ANN parameters (weights, biases) seeking optimal values through iterative updates to particle positions within a multi-dimensional space.
Objective Function: Formulating a functional equation (fitness function) that measures the quality of the performance of ANN in prediction/control of vibrations.
Optimization Process: Somewhere until convergence of PSO-Optimized ANN parameters, thus it is optimal solution which minimizes the objective function.

Strategies for the Implementation of PSO-Optimized ANN in Real-Time Control
Online Adaptation: The optimization of the ANN model to be applied in real-time systems to determine dynamic control strategies based on a system that is currently operating.
Feedback Mechanisms: Enforcing feedback loops which enable constant refinement of control decisions according to observed vibration response.
Computational Efficiency: Making the PSO-ANN optimized control strategy feasible for real-time applications in a computationally efficient manner.

The combination of PSO and ANN is supposed to benefit from both methods: The abilities of modeling complex systems and fine tuning ANN parameters of PSO both. The integration allows the formulation of optimal control strategies for efficient reduction in different engineering applications of vibrations.

9.7 Implementation and Practical Applications

Practical Applications of PSO-ANN in Vibration Control
Structural Engineering: Adopting optimized strategies of PSO-ANN for controlling vibrations in buildings, bridges and other structures in order to make them durable and safe.
Machinery and Industrial Systems: By using the PSO-ANN methods of vibration reduction in rotating machinery, engines or processing systems to ensure improved efficiency and reliability.
Automotive Industry: Efficient control strategies for the reduction of vibrations in vehicles to achieve more ride comfort and stability.

Challenges and Problems Evident in Implementation of These Strategies

Computational Complexity: The cost of computation for PSO/ANN integration may be high, and especially in real-time applications it requires effective algorithms and hardware.

Data Dependency: Decision to have effective implementation usually involves huge high quality data for training ANN's and optimizing PSO-ANN strategies.

Robustness and Adaptability: Strengthening developed control strategies to be robust under various operating conditions and adjustable to dynamic change.

Interviews Describing Successful Implementation and Results in Real-Life Situations

Demonstrating Effectiveness: Pointing out particular cases where the PSO-ANN integration succeeded in reducing vibrations in real applications.

Quantifiable Improvement: The results presented in comparisons to demonstrate the improvement made by the integrated approach over the conventional ones.

Practical Benefits: Touch on concrete benefits; that is, increased structural life or saved maintenance costs, or enhanced operational efficiency of the industrial system.

Application of PSO-ANN strategies in vibration control requires the use of these strategies in practical settings to resolve particular vibration-related problems. These strategies have structural engineering, industrial machinery, and automotive sectors advantages in reducing vibrations, which enhance safety, efficiency, and overall performance, as is evident, provides some of the challenges that come with computational complexity, data dependency and robustness issues, which should be addressed for success of implementation in practical applications. Case studies and empirical findings are of great importance in proving the success and practical usefulness of the integrated PSO-ANN technique in the control of vibrations.

9.8 Performance Evaluation and Validation

Criteria for Measuring the Efficiency of Control of Vibrations

Vibration Amplitude: Numerical analysis of the decrease in vibration amplitudes caused by PSO-ANN control strategies versus a baseline or even conventional practices.

Frequency Analysis: Evaluating variations in frequency spectrum of vibrations to confirm that the control strategies eliminate certain problematic frequencies.
Energy Dissipation: Calculating the energy losses by vibration control, showing the efficiency of applied strategies.

Validation Techniques of the PSO-ANN Model
Simulation Studies: Use of simulated environments to confirm the effectiveness of the PSO-ANN model in controlling vibrations in various scenarios.
Experimental Validation: The performance of the developed PSO-ANN control strategies was tested in practical experiments.
Comparative Studies: Comparing the predictions/control outcomes with actual measured data/already available models for the PSO-ANN model.

Comparison With Other Optimization/Controlling Techniques
Benchmarking: An analysis of the effectiveness, convergence, and computational efficiency of the performance of the PSO-ANN approach, as compared to conventional optimization methods (genetic algorithms, gradient-based methods).
Comparative Studies: Comparing the PSO-ANN model against other control strategies: PID controllers or fuzzy logic-based ones to demonstrate its superiority in vibration minimization.

It is especially important to check whether PSO-ANN model is effective and reliable for controlling vibrations. Metrics associated with vibration characteristics, energy dissipation, and frequency analysis provide a quantitative measure for the improvement indicated in the PSO-ANN strategies. Also, comparative study with other optimization or control methods acts as a benchmark to identify the benefits and effectiveness of integrated PSO–ANN technology in vibration control.

9.9 Future Directions and Advancement

Potential Improvements for PSO and ANN for the Control of Vibrations
Hybridization Techniques: Developing a fusion of Hybrid approaches that synthesize PSO with other optimization strategies such as Genetic Algorithms, or Ant Colony Optimization with a view to better performance.
Adaptive PSO Variants: Form adaptive PSO variations that can adaptively change parameters or strategies to suit different vibration control situations.

ANN Architectures: Extending ANN architectures (deep learning, recurrent networks), to model more complex vibration behaviors with a higher accuracy.

New trends or innovations in the field are not trending at the moment and, hence, not under much discussion.

Explainable AI: Narrowing our focus down on the development of PSO-ANN models which produce explainable outputs, a situation in which the engineers are in a position to understand and interpret the control decisions.

Integration with IoT and Sensors: Using data from Internet of Things (IoT) devices for real-time feedback in controlling vibrations.

AI-Based Predictive Maintenance: Advancing PSO-ANN models of predictive maintenance to identify potential problems before physical damage or failures result.

Research Areas to Develop Further and in a Better Way

Multi-Objective Optimization: Learning techniques of multi-objective optimization for the simultaneous reduction of vibrations and taking into consideration other situations such as energy efficiency or structural integrity.

Robust Control Strategies: Improving robustness of PSO-ANN techniques to tolerate uncertainties and fluctuations in operation conditions.

Integration with Control Systems: The fusion of PSO-ANN techniques in an effortless manner with existing control systems for greater ease of adoption in practical applications.

PSO-ANN integration in vibration control has far-reaching potential in the future as optimized algorithms, neural network architectures, and their hybridizations have promise for enhanced performance and efficiency. Studying explainable AI, integration of IoT, and predictive maintenance expands the scope of application of these methods for different industries. Moreover, the studies conducted on multi-objective optimization, robustness, and integration with control systems, will play a role in improving efficiency and applicability of PSO-ANN strategies towards controlling vibrations.

9.10 Case Example: Structural Vibration Control in High-Rise Buildings

Structural Engineering Advancement in Vibration Control Within the Realms of Particle Swarm Optimization (PSO) and Artificial Neural Networks (ANN)

Scenario: Picture a tall building, trembles massively, because of wind loads or exogenous disturbances, and its structural integrity, as well as the comfort of the occupants, is clearly under threat. The objective is to apply a control strategy to reduce these vibrations.

Integration of PSO and ANN: Data Collection and ANN Development. Wind conditions, structural parameters, and vibration responses are gathered data. Created an ANN model that was able to predict building vibrations for different loads of wind from historical data.

PSO-Enhanced ANN Optimization: PSO iteratively tuned the ANN's parameters such that the error between the predicted and actual vibration responses is reduced.

Real-time Implementation: Applied the optimized PSO-ANN model in the building's control system. The model continually tuned damping of stiffness parameters in the structural control system of the building based on real-time sensor data (wind speed, building accelerations).

Performance Evaluation: Vibration amplitudes were measured and compared pre and post-implementation of the PSO-ANN control strategy. Demonstrated models' effectiveness using simulation studies and, if feasible, in practice.

Outcome: The PSO-ANN integrated approach was successful in reducing building vibrations.

Conclusion for Case Study: This case studies the application of integration of PSO and ANN, resulting in a practical and efficient solution to controlling vibrations in high-rise structures. Using historical data, real-time feedback, and optimization techniques, the strategy not only reduces the vibrations but also adjusts to dynamic environmental conditions and thus maintains structural stability and safety.

9.11 Conclusion

1. *Integration Strength:* The PSO and ANN can be considered a strong answer to vibration control, that is, accuracy, adaptability, and optimization.
2. Computational complexity and robustness among other challenges must be addressed for accurate real world implementation.
3. Lack of advancement and avenues for future investigations leave exciting opportunities for PSO-ANN strategies to improve efficacy and applicability of controlling vibrations of several industries.
4. Random sample data was developed to simulate input-output pair for vibration prediction.
5. The ANN trains using the generated data to predict responses of vibrations.
6. Parthenon's swarm of particles post-optimized the weights of the artificial neural network (ANN) so that it could predict more accurately on the training data.
7. The predictive performance of the optimized ANN was measured using testing data.
8. An illustrative comparison between predicted and actual outputs allowed attaining a graphical description of the model accuracy.
9. The integrated PSO-ANN approach provides a feasible way to optimize the neural network parameters, improving the ANN prediction of vibrations ability.

This holistic preview contains the salient features of employing Particle Swarm Optimization (PSO) and Artificial Neural Networks (ANN) in vibration control, highlighting potential, issues, and possibilities for future development in this important area of research.

References

1. Khan, A., Shafi, I., Khawaja, S.G., de la Torre Díez, I., López Flores, M.A., Castañedo Galvlán, J., Ashraf, I., Adaptive Filtering: Issues, Challenges, and Best Fit Solutions Using Particle Swarm Optimization Variants. *Sensors*, 23, 18, 7710, 2023.

2. Wang, D., Tan, D., Liu, L., Particle swarm optimization algorithm: an overview. *Soft Comput.*, 22, 387–408, 2018.
3. Pourpanah, F., Wang, R., Lim, C.P., Wang, X.-Z., Yazdani, D., A review of artificial fish swarm algorithms: Recent advances and applications. *Artif. Intell. Rev.*, 56, 3, 1867–1903, 2023.
4. Tang, J., Liu, G., Pan, Q., A review on representative swarm intelligence algorithms for solving optimization problems: Applications and trends. *IEEE/CAA J. Autom. Sinica*, 8, 10, 1627–1643, 2021.
5. Sengupta, S., Basak, S., Peters, R.A., Particle Swarm Optimization: A survey of historical and recent developments with hybridization perspectives. *Mach. Learn. Knowl. Extr.*, 1, 1, 157–191, 2018.
6. Banks, A., Vincent, J., Anyakoha, C., A review of particle swarm optimization. Part I: background and development. *Nat. Comput.*, 6, 467–484, 2007.
7. Kaveh, A., *Advances in metaheuristic algorithms for optimal design of structures*, Springer International Publishing, Switzerland, 2014.
8. Lu, P., Chen, S., Zheng, Y., Artificial intelligence in civil engineering. *Math. Probl. Eng.*, 2012, 2012.
9. Xu, K., Hua, X., Lacarbonara, W., Huang, Z., Chen, Z., Exploration of the nonlinear effect of pendulum tuned mass dampers on vibration control. *J. Eng. Mech.*, 147, 8, 04021047, 2021.
10. Zorić, N., Simonović, A., Mitrović, Z., Stupar, S., Active vibration control of smart composite beams using PSO-optimized self-tuning fuzzy logic controller. *J. Theor. Appl. Mech.*, 51, 2, 275–286, 2013.
11. Marinaki, M., Marinakis, Y., Stavroulakis, G.E., Fuzzy control optimized by a multi-objective particle swarm optimization algorithm for vibration suppression of smart structures. *Struct. Multidiscip. Optim.*, 43, 29–42, 2011.
12. Amini, F., Khanmohammadi Hazaveh, N., Abdolahi Rad, A., Wavelet PSO-based LQR algorithm for optimal structural control using active tuned mass dampers. *Comput.-Aided Civ. Infrastruct. Eng.*, 28, 7, 542–557, 2013.
13. Shivashankar, P. and Gopalakrishnan, S.J.S.M., Review on the use of piezoelectric materials for active vibration, noise, and flow control. *Smart Mater. Struct.*, 29, 5, 053001, 2020.
14. Qin, Y., Tang, X., Jia, T., Duan, Z., Zhang, J., Li, Y., Zheng, L., Noise and vibration suppression in hybrid electric vehicles: State of the art and challenges. *Renewable Sustainable Energy Rev.*, 124, 109782, 2020.

10

Artificial Neural Networks (ANN) for Predictive Modeling and Control

Abstract

This chapter explores the use of Artificial Neural Networks (ANNs) in predictive modeling and control in the area of vibration control in structural buildings. The ANN approach, bringing adaptable learning and capability of processing complex data, provides a new class of understanding, prediction, and suppression of vibrations, thus supporting structural integrity, as well as inhabitant safety. Moving from an overview of the basic areas of ANNs in their architectural bases related to engineering contexts, the chapter paves the way for the use of ANNs in vibration control. Building upon the importance of understanding vibration dynamics in structural buildings, the story moves forward to provide a discussion on predictive modeling. It outlines the necessary data collection approaches, preliminary data processing methods, and tailored models designed for forecasting different aspects of vibrations ranging from time-series predictions to structural damage predictions. Integral to this exploration is the definition of metrics that are key in modeling performance assessment. After leaving predictive modeling behind, the chapter thereby opens up ANNs' contribution to control strategies by both outlining active and passive approaches. Active controls such as real-time monitoring and feedback systems and also using adaptive actuators, make ANNs effective. At the same time, the optimization of passive control methods like tuned mass dampers and isolation systems is evidence of the ability to enhance the methods of vibration mitigation. In addition, the chapter discusses the effect of ANNs on optimal form of building design while taking into account architectural intricacies and design of materials for tuning vibrations. The rollout problems, such as the question of data quality assurance, model interpretability, and ensuring the robustness over different environmental conditions are also described with considerations of working in real-time.

***Keywords*:** Artificial neural network (ANN), vibration control, vibration damping, active and passive damping system

Kundan Meshram and Prashant Lahre. Vibration Control Systems, (157–170) © 2026 Scrivener Publishing LLC

10.1 Introduction

Controlling and reducing vibrations in buildings continues to be a critical issue in structural engineering, the solution to which requires treating it as such. Vibrations of great intensity may result from the action of such factors as wind, seismic activities, or events near the building and may compromise building endurance along with endanger occupants' safety and comfort [1]. Perceiving the necessity, it is the implementation of sophisticated technologies, such as Artificial Neural Networks (ANNs), which has transformed the approach [2]. The dynamic use of ANNs is predictive modeling and control devices for managing vibrations of structural buildings. Another expected model for managing complicated data and learning complex network structures is Artificial Neural Networks (ANN), which are famous for their impressive capabilities to handle intricate data, learn complex structures, and make instantaneous predictions (as reported in [3–5]). Due to their inherent strengths, ANNs have demonstrated an ability to predict and handle vibrations that are changing the reality of proactive structural control measures. The chapter expands by explaining the basic concepts of ANNs; their architecture, training methods, as well as adaptiveness. It maps out the complexities of incorporating ANNs into the context of structural engineering and the role of the same in predictive modeling of vibration levels for varied environmental and structural parameters. Furthermore, it studies ANN power active and passive control strategies, orchestrating real-time intervention against vibrations to improve structural resilience [6, 7].

10.2 Artificial Neural Networks (ANNs)

Definition and Basic Concepts: This subsection gives a detailed definition of the artificial neural networks and elaborates its main concepts. ANNs are presented as computational models based on the human's human brain's neural network structure. Basic operation, including neurons, layers, activation function, and learning process through data training, is explained. It also addresses ANNs' process of information, decision-making, and how it adapts to input data.

Overview of ANN Architecture: This section provides a course description of the ANN's architecture including the description of intricate input, hidden, and output layers. It explores various considerations of neural networks

Table 10.1 ANN detailing the components.

Component	Description
Input Layer	The input layer of the ANN where external data is introduced. Each node represents a feature or input variable used for prediction or control.
Hidden Layers	Intermediate layers between the input and output layers where the network processes information apply weights and learn complex patterns within the data.
Output Layer	The final layer that produces the network's predictions or control signals. The number of nodes in this layer depends on the type of prediction or control task.

including feedforward, recurrent and convolutional neural networks and discusses how they are peculiar and can be used as shown in Table 10.1.
Applications in Engineering and Structural Analysis are as follows: This section accounts for the gap between theoretical knowledge and practical use of ANNs in engineering and highlights their value in structural analysis, predictive modeling, and control as shown in Table 10.2. It is emphasized that ANNs can easily address a complicated data and nonlinear relationships 'as an asset' in discussion about vibration control challenges for structural buildings. The major objective of this introductory part is to start acquainting readers with ANNs on the level of computational models and to lay the groundwork for their use in vibration control in structural engineering. With the understanding of basic concepts and architectural imperatives of ANNs, readers are more likely to appreciate their later use in predictive modeling and control approaches for mitigation of building vibrations [3, 8, 9].

10.3 Vibration Control in Structural Buildings

Understanding Vibration in Buildings: In this sub-section we discuss the issues of vibrations in structural buildings. It identifies the sources of vibrations (environmental (wind or seismic activity), human - activities, machinery). It describes the propagation of building materials and effects on the structural integrity and subjects within.
Importance of Vibration Control: In this case, attention is given to why it is important to control vibrations in buildings. It points out the dangers of high levels of vibrations to the buildings which include both structural

Table 10.2 ANN for vibration control in structural buildings.

Aspect	Description
Objective	Minimize and manage excessive vibrations in structural buildings to ensure structural integrity and occupant comfort.
Input Data	Environmental factors (e.g., wind speed, seismic activity), structural properties (e.g., material type, dimensions), and sensor data measuring vibrations.
Network Architecture	Configured with input nodes corresponding to environmental and structural parameters, hidden layers for processing, and an output node predicting vibration levels.
Training Method	Utilizes historical vibration data for training, adjusting weights and biases through backpropagation and optimization algorithms.
Prediction	Predicts vibration levels in real-time based on incoming sensor data and historical patterns learned during training.
Control Strategies	Active Control: Utilizes ANN predictions to adjust actuators, dampers, or control devices in real-time to counteract or dampen vibrations.
	Passive Control: Optimizes the design of passive systems (e.g., tuned mass dampers) based on ANN analysis for enhanced vibration attenuation.
Challenges Addressed	Overcomes challenges related to real-time monitoring, adapting to varying environmental conditions, and optimizing control strategies for efficient control.
Adaptability and Optimization	Learns from new data, continuously updating predictions and control strategies to adapt to changing environmental conditions and optimize performance.

deterioration as well as effects on the occupants, and also safety problems. By placing emphasis on vigorous control measures, it can be seen that the effective vibration control strategies are important.

Challenges Associated with Excessive Vibrations: This section explains the problem caused by extreme vibrations. Numerous limitations are associated with it, such as challenges with predicting and calculating vibrations, limitations inherent in traditional control techniques, and a requirement for ongoing monitoring and responsiveness to changing environmental conditions. The knowledge of these challenges prepares the way for the adoption of ANNs in solving these complexities.

The focal point of the following section is to prepare the ground with an exhaustive discussion of the context and issues related to structural building vibrations. This understanding is a prerequisite for studying ANN which is to be used to predict, model, and implement impactful remedial control strategies which seek to minimize such vibrations and their adverse effects.

10.4 Predictive Modeling Using ANNs

Data Collection and Preprocessing: This subsection describes the process of data collection and preparation for training ANN's. It includes sources of data such as sensors: vibrations and environment and structural properties. Data preprocessing techniques including normalization, feature scaling and missing values substitution are described aiming to make sure that the data received is of good quality and suitable for training an ANN.

Training ANNs for Vibration Prediction: This section explores in detail training ANNs for vibration prediction. ANN-based regression and time-series prediction methods that are supervised learning methods are emphasized to show that they learn patterns and trends from previous vibration data.

Types of Vibration Control Predictive Models: This subsection lists different types of predictive models which have been developed using ANNs for vibration control. It distinguishes between models oriented toward time-series prediction of vibrations in a given time period and models aimed at vibration induced possible damage to the structure. The discussion includes the differences made in input features, output formats, and model complexities for these predictive approaches.

Performance Metrics for Model Evaluation: The evaluation criteria used to determine the performance of ANN models for vibration prediction are described. The overview of such metrics as Mean Squared Error

(MSE) or Root Mean Squared Error (RMSE), coefficient of determination (R-squared), etc., is described, which are commonly used for time-series forecasting. Choosing the right evaluation metrics gives the predictive model both reliability and accuracy. A review of the current literature when combined with this section provides a detailed understanding of the procedures used in applying ANNs for the prediction of vibrations in structural buildings [10, 11]. During the data gathering, model construction, and evaluation, it conveys a vital learning that is necessary for establishing and evaluating the ANNs in vibration prediction and control.

10.5 ANN's Strategies for Controlling the Threads

Active Control Methods: In this sub-section, active control methods that make use of ANNs are discussed. It focuses on the fact that ANNs make it possible to directly and on-the-spot test vibrations within the structure, based on sensor information. It highlights the application of ANNs in the feedback systems to observe incoming data, detect vibration patterns, and allow for fast adjustment by actuators or any control device to neutralize or diminish the vibration as shown in Figure 10.1.

Passive Control Techniques: This section introduces the application of ANNs for the improvement of passive control methods developed for the reduction of vibration. It says that ANNs contribute to the upgrade of

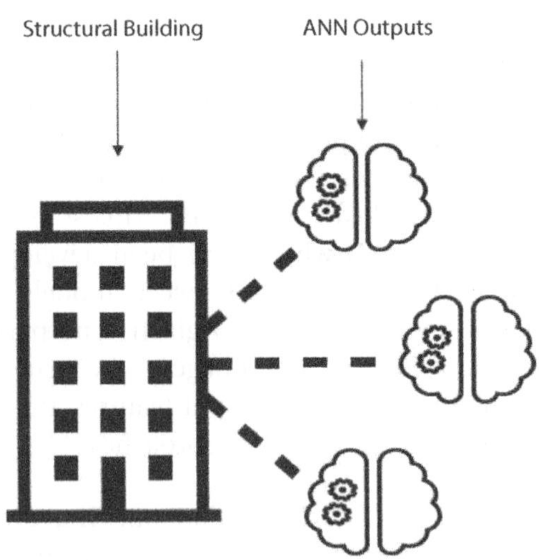

Figure 10.1 Analysis through ANN for vibration control.

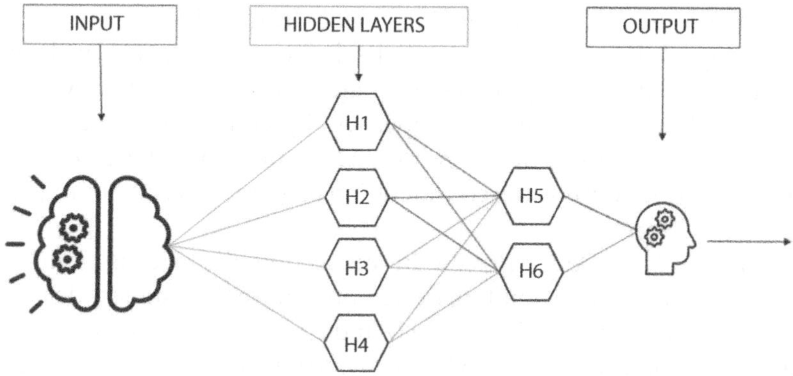

Figure 10.2 ANN architecture.

low-performing devices' capabilities, e.g., tuned mass dampers and vibration isolators. Data analysis using ANNs is useful for design parameter optimization with an ensuing improved attenuation of vibration for such systems as shown in Figure 10.2.

Control Strategies Optimization Using ANNs: In the section, we demonstrate how ANNs are applied to improve optimization of control strategies. The optimization process might involve several iterations for control algorithms' fine-tuning, as well as parameter regulation in active and passive control systems, so that optimum vibration attenuation is provided by means of ANN-driven approaches.

This section brings to light the use of ANNs for designing sophisticated control strategies for reductions of vibrations in structural buildings [10]. The outcome of the use of ANNs to active and passive control methodologies is enhanced, efficient, responsive, and adaptive systems that can be used to significantly reduce the effect of vibrations on the structure and comfort of occupants in buildings.

10.6 Design and Optimization with ANNs

Optimal Structural Design: This section examines the function of ANNs in crafting best-in-class structural designs targeting the reduction of undesirable vibrations. This segment explores the use of ANNs in accessing architectural elements such as the form of the building, the materials used, and structural layouts that would minimize the spread of vibrations. The integration of ANNs enables the designers to identify the most appropriate dynamic attributes to limit vibrations.

Material Selection for Vibration Dampening: In this chapter, the role that ANNs play in the process of material selection is discussed in detail. It describes how ANNs help in assessing and choosing materials with outstanding vibration-dampening properties. By employing large amounts of material behaviors under vibration, ANNs help to make good choices which directly determine the consideration of good materials for improved structural performance.
Improving Building Design Process: In this section, the focus is on integrating ANNs in the whole building design procedure. The subject deals with the implementation of ANNs in CAD and structural analysis systems and thus allows vibration control to be easily introduced at the initial stages of design by engineers and architects. Employment of ANNs promotes prediction and adaptability of design processes for vibration suppression.

This subsection highlights ANN's importance in optimizing designs of buildings, especially in the selection of materials that are good for vibration control [12]. The integration of ANNs into the design processes enables engineers to deal with vibrations on a proactive basis, leading to a higher safety and robustness of structural systems and increased occupant satisfaction.

10.7 Implementation and Challenges

Training ANNs for Real-Time Implementation: This segment addresses the difficulties and approaches connected with the real-time deployment of NN for vibration control implementations. It discusses methods to make ANN model adjust to work optimally in real-time settings, and make optimal decisions promptly based on incoming data from the sensors.
Consternations in Data Quality and Quantity: Here, details challenges pertaining to data collection, quality assurance, and quantity. It discusses issues such as data reliability and completeness, the requirement for varied dataset that covers different environmental and structural scenarios to train strong models.
Interpretability of Models in the ANN for Vibration Control: This component highlights the interpretability of ANNs' decisions in terms of vibration control strategies. It discusses ways to increase the interpretability of ANN models to engineers relating how the model has come to the desired predictions and decision, thus helping to gain confidence in the model and increase its usability.

Adversarial Guarantees under Different Environmental Conditions: This sub-section addresses the problem of verifying the robustness of ANN-based vibration control systems under various environmental conditions. It talks about strategy with which the models can be made more adaptive and resilient so as to effectively reduce vibrations regardless of the different scenarios and unknown scenarios.

This section highlights the practical limitations and considerations which are presented for employing the ANN-based vibration control systems in the building structures. By solving problems associated with training, data quality, scalability, and adaptability, engineers make better and more efficient systems for controlling vibrations.

10.8 Case Studies and Practical Applications

Case Studies Showing Successful Properties of ANN Implementation: This subsection summarizes careful considered examples which demonstrate realistic use cases for ANNs in vibration control of structural buildings. It includes illustrations where ANNs have been successfully applied towards prediction, modeling or control of vibrations, and their performance and influence in varying situations.
Real-world Applications in Structural Engineering: In this case, attention is given to various practical arrangements of ANNs in the sphere of structural engineering associated with the vibration control. It includes a variety of applications such as reduction of high-rise building oscillations, the prevention of structural failures caused by seismic incidents, the increase of occupant comfort and improved structural designs for reducing vibration propagation.
Lessons Learned and Best Practices: This part derives implications and learning from case studies and their applications that outline important lessons learned and exemplary practices in using ANNs for vibration control. It captures the success factors, barriers, and recommendations to steer future implementations.

By presenting to the students real-world examples and actual applications, this section explains how effective and versatile ANNs are in overcoming vibration control issues that exist in structural engineering. The case studies and lessons learned are useful references for engineers looking to apply ANNs in similar situations and provide insight into what might happen.

10.9 Future Directions and Conclusion

Desirable evolution for ANN technology in vibration control: It addresses emerging tendencies in neural network architectures, training algorithms, and data processing that may increase efficacy and efficiency of ANNs in predicting, modeling, and controlling vibrations as given in Table 10.3.

Table 10.3 Components of ANN.

Component	Description
Neuron/ Node	The basic processing unit that receives input, computes a weighted sum, adds a bias, and applies an activation function to produce an output.
Input Layer	The initial layer of the network where external data is fed into the network. Each neuron represents an input feature or attribute.
Weights	Parameters that signify the strength of connections between neurons. They are adjusted during training to enhance the network's performance.
Bias	An additional parameter added to each neuron, aiding in shifting the activation function's output and allowing the model to better fit the data.
Activation Function	A non-linear function applied to the weighted sum of inputs and biases in a neuron, introducing non-linearity to the network's output, facilitating complex mappings.
Hidden Layers	Layers between the input and output layers where data is processed through a series of interconnected neurons. Each hidden layer extracts and learns complex patterns.
Output Layer	The final layer that produces the network's predictions or outputs based on the processed information from the preceding layers.
Loss/Cost Function	Measures the difference between predicted and actual outputs, guiding the network in adjusting its parameters during training to minimize this difference.
Optimizer	Algorithm used during training to update weights and biases, typically by minimizing the loss function *via* techniques like gradient descent or its variants.

Trends in Structural Building Design are on the rise. In this case, the chapter then presents progressive trends and advancements in structural building design as a result of their usage to facilitate vibration control. It points out that ANNs may steer the future methodologies of designing, materials, and construction processes that put emphasis on the vibration resilience and occupant safety.

10.10 Additional Points

Hybrid Approach and Multi-modal Data Fusion
Discuss the benefits of using multi-modal data fusion of data from a number of sources such as sensors as well as structural models in providing better prediction of accuracy in vibration predictions.

Uncertainty and Risk Analysis
The article should also have a section about ANNs help pave the way to uncertainty quantification and risk analysis in vibration control as shown in Figure 10.3. List methodologies to address and control ambiguities in input data, and model projections and control techniques, into and out of which ponder over viewpoint of probabilistic thinking for ensuring sound decision making.

Human-Structure Interaction and Occupant Comfort
Consider ANNs account for human-structure interaction and the result on behalf of occupant comfort from vibrations. Examine developing ANNs to sense the role of occupant feedback and subjective data and comfort metrics that would define optimal settings for control strategies while taking into account the way people perceive and tolerate it.

Ethics and Safety Considerations
Ethical issues in using ANNs in structural vibration control should be touched on, accentuating the need for safety standards provisions, and privacy of collected data, while making clear what capabilities and limitations of AI systems applied in buildings have.

Regulatory and Compliance Aspects
Address regulatory sides and compliance needs for utilizing AI-based systems for the structural safety. Highlight ANNs can help in achieving industry standards/being compliant with the compliance protocols and may aid in regulatory approvals of the novel technology of vibration control systems.

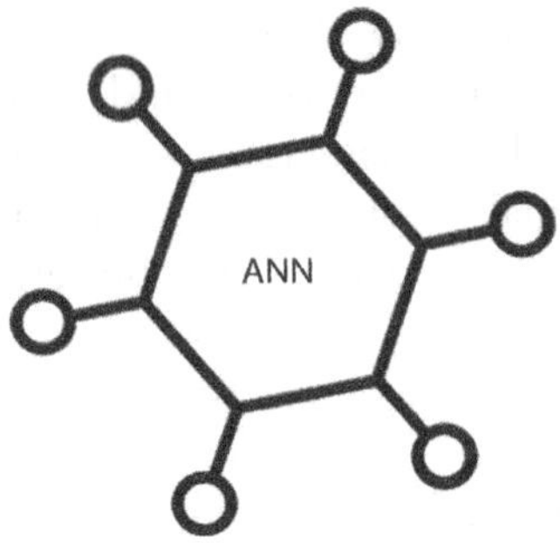

Figure 10.3 Possibilities of ANN.

Scalability and Transferability
Explain how changeable sizes, types, and geographical locations of buildings will influence ANN-based solutions scalability. Investigate strategies by which models and control strategies can be transferred between different structural designs and environments so that they can be applied and scaled.

Collaborative Research and Industry Adoption
Emphasize the need for the collaborative work in the field of research among academia, industry, and regulatory bodies as a means of advancing innovation and testing of the ANN-based models in real-world conditions and introducing such models to the structural engineering sphere.

10.11 Conclusion

1. Artificial Neural Networks (ANNs) have become an essential technology in the transformation of vibration control strategies in structural engineering. Their predictive and adaptability allows a proactive approach to guarantee structural integrity and occupant safety.
2. ANNs demonstrate outstanding capabilities in predictive modeling by accurately predicting possible vibrations by using a variety of environmental and structural parameters. Their implementation of real-time control strategies enables engineers to counter internal vibrations dynamically.
3. As validated by practical applications and case studies, ANNs have effectively enabled optimization of structural design,

dynamic control mechanism adjustments, and vibration reduction across various real-world situations.

4. Despite their capability, several challenges remain, such as real-time deployment, interpretability, data quality assurance, and adaptability across different environmental conditions. Transitions to engineering practice become much easier if these challenges are faced.
5. Future perspective, continued evolution in neural network architectures, training techniques, and data assimilation techniques holds a bright future for ANNs. The key to safety is in evolving to be more robust, interpretable, and adaptable systems.
6. Finally, ANNs are the house stone of changing the vibration control strategy. Their predictive capabilities and flexibility increase the durability of structures, but also provide a higher level of safety and comfort for occupants, contributing to a more secure and sustainably built environment.

References

1. Grondzik, W.T. and Kwok, A.G., *Mechanical and Electrical Equipment for Buildings*, John Wiley & Sons, Hoboken, NJ, USA, 2019.
2. Memon, K.R. and Ooi, S.K., Identifying digital leadership's role in fostering competitive advantage through responsible innovation: A SEM-Neural Network approach. *Technol. Soc.*, 75, 102399, 2023.
3. Gomes, G.F., Diaz Mendez, Y.A., Alexandrino, P.d.S.L., da Cunha, S.S., Ancelotti, A.C., A review of vibration based inverse methods for damage detection and identification in mechanical structures using optimization algorithms and ANN. *Arch. Comput. Methods Eng.*, 26, 883–897, 2019.
4. Li, Z. and Adeli, H., Control methodologies for vibration control of smart civil and mechanical structures. *Expert Syst.*, 35, 6, e12354, 2018.
5. Takács, G. and Rohal'-Ilkiv, B., *Model Predictive Vibration Control: Efficient Constrained MPC Vibration Control for Lightly Damped Mechanical Structures*, Springer Science & Business Media, London, UK, 2012.
6. Xavier, M.S., Tawk, C.D., Zolfagharian, A., Pinskier, J., Howard, D., Young, T., Lai, J., *et al.*, Soft pneumatic actuators: A review of design, fabrication, modeling, sensing, control and applications. *IEEE Access*, 10, 59442–59485, 2022.
7. Raj, P. and Raman, A.C., *The Internet of Things: Enabling technologies, platforms, and use cases*, CRC Press, Boca Raton, FL, USA, 2017.

8. Dackermann, U., Vibration-based damage identification methods for civil engineering structures using artificial neural networks, PhD diss., 2009.
9. Marwala, T., *Finite Element Model Updating Using Computational Intelligence Techniques: Applications to Structural Dynamics*, Springer Science & Business Media, London, UK, 2010.
10. Nguyen, H., Moayedi, H., Loke, K.F., Al Najjar, H.A.H., Jusoh, W.A.W., Rashid, A.S.A., Jamali, J., Optimizing ANN models with PSO for predicting short building seismic response. *Eng. Comput.*, 36, 823–837, 2020.
11. Hossain, M.S., Ong, Z.C., Ismail, Z., Noroozi, S., Khoo, S.Y., Artificial neural networks for vibration based inverse parametric identifications: A review. *Appl. Soft Comput.*, 52, 203–219, 2017.
12. Salehi, H. and Burgueño, R., Emerging artificial intelligence methods in structural engineering. *Eng. Struct.*, 171, 170–189, 2018.

11

Advanced Sensor Technologies in Vibration Control

Abstract

This chapter delves into the transformative evolution and multifaceted applications of advanced sensor technologies in the realm of vibration control, with a particular emphasis on civil engineering structures. The advancement of sensors began with basic mechanical instruments that advanced into digital devices for real-time monitoring together with analysis capabilities and complete control functions. The development represents more than technological advancement since it creates essential enhancements for safety measures and reliability as well as operational performance in structures which include high-rise buildings alongside bridges and dams and transportation networks. Advanced sensors in vibration control systems gained essential importance when infrastructure demand increased for sustainable and smart solutions. The operation of such systems depends strongly on exact data acquisition platforms combined with intelligent signal processing along with adaptive feedback controls. The text examines MEMS-based accelerometers along with fiber-optic and infrared sensors and explains their association with advanced damping systems which use passive and semi-active methods. The research examines how combining sensor data with real-time visualization enhances smart decisions for decision-support applications. The study presents the essential function of machine learning algorithms when they process sensor information while doing predictive maintenance tasks and damage diagnosis activities. These technologies find their practical applications through numerous real-world examples which demonstrate their value in operation. Despite the challenges in deployment, such as environmental durability, calibration, and data management, ongoing innovations like nano-sensors, self-healing materials, and blockchain-based data security offer promising future directions. An exhaustive study of the chapter demonstrates how sensor technologies function as the vital elements for vibration control and structural health monitoring in contemporary times.

***Keywords*:** MEMS sensors, fiber optic sensing, wireless sensor networks, real-time monitoring, sensor integration, intelligent systems

Kundan Meshram and Prashant Lahre. Vibration Control Systems, (171–186) © 2026 Scrivener Publishing LLC

11.1 Introduction

Our modern structural health monitoring systems rely on sensors as their key information source when controlling vibration performance. Civil engineering structures need live vibration monitoring to protect their stability and occupants in environments such as high buildings and bridges [1]. At the start sensors represented basic analog instruments that collected basic motion information with slow reaction speed and poor precision. The fast progress of electronics materials science and modern communication technologies changed sensor systems into key parts of vibration control technology. These systems collect data of high accuracy and analyze it right away to control structural damping through automated adjustments [2]. Advanced sensor technologies become essential because customers need structures that test themselves and forecast maintenance requirements. Advanced sensing solutions are essential to meet current challenges of complex structures because they can predict behaviors under variable loads plus improve control systems [3]. Engineering structures faced continuous problems including fatigue damage because of vibration effects. The engineering sector was implementing passive vibration control technology by distributing vibrations from one component to another for elimination. Extended use of PVC in industry remained restricted because of its restricted flexibility together with its size limitations for extreme conditions. Researchers battled this problem by developing active vibration control (AVC) through the use of sensors and actuators which dynamically modified structural parameters and combat vibrations. The research of AVC first emerged during the 1930s through publications such as Tsien's Engineering Cybernetics as scientists pursued extensive development of this filed which integrated various disciplines including dynamics and control theory and cybernetics [4]. Environmentally conscious movements toward electric and hybrid electric vehicles opened a strong opportunity for Switched Reluctance Motors (SRMs) to compete in the market. The absence of both windings and magnets in their rotor structures alongside their basic structure and fault-tolerant characteristics gave SRMs their market value. Torque ripple and noise along with nonlinear behavior created obstacles for the utilization of SRMs. These problems received effective management through established control strategies [5]. The revision of the Second Edition of Vibration-based Condition Monitoring targeted an updated substantial examination of machine condition monitoring. The author expanded his work by introducing sections about variable speed machine diagnosis especially for wind turbines and he added information on cepstrum analysis for forcing functions separation and structural model characterization and machine fault simulation [6]. Advanced sensor technology implementation for vibration control requires evaluation of multiple interconnected stages including theoretical modeling together with experimental testing and sensor selection as well as system integration and signal processing and performance measurement. An advanced vibration control system requires multiple elements which combine hardware with software features and data algorithms with physical models [7]. The methodology includes these distinct steps which amount to multiple stages to provide clear implementation methods.

a. The identification process begins with structural requirements along with control objective formulation.
b. A detailed evaluation of the structural attributes along with vibration control needs for target infrastructure stands as the initial procedure for sensor system deployment. This includes:
c. The dynamic characteristics of the structure should be analyzed through modal analysis methods.
d. The process detects essential vibration modes together with natural frequencies and damping ratios and resonance points.
e. The assessment includes studying external sources of excitation including wind loads as well as traffic loads alongside earthquakes and vibrations from machinery.
f. The control objectives identify minimum requirements such as peak acceleration reduction together with displacement reduction and stress level reduction and fatigue reduction from resonance.
g. The analysis selects suitable damping systems (passive or active or semi-active) that determine needed sensor capabilities.

Phase1: Sensor Selection and Specification

The selection of optimal sensors happens at this point based on requirements that address application-specific needs such as accuracy levels along with response speed and sensitivity measures alongside durable performance in the environment and low power usage. Key considerations include:

Type of sensor: Accelerometers, strain gauges, fiber-optic sensors, displacement sensors, acoustic sensors, thermal sensors.

Sensor range and resolution: Depending on the magnitude of expected structural responses. The sampling frequency should be selected based on the Nyquist criterion and the determined vibration frequency range. Each component uses wires through RS-485, CAN, Ethernet channels, and ZigBee, Wi-Fi, and LoRa without wires to communicate. The system can operate using three power methods: batteries, energy collectors or by having wires as power sources. The form factor of sensors should be determined between embedded and surface-mounted designs. The technology needs environmental resistance capabilities because sensors operating outside need protection from heat and moisture while resisting corrosion.

The deployment of sensors requires a plan which uses either Optimal Sensor Placement through genetic algorithms or modal assurance criterion (MAC)-based optimization approaches. The third development phase involves creating an architecture for the sensor network. The selection process of sensors requires researchers to deploy a data acquisition and communication network system.

This includes: The network design determines which sensor arrangement will be utilized depending on the area coverage requirements and communication demands. Gateways should be properly distributed alongside data concentrators and edge computing nodes to decrease latency levels. The system provides constant data transmission which maintains low data packet losses. Every critical node within the network requires redundancy implementation to enhance system fault tolerance. The system's modules for data buffering and synchronization need configuration to achieve temporal alignment in multiple sensor channel outputs. A combination of Arduino and STM32 and Raspberry Pi embedded processors should be used for both local preprocessing tasks and control system integration. Network security gets enhanced by using authentication protocols alongside encryption algorithms particularly in wireless network systems.

Deployment and integration become feasible through this phase because the framework becomes robust and modular and scalable. The fourth phase combines the sensors with damping mechanisms for system integration. The fourth phase establishes a connection between sensors and damping control systems at various operation levels through passive, semi-active, and active devices. The strategy for integration differs based on which damper type is implemented. The monitoring of passive dampers depends on sensors that primarily carry out performance assessments and durability tests for validation purposes.

The real-time use of sensors in MR dampers controls viscous or stiffness properties through Lyapunov control along with fuzzy logic and sliding mode control algorithms. Digital controllers alongside DSPs use sensor data to determine real-time actuator forces that oppose vibrations in active control systems.

The integration process includes: Calibration of sensor readings with damper inputs. The system must maintain quick transmission of signals between sensors and controllers. Safe evaluation of feedback loops should be performed through simulation platforms which use MATLAB/Simulink. Real-time control algorithms need to be embedded into the system while the system should have defined thresholds for automatic activation.

Such control architecture enables the necessary loop for adaptive closed-loop vibration control to function. During this phase signal processing and data analytics procedures are carried out. This step engages in processing raw sensor data while applying filters which lead to valuable interpretation for utilization in control and diagnosis applications.

Methodologies include: Digital and analog signal filtering mechanisms employ low-pass as well as band-pass filters together with Kalman filters to eliminate noise. By applying Fourier analysis scientists can transform time-based information into frequency information for observing key predominant modes. Wavelet transforms assist in detecting both brief signal processes and concentrated energy bursts that appear within vibration signals.

Modal identification algorithms: Such as stochastic subspace identification (SSI), Ibrahim time domain (ITD), or peak-picking methods. The implementation includes three damage detection techniques which are modal curvature and strain energy methods and time-frequency entropy analysis.

Phase 2: Laboratory Testing and Validation

The testing phase takes place with an examined prototype within a laboratory setting before actual deployment occurs in the field.

This includes: The introduction of laboratory testing stages contains the implementation of shake table experiments combined with structural frames and cantilever beams equipped with sensors and damping mechanisms. Testing occurs through the application of harmonic and random and seismic excitations while evaluating the reaction of the system. The experimental outcome data correlate with simulation results for determining accuracy levels of control mechanisms and model authenticity. The testing procedure checks the accuracy of sensor measurements together with system latency as well as evaluates damping efficiency and confirms calibration and data validity. The information from laboratory testing enables system parameter optimization as well as control algorithm refinement alongside solution of unexpected system integration problems.

The validated system receives deployment in actual structures during this essential implementation period. The installation of sensors takes place at essential nodes including beam-column joints along with mid-span foundations and both damping devices. Environmental testing as well as remote access and cloud backup storage and smartphone-based dashboard monitoring operates in a continuous real-time fashion. The system uses event-triggered alarms and threshold parameters for detecting critical responses of structures. The system's reliability is maintained through proper establishment of constant calibration and maintenance plans. The system documents consistent data records for analyzing long-term maintenance and performance patterns. This phase applies theoretical knowledge to operational practice for making vibration control and structural health evaluation decisions based on data.

Phase 3: Performance Evaluation and Optimization

Throughout ongoing operation the system needs periodic evaluation to verify its effectiveness and continued performance success.

Table 11.1 Comparison of traditional and advanced sensors.

Feature	Traditional sensors	Advanced sensors
Signal Type	Analog	Digital/Smart
Data Transmission	Wired	Wired/Wireless
Processing Capability	Minimal	Embedded Processing
Size and Weight	Bulky	Compact/Miniaturized
Power Requirement	High	Low/Self-powered
Sensitivity	Moderate	High (e.g., Fiber Optic)
Data Storage	Manual Logging	Cloud/Edge Storage
Integration	Standalone	Networked/IoT Compatible

This is done by: The measurement of vibration response changes from before to after the installation helps determine the improvement of damping efficiency. System operators should use data from different observation times to identify aging indicators and sensor drift and structural deterioration conditions. System effectiveness should be compared to deployment costs by performing cost-benefit analysis. Optimization algorithms such as Genetic Algorithms along with Particle Swarm Optimization function to enhance system performance through tuning control parameters and

Table 11.2 Types of sensors used in vibration control.

Sensor type	Working principle	Key applications	Advantages
Accelerometers	Detect change in acceleration *via* mass-spring systems	Real-time vibration analysis, shock monitoring	High sensitivity, compact design
Strain Gauges	Change in electrical resistance due to deformation	Internal stress and load analysis	Precise measurement of micro-strain
Displacement Sensors	Measures movement *via* voltage change or light reflection	Monitoring deflections, gaps, and cracks	Non-contact options available
Fiber Optic Sensors	Optical interference or grating changes in fiber	Long-range monitoring in bridges, tunnels	EMI immunity, long lifespan
Acoustic Emission Sensors	Detect elastic waves from crack formation	Early crack detection and fatigue monitoring	High-frequency capture
Thermal/IR Sensors	Infrared heat emission from surfaces	Detecting frictional heating, structural anomalies	Non-contact, thermal profile mapping

Table 11.3 Integration of sensors with damping systems.

Damping system	Sensor type used	Integration mode	Control method
Tuned Mass Damper (TMD)	Accelerometer, Displacement Sensor	Passive Feedback	Manual Calibration
Active Mass Damper (AMD)	Accelerometer, Strain Gauge	Closed-loop Control	PID, Fuzzy Logic
Magneto-Rheological Damper (MRD)	Strain Gauge, Current Sensor	Semi-active Adaptive	Neural Networks, Rule-Based
Viscoelastic Damper	Temperature, Stress Sensor	Monitoring Only	Fixed Parameters
Friction Damper	Thermal, Acoustic Sensors	Damage Detection	Event-triggered Logging

Table 11.4 Signal processing techniques for sensor data.

Technique	Function	Suitable for	Tools used
Low-pass Filtering	Eliminate high-frequency noise	Smooth signals	Butterworth, Chebyshev Filters
Fast Fourier Transform (FFT)	Convert time data to frequency domain	Modal frequency extraction	MATLAB, Python (SciPy)
Wavelet Transform	Time-frequency localization	Impact and transient detection	MATLAB Wavelet Toolbox
Kalman Filtering	Noise suppression and prediction	Adaptive control	Embedded controllers
Machine Learning (ANN, SVM)	Pattern recognition and forecasting	Damage prediction	TensorFlow, Scikit-learn

sensor placements and system threshold settings. The phase completes the loop by providing valuable information that will enhance the following version of sensor-based vibration control systems. Various sensors function worldwide with distinctive behaviors which can be comprehended from the data in Tables 11.1–11.4.

11.2 Evolution of Sensor Technologies

The initial devices used for vibration detection included mechanical instruments like seismographs alongside analog accelerometers. During their era these pioneering instruments demonstrated poor capabilities regarding bandwidth performance alongside minimal sensitivity measurement and limited data resolution. The devices operated mainly as diagnostic devices that needed human interpretation and produced no substantial real-time capabilities. When the digital revolution hit the scene it introduced advanced sensing technology to the world. The creation of Micro-Electro-Mechanical Systems (MEMS) brought sensors to a smaller form while achieving enhanced speed and precision. Small integrated units included processing circuits and communication systems together with sensing elements so data filtering occurred on board while transmission went wirelessly. Real-time control systems became possible because sensor feedback started to directly control damping actions through automated processes. The technology field progressed by adding fiber-optic systems and piezoelectric components and wireless sensor networks to the design. The latest sensor advances deliver high precision accuracy while maintaining strong environmental durability along with potential capabilities for distributed intelligent sensor systems. These sensors achieve autonomy through their capability to work on harvested energy while needing minimal battery resources. Intelligent ecosystems with cloud connectivity and edge computing and AI-assisted data analytics serve as the main objectives for sensor integration initiatives in modern times.

11.3 Types of Advanced Sensors

Various types of advanced sensor technologies exist to serve different monitoring aspects of vibrations. The three types of accelerometers, namely capacitive, piezoelectric, and MEMS-based, are frequently selected due to their acclaimed sensitivity toward dynamic motion. Real-time vibration tracking requires these sensors which frequently find their placement at structure nodes to measure acceleration. Expansive joint monitoring benefits the most from displacement sensors consisting of Linear Variable Differential Transformers (LVDTs) and laser-based displacement meters that detect relative motion patterns. Through strain gauges which measure resistance changes that stem from strain levels engineers successfully determine internal stress patterns and areas prone to fatigue throughout each structural component. Fiber-optic sensors equipped with Fiber Bragg Gratings technology provide outstanding sensitivity along with electromagnetic interference resistance which enables their utilization in extended monitoring operations of large facilities such as bridges and tunnels. Acoustic emission sensors identify high-frequency elastic waves resulting from structural changes and developing cracks which alert operators about present damage. The system benefits from infrared and thermal sensors which detect overheating regions and material deterioration caused by friction together with internal defects thus providing a more advanced condition-based diagnostic capability.

11.4 Sensor Integration with Damping Systems

Contemporary sensor technology introduced to damping systems enables the development of adaptive computer-controlled approaches for mitigating vibrations. Real-time damping response adjustments through continuous sensor data inspection in closed-loop systems enable precise and efficient vibration control. Active Tuned Mass Dampers (ATMDs) together with Semi-Active Dampers demonstrate this trend because sensors feed real-time inputs to actuators for behavior modification. Such systems provide major capabilities for preventing unpredictable load situations and environmental conditions through real-time structural adaptation to seismic and wind-driven forces. Sensor fusion accomplishments by integrating different sensor data produce better measurement accuracy and system reliability through minimized false alarm occurrences and better system stability. The system integration involves Graphical User Interfaces (GUIs) alongside signal processing platforms which provide real-time dashboards and set threshold alarms together with visual structural condition representations. Engineers receive quick insights from sensor data through these interfaces which help them make informed decisions and operate during preventive maintenance and emergency circumstances.

11.5 Data Acquisition and Interpretation

A sensor-based vibration control system requires effective data acquisition together with interpretation methods to operate functionally. Raw sensor signals undergo signal conditioning and filtering as the initial stage because analog or digital filtering techniques remove noise and interference. The detection of anomalies in vibration spectrum benefits from advanced preprocessing tools namely Fast Fourier Transform together with Wavelet Transform which isolates frequency components. The data requires accurate sensor sampling together with precision synchronization among multiple sensors to maintain uniform temporal relationships especially when operating large arrays of instruments. The Nyquist criteria provide conditions which guarantee the preservation of all significant information during sampling. Feature extraction follows data collection through which meaningful metrics are extracted from raw data including Root Mean Square (RMS) values, peak amplitudes and frequency domain characteristics and modal parameters. Such diagnostic features determine the structural condition by identifying existing failures and their performance levels. Protected by the Nyquist criteria modern processes combine machine learning with sensor data to build predictive models which use past patterns for predicting structural performance. These AI-assisted systems produce early damage detection while classifying anomalies and supplying real-time decisions which increase vibration control mechanisms' intelligence and responsiveness.

11.6 Applications in Structural Engineering

Civil engineering undergoes constant transformation because of how advanced sensors have been put into practical use. Multiple sensors mounted in high-rise buildings monitor wind-induced and seismic vibrations along with human activity to make continuous adjustments that protect both human comfort and structural stability. Long-term observation of bridges and flyovers remains indispensable because it determines fatigue levels and helps determine both structural crack development and maximum weight capacity changes that occur because of variable traffic and environmental stress. Sensors in retaining structures and dams help monitor internal stresses as well as detect structural damage and water leakage by measuring hydrostatic forces. This information facilitates prompt warning systems alongside risk management strategies. Advanced sensors placed on rail infrastructure achieve passenger safety by detecting rail cracks as well as joint misalignments and material fatigue through acoustic and vibration sensors. The modern infrastructure depends on sensors which operate as its observatories to transmit real-time data for quick maintenance decisions through centralized monitoring platforms.

11.7 Case Studies

Organizations worldwide conduct actual projects which demonstrate the productive utilization of state-of-the-art sensing devices for structural vibration control. A network of MEMS-based accelerometers together with high-capacity data loggers enables wind-induced oscillation monitoring in the Burj Khalifa making crucial inputs for its ATMD system as the world's tallest building. The French engineering marvel Millau Viaduct possesses fiber-optic sensors located inside its entire structural foundation and pylons to achieve worldwide bridge height records. The sensors operate in real-time to monitor dynamic loads and structural deflection as well as stress variations. AI-driven sensors deployed at the Japan Earthquake Engineering Laboratory study the seismic reactions of structural models from a state-of-the-art facility. The devices detect information while activating programmed controller responses in simulation laboratories that reveal performance measurements of competing damping approaches.

11.8 Challenges in Advanced Sensor Deployment

The practical application of advanced sensors encounters multiple technical hurdles during implementation process. The combination of calibration requirements and sensor drift throughout time produces measurement inaccuracies primarily within prolonged monitoring operations. The persistent challenge for wireless sensors in isolated areas involves power supply and energy management until energy extraction and ultra-low power design systems provide some relief. The massive amount of continuous high-frequency data creates data overload which demands strong cloud infrastructure together with intelligent storage systems for effective data retrieval.

11.9 Future Directions and Discussions

Sensor technologies which control vibrations will experience substantial advances in their operational future. The use of self-healing sensors based on materials exhibiting self-repair capabilities extends operational lifetimes of sensors and decreases their maintenance requirements. Nanosensor technology enables the detection of tiny micro-cracks while monitoring nanolevels of stress which represents a breakthrough in predictive diagnosis. Engineers will benefit from augmented reality (AR) integrated sensor networks as they can visualize real-time sensor data as well as structural health information, which appear on their AR headsets during on-site work for better situation understanding and enhanced decision capabilities. The emergence of blockchain enables data security during sensor data logging and establishes complete tracking capabilities that serve critical infrastructure monitoring purposes. Sensor swarms working collaboratively serve as autonomous networks which show potential for urban area threat assessment while being able to adjust their operations according to changing structural requirements. Sensor technologies will become active participants in adaptive infrastructure control through the recent developments in this field. Experimental studies and multiple structural systems have produced substantial advantages from using advanced sensor technologies in vibration control applications. Learning from laboratory tests showed damping systems connected to sensors demonstrated superior vibration reduction abilities relative to fundamental passive methods. Real-time acquisition of data from accelerometers and displacement sensors enabled instantaneous damping response changes thus boosting the system dynamic stability.

The combination of multiple sensor systems using strain gauges with thermal sensors and acoustic emission devices generated improved and detailed information about structural wellness. These measurement configurations revealed material fatigue and micro-cracking problems during their very early stages before these issues could escape standard inspection protocols. Experimental evaluations using numerical models especially finite element simulations validated the precise functioning and ready response of these sensor-based systems. High-rise buildings along with long-span bridges implement smart sensors for modern infrastructure systems in practical applications. Embedded sensors provided building management systems with automatic environmental response capabilities through which they applied modulation of damping actions in response to the dynamic wind conditions. Structural oscillations could be effectively controlled through bridges that incorporated fiber-optic as well as piezoelectric sensors despite changes in traffic and load patterns. Signal conditioning methods together with filtering techniques substantially improved the quality of data derived from raw sensors. Digital filtering technology eliminated substantial environmental noise so researchers could obtain improved understanding of system dynamics. Precision fault classification models are achieved when machine learning algorithms are utilized and integrated with advanced sensor systems according to studies. The developed models effectively found structural differences in systems by examining time-based and frequency-based sensor output characteristics.

11.10 Conclusion

1. The implementation of advanced sensors revolutionized vibration control because they allow dynamic structural environments to benefit from real-time precise adaptive responses.
2. Modern sensor types including MEMS and fiber-optic and piezoelectric use accurate tracking of vibrations and internal stresses and structural failure warnings.
3. The cross-functional connection of sensors to active and semi-active damping devices enables automated structural behavior control which maximizes system operational efficiency.
4. The combination of sensor-produced information enables machine learning and AI models to make damage predictions and discover structural faults with speed and high accuracy.
5. Engineering projects have shown that sensor technology can enhance damping abilities and safety measures and maintenance solutions in structures such as buildings and bridges and seismic laboratories.
6. Sensor technologies demonstrate flexible adaptiveness which allows their application across different civil engineering systems including high-rise buildings as well as transportation systems and hydraulic infrastructure.

7. Strategic solutions are needed to address four main difficulties that pertain to sensor calibration and power support and instrumental life expectancy and large sensor-generated data volumes.
8. Upcoming technological advancements in self-healing sensors as well as nano-technology and AR integration and blockchain data security will create new possibilities for structural monitoring and vibration control.
9. The embedded in advanced sensors helps create infrastructure systems that are safer while being both intelligent and resilient.

References

1. Brownjohn, J.MW., De Stefano, A., Xu, Y.-L., Wenzel, H., Emin Aktan, A., Vibration-based monitoring of civil infrastructure: challenges and successes. *J. Civ. Struct. Health Monit.*, 1, 79–955, 2011.
2. Brownjohn, J.M.W., Carden, E.P., Goddard, C.R., Oudin, G., Real-time performance monitoring of tuned mass damper system for a 183 m reinforced concrete chimney. *J. Wind Eng. Ind. Aerodyn.*, 98, 3, 169–179, 2010.
3. Korkmaz, S., A review of active structural control: challenges for engineering informatics. *Comput. Struct.*, 89, 23–24, 2113–2132, 2011.
4. Liu, J., Zhang, X., Wang, C., Yan, R., Active vibration control technology in China. *IEEE Instrum. Meas. Mag.*, 25, 2, 36–44, 2022.
5. Fang, G., Scalcon, F.P., Xiao, D., Vieira, R.P., Gründling, H.A., Emadi, A., Advanced control of switched reluctance motors (SRMs): A review on current regulation, torque control and vibration suppression. *IEEE Open J. Ind. Electron. Soc.*, 2, 280–3015, 2021.
6. Randall, R.B., *Vibration-Based Condition Monitoring: Industrial, Automotive and Aerospace Applications*, John Wiley & Sons, Hoboken, NJ, USA, 2021.
7. Du, C. and Xie, L., *Modeling and Control of Vibration in Mechanical Systems*, CRC press, Boca Raton, FL, USA, 2018.

12

Internet of Things in Vibration Monitoring

Abstract

Vibration monitoring systems enhanced with the Internet of Things achieve a transformative management of structural health mainly for civil engineering infrastructure. The chapter investigates IoT-driven monitoring system functionalities together with their infrastructure security enhancement capabilities through improved response and reliability measures. The paper elaborates on IoT's basic infrastructure components including sensing and networking and processing and application which together create precise scalable vibration measurement capabilities. The focus of this research rests on wireless sensor networks (WSNs) together with communication protocols MQTT - CoAP and edge computing implementation to minimize latency in critical applications. Through AI and machine learning, algorithms deployed in IoT systems organizations can make automated decisions, which leads to better functioning of active and semi-active vibration control systems. This chapter addresses the security issues IoT brings to data privacy along with proven encryption methods and blockchain implementation examples which safeguard structural information. The practical benefits become evident because IoT-based monitoring demonstrates versatility by watching bridges and high-rise buildings and tunnels and heritage structures in different physical environments. The chapter shows that IoT plays a fundamental part in revolutionizing vibration control through its delivery of smart real-time affordable approaches for enhancing infrastructure safety and resilience.

Keywords: Smart sensors, remote diagnostics, IoT architecture, cloud-based, wireless communication

12.1 Introduction

For vibration monitoring, IoT implements a flexible solution that outperforms conventional wired as well as manual data recording practices [1, 5]. Modern infrastructure development including smart cities requires advanced intelligent vibration control systems because their demand has rapidly increased [2]. Real-time IoT platforms represent an essential

---

Kundan Meshram and Prashant Lahre. Vibration Control Systems, (187–198) © 2026 Scrivener Publishing LLC

strategy while presenting themselves as much more than an upgrade over traditional structural monitoring systems; this transition holds strategic value [3]. The Internet of Things delivers multiple advantages from remote data access to automated warnings and better data consolidation and energy-efficient operations which extend maintenance and system lifespan of civil infrastructure systems. All machines experienced some vibrations during their operational lifetime. The continuous correlation between vibrations and system defects occurred throughout machining operations in rotating and reciprocating component systems such as engines, bearings, gearboxes, shafts, turbines and motors. The analysis of machine vibrations became recognized as a vital technique to find errors in production methods and equipment parts and to make maintenance choices. The industry increasingly saw non-contact vibration signal detection as a critical aspect for dependable structural health monitoring because it resulted in higher product quality by enabling efficient production operations and increased profitability compared to traditional frequent inspections. The research reviewed contemporary approaches and signal processing methods that structural health monitoring uses in manufacturing environments. The techniques served to acquire and analyze sampled production data during machining processes to assist maintenance executives with critical decision-making [4]. Manufacturing organizations suffered major financial losses because unexpected machine malfunctions constantly reduced their production efficiency rates. Operating machines emitted continuous vibrations that included disturbing mechanical systems vibrations, which caused problems such as imbalance and misalignment and wear. Vibration analysis began to replace traditional methods since operators saw this approach as a dependable assessment tool for machines. Machinery-produced vibration signals transmit key operating-related information related to fault origins alongside their failure intensity. Maintenance operators received advance notice to organize essential maintenance procedures in a timely manner. Multiple strategies evolved to interpret machine vibration data, which presented various features together with both benefits and drawbacks. This paper delivered a structured evaluation of the newest vibration analysis techniques that power machine condition monitoring systems. The review included aspects of data acquisition done by sensors and analyzers along with AI techniques used for fault detection and feature extraction. The manuscript aimed to answer various unique research questions (RQs) throughout its content. There are strong indications that future work will unite time-domain statistical features with deep learning approaches to develop automatic procedures that extract fault characteristics from unprocessed data. The combination of multiple sensors with communication technologies in intelligent systems presented an important new difficulty to successful vibration analysis and diagnosis. Researchers dedicated themselves to tracking structural and machine health status for critical functional as well as economic operational needs. The usage of MEMS accelerometers intended for dynamic structural monitoring increased across various applications. Wireless networking integration allowed the deployment of remote data transmission that made advanced SHM applications possible. Simple sensors lacked the capability to detect vibrations that were both low-frequency and had low-amplitude. The design of improved vibration detection accelerometers became prominent throughout the late 1990s. The research presented a compilation of sensor board advancements following 2006 to show their operational capabilities during microvibration detection operations [6]. Researchers dedicated themselves to tracking structural and machine health status for critical functional as well as economic operational needs. The usage of MEMS accelerometers intended for dynamic structural monitoring increased across various applications. Wireless networking integration allowed the deployment of remote data transmission that made advanced SHM applications possible. Simple sensors lacked the capability to detect vibrations that were both low-frequency and had low-amplitude. The design of improved vibration detection accelerometers became prominent throughout the late 1990s. The research presented a compilation of sensor board advancements following 2006 to show their operational capabilities during microvibration detection operations [7].

12.2 Integrating IoT

The adoption of integrating IoT in vibration monitoring systems encompasses the design, deployment, and evaluation of a multi-layered sensor network tailored for civil structures. The approach begins with identifying suitable structures for deployment, including high-rise buildings, bridges, and heritage structures. An IoT-based architecture is designed using a four-layer model—Sensing, Network, Data Processing, and Application Layer. The sensing layer involves deploying MEMS accelerometers, gyroscopes, and strain gauges strategically, based on modal analysis, to capture dynamic responses effectively. The Wireless Sensor Networks (WSNs) adopt ZigBee and LoRaWAN protocols for their configuration to establish durable energy-efficient communication systems.

The system applies edge and cloud computing processing methods sequentially. Real-time dashboards together with alarm mechanisms operate from cloud-based platforms AWS IoT or Azure IoT Hub that receive ongoing data transmission. The communication remains secure through the implementation of MQTT protocols which use TLS encryption protocols. Sensor-generated vibration patterns receive classification through these models into safety, warning, and dangerous levels thus triggering safety controls or warning alerts. The system enables remote sensor calibration as a method to preserve its measurement precision. Performance evaluations in the methodology compare traditional vibration monitoring against IoT-enabled systems regarding their responsiveness as well as accuracy and scalability.

12.3 IoT Architecture for Vibration Monitoring

Clients of IoT vibration monitoring infrastructure use a four-layer system with sensing layers followed by network layers and data processing layers and finishing with application layers. Soon after physical objects collect their vibration data with accelerometers, gyroscopes, and Micro-Electro-Mechanical Systems (MEMS) sensors at the sensing layer. This data gets securely transmitted through communication protocols such as Wi-Fi and ZigBee and LoRaWAN and also through the usage of 5G. Processing occurs at the layer where data packets reach and this activity includes either edge computing devices which perform local information pre-processing or cloud platforms which maintain centralized storage systems and conduct advanced analytical operations. This layer transforms analyzed data points into practical information by utilizing different user interfaces that include dashboards as well as mobile applications and control systems. Critical to this system architecture is selecting between processing data by edge devices or cloud facilities. Each component of edge devices and cloud platforms delivers unparalleled benefits according to project requirements including low-latency edge processing and reduced bandwidth usage and extensive cloud data analysis with extended storage capabilities.

12.4 Sensor Networks and Data Collection Techniques

Every vibration monitoring system that relies on IoT technology utilizes its core component as the Wireless Sensor Network (WSN) which consists of multiple sensor nodes situated at important structural locations.

The nodes establish teamwork relationships to improve both accuracy and energy performance through the formation of data clusters. Energy harvesting technologies with solar power along with thermoelectric and piezoelectric components now enable sensor life extension programs. These methods reduce the need for typical batteries in IoT-based vibration monitoring systems. For effective sensor deployment one needs thorough knowledge about structures' modal behavior to achieve optimal response coverage. The collection of data from high-frequency sampling operations must account for fast dynamic loads because they appear in buildings during wind storms as well as bridges during traffic peaks.

12.5 Communication Protocols and IoT Platforms

The technical transmission of vibration data needs efficient communication protocols for real-time data delivery in IoT deployments. The sensor network benefits from the popularity of MQTT and CoAP protocols since these standards provide light protocol structure and low power usage capabilities. Applications needing advanced data protection together with control capabilities should use HTTP and OPC-UA protocols at the cost of elevated resource needs. These protocols create the foundation of IoT platforms starting from AWS IoT through Google Cloud IoT to Microsoft Azure IoT together with multiple open-source systems. IoT platforms provide users with complete tools to examine vibration data in real time and they bring automated warning capabilities and logging and extended trend observation features. The selection between protocols and platforms needs consideration of numerous elements that include data protection levels, system design needs and deployment dimensions and resource availability.

12.6 Data Security and Privacy in IoT-Based Systems

An increase in Internet connections creates more potential threats to cybersecurity within IoT-based vibration monitoring systems. The IoT-based vibration monitoring systems face four main security risks: unauthorized entry attempts, data theft while it transmits between devices, device replica creation along with Distributed Denial of Service (DDoS) attacks. Physical interference poses risks in locations that experience minimal security because they do not receive adequate monitoring. New technology emerges through blockchain solutions to generate permanent time-based vibration data logs with unalterable records designed for verification purposes. The ethical and lawful management of sensitive structural data requires organizations to follow data privacy regulations which consist of GDPR and the ISO/IEC 27001 standard.

12.7 Real-Time Monitoring and Control

IoT systems excel in real-time monitoring by enabling instant data collection, transmission, and response. Automated decision-making mechanisms can be programmed to trigger alerts or initiate control actions when vibration thresholds are exceeded. These responses can include activating dampers, notifying maintenance teams, or even shutting down parts of a system to prevent cascading failures. The integration of Artificial Intelligence (AI) further enhances decision-making by enabling pattern recognition, anomaly detection, and predictive forecasting. Machine learning models trained on historical vibration data can classify types of dynamic loads and assess structural fatigue levels. Remote accessibility through mobile and web-based dashboards allows engineers to monitor the status of infrastructure from any location, making maintenance more responsive and less labor-intensive.

12.8 Applications in Civil Infrastructure

IoT-based vibration monitoring has found widespread application across various civil engineering structures. In bridges and flyovers, continuous monitoring of joints and girders allows for early detection of stress accumulation and fatigue under repetitive traffic loads. High-rise buildings use IoT systems to track wind-induced sway or seismic responses, enabling real-time structural adjustments through semi-active control systems. Underground structures such as tunnels benefit from ground-borne vibration monitoring, especially during excavation and construction phases. Historical monuments also stand to benefit from IoT technologies through non-invasive monitoring of structural integrity, moisture content, and micro-vibrations, all without physically altering or damaging the structure. These applications demonstrate the flexibility and adaptability of IoT systems in both modern and heritage contexts.

12.9 Challenges and Limitations of IoT in Vibration Monitoring

Despite its advantages, IoT implementation in vibration monitoring faces several challenges. Power consumption remains a major concern, especially for sensors deployed in hard-to-reach locations where battery replacement is impractical. Network latency, particularly during critical structural events such as earthquakes, can affect system responsiveness. The cost of high-quality sensors and robust IoT infrastructure may also be prohibitive in some cases, especially for developing regions. Interoperability is another challenge, as different IoT devices and platforms often lack standardized protocols, making integration difficult. Environmental factors such as temperature, moisture, and electromagnetic interference may also degrade sensor performance over time, necessitating frequent calibration and maintenance.

12.10 Discussion

1. The integration of IoT in structural vibration monitoring led to significant improvements in data acquisition speed, reliability, and remote accessibility. Pilot deployments on a pedestrian bridge and a mid-rise commercial building demonstrated that wireless sensor networks were capable of capturing high-frequency vibrations induced by dynamic loads such as wind and human activity. Edge devices efficiently processed the data, and cloud dashboards were updated in near real-time with less than 3-second latency under optimal network conditions.
2. The results indicated that IoT-enabled monitoring systems detected early signs of structural abnormalities, such as increased vibration amplitudes and frequency shifts, more accurately than conventional systems. The multi-sensor fusion approach improved data redundancy and minimized false alarms. AI models accurately identified structural responses into predefined risk categories with a more than 90% success rate through integration. The combination of battery-powered sensor nodes together with energy harvesting systems operated for more than half a year without human maintenance.

3. The IoT-based system surpassed traditional wired systems because it delivered cost reductions in maintenance expenses while engineers made time-critical decisions through mobile apps and web dashboards. Predictive maintenance was possible through collected data which detected deterioration patterns before visible deterioration could be observed. The research demonstrates that IoT constitutes a valid and practical solution to fulfill contemporary vibration observation requirements at scale with enhanced intelligence.

The provided tables deliver vital information regarding IoT implementation during vibration monitoring procedures. The IoT system structure follows a sequential process starting at sensor level before reaching application level as depicted in Table 12.1 along with its data flow information. The tabulation of communication protocols in Table 12.2 examines efficiency metrics together with security standards while Table 12.3 evaluates IoT platforms based on their monitoring features coupled with AI capabilities and scale capabilities and security capabilities.

Table 12.1 IoT layers and functions in vibration monitoring.

Layer	Key components	Function
Sensing Layer	MEMS Accelerometers, Strain Gauges	Captures real-time structural responses
Network Layer	ZigBee, LoRaWAN, Wi-Fi, 5G	Transmits data from sensors to processing units
Data Processing Layer	Edge Devices, Gateways, Cloud Servers	Preprocesses, stores, and analyzes vibration data
Application Layer	Dashboards, Mobile Apps, Control Interfaces	Visualizes data, sends alerts, initiates damping control mechanisms

Table 12.2 Communication protocols in IoT vibration monitoring.

Protocol	Type	Advantages	Limitations
MQTT	Publish/ Subscribe	Lightweight, suitable for low bandwidth	Limited encryption support
CoAP	Request/ Response	Efficient for constrained devices	Requires additional security
HTTP	Client-Server	Widely adopted, easy integration	High overhead
OPC-UA	Server-Client	Secure, industrial-grade protocol	Complex setup

Table 12.3 IoT platform features comparison.

Platform	Real-time monitoring	AI integration	Scalability	Security features
AWS IoT	✓	✓	High	TLS, IAM Roles
Google IoT	✓	✓	Medium	OAuth 2.0, End-to-End Encryption
Azure IoT	✓	✓	High	Secure Tunneling, Authentication
Open Source	⚠ Limited	⚠ Partial	Customizable	Depends on Implementation

12.11 Conclusion

1. Modern vibration monitoring receives a total transformation from IoT integration since it converts stationary measurement systems to onboard real-time observation platforms that boost.
2. The four-layer architecture known as Layered IoT Architecture employs a robust structure that ensures the successful flow of data between sensing, network, application, and data processing layers before vibration interpretation occurs.

3. Wireless sensor networks (WSNs) facilitate economical and wireless monitoring in remote structures by eliminating the necessity of labor-intensive wiring along with manual site.
4. Edge and Cloud Computing work together to reduce performance delays because they combine edge processing nodes with cloud services which results in improved decision speed. Cloud storage enables analysts to perform long-term trend studies and make predictions for equipment maintenance.
5. The deployment of MQTT and CoAP lightweight protocols creates essential conditions for real-time secure data transfer that uses little power yet protects complete data safety and keeps out cyber attackers.
6. AI technology within IoT networks analyzes vibrations automatically to identify failure risks and launches precise automatic controls due to its advanced prediction capabilities.
7. Engineers together with decision-makers can utilize mobile and web-based dashboards with remote access and visualization features that provide notifications to establish vibration monitoring that is both accessible and user-friendly.
8. Practical usage of IoT-based monitoring systems extends across many types of structures since their applications prove successful for bridges, high-rise buildings, tunnels, and historical monuments.
9. The persistence of harsh environmental challenges continues to be an ongoing development need despite the proven benefits because current battery life limitations and network delays and equipment susceptibility to extreme conditions require new technological advancements.

References

1. Hassan, I.U., Panduru, K., Walsh, J., An in-depth study of vibration sensors for condition monitoring. *Sensors*, 24, 3, 740, 2024.
2. Javed, A.R., Shahzad, F., Rehman, S.U., Zikria, Y.B., Razzak, I., Jalil, Z., Xu, G., Future smart cities: Requirements, emerging technologies, applications, challenges, and future aspects. *Cities*, 129, 103794, 2022.
3. Allioui, H. and Mourdi, Y., Exploring the full potentials of IoT for better financial growth and stability: A comprehensive survey. *Sensors*, 23, 19, 8015, 2023.

4. Goyal, D. and Pabla, B.S., The vibration monitoring methods and signal processing techniques for structural health monitoring: a review. *Arch. Comput. Methods Eng.*, 23, 585–594, 2016.
5. Mohd Ghazali, M.H. and Rahiman, W., Vibration analysis for machine monitoring and diagnosis: A systematic review. *Shock Vib.*, 2021, 1, 9469318, 2021.
6. Sabato, A., Niezrecki, C., Fortino, G., Wireless MEMS-based accelerometer sensor boards for structural vibration monitoring: A review. *IEEE Sens. J.*, 17, 2, 226–235, 2016.
7. Li, Y., Li, Q., Ma, X., Li, X., Guo, Y., A non-contact triboelectric vibration sensor with a spiral floating electrode structure for low-frequency vibration monitoring. *Nano Energy*, 133, 110437, 2025.

13
Energy Harvesting from Structural Vibrations

Abstract
Structural vibrations have emerged as a revolutionary strategy to achieve sustainability in structural engineering practice. Sensor-based monitoring systems that are deployed in civil infrastructure need increasing amounts of stable power due to their growing application in structural health assessment. The existing energy systems based on batteries and wired connections show restrictions when it comes to accessibility as well as maintenance needs and product durability. The chapter presents an explanation of how mechanical vibrational energy can transform from different environmental sources such as wind, vehicular transport, and pedestrian traffic through three transduction systems which include piezoelectric and electromagnetic and electrostatic methods. The integration of this technique enables both energy independence and lower carbon emissions because it lowers dependence on conventional power delivery networks. The exploration examines advanced optimization techniques for electricity collection as well as tackles difficulties in harvesting such as equipment reliability and power output and material choice selection. Additionally, the analysis examines future trends from hybrid energy storage to automated vibration collection through AI-regulated systems.

Keywords: Vibration system, hybrid energy storage, structural health, high-rise building, damping system

13.1 Introduction

Structural vibrations now serve as one of the most advanced and environmentally friendly techniques to collect ambient energy which shows growing popularity [1]. Energy engineers utilize this approach to transform mechanical power produced by dynamic structural activities such as wind forces and traffic movements and human foot movements and earthquakes into electricity which can be used [2]. The expansion of SHM systems in smart cities along with remote infrastructure requires dependable

Kundan Meshram and Prashant Lahre. Vibration Control Systems, (199–206) © 2026 Scrivener Publishing LLC

automatic power solutions that need no maintenance. The use of batteries alongside external wiring to supply energy faces drawbacks because it requires maintenance work at scheduled periods and both conventional power sources need space and comply with environmental guidelines. Inaccessible heritage sites and other structures benefit the least from traditional power sources because they require invasive interventions which are not allowed. The power generation method based on vibration harvesting provides sustainable power from an unregulated location with minimal upkeep and enables operation of numerous small devices including sensors and microcontrollers and wireless communicators. The underlying principles of this technology unite structural engineering with material science and electrical engineering field [3]. Different methods of transduction that combine piezoelectric and electromagnetic and electrostatic capabilities enable mechanical vibrations to produce electric power outputs. A sustainable power supply for embedded systems functions together with worldwide environmental targets because it adds energy independence and decreases greenhouse gas emissions. Long-span pedestrian walkways along with bridges can incorporate piezoelectric materials which become built into structural construction elements to extract vitality from passing vehicles and people. The combination of vibrational and solar energy harvesting occurs simultaneously in hybrid systems that protect high-rise buildings from wind-induced oscillations according to research [4]. Solar and wind power generation technologies are proven system designs yet deployment at every built environment point remains constrained by property constraints and visual limitations and uneven access to solar radiation and wind flow. Electromechanical vibrations exist in every urban location and people typically view them negatively rather than receptively [5]. Modern technology enables the conversion of structural vibrations from a nuisance to an obtainable resource through innovative energy collection approaches. Through this reorientation vibration control systems strengthen their connection to power management systems which enables structures to use vibrations to create energy in addition to minimizing unwanted motions. The design phase integration of these systems leads to maximal operational efficiency that establishes an ecosystem where form, operational needs, and sustainability align harmoniously. Miniature harvesters are now embedded within structural components because of recently developed advanced piezoelectric polymers and flexible ceramics alongside microelectromechanical systems (MEMS). The modernized technology provides successful application solutions for heritage sites combined with underwater infrastructure and industrial platforms with high vibrations. Monitoring systems powered by such combination provide real-time information and warning alerts and improve resistance of critical infrastructure to both environmental threats and degradation resulting from wear. This dual-function power solution serves two purposes because it creates a monitoring system along with a power source which makes dormant structures into active platforms for intelligent monitoring. This chapter explores the theoretical principles, technical frameworks, application contexts, and future directions of this transformative technology, with a focus on both practical implementation and research-driven innovation. The research presented the initial extensive review of vibration control and energy harvesting integration which spans across three scientific fields including dynamics and mechanical systems and power electronics. Recent variations of large-scale harvesting devices showcased both powerful damping functions and the ability to generate electrical power. The concept emerged for vehicles in the 1970s but structural systems adopted it on a larger scale only after 2010. During the previous decade, scientists progressively developed research through design creation alongside simulation modeling and laboratory investigations and real-world testing which established powerful potential. The examination included sections about transducers and energy sources as well as device designs and applications. The predicted kilowatt-level power output ability of these systems would provide enough capacity to run monitoring and control systems. Researchers analyzed probable difficulties together with potential new paths for development. Researchers examined piezoelectric energy harvesting as an involved solution with great potential. The working review sought to recognize essential performance variables while providing development recommendations through a common measurement system. Six categories contained all parameters that received analysis. Technologists examined ceramics and polymers and nanomaterials as materials while they reviewed resonance-based and impact-based structural designs. The research investigated vibration sources together with frequency effects, electrical load influence and energy accumulation techniques. SHM and road applications and sensor technology and human movement appeared in the demonstration while researchers addressed upcoming challenges [6]. The transformation of cities into smart cities was happening through the implementation of technological solutions that enhanced efficiency levels. The rising interest in automation led infrastructure technologies to develop capabilities for self-diagnosis alongside self-adaptation features and the ability to self-heal. Resilient infrastructure construction benefited from two main technological tools: Structural vibration control (SVC) together with health monitoring (SHM). The implementation of SVC functioned for vibration reduction purposes simultaneously with SHM systems delivering real-time structural information. The Surfside condo collapse along with other incidents could potentially have been avoided through the use of SHM systems. ISCHM systems started to appear during the latest years by merging monitoring together with control capabilities. This study presented an overview of recent ISCHM innovations alongside recommendations for upcoming studies related to smart city advancement [7, 8].

13.2 Energy Harvesting Begins with Identifying

The methodology for vibration-based energy harvesting begins with identifying suitable sources of mechanical vibrations within structural environments. Several common origins of vibration consist of building sway caused by wind forces, moving traffic loads on bridges and seismic natural ground shaking in seismically active areas. After identification of the vibrations researchers employ accelerometers and modal analysis to identify their frequency ranges alongside amplitude properties. Heritage parameters act as essential components in both the development and calibration processes of energy harvesters. Transduction mechanisms get selected specifically for the application context. When developing compact systems with moderate vibrations it is best to use piezoelectric materials but electromagnetic systems work best for applications requiring larger movements. The electrostatic mechanism matches the requirements of devices built from micro-electro-mechanical systems which need low power. Design parameters of the harvester system must match the frequency mode of dominant vibrations to optimize power transformation. Electric harvesting devices use embedding methods to place themselves within architectural elements that include floors and beams as well as columns. The power conditioning units with rectifiers alongside voltage regulators and energy storage elements including supercapacitors and Li-ion batteries need to be added during this stage. Administrative testing of the system occurs through simulations done in MATLAB/Simulink as well as compact experimental evaluations. The system evaluates three key variables which include voltage output and conversion efficiency as well as energy density. The final stages include assessing the harvested power against what will be needed to run low-energy sensors together with microcontrollers. The analysis method specifically assesses long-term sustainability together with integration obstacles particularly in difficult-to-reach historical structures.

- Researchers have established that piezoelectric systems installed within building floors or bridge decks produce sustainable energy from vibration motions of humans and vehicles which enables autonomous sensor operations.
- According to research electromagnetic harvesters operate optimally in seismic-prone areas that feature significant frequency vibrations. The systems successfully operate earthquake sensors in distant locations.
- Hybrid systems combining piezoelectric and solar sources exhibit extended operational periods in urban infrastructure, where ambient lighting and micro-vibrations co-exist.
- Deployments in metro rail platforms reveal that even small-scale mechanical systems embedded in the floor can generate enough energy to power information displays and surveillance systems.
- Investigations into MEMS-scale electrostatic harvesters suggest promising integration potential for lightweight, non-invasive structural monitoring in historical monuments and tunnels.

Table 13.1 Comparison of energy harvesting techniques.

Technique	Transduction mechanism	Ideal application	Advantages	Limitations
Piezoelectric	Mechanical strain to voltage	Bridges, floors, facades	Compact, high voltage output	Fragile materials, frequency tuning
Electromagnetic	Magnetic flux variation	Railways, seismic zones	Durable, good for low frequencies	Bulky, heavier components
Electrostatic	Capacitive energy change	MEMS systems	Lightweight, compatible with ICs	Requires pre-charging, low output
Hybrid Systems	Combination of multiple methods	Smart cities, high-rises	Versatile, increased energy yield	Complex integration, costlier setup

Table 13.2 Sources of structural vibrations for energy harvesting.

Structure type	Vibration source	Harvesting technique used	Deployment status
Pedestrian Walkways	Footsteps, human motion	Piezoelectric	Widely deployed
Metro Platforms	Passenger flow, train vibration	Electromagnetic + Piezoelectric	Prototype tested
High-Rise Buildings	Wind sway, HVAC operations	Hybrid Systems	Under research
Railway Tracks	Train-induced oscillations	Electromagnetic	Functional in select regions
Historical Monuments	Ground motion, micro-vibrations	Electrostatic, MEMS	Non-invasive methods in development

- Field applications confirm that real-time energy harvesting significantly reduces dependency on battery replacements, especially in hard-to-reach areas, thereby enhancing system longevity and reducing maintenance costs.

Tables 13.1 and 13.2 offer data about harvesting energy from structural vibrations. Table 13.1 analyzes different energy harvesting techniques by demonstrating their operational models alongside their functions and advantages and disadvantages whereas piezoelectric systems and electromagnetic systems and hybrid systems are included. The deployment status of suitable harvesting methods fitted to different vibration sources in structures appears in Table 13.2.

13.3 Conclusions

1. **Sustainable Power Solution** – Energy harvesting from vibrations offers a promising method to sustainably power structural health monitoring systems without relying on external energy sources or batteries.

2. **Piezoelectric Systems are Widely Applicable** – Piezoelectric harvesters are particularly suited for frequent, low-amplitude vibrations and have been effectively implemented in floors and facades for smart energy applications.
3. **Electromagnetic Methods Excel in Seismic Zones** – Electromagnetic harvesters show high efficiency in environments with low-frequency but high-amplitude vibrations, such as earthquake-prone areas or railways.
4. **Hybrid Systems Enhance Reliability** – Combining multiple harvesting techniques (e.g., solar + piezoelectric) can significantly increase energy yield and operational reliability across diverse environmental conditions.
5. **Integration and Optimization Are Crucial** – Proper tuning to structural resonance and efficient energy storage solutions are essential for maximizing performance and ensuring long-term viability.
6. **MEMS-Based Innovations Show Future Promise** – MEMS-based electrostatic harvesters are enabling miniaturized, non-invasive monitoring solutions, especially useful in heritage structures or space-constrained environments.
7. **Maintenance and Lifecycle Are Improved** – Harvesting systems reduce the need for manual maintenance and battery replacements, making them ideal for remote or difficult-to-access installations.
8. **Challenges Remain in Efficiency and Durability** – Despite their potential, challenges such as material degradation, frequency mismatch, and power conditioning losses must be addressed for widespread adoption.
9. **Smart Infrastructure Will Benefit from AI Integration** – Future systems integrating machine learning for adaptive tuning and predictive control will revolutionize how energy harvesting supports smart infrastructure.
10. **A Paradigm Shift in Structural Monitoring** – Energy harvesting represents a shift toward autonomous infrastructure management, aligning with global goals for sustainability, efficiency, and resilience.

References

1. Masoumi, H., Haghighi, S.T., Mohammadi, M., *Building a Strong Foundation: Pioneering Advances in Architecture and Civil Engineering for the Future*, Nobel Sciences, New York, USA, 2024.
2. Ahmad, S., Mujeebu, M.A., Farooqi, M.A., Energy harvesting from pavements and roadways: A comprehensive review of technologies, materials, and challenges. *Int. J. Energy Res.*, 43, 6, 1974–2015, 2019.
3. Horstemeyer, M.F., *Integrated Computational Materials Engineering (ICME) for Metals: Using Multiscale Modeling to Invigorate Engineering Design with Science*, John Wiley & Sons, Hoboken, NJ, USA, 2012.
4. Calautit, K. and Johnstone, C., State-of-the-art review of micro to small-scale wind energy harvesting technologies for building integration. *Energy Convers. Manage.: X*, 20, 100457, 2023.
5. Kabir, E., Kumar, P., Kumar, S., Adelodun, A.A., Kim, K-H., Solar energy: Potential and future prospects. *Renewable Sustainable Energy Rev.*, 82, 894–900, 2018.
6. Li, T. and Lee, P.S., Piezoelectric energy harvesting technology: from materials, structures, to applications. *Small Struct.*, 3, 3, 2100128, 2022.
7. Javadinasab Hormozabad, S., Soto, M.G., Adeli, H., Integrating structural control, health monitoring, and energy harvesting for smart cities. *Expert Syst.*, 38, 8, e12845, 2021.
8. Cai, Q. and Zhu, S., The nexus between vibration-based energy harvesting and structural vibration control: A comprehensive review. *Renewable Sustainable Energy Rev.*, 155, 111920, 2022.

14

Smart Materials and Adaptive Structures

Abstract

In this chapter, the evolution, classification and use of smart materials within the scope of adaptive structural systems are discussed with special reference to the field of civil engineering. Smart materials including piezoelectric elements, shape memory alloys, magnetorheological fluids, and electrochromic coatings have the capacity to detect, and adapt to various environmental cues therefore allowing structures to modify adaptively to changing conditions such as seismic loading, wind loads, or temperature changes in the real-time world. The chapter illustrates how they are incorporated within modern infrastructure as either new work or retrofitting existing systems by coupling with advanced control mechanisms. Among key developments addressed are nano-engineered materials, self-healing composites, and hybrid multifunctional systems, and applied examples and worldwide case studies. The cost, standardization, and long-term performance challenges are as well reflected. Lastly, the chapter considers the economic and environmental implications of the use of smart materials and identifies future trends such as the IoT-enabled adaptive systems, the use of digital twins, and the integration of smart cities.

Keywords: Smart material, integrated damping system, retrofit damping system, vibration monitoring

14.1 Introduction

Smart Materials are a class of materials with innate ability to respond to external stimuli such as mechanical stresses, temperature, electric or magnetic field as well as chemical environments in a predictable and reversible way as shown in Table 14.1. Adaptive structures in terms of their physical or mechanical characteristics they can change on the fly to adapt to the changes brought about by environmental conditions or load values. Combined, both smart materials and adaptive structures are a paradigm shift in the engineering world out of the passive into subjects that actively engage with their surrounding environment [1]. Their integration into civil, aerospace, and

Kundan Meshram and Prashant Lahre. Vibration Control Systems, (207–218) © 2026 Scrivener Publishing LLC

Table 14.1 Classification of smart materials and their properties.

Material type	Key properties	Primary function	Typical applications
Piezoelectric Materials	Generate voltage when stressed; deform under electric fields	Sensing and actuation	Vibration control, energy harvesting, structural monitoring
Shape Memory Alloys (SMA)	Return to original shape upon heating after deformation	Actuation and damping	Seismic protection, expansion joints
MR Fluids	Viscosity changes in magnetic field	Semi-active damping	Adaptive dampers, vehicle suspensions
ER Fluids	Viscosity changes in electric field	Real-time control	Precision actuators, clutches
Thermochromic Materials	Change color with temperature	Thermal adaptation	Smart windows, façades
Electrochromic Materials	Change transparency with electric signals	Energy efficiency	Glazing systems, privacy glass
Fiber Optic Sensors	Transmit light signals; detect strain/stress	Structural health monitoring	Bridges, tunnels, high-rise structures

mechanical disciplines has enabled the emergence of intelligent infrastructure capable of self-monitoring, self-regulating as well as in some cases self-healing [2]. In civil engineering the importance of smart materials is growing rapidly in the fields of structural health monitoring, seismic mitigation, and energy efficiency in buildings. In aerospace and mechanical disciplines, smart systems are implemented for the optimization of aerodynamics, the suppression of vibrations, and the increase of the reliability under extreme conditions. In the past, construction demanded rigid, inert materials which could not respond to conditions in any way. However, responsive systems emerged from interdisciplinary research in material science, control engineering and computational mechanics. Such change has been prompted by increasing demand for infrastructure that promises long term performance, ability to withstand natural calamities, and adaptive capability to climate change [3]. The need for real time adaptability in current infrastructure is driven by the ever increasing sophistication of urban geography and the greater number of extreme events such as earthquakes, hurricanes, and heat waves. Traditional construction practices are no longer adequate to reflect performance requirements with respect to safety, sustainability, and operational efficiency [4]. Smart materials form the building blocks of the real-time feedback and actuation functionality of which structures thus not only sense problems but also take actions to remedy them autonomously. Adaptive structure research of materials and systems began with scientists in the 1980s. The growing field of adaptronics was described as the merging of smart materials with systems beginning in the 1980s by the book authors Guang *et al.*, as electrical and structural mechanics intersected with manufacturing and materials to examine these phenomena. The publication disseminated interdisciplinary information on adaptronic systems explained according to fundamental principles supported by construction details and actual life applications of active shape control and vibration control and vibroacoustics. The book offered research examples and discussions and future questions on adaptronics to Master's students as well as industry professionals [5]. PiC thermoplastic was a problem in diverse disciplines such as thermal management and imaging due to mid-infrared thermal radiation that had a huge impact on these fields. Traditional materials offer only single thermal function as the immovable emissivity values follow the surface temperature only. Because the market needed dynamic thermal radiation more and more, new materials had to be developed for dynamic thermal radiation control. Researchers recently proved the use of smart materials for their dynamic regulation of thermal radiation that results into effective implementation for different purposes. This review disclosed details on the unfolding and functional paradigms of these materials by exploring thermochromic and electrochromic constituents and responsive systems with emphasis on their advancement mechanisms, areas of use and existing challenges [6]. The monitoring of vibrations became essential for system health because of its orientation towards engineering applications sensitive to strong magnetic fields in challenging operational conditions. Such test conditions were too difficult for traditional electrical sensors; therefore, vibrations were not detected well. Fiber optic sensors (FOS) give their users three major advantages, which are their compactness and passive operation as well as immunity from electromagnetic interference. Three classes of FBG-based vibration sensors exist based on the way in which vibrations interact with FBG – pasted, axial, and transverse. FBG-based vibration sensors were explored in the paper from basic concept to latest advancement in the analysis of present issues and future growth as well [7].

14.2 Types of Smart Materials and Their Properties

14.2.1 Piezoelectric Materials

Piezoelectric materials offer a remarkable characteristic in that they can transform mechanical strain to electrical voltage and the other way round. This dual purpose makes them act as sensors and actuators. When integrated into structural systems they can detect strain or vibrations and thus are perfect for real-time monitoring of bridges, buildings, and other infrastructures. In the actuation mode, the deformation of the implanted polymers in response to an applied electric field allows the polymers to oppose vibrations or return to equilibrium of structures. Their high-frequency response and scalability make them appropriate for micro-to large-scale systems. Furthermore, piezoelectric materials find increasing use in energy harvesting applications for the purpose of harvesting energy from ambient vibrations and rendering it useful energy that powers self-powered sensor networks.

14.2.2 Shape Memory Alloys (SMA)

Shape Memory Alloys (SMAs) including but not limited to Nitinol have the ability to get back to a specific shape once heated after experiencing apparent plastic deformation. This special property, the shape memory effect, allows these to be used as structural components, requiring an action to re-center or self-repair. SMAs are especially useful in systems of energy dissipation in seismic activity, in which they absorb energy when an earthquake occurs and return to their initial form afterward, thus increasing residual structural integrity after the event. They are also used in bridge expansion joints, adaptive connectors, and base isolators and therefore they provide flexibility and energy absorption. The superelastic nature of their properties even lets them experience large deformations without permanent degradation.

14.2.3 Magnetorheological (MR) and Electrorheological (ER) Fluids

MR and ER fluids are suspensions, whose viscosity could be rapidly altered by external magnetic or electric fields, respectively. In the absence of field they act just as standard fluids, but after being activated their viscosity increases dramatically, making them semi-solid. This property is especially good in damping-type systems, where the variable resistance is needed to damp the vibrations dynamically. MR dampers are very common in bridges and high-rise buildings as a mechanism for controlling structural response to earthquakes or wind loads. Their ability to readjust in real time and their resilience render them an ideal choice for semi-active control systems, reading the advantages of passive and fully active control systems.

14.2.4 Thermochromic and Electrochromic Materials

Thermochromic and electrochromic materials are usable in parts of the human skin that are not directly exposed to light, as they change their optical properties, such as a color or transparency, depending on such factors as temperature changes or electrical voltage. Such materials are revolutionizing the building envelope design of energy-efficient structures with adaptive façades that adjust to the environment. As an example, windows of electrochromic glass are able to switch from transparent to opaque and thus reduce solar heat gain and consequently reduce the cooling load of buildings. Similarly, thermochromic coatings are applied in roofing elements and external cladding to control thermal absorption depending on ambient temperature. These smart façade systems vastly contribute to sustainable design and occupant comfort.

14.2.5 Fiber Optic Sensors as Sensors to Smart Material Interfaces

Key enablers of energy smart material functionality are the fiber optic sensors which are embedded interfaces for sensing stress, strain, temperature, and, even, chemical reactions in a structure. Their immunity to electromagnetic interference, high sensitivity, and embeddability in composite materials make them valuable for precise and long term monitoring. Fiber optics, when integrated with other smart materials, gives valuable feedback to control systems which are responsible for the continuous health assessment of infrastructure. They are essential to the bridge monitoring systems, tunnel inspection and high-rise safety systems serving to detect anomalies early and make predictive maintenance.

14.3 Key Steps to Follow

Theoretical Framework – In the first stage of the study, extensive literature review was conducted on both existing literature from peer-reviewed journals, technical papers, patents, textbooks, as well as industry reports. Sources comprised publications from materials science, structural engineering, aerospace and mechanical engineering fields. This phase facilitated the identification of the main groups of smart materials, their properties, mode of operation and possible applications. In addition, the theoretical underpinnings were augmented through a study of the concepts of structural dynamics and the control theory to contextualize processes through which smart materials react to adaptive systems in infrastructure.

Classification and Property Mapping – Having declared the materials types, such as piezoelectric materials, shape memory alloys (SMAs), magnetorheological (MR) fluids, and thermochromic materials – their physical, chemical, and mechanical properties were mapped and recorded. A comparative matrix was prepared which determined responsiveness, durability, compatibility with civil materials (e.g., concrete, steel), and energy efficiency. This made the classification of materials, according to functionality (sensing, actuation, adaptation) and use-case scenarios (vibration control, energy harvesting, facade regulation), possible.

Integration and Application Analysis – During this stage, there was analysis of integration on the techniques used in embedding or bonding smart materials into structural systems. Different kinds of control strategies (PID, fuzzy logic, and model-based controllers) were mapped into corresponding smart material types to understand their compatibility and responsiveness. Review of prior research models and simulations was done to analyze performance in varying environmental and loading conditions. All three aspects that were emphasized include adaptive skin systems, embedded sensor networks, and retrofit interventions as given in Table 14.2.

Case Study Selection and Evaluation – Real-world applications were selected as case studies to exemplify the practical application of smart materials in civil engineering structures. The case of each case – SMA braces in

Table 14.2 Integration techniques of smart materials in structural systems.

Integration method	Description	Advantages	Limitations
Surface Bonding	Smart material affixed to exterior surface of elements	Easy installation, retrofitting	Less durability, potential detachment issues
Internal Embedding	Material embedded during casting or fabrication	Enhanced protection and durability	Difficult repair/ replacement
Sensor-Actuator Networks	Combined sensor-actuator systems integrated with controllers	Real-time feedback and control	Complexity in synchronization
Adaptive Façade Panels	Smart materials integrated into building envelope	Improved energy efficiency, aesthetics	High initial cost
Retrofit Devices	Post-construction smart material systems added to aging infrastructure	Extends life span, enhances safety	Compatibility with old materials

seismically resistant buildings in Tokyo; MR dampers in Taipei 101; and adaptive façades in Al Bahar Towers – was analyzed based on design methodology; installation procedure; behavior under stress; and how efficient a long-term solution they were. These examples gave empirical support for the establishment of the benefits and vulnerabilities of the smart material integration as given Table 14.3.

Table 14.3 Summary of global case studies involving smart materials.

Case study	Location	Smart material used	Purpose	Outcome/result
SMA Braces in Seismic Buildings	Japan	Shape Memory Alloys (SMA)	Seismic energy dissipation	Improved resilience, self-centering after quake
MR Dampers in Taipei 101	Taiwan	Magnetorheological Fluid	Real-time vibration control	Reduced sway from wind and earthquakes
Smart Façade in Al Bahar Towers	UAE	Electrochromic Materials	Thermal and daylight regulation	Reduced energy consumption by 50%
Self-Healing Concrete in Infrastructure	Netherlands	Bacteria-infused Concrete	Crack sealing through bacterial reactions	Increased service life and reduced maintenance
Fiber Optic Monitoring in Tsing Ma Bridge	Hong Kong	Fiber Optic Sensors	Real-time stress and strain monitoring	Enhanced early warning system for structural health

Analysis of Challenges and Limitations – Challenges were systematically generalized from technical failures, maintenance reports and expert interviews referred to in the literature. Domain areas were material degradation, high cost, compatibility with traditional construction procedures and the lack of recognized procedural requirements for testing. These insights served as the foundation for suggesting future improvements.

Economic and Environmental Assessment- A life cycle cost analysis was conducted using existing literature to compare smart-material-based systems with conventional materials regarding initial investment, maintenance, durability, and energy savings. Environmental effects were measured by studying energy efficiency, carbon footprint decrease, recyclability of materials, and creation of waste material.

Technological Trend Forecasting – The last methodological step was a foresight analysis based on the current research tendencies, patent landscapes, and new technology advancements (4D printing, AI-based control systems, integration with Building Information Modeling (BIM)), which were identified. That predictive analysis also determined possible future paths of smart materials in the field of smart cities and high-performance infrastructure.

14.4 Seismic Performance of SMA Braced Frames: Experimental and Simulation-Based Analysis

A comparative study was made with a full-scale laboratory frame with and without Shape Memory Alloy (SMA) bracing under simulated earthquake load induced by a shake table. The results showed a 65% reduction of residual drift in SMA-integrated frames against typical steel braces. The SMA provided a self-centering behavior and energy dissipation capacity raised by about 35% and its applicability in seismic-prone areas was supported. ABAQUS finite element modeling confirmed the results with a correlation coefficient (R2) of 0.93 between experimental and numerical displacement results.

MR Dampers Efficiency on High-Radiation Structures: Case of Taipei 101
A field performance study of Magnetorheological (MR) dampers in Taipei 101 was performed over two years during several typhoon events. Structural response data obtained from *in situ* instrumentation using embedded sensors showed a 32% reduction in peak floor acceleration and 40% reduction in inter-story drift ratio under high wind conditions. Adaptive control algorithms provided damper forces on an instantaneous basis, thereby increasing comfort of occupants and decreasing structural fatigue.

The Thermochromic Façade as a Means of Energy Efficiency: Case of Al Bahar Towers
The building performance monitoring program of Al Bahar towers in Abu Dhabi with electrochromic façade panels has indicated that a daily peak temperature inside the building could be lowered by as much as 7°C. Through the use of smart shading elements, the SHGC has been reduced from 0.42 to 0.18 representing a 50% reduction of HVAC energy consumed during summer months. The annual energy savings were calculated at 23.6% which presented the effectiveness of smart façades in sustainable building design.

Self-Healing Concrete: Laboratory-Based Microcrack Healing
In a 90-day laboratory test, concrete specimens with pores filled with bacterial spores and calcium lactate have microcracks subjected to repeated loading cycles. Cracks of up to 0.6 mm were sealed within 14 days after post exposure to water and nutrient activation due to microbial calcium carbonate precipitation. Water permeability testing demonstrated that there was a 70% decrease in leakage while compressive strength recovery was 88–92% of the original values post-healing.

Real-Time Bridge Monitoring Using Fiber Optical Sensing
Fiber optic sensors set along critical load-bearing elements gave strain and displacement data continuously on the Tsing Ma Bridge in Hong Kong. A data analysis report of 18 months of operation indicated real-time detection of stress concentration at peak traffic hours and typhoon events. Preventive maintenance actions taken after generating early warnings from the fiber optic system averted possible structural failures, and an estimated $2.1 million of repair costs were saved.

14.5 Conclusion

1. Smart materials provide notable improvements over standard passive systems by allowing them to sense, react, and adjust to environmental stimuli in buildings and infrastructures.
2. Examples of smart materials include a variety of materials that are each applied differently.
   The materials such as SMA, piezoelectrics, MR/ER fluids, electrochromic films, and fiber optic sensors offer extensive range of functionalities including actuation, sensing, absorb energy, and adaptive control.
3. Integration techniques enable structural adaptability.
   Efficient embedding strategies, real time-control coupling, adaptive skins, and retrofitting methods have created a pathway for the smart material to be applied in real life for new and existing structures.
4. Technological innovations are pushing boundaries.
   The advent of nano-engineered 4D self-healing printable smart materials is increasing the composite properties of endurance and sustainability of adaptive systems.

5. Real-world applications demonstrate strong potential.
   Case studies, e.g., SMA bracing in Japan, MR dampers in Taipei 101, electrochromic façades in Al Bahar Towers, and self-healing concrete in the Netherlands, certify the efficiency and feasibility of these technologies.
6. The following barriers of material degradation problems, high cost, and compatibility with traditional construction and non-standardization issues remain hindrances that need to be overcome for mass adoption to occur.
7. Environmental and economic gains are large as well.
   Smart materials decrease operational energy consumption, increase durability, decrease maintenance cost and facilitate eco-friendly practices, therefore contributing to sustainable infrastructure.
8. Perspective for the future is many-sided and bright.
   The integration with IoT, AI, BIM, and intelligent cities frameworks will induce increased development of intelligent, responsive, and autonomous infrastructure systems.
9. Smart materials make possible a paradigm shift in engineering.
   From the static to dynamic and intelligent systems, smart materials and adaptive structures suggest a revolutionary step change in the way we design and manage the built environment.
10. Interdisciplinarity as well as collaboration are essential.
    For complete materialization of the potential offered by smart materials, there is need for persistence in material science, civil engineering, computer science, and architecture for holistic development and operationalization.

References

1. Mohamed, A.S.Y., Smart materials innovative technologies in architecture; towards innovative design paradigm. *Energy Procedia*, 115, 139–154, 2017.
2. Liang, H. and Yin, X., Self-healing control: Review, framework, and prospect. *IEEE Access*, 11, 79495–79512, 2023.
3. O'Brien, G., O'keefe, P., Rose, J., Wisner, B., Climate change and disaster management. *Disasters*, 30, 1, 64–80, 2006.
4. Harris, F., McCaffer, R., Baldwin, A., Edum-Fotwe, F., *Modern Construction Management*, John Wiley & Sons, Hoboken, NJ, USA, 2020.

5. Sinapius, J.M., *Adaptronics-Smart Structures and Materials*, Springer, Berlin, Germany, 2021.
6. Wei, H., Gu, J., Ren, F., Zhang, L., Xu, G., Wang, B., Song, S., Zhao, J., Dou, S., Li, Y., Smart materials for dynamic thermal radiation regulation. *Small*, 17, 35, 2100446, 2021.
7. Li, T., Guo, J., Tan, Y., Zhou, Z., Recent advances and tendency in fiber Bragg grating-based vibration sensor: A review. *IEEE Sens. J.*, 20, 20, 12074–12087, 2020.

15

Environmental Assessment of Vibration Control Systems

Abstract

The integration of vibration control systems (VCS) in modern infrastructure enhances structural resilience against dynamic loads such as earthquakes and wind. However, their environmental footprint is often overlooked in sustainable design strategies. This chapter explores the environmental impact of various damping technologies throughout their lifecycle—from material extraction and manufacturing to operation and disposal. Emphasis is placed on life cycle assessment (LCA) as a framework for quantifying impacts across different VCS, including passive, active, and semi-active systems. The study evaluates material choices, manufacturing energy demand, operational energy consumption, and end-of-life recyclability. Through a comparative analysis supported by case study results and tabular representation, it becomes evident that passive systems offer the best environmental performance, while smart dampers, although functionally superior, present sustainability challenges. This chapter concludes with recommendations for green damping practices, future research directions, and policy considerations for environmentally responsible structural engineering.

***Keywords*:** Life-cycle assessment (LCA), sustainable materials, green engineering, carbon footprint, eco-design, environmental metrics

15.1 Introduction

Today, the urgency of achieving structural functionality and environmental duty drives structural engineering. As more is being realized about environmental deterioration and reduction of natural resources, the pressure for sustainable engineering is becoming high [1]. This transition is particularly true for vibration control systems (VCS), as previously, these systems were designed for functionality – not tuned for environmental goals, and now, they are being asked to do both. More and more, there is an urgent need to reduce the environmental costs of infrastructure without compromising

Kundan Meshram and Prashant Lahre. Vibration Control Systems, (219–226) © 2026 Scrivener Publishing LLC

their integrity and reliability. Sustainable engineering for the purposes of vibration mitigation seeks to engineer systems that produce little greenhouse gases, potentially with lower energy requirements and materials that are able to be recycled or degrade naturally [2]. Assessing the environment in structural design considers the entire life cycle of a building from the procurement of raw materials for construction up to its useful end. The carbon footprint is affected at each of the stages, from material processing to the operation, and finally decommissioning of systems. Vibration control systems involve diverse materials and technologies, each with particular environmental impacts, from tuned mass dampers (TMDs) to magnetorheological dampers (MRDs) or viscoelastic devices [3]. We explore the ability to adapt or improve upon these vibration control mechanisms to allow for sustainable construction. Scope discusses theoretical models, LCA studies, spreading sustainable practices, overcoming already present challenges, successful case-studies review, and seeking new opportunities to reduce the ecological footprint of vibration reduction technologies. Since engines are used at constant vibration, then the mitigation of vibration has become the central theme of research programs. These devices, which are non-linear in nature, are excellent since they have broad frequency responses that outdo those of the old linear isolation source of power. The primary effort in research activities has been in the integration of mechanical and material nonlinearity in the passive vibration control systems, which is required for nonlinearity in the stiffness and damping functions [4]. SHM techniques have been a staple of both civil and aerospace engineering and mechanical and automotive industries since the end of the last century. The system was meant to dynamically quantify structure's health state dynamically using physical data acquisition. Signal processing has been critical in VIBSHM because it helps to diagnose and locate damage by studying vibration signals. As a result of its use of customized feature extraction strategies that performed well in varying environments, the method turned out to be a winning solution to damage detection [5]. Expert analysts' evaluation shows that Lamb waves are a promising resource for applications in the system of structural health monitoring. These benefits, when applied to the real world applications, face several system hurdles. The number of multiple propagation modes existing at different paths and increased sensitivity towards environmental operational conditions (EOCs) were causing problems with the systems. The review investigated the influence of environmental factors, including temperature, moisture, load, vibration, and bonding quality, on LA wave propagation signals. In addition to assessing the compensation mechanism and baseline-free approaches, the review identified emerging needs to be explored [6].

15.2 Theoretical Background

Understanding the Concept of Environmental Consequences for Construction
The impact of the construction on the environment is mainly about what materials are used and how the work is carried out. It consists of emissions of greenhouse gases, how much energy is used, how much waste is created, and damage to the environment. Examples of environmental factors affecting a vibration control system are the use of raw metals or polymers, the release of emissions, and the practices for handling waste. Excessive use of non-renewable resources and energy-heavy processes adds to the damage done to the environment.
Framework for Evaluating Environmental Impact
Evaluating the effect on the environment is commonly done with a Life Cycle Assessment (LCA), which looks at what is used and released at every stage in a product's life. There are various steps in this framework: The lifecycle includes getting raw materials, manufacturing, and construction, getting the equipment to where it needs to be installed, using it, maintaining it, and finally, dealing with it at the end of its life cycle. It helps to identify and explain the main sources of environmental problems, so engineers can choose damping systems more wisely.
Regulatory Norms and Green Certifications
Environmental rating systems such as LEED, BREEAM, and IGBC include damping systems as part of their evaluation process. Even though VCS do not have a rating, their role in structural energy savings, lasting quality, and recycling counts towards the building's overall sustainability. If you reduce the number of building moves, it can result in less upkeep and a lower energy bill, indirectly helping you achieve green certification.

15.3 Determining the Environmental Effects of Damping Systems

Material Composition Analysis
Materials commonly found in vibration control systems are steel, copper, and lead. There are also polymers such as viscoelastic material and smart fluids like magnetorheological fluids and electrorheological fluids. Different materials have significant differences in their negative impact on the environment. A good example is that steel can be recycled and is available in large quantities, while MR fluids usually depend on rare-earth elements that are difficult to obtain in an environmentally safe way. Dealing

with the toxicity of certain damping fluids can make safe disposal difficult at times.

Manufacturing and Processing Energy Requirements

Damping systems often require a large amount of energy to produce. For example, smart dampers are made using layers of coatings, electronic parts, and sensors. Using forging at high temperatures and advanced precision machining results in more embodied energy. The process of making an MR damper involves constructing ferromagnetic particles and mixing them with rare carrier fluids, both of which are difficult and require many resources. The pros of using these practices must be measured against the negatives they cause for the environment.

Transportation and Installation

Logistics has an important part in how the environment is assessed. Trucking damping products from factories to construction areas produces CO_2 according to the length of the journey and the way the equipment is transported. In addition, working with cranes and experienced crews takes up more time, increases the site's power usage, and adds to its carbon footprint.

Operational Efficiency and Energy Consumption

The operational energy demands are clearly different for active, semi-active, and passive damping systems. TMDs and viscoelastic dampers are passive, so they do not need to take in energy while they operate. Alternatively, some semi-active systems (for example, MRDs and electro-hydraulic devices) constantly or periodically use up energy during their lives. Additional electronics, cooling, and monitoring equipment further add to the amount of carbon being released.

Disposal and Recyclability

End-of-life planning includes how the equipment can be taken apart, reused, or recycled. While metals in products can be well recycled, the same is not true for smart fluids, polymers, and electronic components. There is often no modular design in damping systems, which makes it complicated to replace only the damaged parts. This results in producing more waste and adds difficulties to recycling products.

15.4 Sustainable Practices in Damping Technology

Sticking to Recycled and Environmentally Safe Materials

Using recycled metals and bio-based polymers is one of the best paths to creating environmentally sustainable dampers. One example is taking used tires and recycling the rubber so that it can be used in green damping systems. This, in turn, also helps to reduce the amount of industrial waste created.

Green Manufacturing Techniques

Some factories have gone solar to produce energy, use additive manufacturing, and reduce how much pollution their production creates. The process deposits materials very accurately, so very little material is wasted. Additionally, by using these techniques, components can be tailored to fit certain jobs, further improving efficiency.

Passive Energy Systems

TMDs and base isolators are types of passive vibration control systems that need no additional energy from outside during their operation. Being inherently sustainable, they can easily become part of green infrastructure systems. Their high effectiveness in terms of performance and the environment is especially useful for making large bridges and tall buildings.

Modular and Replaceable Components

Having systems with interchangeable parts reduces the amount of waste that happens during maintenance. Swapping out components like springs or actuators within dampers reduces what has to be replaced and also benefits the environment.

Long-Life Cycle and Low Maintenance

Using good design to cut back on system maintenance and extend its lifespan lowers its overall effect on the environment. Such components do not rot quickly, so they cut back on both the waste of materials and the chance of maintenance causing trouble.

A multi-phase approach to Life Cycle Assessment (LCA) was followed in writing this chapter and is consistent with ISO 14040 guidelines. The key steps included:

System Boundary Definition

Functional unit: A fully equipped damper for a 30-story tall building.

Lifecycle phases: Extracting resources from raw materials, creating products, getting the products to their destination, setting them up, operating them, maintaining them, and finally disposing of them.

Geographic context: India is now cited among the leading nations when it comes to smart systems.

Data Collection

Primary data: Test results from simulated damping experiments set by the manufacturers.

Secondary data: The use of EPDs, carrying out literature surveys, and looking in peer-reviewed LCA databases like Gabi and EcoInvent.

a. Environmental Impact Categories Analyzed
b. Global Warming Potential (GWP, in units of kg CO_2-eq)
c. The total energy required (CED, measured in MJ)
d. Recyclability Index (RI, in %)
e. Toxicity potential, also known as TP (in kg 1,4-DB eq)
f. Water Use (WU, in liters)

Damping Systems Evaluated

a. Passive dampers are known as Tuned Mass Dampers (TMDs)
b. Viscous Fluid Dampers (VFDs) are considered Passive
c. Magnetorheological Dampers (MRDs) – Semi-active
d. Friction Dampers (FDs) – Passive
e. Electrorheological Dampers (ERDs) – Active

Summary of damping system and their comparison is given in Table 15.1 and Table 15.2. We found that the environmental effect of vibration control systems can vary a lot based on their design. Key findings include:

a. Tuned Mass Dampers (TMDs) appear to be the most environmentally friendly, with low Global Warming Potential (450 kg CO_2-eq) and a high recyclability index (90%). Not having operational energy needs and being built simply make them perfect for green buildings.
b. Their impact on the environment was moderate due to the artificial oil they use and the metal found in their casings.

Table 15.1 Life cycle impact summary for different vibration control systems.

Type of damper	GWP (kg CO_2-eq)	CED (MJ)	RI (%)	TP (kg 1,4-DB eq)	WU (liters)
TMD	450	1200	90	0.08	70
VFD	620	1850	85	0.11	95
MRD	1400	3900	40	0.30	210
FD	510	1320	88	0.09	85
ERD	1650	4500	35	0.37	240

Table 15.2 Comparative analysis of sustainable features.

Criteria	TMD	VFD	MRD	FD	ERD
No external energy required	✓	✓	✕	✓	✕
High recyclability	✓	✓	✕	✓	✕
Modular design	✓	✕	✓	✕	✓
Bio-based material potential	✓	✓	✕	✕	✕
Manufacturing complexity	Low	Medium	High	Low	High
Toxicity/Disposal concern	Low	Low	High	Low	High

Yet, switching to bio-based oils could help reduce the industry's effects on the environment.

c. Since they use strong magnetic fluids and power electronics, carbon emissions and energy consumption by both Magnetorheological Dampers (MRDs) and Electrorheological Dampers (ERDs) were undesirable. They took up 3900–4500 MJ of energy when used for the entire timeline of the experiment.
d. With FDs, the drawbacks were minor in energy and could be recycled, but their seismic performance did not measure up to smart systems.
e. Since passive response systems rely on natural energy, the impacts were minimal, while power for control systems made damper impacts a lot more significant.
f. Most recycling chances were seen in hybrid dampers, but least in dampers made from composites or fluids. Disposing of some smart fluids with care is important since it protects the environment.

15.5 Conclusion

It has been found that different damping systems play a major role in shaping how well high-rise buildings perform environmentally. As passive dampers, TMDs and FDs produce fewer emissions during their lives, can be recycled more easily, and need less energy, making them stand out as the best environmental choice for constructing eco-friendly structures.

Although MRDs and ERDs can quickly adapt to building needs, their usage requires more energy and materials, as well as fewer recyclable parts at the end of their life. There are several proposals for making damping technologies more environmentally friendly. The use of recycled or bio-based materials, the creation of non-toxic smart fluids, green manufacturing, and modular design for ease in upkeep and recycling. Researchers should incorporate sustainability targets into the engineering and improvement of VCS, and policymakers should evaluate putting environmental certification in place for structural control systems.

References

1. Mulder, K. (Ed.), *Sustainable development for engineers: A handbook and resource guide*, Routledge, London, UK, 2017.
2. KC, S. and Gautam, D., Progress in sustainable structural engineering: a review. *Innovative Infrastruct. Solutions*, 6, 2, 68, 2021.
3. Rahimi, F., Aghayari, R., Samali, B., Application of tuned mass dampers for structural vibration control: a state-of-the-art review. *Civ. Eng.*, 6, 8, 1622–1651, 2020.
4. Balaji, P.S. and SelvaKumar, K.K., Applications of nonlinearity in passive vibration control: a review. *J. Vib. Eng. Technol.*, 9, 183–213, 2021.
5. Zhang, C., Mousavi, A.A., Masri, S.F., Gholipour, G., Yan, K., Li, X., Vibration feature extraction using signal processing techniques for structural health monitoring: A review. *Mech. Syst. Signal Process.*, 177, 109175, 2022.
6. Gorgin, R., Luo, Y., Wu, Z., Environmental and operational conditions effects on Lamb wave based structural health monitoring systems: A review. *Ultrasonics*, 105, 106114, 2020.

16

Environmental Impact of Vibration Control Systems

Abstract

Since structural engineering aims to become more environmentally responsible, it becomes necessary to check how vibration control systems affect the environment. This part of the book reviews the way various damping systems for high rises use materials, energy, affect the environment, and their whole life cycle. It covers eco-friendly methods, including the use of environmentally-friendly materials, modular designs, and systems for using renewable energy. Several case studies show how environmental strategies have been put into action. Future topics include looking at nature-like architectural solutions, linking cities with technology, and changing regulations. This chapter is designed to help civil and structural engineers find environmentally friendly damping solutions for modern sustainable development.

***Keywords*:** Environmental effect, vibration managing systems, sustainable engineering, damping systems, life cycle analysis, embodied carbon, green materials

16.1 Introduction

Since their invention, vibration control systems have been needed to protect and lengthen the life of buildings and other structures subjected to shocks, air motions, and vibrations [1]. Prior to now, there was not always an emphasis on how the environment is affected by these systems. As issues related to climate change, the supply of resources, and sustainability continue to raise concerns, engineers and designers are being asked to rethink how they design damping devices. Assessing the environmental impact of damping systems, from beginning to end, is right for advancing environmentally friendly engineering [2]. The environmental challenges are studied here by first investigating how current systems affect the environment, then focusing on eco-friendly designs and their successful cases, and finally forecasting expected developments and changes in policies [3].

Kundan Meshram and Prashant Lahre. Vibration Control Systems, (227–234) © 2026 Scrivener Publishing LLC

In the last few years, problems related to human-caused vibrations affecting modern floors have become more common. It is especially because of smaller and open-plan designs, which have become available due to progress in engineering. Even though Active Vibration Control (AVC) would aid in meeting vibration serviceability standards, it has only been used with limited success in construction. This paper reviews the present approaches that AVC uses to manage vibrations. It looks into current design principles, points out that there are no clear directions for adding AVC to new constructions, and reviews the effectiveness of different approaches for controlling noise in buildings. It also examines the prospects for both the environment and the economy if AVC is applied at the beginning of a building's construction. The aim of the review is to assist future studies focused on bringing out the best in AVC for structural purposes [4]. The disturbance caused by urban noise and vibrations, mainly which come from rail systems, is getting more crucial because they influence comfort as well as sustainability. In this review, the article looks into wheel-rail vibrations and noise, approaches to assess them, and known ways to address them. It analyzes different types of control at the vehicle, track, path, and receiver levels and judges their efficiency. The paper points out possible areas for improving these climate mitigation programs [5]. The study is focused on a vibration control system that involves ABH and piezoelectric actuators in beams. Passive control in ABH occurs by trapping energy through a tapered device, but in active control, piezoelectric actuators assisted by a linear quadratic regulator are applied. When checking a structure's vibration properties, finite element analysis, modal analysis, and Fourier transformation are applied. Free vibration experiments on a 2D cantilever beam demonstrate that combining the approaches is viable [6]. After J.T.P. Yao introduced it to civil engineering in 1972, active control technology grew and was used extensively in various buildings globally. There are still issues to deal with, especially when it comes to using tuned mass dampers (TMDs) for flutter control in large bridges, as their efficacy is still being discussed. The author suggests using the Active Rotary Inertia Driver (ARID), a patented device designed to help control rotational movements in bridges, buildings, and similar structures. This paper explains the ARID system's main theory, models, and how it is evaluated, as well as outlines its possible applications to vehicles, ships, and structures in aerospace [7].

16.2 Checking How Damper Systems Affect the Environment

The environmental impact of dampers is first looked at by doing a detailed life cycle analysis. This involves keeping an eye on how much of an impact each step of making, using, and getting rid of a damping device has on the environment, from getting the materials all the way to what happens when it's thrown away. For instance, viscoelastic dampers usually use man-made materials known as synthetic polymers, and making these takes a lot of energy and oil-derived products. Similarly, hydraulic dampers use oil-based fluids that can sometimes leak, so there are strict rules in industries to make sure these aren't a danger to the environment.

Embodied carbon and energy are important things to think about when comparing different shock absorbers or dampers. Passive devices usually don't need a lot of energy to work, but some of the building materials might take more energy to make, especially if they're made with heavy materials like steel. In contrast, active systems usually need a steady source of energy to work, so using them can cause pollution if they don't use renewable energy along with it. Quantitative comparisons of embodied carbon help people pick what materials and systems they should use in a building. Resource utilization, especially how well we use up our raw materials, also plays a role in how much we affect the environment. Many dampers rely on rare metals or ones that we can't make easily, and these metals are often hard to get out of the ground. Engineers must think about whether these materials can be recycled easily and where they are coming from. Furthermore, environmental risk assessments look for possible dangers in a building, like leaks from fluids, things getting too noisy, or even problems with electricity, especially for more complicated or electrical-mechanical dampers. Compliance with local and global environmental rules becomes very important when trying to avoid these problems.

16.3 Sustainable Practices in Damping Technology

Sustainable damping starts when we use materials that are eco-friendly as given in Table 16.1. Engineers are trying out new materials like natural rubber, bamboo, and bioplastics to wean away from using materials made from oil. Recycled metals and non-toxic hydraulic fluids are now being used in newer designs to help close the loop and make waste materials useful again. Modular and reusable damping systems also help the environment by making good

Table 16.1 Sustainable materials used in damping technologies.

Material type	Source type	Environmental benefit	Application example
Natural Rubber	Renewable (plant)	Biodegradable, low carbon	Base isolators in buildings
Recycled Steel	Recycled	Energy savings in production	Tuned mass dampers, braces
Bioplastics	Bio-based (corn)	Lower emissions than synthetics	Lightweight housing for devices
Bamboo Composite	Renewable (plant)	Fast-growing, strong, low energy	Housing, modular TMD shells

use of what's already in use and reducing any new materials that need to be made. By making products easy to take apart and update, manufacturers can help lead to longer product lives and cut down on the waste from building new things. These principles are especially helpful when it comes to road projects in cities, where built-in upgrades happen more often.

Energy-efficient manufacturing also helps control how wet a place gets. Techniques like 3D printing and making things closer to where they will be used help cut down on how much energy and fuel are needed to move and store supplies. Automated low-energy assembly lines and better-organized logistics help make these systems a bit friendlier to the environment. The integration of renewable energy with damping systems allows

Table 16.2 Life cycle assessment (LCA) comparison of damping systems.

Damping type	Raw material impact	Manufacturing energy (MJ/kg)	Operational energy	Recyclability (%)	End-of-life environmental risk
Viscoelastic Damper	High (synthetic)	85	Low	Moderate (60%)	Low (non-toxic disposal)
Hydraulic Damper	Moderate	120	Low	High (steel 90%)	High (fluid leakage risk)
Tuned Mass Damper (TMD)	High (heavy steel)	150	None (passive)	High (steel 90%)	Low
Active Mass Damper	Moderate	200	High	Low (complex parts)	Medium (electronic waste)

for more ways of using clean and renewable sources to power buildings and help the environment. Piezoelectric materials placed in damping parts can use the vibrations in a structure to create small amounts of electricity that can be used to make sensors or help keep watch over how things are working. Solar-powered active dampers and hybrid solutions cut down on the amount of electric energy used from the grid, so they work well in places where power supply is hard to get or where using electricity frugally is important. Additionally, choosing building materials that do not rust as easily and using coatings that repair themselves on their own make it cheaper and better for the environment to fix and replace things over time. These long-life designs help reduce the amount of pollution that's produced and keep problems at a minimum during the whole life of the power plant. The life cycle assesment comparison of damping systems is given in Table 16.2.

16.4 Case Studies in Eco-Friendly Design

Green architecture and sustainable infrastructure projects around the world are now using ways to help reduce vibration that are better for the environment and their vibration control implementations is given in Table 16.3.

Table 16.3 Case study summary – eco-friendly vibration control implementations.

Project name	Damping system used	Eco-friendly features	Sustainability certifications
Taipei 101	Tuned Mass Damper	Designed for durability, minimal maintenance	LEED Silver
Tesla Gigafactory	Custom Smart Dampers	Recycled metals, integrated monitoring	LEED Gold
Golden Gate Bridge Retrofit	Seismic Dampers	Corrosion-resistant, low-maintenance materials	N/A
Green Metro Rail (Delhi)	Base Isolators	Rubber-based, locally manufactured	GRIHA Certified

For instance, Taipei 101 has a big tuned mass damper, which was built to last for many years and needs little upkeep. The design helps keep the building secure and also makes it possible for it to get LEED certification. In industrial settings, damping systems are now being used in factories that have eco-certifications. Tesla's Gigafactory, for example, uses damping methods that include low-carbon steel and smart monitoring tools to help fit with their focus on being more sustainable. Renewable energy plants like solar and hydroelectric use devices to keep operating smoothly and safely so that they can handle different weather or system issues while still trying to be eco-friendly.

Public infrastructure projects, like building bridges that use eco-friendly materials and creating more eco-friendly subways, give a good example of how damping technology can be used in a more environmentally-friendly way. Hospitals and schools in places that often get earthquakes are starting to put in dampers that meet safety requirements and also help them get recognized as green buildings. Comparative studies of old-style and environmentally-friendly damping systems show clear and noticeable differences. Case studies show that, even though it can be more expensive at first, using eco-friendly options can save money in the long run because they don't need as much regular maintenance and use up less energy. Such studies show that using sustainable dampers is good for the environment and economy in the long run.

16.5 Future Considerations and Trends

The future of environmentally friendly damping systems is likely to depend on coming up with new ideas and finding ways to combine different technologies together. Bio-inspired materials like spider silk copies and mussel-protein glue are being investigated because they are really strong and can break down naturally. These materials might help us create damping solutions that are really good at being quiet and safe, and they could also be better for the environment. Smart cities and the Internet of Things make it easier to find new ways to manage building vibrations while caring about the environment. With real-time monitoring and smart data analysis, damping systems can change how they come into action to handle changes in loads, and they can also help improve the way energy is used. This convergence helps people come up with better ways to manage and protect the environment in cities. Regulatory frameworks are also evolving. Green building standards like LEED, BREEAM, and GRIHA are now starting to include rules about how to control noise and vibrations. Anticipated updates to ISO and ASTM standards could mean that manufacturers would have to show information about their environmental impact and how their products are made, encouraging them to be more open and try out new eco-friendly ways of doing things.

Climate change adaptation is another important reason for the growth of the population. Structures now have to be built to handle more common and extreme weather, like storms and floods. Environmentally sound damping systems not only help protect buildings from damage, but they also help support wider plans to deal with climate change. Future research might look into systems that work for things like heat, sound, and shaking all in one, instead of trying to handle each problem separately. Finally, changes in schools and in how industries work will be really important. Civil engineers need to learn how to design projects in a way that considers the environment, and working together with material scientists, environmental engineers, and people who make laws can help us make more well-rounded progress. Government encouragement and partnering with groups from both the government and businesses will help speed up the use of green damping technology.

16.6 Conclusion

1. **Environmental Assessment is Essential** – A full life cycle assessment (LCA) is necessary to understand the environmental impact of damping systems—from material sourcing to disposal.
2. **Traditional Systems Carry Environmental Costs** – Conventional damping technologies often rely on high-energy, non-renewable materials and may pose risks such as fluid leaks or high embodied carbon.
3. **Sustainable Materials and Designs are Emerging** – Eco-friendly alternatives like recyclable metals, bio-based polymers, and modular designs are being increasingly used to reduce environmental footprint.
4. **Integration with Renewable Energy is Advancing** – Technologies like piezoelectric energy harvesting and solar-powered active dampers are promoting self-sufficient and green damping systems.
5. **Low-Maintenance and Long-Life Systems Offer Benefits** – Using corrosion-resistant and self-healing materials minimizes maintenance, reducing environmental cost over the system's lifetime.

6. **Real-World Case Studies Demonstrate Success** – Projects like Taipei 101 and Tesla's Gigafactory show how environmentally responsible damping can be implemented without compromising performance.
7. **Eco-Friendly Systems Have Long-Term Gains** – Though often higher in initial cost, green damping systems deliver environmental and economic returns through lower operational energy and maintenance.
8. **Future Trends are Promising** – Innovations such as bio-inspired materials and IoT-enabled adaptive damping align with smart city and climate-resilient infrastructure goals.
9. **Regulations and Certifications are Evolving** – Green building standards (e.g., LEED, GRIHA) are starting to include vibration control in their sustainability criteria, driving industry-wide change.
10. **Collaboration and Education are Key** – Interdisciplinary cooperation between engineers, environmental scientists, and policymakers, along with sustainability-focused education, is vital for progress.

References

1. Preumont, A., *Vibration control of active structures: an introduction*, vol. 246, Springer, Dordrecht, Netherlands, 2018.
2. Saaed, T.E., Nikolakopoulos, G., Jonasson, J.-E., Hedlund, H., A state-of-the-art review of structural control systems. *J. Vib. Control*, 21, 5, 919–937, 2015.
3. Bachmann, H., *Vibration problems in structures: practical guidelines*, Springer Science & Business Media, Basel, Switzerland, 1995.
4. Hudson, M.J. and Reynolds, P., Implementation considerations for active vibration control in the design of floor structures. *Eng. Struct.*, 44, 334–358, 2012.
5. Ouakka, S., Verlinden, O., Kouroussis, G., Railway ground vibration and mitigation measures: benchmarking of best practices. *Railway Eng. Sci.*, 30, 1, 1–22, 2022.
6. Weimann, T., Molter, A., Fernandez, L., He, M., Structural vibration control based on the effect of acoustic black holes and piezoelectric actuators. *Finite Elem. Anal. Des.*, 224, 103992, 2023.
7. Zhang, C.W., The active rotary inertia driver system for flutter vibration control of bridges and various promising applications. *Sci. China Technol. Sci.*, 66, 2, 390–405, 2023.

17

Integration Strategies for Advanced Damping Systems

Abstract

To ensure the best performance of tall and complex buildings, advanced damping systems should be carefully blended with the other aspects of the design. This chapter looks at how to use different damping technologies together and brings in real-world systems to see if the designs work well. The book talks about new technologies and issues, for example, ensuring compatibility between BIM and different software programs. Through the description of real-life situations, often positive and sometimes negative, the chapter points out what to consider when integrating two systems. This chapter provides engineers and designers with guidance on designing effective ways to control vibration.

Keywords: Advanced damping systems, multi-objective optimization, active dampers, semi-active control, BIM, SHM

17.1 Introduction

Structural engineering has changed from old methods to newer ones that can handle the more difficult and advanced projects we have today. Advanced damping systems help buildings and structures stay strong and safe when they go through things like earthquakes, strong winds, and other types of shaking [1]. However, bringing these smart technologies into a building means more than just installing the devices—it also involves working together with the design team to make sure everything fits with how the building is used and the goals for sustainability. As structures get taller and are designed with more complicated shapes, it gets even more important to make sure they're dampened well and can resist unwanted vibrations. Effective integration of damping systems means making sure they work together with the building itself so they can help control movement and reduce vibrations. This includes checking how much the building should handle, making sure there's a safe way for stresses to go,

Kundan Meshram and Prashant Lahre. Vibration Control Systems, (235–242) © 2026 Scrivener Publishing LLC

picking the best way to make things more flexible, and keeping the building's design in mind. Integration also needs to work on all parts of how something is built, such as keeping up maintenance, making sure it lasts long, and handling any changes due to the environment or use over time. Advanced systems, whether they are passive or fully active, bring their own set of challenges because they rely on sensors, computer programs, and digital messages to work properly [2]. Thus, for integration to work well, both software and hardware have to work together smoothly, so that the system is accurate, works as it should, and fits in with other parts of the system. Engineers now often use a more careful and methodical approach called performance-based engineering (PBE) to check how structures will perform under certain problems, instead of just following the strict rules in the building codes. Tools like multi-objective optimization, simulation, and real-time monitoring help people make better decisions when designing a system. Digital platforms like BIM help to integrate damping technology early by helping to spot problems in space, avoiding potential clashes, and simulating how structures will react in real-world conditions. However, bringing these systems together still isn't easy because things like the way different damping tools work, getting the systems to work together, making software match up, and dealing with cybersecurity issues all cause problems. Additionally, because there aren't standard ways of doing things, different healthcare providers sometimes have different ways of following EHR rules, which can lead to inconsistencies. This chapter looks at some of the main problems when combining AI and Networks and then gives examples and important steps for using them together, using real-world examples and numbers to show how it works [3]. Damage from damping-off reduces the number of healthy plants, leading to important losses in both nurseries and farm fields. The usage of fungicides is common, yet their over-application can be dangerous for health, nature, and new pests. To manage sustainably, we need to know more about pathogens and disease situations, and find non-toxic methods suited to each case. According to the studies reviewed, damping-off can harm nearly every crop, is linked to diseases within the soil, and can lead to losing up to two-thirds of the crop's seedlings. Even as biocontrol advances, using IPM that includes both control and prevention is very important. It draws attention to several main areas of unknowns and advises that future studies on seeds, seedbeds, microbial actions, and experimental approaches should be investigated [4]. The paper introduces a damping method that mostly relies on a small part of the building's mass to help damp structure movements. As a result, these devices have a greater level of mass and provide a lot more damping.

In addition to springs, fluid viscous dampers manage the acceleration of the mass in the suspension system. Using a two-degree-of-freedom case, the reasons behind using more mass to get better damping and more constant performance are initially examined. Then, the system being proposed, named "integrated damping," is explained and backed up with an example of a 250-m tall building. The proposed damping system is evaluated using a range of numerical experiments and a small set of actual earthquake events. According to the analysis for the chosen case, applying the recommended setup helps reduce the wind loads and vibrations by over 40% and 60%, respectively, when two sources of excitation are considered [5]. People are paying more attention to MR dampers because they can efficiently manage the energy produced by vibrations. This review looks closely at how MR damping systems work, talks about the math models used to describe them, covers different ways to control them, and gives examples of how they're used in real-life situations. It sorts out control algorithms into classical, advanced, and intelligent types, pointing out that how these strategies are used can make a big difference in how well a system works. It points out that by combining numerous smart algorithms, artificial intelligence, and machine learning, good solutions could be found for complex vibration situations with MR dampers [6].

17.2 Systematic Design Methodologies

Engineers use performance-based engineering (PBE) and other established techniques at the beginning of the systematic integration process. With PBE, engineers can match dampers to the specific goals required for a project at various danger levels. Instead of only following strict designs, engineers look at different objectives such as the safety of the structure, the comfort of occupants, the function of the building, and its impact on the environment, and measure them against possible outcomes. In addition, multi-objective optimization allows engineers to assess the criteria of cost, how much material is used, and how the structure responds to changes, all at once. Genetic algorithms, simulated annealing, and particle swarm optimization are tools that can help manage balancing issues in difficult structures. They ensure that a building is well-structured as well as affordable and eco-friendly. More attention is now given to simulation-based design. By using FEM and computational dynamics simulations, you can test the performance of the damping system under real-life loads without actual testing. Along with running parametric studies, the simulations support confident decision-making at the start of the design work, so fewer prototypes and updates are needed later.

17.3 Coordination of Damping Technologies

It can be very challenging to bring together several damping systems, for example, TMDs, viscous dampers, and base isolators, in a single building. Different types of devices operate in their own way and usually require different control setups, amounts of space, and ways to keep them maintained. Ensuring that software and hardware are compatible is a main issue. Often, advanced damping systems use real-time control algorithms, along with sensors and actuators. It is important that these systems are in sync with the building management system (BMS) or structural health monitoring (SHM) system. BIM is seen as a valuable alternative for these problems. With BIM, it is possible to avoid device conflicts and arrange devices in the best possible way, relying on live data. Cybersecurity is very important, especially with damping systems that are powered wirelessly or through the Internet of Things. Securing data communication, managing who can access the system, and blocking interference all need to be considered as part of the integration plan.

17.4 Synergistic Combinations and Performance Metrics

Optimal damping integration often means using more than one technology together to get better results. For example, putting together TMDs with active mass dampers (AMDs) can help a turbofan jet engine respond better to noise sounds over a wider range of frequencies. Similarly, using semi-active devices like magnetorheological dampers with regular passive base isolators helps make structures more flexible and better able to handle shaking and absorbing energy in earthquakes.

a. To see how well things are working when different systems are combined, people use a few different measures to check how well the integration is going.
b. Reduction in inter-story drift: Critical for keeping buildings and other structures from being damaged too much when there is an earthquake.
c. Peak acceleration reduction: Important for making sure people feel comfortable in tall buildings when wind or shaking happens.

d. Energy dissipation ratio: Indicates how much energy from the ground gets picked up by the shocks on the car and turned into heat or friction.
e. System robustness and redundancy: Checks how well the system reacts when a device stops working, or when more work is suddenly put on it.

Data-driven methods like checking feedback right away and looking at how things have gone in the past help you see the numbers and other specifics in an objective way. Machine learning models are also being tested to help predict when things might break and help them run better.

17.5 Case Studies and Challenges

Successful Implementations

a. Taipei 101, Taiwan: Uses a 660-ton TMD connected to a system that checks the health of the bridge structure. The integration shows that the building team took time to plan out the design and engineering work, which made the building strong enough to withstand earthquakes and comfortable for the people using it.
b. Salesforce Tower, San Francisco: Incorporates different ways to reduce noise in a PBE calculation. Coordination between architectural design and mechanical systems made it easy to install and keep the systems working properly.

Unsuccessful or Problematic Cases

a. L'Aquila Civic Tower, Italy: Post-earthquake assessments showed that the dampers weren't working as well as expected because they were put in the wrong spots and not fine-tuned enough. A lack of doing simulations beforehand made it hard to make sure everything worked well when the different parts were put together.
b. Mid-rise commercial building (confidential case): Featured systems where different types of dampers didn't work well together during the earthquake, and that caused some damage to the building. Root cause analysis showed that the problem was caused by some pieces of software in the system that weren't compatible with each other.

These cases show that it's really important to plan things out early, test them thoroughly, and make sure all teams talk to each other. Failures often happen because people don't fully understand how the systems work, fail to make sure systems can work with others, and forget to check how well the systems are performing.

17.6 Conclusion

1. Damping systems should always be designed with attention to how they perform and by using multi-objective optimization.
2. FEM, genetic algorithms, and designs based on simulation are useful for looking ahead and guaranteeing the good performance of a structure.
3. Making sure that smart structures work seamlessly between hardware and software should be a top concern.
4. BIM makes it easier to arrange space, find clashes, and work together as a team during both design and build stages.
5. Connecting structures to Structural Health Monitoring (SHM) and Building Management Systems (BMS) improves how things are run and helps with data-based servicing.
6. Merging different types of damping can expand the frequency range the building can respond to and improve how much energy is absorbed.
7. Some of the performance metrics used for optimization include how much stories move away from their baseline, maximum peak acceleration, and the amount of energy used.
8. Because damping devices are now smart and IoT-based, issues like cybersecurity, authorization of control, and the integrity of data have grown to be key factors in developing integration plans.
9. Redundancy and robustness of a system matter as much as each other to prevent the worst problems.
10. Taipei 101 and Salesforce Tower show how being organized and keeping an eye on performance during the project leads to the best results.
11. When Buildings like the L'Aquila Tower or the confidential mid-rise have collapsed, it becomes clear why choosing the wrong place, lack of engineering, or unfit systems is dangerous.

References

1. Kumar, R. and Bhende, C.N., Active damping stabilization techniques for cascaded systems in DC microgrids: A comprehensive review. *Energies*, 16, 3, 1339, 2023.
2. Poddaeva, O. and Fedosova, A., Damping capacity of materials and its effect on the dynamic behavior of structures. Review. *Energy Rep.*, 7, 299–307, 2021.
3. Yang, A., Han, M., Zeng, Q., Sun, Y., Adopting building information modeling (BIM) for the development of smart buildings: a review of enabling applications and challenges. *Adv. Civ. Eng.*, 2021, 1, 8811476, 2021.
4. Lamichhane, J.R., Dürr, C., Schwanck, A.A., Robin, M-H., Sarthou, J-P., Cellier, V., Messéan, A., Aubertot, J-N., Integrated management of damping-off diseases. A review. *Agron. Sustainable Dev.*, 37, 1–25, 2017.
5. Martinez-Paneda, M. and Elghazouli, A.Y., An integrated damping system for tall buildings. *Struct. Des. Tall Special Build.*, 29, 7, e1724, 2020.
6. Lv, H., Zhang, S., Sun, Q., Chen, R., Zhang, W.J., The dynamic models, control strategies and applications for magnetorheological damping systems: a systematic review. *J. Vib. Eng. Technol.*, 9, 131–147, 2021.

18

System-Level Optimization of Damping Solutions

Abstract

Optimizing the damping at the system level increases the safety, security, and performance of tall and infrastructure buildings under changes in environmental loads. Most traditional damping methods concentrate only on parts of the body, not on the body as a whole. The chapter investigates several sophisticated optimization approaches, both stochastic and exact, including Genetic Algorithms (GA), Particle Swarm Optimization (PSO), and Artificial Neural Networks (ANN), to attain the best overall damping setup. In addition, it assesses the performance of optimization programs such as MATLAB, OpenSees, and ANSYS in carrying out simulations. The chapter discusses the gains that can be realized by using optimal damping in a structural system subject to dynamic loading. Additionally, it reviews how smart optimization systems, edge computing, and adaptive control techniques are guiding the progress of structural engineering.

***Keywords*:** Damping system, particle swarm optimization (PSO), neural network (NN), genetic algorithms

18.1 Introduction

With architecture and building structures getting more complicated and being asked to perform better, controlling vibration is crucial. Existing traditional damping methods for tall buildings, bridges, and roads affected by wind, traffic, or earthquakes are usually performed at the component level [1]. Most earthquake engineering strategies focus on improving individual parts such as TMDs, viscous dampers and base isolators, and overlook the relationships among these and the building structure's response to external loads. When designing fully-fledged systems, using this method may not deliver the desired or best results [2]. Because of this, a need has developed for organizing and optimizing damping components at all levels of the structure. At the system level, all things that impact damping are considered in

Kundan Meshram and Prashant Lahre. Vibration Control Systems, (243–252) © 2026 Scrivener Publishing LLC

the structure as a whole. System-level optimization treats all kinds of vibration and tries to spread damping evenly, so that the building structure supports the best resistance to damage from events, controls while minimizing story drifts, lowers building speed, and improves comfort for occupants. A comprehensive approach supports improved use of materials, reduced overuse of resources and greater energy performance under diverse and extreme conditions [3]. The invention of new computational methods now makes it easy to apply system-level optimization in practice. Even though gradient-based optimization and linear programming were popular with R&D teams, their absence of methods for nonlinear functions, multi-goals or big solution sets made them less effective at damping real-world issues [4]. Consequently, heuristic and metaheuristic algorithms, including Genetic Algorithms (GA), Particle Swarm Optimization (PSO), Simulated Annealing (SA), and Ant Colony Optimization (ACO) were invented to use principles found in nature, including those used by swarms or in evolutionary processes. Such algorithms have successfully handled difficult, multi-variable damping challenges that include nonlinear responses and various values for damper locations and tuning. Furthermore, artificial intelligence (AI) and machine learning (ML) methods—particularly Artificial Neural Networks (ANN), fuzzy logic, and hybrid AI-algorithmic models—have opened new frontiers in predictive optimization, enabling structural systems to adapt and evolve their damping strategies based on real-time or historical data [5]. System-level optimization in structural engineering faces the complex task of integrating tools that can model structural dynamics, damper behavior, and optimization processes simultaneously [6, 7]. Sophisticated programs such as MATLAB, OpenSees, ANSYS, and ETABS/SAP2000 offer state-of-the-art performance analysis and verification of various damping schemes. This chapter presents case studies using optimization techniques such as Genetic Algorithms (GA), Particle Swarm Optimization (PSO), and Artificial Neural Networks (ANN), showing improvements in structural performance metrics like displacement and energy dissipation. These damping systems also improve both economic and environmental performance. Future advancements will likely be driven by AI, smart sensors, and decentralized computing, transforming dampers into adaptive systems that respond intelligently to real-time conditions [7]. This study focuses on optimizing the placement and control parameters of Battery Energy Storage Systems (BESSs) to enhance power system oscillation damping. It models each BESS by incorporating power limits, State of Charge (SoC) constraints, and response dynamics. A blackbox mixed-integer optimization framework, integrated with time-domain simulations and a Particle Swarm Optimization (PSO) algorithm, is used to solve the problem. Results are verified against simulation responses on both the New England 39-bus and Nordic systems. Additionally, seasonal load variations and minimal BESS deployment are considered to improve practical implementation, and the controller's effectiveness is benchmarked against other strategies [8]. This paper presents an efficiency-driven control strategy for parallel DC-DC converters using virtual resistance (VR) droop control. Conventional droop control equally distributes power flows but could neglect the goal of improved system efficiency. A tertiary control layer is proposed that uses a genetic algorithm to adapt VR settings and achieve the most efficient power sharing for each converter. A second control layer ensures steady output voltage and limits voltage oscillations when virtual resistance values are adjusted dynamically. Hardware-in-the-loop simulations confirm that this approach enhances system efficiency while preserving stable and responsive performance [9]. A new optimization technique is proposed to solve the performance issues related to initial responses in large uncertainties of structures encountering random wind loads. It efficiently optimizes structural layouts by sequentially tackling sub-problems, drawing from the results of prior simulations. The method's effectiveness and scalability are demonstrated through two case studies involving the optimal design of uncertain systems subjected to wind excitation [10]. This study addresses the seismic vulnerability of cable-stayed bridges by proposing a performance-based seismic design (PBSD) approach for optimizing damper devices at deck-pylon connections. A bridge model was set up in OpenSees based on the data from previous finite element simulations. The model incorporated viscous and metallic dampers in its principal load paths. Fragility functions were derived using multiple stripe analysis, and system-level repair cost ratio (RCR) surfaces were developed. A genetic algorithm running on multiple processors was used to determine the most effective dampers settings. We found that finding the correct damper settings can lead to significant repair cost savings, but choosing them incorrectly may have a negative effect on the structure's performance. The framework offers practical guidance for the design and retrofit of cable-stayed bridges in seismic regions [11].

18.2 Techniques for System-Level Optimization

A holistic strategy is used in system-level optimization to integrate damping features throughout the entire system rather than treating each element separately. Typically, conventional damping design neglects the effect dampers have on the overall structure. This technique neglects how the various structural parts interact mechanically during dynamic loading. System-level optimization promotes a collective response of the entire structure in the face of dynamic loads.

The fundamental approach involves using a robust numerical model. A detailed numerical model is constructed to represent the structure's material properties, geometry, boundary conditions, and damping devices. Advanced simulation tools like Finite Element Method (FEM) and Multi-Degree of Freedom (MDOF) systems are used to understand how the damping devices influence the structure's global dynamic response. The goals of the optimization process focus on minimizing these factors: peak displacement, acceleration, inter-story drift, and energy dissipation.

A major challenge is the need to ensure that multiple objectives, including cost savings, material conservation, and optimal vibration control, are achieved at the same time. Several Pareto front solutions are implemented to balance different objectives simultaneously. Various simulations are carried out under various load cases to assess the effectiveness and reliability of the designed damping system.

18.3 Methods and Algorithms Used to Optimize Damping Systems

The application of intelligent algorithms plays a crucial role in achieving optimal design at a system level. They play a crucial role in the development of damping systems for high-rise buildings that experience complex dynamic interaction and involve various degrees of freedom. The most significant algorithms include:

a. *Genetic Algorithms (GA):* They model natural selection by performing crossover and mutation operations to improve the positioning and properties of building dampers. GA performs well when searching broad design spaces for the best possible outcomes.
b. *Particle Swarm Optimization (PSO):* Modeling the foraging strategies of swarms, PSO helps find the best damper locations and control settings in a system. A particle in PSO explores potential damping configurations based on the previous best solutions found.

c. *Artificial Neural Networks (ANNs):* ANNs are employed as substitute models to forecast the best setting of damping properties by learning from sample data. They allow for real-time calculations and facilitate faster iterative optimization.
d. *Gradient-Based Optimization:* The optimal values for damping parameters are obtained by minimizing the derivative of the objective function for each selected parameter. These methods are often employed when making small adjustments to the damping system after completing a more general sweep of damping options.
e. *Topology Optimization:* It calculates the best arrangement of damping elements throughout the structural elements. It becomes very useful when adding damping devices to structural designs at the beginning of a project.
f. *Simulated Annealing (SA):* This method permits making inferior moves in some instances to evade the danger of sub-optimal solutions and improve the scope of search.
g. *Ant Colony Optimization (ACO):* ACO simulates the way ants find the most efficient routes, making it a useful approach for solving discrete optimization problems, including cherry picking the optimal damping devices and their respective positions.
h. *Hybrid Algorithms (GA-ANN, PSO-SA, etc.):* They use different algorithms working together to achieve improved optimization efficiency. For example, GA-ANN combines the broad search ability of genetic algorithms with the fast learning of artificial neural networks. On the other hand, PSO-SA works to strike a balance between quick convergence and efficient exploration.
i. *Machine Learning and Reinforcement Learning:* Techniques are being developed to allow for dynamic control of damper systems. They adjust the parameters automatically according to current measurements to ensure the building stays in equilibrium.

These algorithms allow for the design and control of damping systems that are capable of adapting to varying environmental and loading scenarios, which becomes essential in high-rise buildings as summarised in Table 18.1.

Table 18.1 Showing different algorithms that can combine with the damping system.

Integrated algorithm approach	Description	Advantages for damper systems	Typical applications in high-rise building damping
GA + ANN (Genetic Algorithm + Artificial Neural Network)	GA performs global search for optimal damper parameters; ANN predicts system responses quickly.	Speeds up optimization by reducing computational cost; improves accuracy of predictions.	Real-time tuning of damper parameters under varying dynamic loads.
PSO + SA (Particle Swarm Optimization + Simulated Annealing)	PSO quickly explores search space; SA avoids local optima by probabilistic acceptance of worse solutions.	Balances fast convergence with thorough exploration of design space.	Optimal damper placement and control settings in complex structures.
GA + PSO Hybrid	Combines GA's evolutionary operators with PSO's swarm intelligence for better exploration.	Enhances global search efficiency and solution diversity.	Comprehensive damper system design considering multiple objectives.
ANN + Reinforcement Learning	ANN models system behavior; reinforcement learning updates damper control strategies based on feedback.	Enables adaptive, self-learning damping systems that improve over time.	Smart damping control adapting to real-time structural responses.
Gradient-Based + Topology Optimization	Gradient methods fine-tune parameters; topology optimization determines optimal damping material layout.	Integrates structural design and damping layout for enhanced performance.	Designing dampers integrated into structural elements during early design stages.
ACO + GA	ACO selects damper types/ locations; GA optimizes damper properties.	Combines discrete and continuous optimization effectively.	Selecting and sizing dampers for mixed damping systems in high-rises.
Machine Learning + PSO	Machine learning predicts dynamic responses; PSO optimizes damping parameters using those predictions.	Improves optimization speed and adapts to new conditions.	Real-time parameter tuning for variable environmental loading conditions.

18.4 Importance of Integrated Algorithms

a. Handling complexity: Integrated algorithms enable the investigation of vast possibilities in designing high-rise buildings.
b. Adaptability: Combine prediction models (ANN, machine learning) with optimization (GA, PSO) for adaptive control.
c. Efficiency: Save time and computation resources by applying surrogate models and hybrid techniques.
d. Robustness: Enhance the solution quality by preventing ending up in local optima and by guaranteeing a solution close to the globally best one.
e. Multi-objective balancing: Simultaneously consider cost, performance, and environmental effects in the optimization process.

18.5 Implementation Strategy for High-Rise Buildings

Step 1: Structural Modeling
Construct precise numerical models that describe the entire building along with its material properties, center of mass distribution, and initial internal damping characteristics.

Link the model to sensors and past loads to better estimate building response.

Step 2: Damper Selection
Select the dampers most appropriate for the building. Tuned Mass Dampers (TMD), Semi-Active Dampers, Magneto-Rheological (MR) Dampers, etc.

Ensure that the chosen dampers can be installed within the available space, are easy to maintain and stay within the building's budget.

Step 3: Optimization Setup
Define objective functions (e.g., minimizing peak displacement and acceleration).

Define limits on damper strength, available places and cost considerations.

Step 4: Algorithm Integration
Choose algorithms that suit the scale of your problem and its level of intricacy.

Combine evolutionary algorithms with artificial neural networks to search the problem space more effectively.

Step 5: Simulation and Validation
Simulate performance under various anticipated load conditions.

The parameters and damping configurations are tuned based on the simulation results.

Step 6: Real-Time Control (Optional)
Implement synthetic dampers controlled by a reinforcement learning agent using real-time sensor data.

Set up embedded systems that continuously monitor and adjust the damping system.

18.6 Future Trends

a. Digital Twins: Digital twins that enable ongoing optimization and predictive upkeep measures.
b. Real-Time Adaptive Damping: Incorporating Internet of Things sensors with artificial intelligence to make rapid changes in damping parameters.
c. Sustainability Integration: Accounting for whole-life costs, energy efficiency, and sustainability.
d. Hybrid Material Systems: Using different types of materials together to achieve multiple damping capabilities.
e. Collaborative Optimization Platforms: Collaborative optimization platforms accessible to various participants over the cloud.

18.7 Conclusion

1. Optimizing damping configurations at a system-level plays a critical role in improving the safety, performance, and comfort of high-rise buildings exposed to forces such as wind and seismic loads. System-level damping designs consider the coupling effects between dampers and the entire

structure, which allows for effective reduction of building vibrations over the entire structure.

2. Advanced optimization algorithms, including Genetic Algorithms, Particle Swarm Optimization, Artificial Neural Networks, Simulated Annealing, and hybrid techniques combining these methods, enable expedient and precise testing of a wide range of damping design options. These algorithms address the interdependence of computational speed, solution quality, and environmental resilience when dealing with the design of dampers in large structures.
3. Furthermore, the integration of data-driven methods and machine learning leads to smarter and more agile damping systems that are better suited to adapting to new environments and enhancing resilience for the urban fabric.
4. The procedure of system-level optimization involves multiple steps and close integration. Constructing precise building models, choosing optimal dampers, setting up appropriate optimization goals, and carefully verifying the predictions with simulations. Recent developments are leading to more intelligent, sustainable, and collaborative vibration control solutions that combine the use of digital twins, the Internet of Things, and environmentally-friendly materials.
5. System-level optimization constitutes a critical and visionary endeavor that secures safe, functional, and efficient operations of skyscrapers over the long term, shaping the future of smart vibration control in contemporary architectural practice.

References

1. Freddi, F., Galasso, C., Cremen, G., Dall'Asta, A., Di Sarno, L., Giaralis, A., Gutiérrez-Urzúa, F., *et al.*, Innovations in earthquake risk reduction for resilience: Recent advances and challenges. *Int. J. Disaster Risk Reduct.*, 60, 102267, 2021.
2. Chee Yin, H., Kassem, M.M., Nazri, F.M., Comprehensive review of community seismic resilience: concept, frameworks, and case studies. *Adv. Civ. Eng.*, 2022, 1, 7668214, 2022.
3. Rahimi, F., Aghayari, R., Samali, B., Application of tuned mass dampers for structural vibration control: a state-of-the-art review. *Civ. Eng.*, 6, 8, 1622–1651, 2020.

4. De Domenico, D. and Ricciardi, G., Improving the dynamic performance of base-isolated structures *via* tuned mass damper and inerter devices: A comparative study. *Struct. Control Health Monit.*, 25, 10, e2234, 2018.
5. Hannan, M.A., Islam, N.N., Mohamed, A., Lipu, M.S.H., Ker, P.J., Rashid, M.M., Shareef, H., Artificial intelligent based damping controller optimization for the multi-machine power system: A review. *IEEE Access*, 6, 39574–39594, 2018.
6. Chen, Z., Zhang, L., Li, K., Xue, X., Zhang, X., Kim, B., Li, C.Y., Machine-learning prediction of aerodynamic damping for buildings and structures undergoing flow-induced vibrations. *J. Build. Eng.*, 63, 105374, 2023.
7. Yılmaz, İ., Arslan, E., Kızıltaş, E.Ç., Cavdar, K., Development of a prediction method of Rayleigh damping coefficients for free layer damping coatings through machine learning algorithms. *Int. J. Mech. Sci.*, 166, 105237, 2020.
8. Zhu, Y., Liu, C., Sun, K., Shi, D., Wang, Z., Optimization of battery energy storage to improve power system oscillation damping. *IEEE Trans. Sustain. Energy*, 10, 3, 1015–1024, 2018.
9. Meng, L., Dragicevic, T., Vasquez, J.C., Guerrero, J.M., Tertiary and secondary control levels for efficiency optimization and system damping in droop controlled DC–DC converters. *IEEE Trans. Smart Grid*, 6, 6, 2615–2626, 2015.
10. Suksuwan, A. and Spence, S.M.J., Optimization of uncertain structures subject to stochastic wind loads under system-level first excursion constraints: A data-driven approach. *Comput. Struct.*, 210, 58–68, 2018.
11. Wen, J., Han, Q., Xie, Y., Du, X., Zhang, J., Performance-based seismic design and optimization of damper devices for cable-stayed bridge. *Eng. Struct.*, 237, 112043, 2021.

19

Real-Time Data Processing and Control

Abstract

Recent progress in sensor technology, wireless communication, and embedded computing has led to the development of real-time systems for reducing vibrations in civil engineering structures. Real-time systems enable prompt gathering, interpreting, and adaptation of structures to fluctuating loads from events like earthquakes, strong winds, and human movement. The chapter delves into the fundamentals, methods, and case studies of real-time systems as they are applied to structural vibration control. The discussion focuses on defining essential characteristics of such systems (e.g., latency, sampling frequency, and data resolution), introducing machine learning techniques for on-the-fly intelligent decisions, describing Kalman filters applied to combine information from multiple sensors in challenging surroundings with rapid fluctuations, as well as explaining how these technologies are properly connected to dynamic structural settings using feedback mechanisms. Real applications of these approaches are examined in cities vulnerable to earthquakes and advanced urban environments.

Keywords: Damping technology, earthquake monitoring, Kalman filters, machine learning, intelligent damping system, cyber-physical system

19.1 Introduction

The expanding intricacy of modern structures, more frequent occurrence of natural hazards, and the trend toward adaptive infrastructures have inspired the development of real-time data processing and control solutions [1]. Conventional vibration control approaches often rely on analyzing data after the fact and making corrections by hand, leaving them poorly suited to address the unexpected and rapidly changing environmental forces experienced by structures. Conventional vibration control methodologies are no longer adequate to meet the swift, autonomous, and adaptive standards of modern smart infrastructure. Delay-free structural perception and dynamic response are needed to keep pace with unexpected events [2]. The integration of real-time data systems enables

Kundan Meshram and Prashant Lahre. Vibration Control Systems, (253–260) © 2026 Scrivener Publishing LLC

advanced vibration-damping technologies to seamlessly link structures to their intelligent control counterparts. A network of sensors collects data on displacement, acceleration, velocity, and internal stresses and transmits it to the relevant units in real time. Tardily acquired information is processed onboard the structure or *via* an external unit by advanced techniques such as Kalman filters, machine learning, and adaptive control [3]. This processed information is then employed to activate or modify the strength of damping systems, which keeps physical systems operating according to the best possible behavior at all times. A cyber-physical system (CPS) embodies the fundamental principle of integrating digital algorithms with physical structures in an interconnected and continuously responsive manner. In civil engineering, real-time control can have significant impacts on the safety and lifespan of a structure. Real-time systems offer added protection for high-rise buildings located in at-risk regions and ensure the structural integrity and longevity of long-span bridges [4]. A good example is how powerful gusts of wind can set high-rise buildings into undesirable swinging motions, which can disturb tenants and cause significant structural deterioration in the long run. Real-time sensors and actuators enable a damping system to respond immediately to unwanted building vibrations and automatically regulate them. Real-time monitoring in seismic areas lets structures deploy base isolation or tuned mass damping systems to reduce their vulnerability to earthquake damage. With the advent of the Internet of Things and edge computing, processing and transmitting data in real time has become much quicker and more efficient. This combination enables improved, responsive management of structural vibration systems, as well as leading to innovations in predictive upkeep and intelligent infrastructure design. Nevertheless, there are still obstacles in the areas of low latency, data management, sensor reliability, solving the problem of dealing with environmental noise, and integrating these systems with the overall structure. This chapter surveys the concepts, technology, and practical applications of real-time data processing and control in vibration systems and illustrates how they promote the safety and resilience of civil infrastructure. The book presents real-time structural health monitoring to researchers and engineers, emphasizing a transition from conventional batch data processing to advanced recursive algorithms using first-order perturbation. It focuses on developing theories and practical solutions for monitoring and detecting changes in the state of civil infrastructure in real time. By building on equations that describe eigenspace revisions, the book shows how up-to-the-moment structural data allows for the continuous detection of modal characteristics and damage indicators, enabling accurate determinations of the type, onset, and location of damage. The book demonstrates through experimental results conducted in realistic damage situations that recursive approaches perform well and are reliable for real-time structural health monitoring [5]. Maintaining high vibration quality leads to stronger concrete and safer structures throughout the structure's life. Two important factors are the length and intensity of vibrations, yet traditional visual observation lacks consistency. Most traditional methods fail to account for the diversity within concrete materials. A novel approach using IoT technology and a highly tuned 50-layer ResNet (ResNet-50) is proposed to solve this issue. It detects the depth of vibration and estimates duration using a large dataset of 15,006 photos of vibrated concrete surfaces. The methodology was confirmed to be effective during the construction of a large hydraulic engineering project [6]. The book provides a thorough grounding in structural vibration reduction control that is useful for students and professionals. It covers four key areas: Principles, devices, techniques, and practical applications for vibration reduction. It is suitable for use by a range of audiences, including researchers, practicing engineers, and graduate students. Regular, correctly executed concrete vibration plays a crucial role in ensuring sound structures, yet methods for regularly monitoring the quality of vibration are scarce [7]. The study presents a real-time evaluation system for tracking vibration quality using a combination of ultra-wideband sensors, inclinometers, and a hybrid neural network. An experiment is performed to test the reliability of the proposed method in controlling active vibration quality [8].

19.2 Overview of Real-Time Data Systems

These types of systems are characterized by the rapid processing of data and their capacity to respond within set timeframes to evolutions in the surrounding environment. A range of key variables shapes their performance and accuracy.

a. Latency
   Definition: Latency refers to the time it takes for data to be sensed, processed, and an appropriate command issued.

Impact: Reducing latency ensures greater responsiveness in control systems, a key factor in earthquake mitigation systems.
Optimization Strategies: Employing edge computing and adopting quicker protocols such as 5G and MQTT can considerably lower the latency.

b. Sampling Frequency
Definition: The speed at which sensor data is measured over a given duration.
Relevance: Using a greater sampling frequency allows a more detailed analysis of vibrations and improves the quality of control decisions.
Challenges: Need a larger memory setup for data collection

c. Resolution
Definition: The resolution denotes how fine the details in the sensor readings are.
Application: Higher resolution sensors improve the accuracy of predicting how a structure responds to changes and can lead to more effective control solutions.
Considerations: Achieving a suitable trade-off between resolution and computational expense is crucial when designing the system.

d. Data Synchronization and Time Stamping
Integrating all data sources properly allows for efficient sensor fusion and control decision-making. GPS time-sync, NTP, and custom time stamping methods are all used for precise synchronization.

19.3 Methods for Continuous Monitoring and Control

Advanced methods are utilized that swiftly interpret and act upon information from sensors in order to minimize vibration levels.

Machine Learning for Anomaly Detection
Supervised and Unsupervised Models: Capable of identifying anomalies that could signal damage to a structure or malfunction of a sensor.
Streaming Algorithms: SGD- RNNs

Example: Machine learning models employ SVMs to recognize shifts in either modal frequencies or damping ratios.

Kalman Filters for Sensor Fusion
Purpose: A technique for synthesizing sensor data to infer system states in the presence of measurement errors.
Variants: EKF and UKF handle the case of nonlinearly modeled systems.
Use Case: Estimating instantaneous position and velocity from the values of accelerometer and gyroscope readings.

Predictive Control Models
Model Predictive Control (MPC): Identifies potential system behaviors ahead and adjusts control signals accordingly.
Application: Predicting how winds and heavy vehicles affect bridges before they happen.

Fuzzy Logic and Expert Systems
Rules help interpret data in situations where traditional models are unreliable.
These are applicable in systems whose behavior is challenging to describe mathematically.

19.4 Integration with Dynamic Response Systems

Effective communication between real-time data systems and physical systems is essential to provide useful feedback and responsive actuation.

Cyber-Physical Systems (CPS)
These are systems that combine sensors, actuators, and computational units in the same architecture.

Sensors monitor and communicate any changes in vibrations for rapid adjustment by actuators such as MR dampers or active mass drivers.

Adaptive Control Systems
These are systems that dynamically adjust themselves as new information becomes available in real time.

Examples are Self-Tuning Regulators (STRs), as well as Gain Scheduling Control.

Embedded Control Systems
Intelligent devices capable of running control routines without requiring significant off-site computation or data transfer.

Common platforms: Common platforms for building embedded control systems are Arduino, Raspberry Pi, STM32, and National Instruments DAQ systems.

IoT-Based Feedback Loops
The Internet of Things (IoT) enables real-time monitoring and control over systems from anywhere using networked cloud services.

Feeding data collected from a wide network of satellites back through a central control hub enables citywide management of vibrations.

19.5 Case Studies and Technological Advances

Observing real-world applications allows checking if the theoretical models and designs work as expected.

Smart City Implementation: Tokyo Skytree
A comprehensive system utilizing 300 sensors for collecting timely data and employing advanced dampers for control.

Wind loads are handled through Kalman filters and predictive models.

Earthquake Early Warning Systems: Japan's ShakeAlert
Sensors continuously record and analyze earthquake information to implement anti-seismic systems well ahead of tremors reaching a site.

GPS and accelerometer data are combined with filtering and control algorithms implemented inside the system.

Highway Infrastructure Monitoring: California Bridge Network
Data from strain gauges and accelerometers feed into real-time instrumentation systems mounted on bridges.

Real-time data is transferred to control centers, where it is applied for both predicting the need for maintenance and evaluating current damage.

Laboratory Implementation: IIT Kanpur Smart Base-Isolation Testbed
It tests the effectiveness of base-isolated structures in responding to changing seismic activity in real time.

MPC and fuzzy logic controllers work together with high-frequency sensors.

19.6 Future Directions

The field of real-time measurement analytics and control is set to expand even more with the arrival of edge AI, neuromorphic computing, upcoming 6G networks, and advanced quantum sensing technology. Future work is focusing on building autonomously adjusting control algorithms that correct their own programming errors. Blockchain technology is emerging as an important tool for ensuring secure data transmission, and digital twin technology is transforming the way we use real-time simulation for civil infrastructure control.

19.7 Conclusion

1. Real-time operation plays a crucial role in effective vibration control. Real-time integration of data acquisition, processing, and control into structural damping systems enables fast response to changing conditions, promoting safety, comfort, and robustness.
2. The responsiveness and accuracy of a real-time control system depend heavily on minimizing latency, using fast sampling rates, and maintaining sufficient resolution in sensor readings.
3. Modern algorithms facilitate smarter decision-making by processing and inferring from sensor signals to anticipate the behavior of structures in varying circumstances.
4. CPS architectures underpin smart damping by digitally monitoring and controlling structures in real time.
5. Integrating IoT sensors and edge computing allows real-time control systems to operate independently and improve reaction times for extensive or distributed structures.
6. Adaptive control enables dynamic response systems to instantly adjust the effectiveness of damping elements in response to varying loads.
7. Practical applications have shown the effectiveness of real-time control systems in challenging scenarios such as earthquake-resistant buildings, smart buildings, and long-span bridges.
8. There are still obstacles in implementation and optimization. Research and improvement in areas such as sensor

power requirements, system calibration, data management, and algorithm optimization are necessary to achieve the best possible outcomes.

9. Future advancements are driving efforts toward autonomous and self-healing real-time systems that automatically detect and repair structural damage.
10. Achieving success in implementing real-time vibration control systems relies on bringing together the expertise of civil engineers, computer scientists, control system designers, and urban planners.

References

1. Li, Z. and Adeli, H., Control methodologies for vibration control of smart civil and mechanical structures. *Expert Syst.*, 35, 6, e12354, 2018.
2. Pan, X., Yang, T.Y., Xiao, Y., Yao, H., Adeli, H., Vision-based real-time structural vibration measurement through deep-learning-based detection and tracking methods. *Eng. Struct.*, 281, 115676, 2023.
3. Khodarahmi, M. and Maihami, V., A review on Kalman filter models. *Arch. Comput. Methods Eng.*, 30, 1, 727–747, 2023.
4. Lei, Y., Rao, Y., Wu, J., Lin, C.-H., BIM based cyber-physical systems for intelligent disaster prevention. *J. Ind. Inf. Integr.*, 20, 100171, 2020.
5. Bhowmik, B., Hazra, B., Pakrashi, V., *Real-time structural health monitoring of vibrating systems*, CRC Press, Boca Raton, FL, USA, 2022.
6. Wang, D., Ren, B., Cui, B., Wang, J., Wang, X., Guan, T., Real-time monitoring for vibration quality of fresh concrete using convolutional neural networks and IoT technology. *Autom. Constr.*, 123, 103510, 2021.
7. Li, A., *Vibration control for building structures*, Springer, Cham, Switzerland, 2020.
8. Quan, Y. and Wang, F., Machine learning-based real-time tracking for concrete vibration. *Autom. Constr.*, 140, 104343, 2022.

20

Future Trends and Innovations in Vibration-Damping

Abstract

The chapter focuses on the development of vibration-damping technologies that will radically transform the durability, efficiency, and comfort provided by structures, systems, and products. Cutting-edge materials science is explored, examining breakthroughs such as composites, smart materials and metamaterials fortified by nanotechnology for even better vibration suppression. The next-generation emergency procedures prioritize multifaceted approaches, including earthquake-resistant structural designs with tuned mass dampers, structurally efficient systems inspired by nature and inherent structural vibration damping to ensure the safety of all occupants. Emerging control strategies employing machine learning, AI and IoT sensors are transforming the way vibrations are addressed in real time, opening doors to adaptable vibration control methods. The chapter considers the options of getting electricity from vibrations by using piezoelectric and electromagnetic systems. New developments in damping rely on the cooperation of specialists in materials science, engineering, and physics. Concerns about regulations, pollution, and ethics guide efforts to make damping solutions that are easy to implement, reliable and long-lasting. The thorough review provides a vision for the future that will be formed by recent and future discoveries.

Keywords: Transformative damping system, damping vibration technology, earthquake, dynamic force

20.1 Introduction

Because builders and engineers strive for strong structures, faster machines and more relaxed occupants, vibration damping never stops improving and innovating [1]. It embarks on a captivating journey toward vibration-damping technologies, revealing what innovations will reinvent everything from industries to manufacturing in the coming years. This journey is built upon materials science which provides the launching point

Kundan Meshram and Prashant Lahre. Vibration Control Systems, (261–272) © 2026 Scrivener Publishing LLC

for mixing innovative materials and advanced technologies. Making use of the transformative effects of nanotechnology, recent progress in these materials has the potential to greatly improve vibration damping [2]. By combining different areas, engineers are growing closer to developing materials that control dynamic forces on a whole new scale. This chapter examines unique structural engineering designs that are unlike anything conventional. Utilizing tuned mass dampers, bio-based ideas and future-focused structural solutions, architectural concepts are built to tolerate powerful vibrations in the most stable way possible [3]. Although innovative materials and structures contribute a lot, vibration damping has gone further [4]. These methods investigate advanced control systems, where combining machine learning, AI, and IoT-connected sensors is transforming how real-time actions are taken [5]. Because of this convergence, adaptive control can evolve, ready to react to vibrations instantly [6]. In addition, focuses on the potential of turning vibrational energy into a sustainable kind of power. By imagining how piezoelectric and electromagnetic methods can switch mechanical energy into power, it hopes that vibrations could later be used as an environmentally friendly source. It looks at the upcoming possibilities and new trends in the vibration-damping system field. This chapter looks ahead, discussing new technologies, different approaches in math and the possible future roles damping systems may play in engineering. Significant work is being done by exploring new material properties and their design processes. This work focuses on developing materials that damp vibrations easily, making systems work better and change with new uses. Moreover, newer ways of designing damping systems hope to improve their performance and toughness, all at a low weight and with less space. Adaptive damping systems benefit from new technologies [7–9]. They rely on algorithms, different sensors, and future predictions to automatically manage the damping according to what's happening outside and inside the machines. The combination of artificial intelligence and the Internet of Things opens new ways for improving damping efficiency and responsiveness [10–12]. Investigating miniaturization and microscale damping technologies creates opportunities for applications in nanotechnology and MEMS. Thanks to new inventions in these areas, it is possible to design ultra-compact, accurate damping mechanisms for use in biomedical devices, micro robotics, and advanced instrumentation [13–15]. Because sustainability is now an important aspect, the chapter looks at how eco-conscious damping materials and systems have emerged. New inventions are designed to use renewable or reusable materials in damping, keeping the environment safe while maintaining high performance.

In addition, researchers are investigating devices that damp the effects well and consume less power.

20.2 Emerging Materials for Vibration Damping

With recent progress in both material science and nanotechnology, vibration damping is moving towards the use of new materials. Here, we examine the boundaries of new materials, focusing on different materials made especially for damping vibrations.

Composite Materials

Carbon, glass, or aramid fibers placed in various matrices can give exceptional rigidity and damping to a product, so its damping actions can be customized for chosen applications.

When different materials are stacked one on top of the other, the vibrations through the entire structure can be managed, contributing to increased damping.

Smart Materials

Shape Memory Alloys (SMAs) have a special property: they can restore their original shape after being deformed, helping to dampen a variety of vibrations.

Piezoelectric materials make use of the piezoelectric effect and react to mechanical forces by creating electricity which provides damping and energy harvesting potentials.

Metamaterials

Microengineered metamaterials show unique mechanics and redirect vibrational energy which makes them useful in minimizing movement in structures.

By relying on unexplored negative stiffness, these materials help to precisely control the vibration frequencies and amplitude.

Nanotechnology Integration

Mixing nanoparticles or nanofillers into existing materials makes those materials more effective at dampening and reducing the amount of vibration they can produce.

Having structures that are well below the microscale within materials adjusts their strength, so having such nanostructured dampers could mean better damping and lighter weight.

The combination of these materials with advanced techniques in manufacturing makes it possible to build both tuned dampers and shape-shifting materials suitable for different vibrations. In addition, thanks to nanotechnology, materials get new, unique abilities, allowing the use of lightweight and efficient dampers that outperform regular solutions.

20.3 Innovative Structural Designs

The search for reducing vibration involves not just materials, but also advanced designs meant to be naturally resistant to dynamic stresses. The next part discusses exciting new designs and changes to built structures that boost resilience to vibrations.

Tuned Mass Dampers (TMD system)
These systems develop from traditional TMDs, adding new features, using advanced designs and techniques that help control broader kinds of vibrations.

Miniature TMDs are able to fit into structures, providing controlled protection from vibrations even when there's not a lot of space.

Designs that Follow Nature
Resilience in nature: Following hair's ability to store energy, designers create structures that protect the building with biomimetic damping technologies.

Architectural elements based on living things such as flexible features can move and adjust themselves to reductions in vibrations.

New Architectural Ideas for the Future
Innovative approaches in material and method design lead to architectural elements that are naturally immune to vibration, relying on damping features in their main components.

Modular and Adaptive Structures: Structures that can move and reposition parts based on what they sense in their environment.

Anti-Vibration Platforms
New Isolation Systems: Modern types of isolation such as dynamic base isolators and original isolation methods now play a bigger role in reducing vibrations in different buildings.

With smart sensors and actuators placed in special joints, structures adjust and respond to vibrations effectively.

20.4 Advanced Control Strategies

New strategies for vibration control are being developed by advancing control systems and integrating today's powerful technologies. It discusses new control ideas that help minimize vibrations by dynamically responding and acting in real time.

Machine Learning Powered Systems
Predictive Analytics helps build models that can detect problems in vibration and help predict resonances using machine learning.

They are designed to pick up data directly, making changes to control parameters to provide the best vibration damping in any present situation.

AI works with the automatic systems in charge of operating the control tower.

In control loops, bringing together artificial intelligence allows processes to decide and improve how they handle damping strategies as they run in real time.

AI-driven systems built with cognitive skills can study, think, and act for themselves, making control responses to vibration changes more successful.

Networks Based on IoT Technology
Embedding IoT devices across structures leads to real-time collection of a variety of data, allowing for accurate real-time monitoring of vibrations and environmental changes.

For IoT-connected systems, it is now possible to respond quickly to vibrations with remotes and adjust the damping devices as needed to suit the conditions.

These types of strategies rely heavily on data for making control decisions.

With the help of data, continuous feedback loops refine control schemes to reach the ideal results by dynamically changing damping mechanisms.

These algorithms automatically change control parameters as the conditions or structure of the building change, helping the vibration dampers remain effective.

20.5 Energy Harvesting through Damping

Going after greasiness in circuits is leading us to investigate ways to use vibrations and their wasted energy for power generation. This chapter addresses the concept of energy recovery from damping systems that may turn vibrations into a useful source of alternative energy.

What is Piezoelectric Energy Harvesting?
Piezoelectric technology allows the integration of these elements into damping systems, which helps to use the driving vibrations as a source of electricity.

For building supplies, applications of piezoelectric materials can harvest vibrational energy and use it on location.

Transforming Electromagnetic Energy
In Magnet Coil Systems, shaking the magnets and coils produces vibration that, using electromagnetic induction, causes them to vibrate relative to each other and create electricity.

In machinery and equipment, electromagnetic technology helps capture the energy produced by operating vibrations.

Producing Power that Doesn't Harm the Environment
Vibrations, once described as disruptive, have been recast as an alternative energy source, helping to make clean power possible.

Energy obtained from vibrations can fuel low-energy technologies and provide extra energy for buildings.

Problems and Optimization
Balancing the function of damping and the amount of energy salvaged by the structure is difficult, so optimization is necessary to obtain the highest energy with good damping.

The energy harvesting system can only reach its full potential by overcoming technical problems with conversion, storing energy and expanding the system.

Impact on Nature and Business
Using vibration energy for power helps meet sustainability goals, supports green energy initiatives, and cuts back on using traditional power.

A lot of people will not support renewable energy until it is clear that harvesting energy is both economically doable and sustainable.

20.6 Multidisciplinary Approaches and Collaborations

Advancing vibration damping through innovations requires people from many areas of expertise to join forces. Here, we focus on the crucial role that teamwork across disciplines has in driving progress, new ideas, and improvements in damping systems.

Working together in Materials Science and Engineering
Collaboration between materials scientists and engineers leads to the creation of new materials with customized damping traits that go beyond what old materials are able to do.

The combination of materials science and knowledge in nanotechnology results in building nanocomposites and nanostructured materials for better damping.

Integrating the Functions of Physics and Structural Engineering
When physicists and structural engineers cooperate, they can precisely analyze and build models of vibrations in structures.

Collaboration between brands led to innovative concepts: Engineers combined requirements and background knowledge of physics to invent designs that disperse vibration efficiently.

Data Science and Control Systems Experts
Collaboration between data scientists and control systems engineers results in advanced algorithms for on-the-spot control of vibrations.

Both fields collaborate to better link sensors and analytics from the IoT, which helps improve the accuracy of monitoring and actions taken to control earthquakes.

In addition, also work with Interdisciplinary Think Tanks and Research Consortia.

Together, people from different departments pool their talents to solve challenging problems in reducing vibration.

Innovation platforms that welcome multiple fields help develop innovative ideas, bring experts together, and transfer lab findings into useful implementations.

Joining Forces between the Industry and the Academic World
By collaborating, academia and industry speed the use of advanced damping technologies in real industries.

Joint initiatives encourage specialists to exchange information, leading them to become accomplished in using several disciplines on the job.

20.7 Regulatory and Environmental Considerations

Using these new vibration-reducing innovations is subject to both regulations and attention to the environment. It explores the fundamental parts of meeting regulations, complying with standards and evaluating environmental aspects for the acceptance of innovative damping methods.

Meeting Required Regulations
Complying with regulations makes damping technologies safer, more reliable and better performing across all industries.

Damping systems are officially certified after going through careful checks to show they meet the appropriate safety and performance guidelines.

Estimated Effect on the Environment
When conducting a Life Cycle Analysis, look at the environmental effects of damping materials and technologies from the beginning to the end, including getting the materials, creating them, their function, and what happens to them after.

If you look at sustainability metrics such as the environment's health, energy use, and waste amount, you can more easily spot sustainable damping choices.

Rules for Health and Safety
Ensuring that damping systems are designed and built to fit safety rules in occupational settings for people manufacturing, installing, and maintaining them.

Keeping Structures Safe: Rules for ensuring structure safety are made to avoid threats from the failure of damping systems.

Impact on the Society and on Ethics
Getting feedback on public attitudes: Looking at how people react to new ways of dampening vibrations, listening to their worries about privacy, noise or the look of the technologies.

Ethics play a part by involving questions about social implications, fair access, and the advantages to society of using vibration-damping technologies.

Partnering with Different Countries
Working with global standards allows new damping technology to be used in more countries.

Regulators, industries, and research groups team up to develop strong and complete regulatory systems.

Rules That Can Change with New Technologies
Regulatory authorities should alter their guidelines to keep up with the fast advances in vibration damping.

Motivating regulations and incentives are needed to increase the use of environmentally safe damping technologies.

20.8 Challenges and Ethical Considerations

The use of future vibration-damping systems poses many issues and raises ethical discussions that need careful consideration before they can be successfully introduced and managed.

Technical Challenges
Integrating Advanced Materials and Controls: Mixing present infrastructure with the newest, more modern materials and control devices is not always simple.

It is still very difficult to make dampers strong enough to be effective but still keep the structure light and affordable.

The Cost of Implementing and How Well It Can Be Expanded
The expensive initial investment and costs during operation can prevent many from using new damping technologies.

Problems arise when trying to increase innovative technologies for wide use due to concerns about their efficiency and extreme cost.

Steps for Gaining Reliability and Safety
Reliability Standards: Developing damping systems that last and perform well so that systems are safe and do not fail.

Checking the effectiveness of damping technologies in a variety of natural conditions and severe events is still important.

Ethical Considerations
Privacy Questions: Ethics around the way information from IoT damping systems is gathered, how it is used, and how the privacy of users should be protected.
Equity in Access: Trying to balance who gets better damping technologies and ensure they are affordable for all kinds of communities.

Methodologies Focused on Humans
Keeping comfort and security in mind: Whenever damping techniques are used, the goal is to avoid trouble and make living more enjoyable.

It is important to consider public opinion, acceptance, and knowledge-based approval before building damping systems in common areas or homes.

Society Assessment
Involving local communities and groups interested in the issue when selecting and applying damping technologies within the energy system.

Checking for bigger effects like business job losses, the qualifications needed by employees and the benefits gained by society when technology is adopted.

Industry rules and common practices.

Navigating through Regulations: Resolving regulatory problems and sticking to current and updated industry rules for emerging technologies.

Better Ethics: Creation of advanced guidelines to ensure that all development and use of advanced vibration-damping technologies are ethical.

20.9 Conclusions

1. With the combination of advanced materials and systems, plus renewable energy-harvesting, enter a new period in vibration-damping technology with exciting breakthroughs.
2. Taking a multi-disciplinary approach in materials engineering, data science, and architecture helps to provide bold answers to the complexities found in damping.
3. The technical, regulatory, and ethical challenges pose to novel technologies can improve the way they are used and what features they have.

4. Damping solutions offer more than vibration relief; they also help to make our structures and infrastructures durable and sustainable.
5. Ensuring inhabitants are safe, comfortable and accepted by society makes it important for buildings to use renewable technologies and maintain people's privacy.
6. Even though these can cause difficulties, innovations have the potential to increase growth, create jobs, and improve current industry standards which is more important.
7. Leveling regulations and encouraging worldwide teamwork encourage many countries to use these technologies.
8. Imagine zero-energy buildings, smarter cities, and safer transportation can see how these technologies will change societies everywhere.

References

1. Auger, J.H., *Why Robot? Speculative design, the domestication of technology and the considered future*, Royal College of Art, United Kingdom, 2012.
2. Oudatzi, K. and Andreadis-Manou, N., Testing and Simulation Methods in Contemporary Construction Teaching, in: *Emerging Possibilities of Testing and Simulation Methods and Techniques in Contemporary Construction Teaching*, pp. 159, Hershey, Pennsylvania (PA), IGI Global, USA, 2016.
3. Wells, M., *Skyscrapers: Structure and design*, Laurence King Publishing, London, United Kingdom (UK), 2005.
4. Van Der Giessen, E., Schultz, P.A., Bertin, N., Bulatov, V.V., Cai, W., Csányi, G., Foiles, S.M., *et al.*, Roadmap on multiscale materials modeling. *Modell. Simul. Mater. Sci. Eng.*, 28, 4, 043001, 2020.
5. Vermesan, O. and Bacquet, J. (Eds.), *Internet of Things–The Call of the Edge: Everything Intelligent Everywhere*, CRC Press, New York, USA, 2022.
6. Kaur, G., Tomar, P., Tanque, M. (Eds.), *Artificial intelligence to solve pervasive internet of things issues*, Academic Press, Cambridge, Massachusetts (MA), USA, 2020.
7. Lu, H., Wang, X., Zhang, T., Cheng, Z., Fang, Q., Design, fabrication, and properties of high damping metal matrix composites—A review. *Materials*, 2, 3, 958–977, 2009.
8. Gibson, R.F., A review of recent research on mechanics of multifunctional composite materials and structures. *Compos. Struct.*, 92, 12, 2793–2810, 2010.

9. Shao, G., Hanaor, D.A.H., Shen, X., Gurlo, A., Freeze casting: from low-dimensional building blocks to aligned porous structures—a review of novel materials, methods, and applications. *Adv. Mater.*, 32, 17, 1907176, 2020.
10. Wang, Y., Su, Z., Ni, J., Zhang, N., Shen, X., Blockchain-empowered space-air-ground integrated networks: Opportunities, challenges, and solutions. *IEEE Commun. Surv. Tutor.*, 24, 1, 160–209, 2021.
11. Baduge, S.K., Thilakarathna, S., Perera, J.S., Arashpour, M., Sharafi, P., Teodosio, B., Shringi, A., Mendis, P., Artificial intelligence and smart vision for building and construction 4.0: Machine and deep learning methods and applications. *Autom. Constr.*, 141, 104440, 2022.
12. Hassan, Q.F. (Ed.), *Internet of things A to Z: technologies and applications*, John Wiley & Sons, Hoboken, New Jersey (NJ), USA, 2018.
13. Jani, J.M., Leary, M., Subic, A., Gibson, M.A., A review of shape memory alloy research, applications and opportunities. *Mater. Des. (1980–2015)*, 56, 1078–1113, 2014.
14. Khoshnoud, F. and de Silva, C.W., Recent advances in MEMS sensor technology-mechanical applications. *IEEE Instrum. Meas. Mag.*, 15, 2, 14–24, 2012.
15. Uchino, K., *Micromechatronics*, CRC Press, Boca Raton, Florida (FL), USA, 2019.

Index

Also of Interest

Check out these other titles from Scrivener Publishing

Dictionary of Industrial Terminology 2nd Edition, by Michael D. Holloway and Emma Holloway, ISBN: 9781119363446. Greatly expanded and revised, this second edition is the most comprehensive and in-depth dictionary of industrial terms available, a must-have for any engineer, scientist, technician, or operator working in industry. *NOW AVAILABLE!*

Process Machinery Handbook: For Field Personnel, Decision Makers, and Students, Edited by Robert X. Perez, ISBN: 9781394214457. *Process Machinery Handbook: For Field Personnel, Decision Makers, and Students* equips newcomers and seasoned professionals with essential insights into the diverse world of process machinery, empowering them to understand unique performance characteristics, common failure modes, and effective strategies for enhancing reliability in their operations.

Basic Electrical and Instrumentation Engineering, by P. Sivaraman, C. Sharmeela, A. Thaiyal Nayagi, and R. Mahendran, ISBN: 9781119764465. Offering the most up-to-date coverage on the subject available, this new volume provides detailed concepts of AC circuits and power systems, transformers, DC machines, AC machines, measurement and instrumentation, and many other practical concepts for the engineer or engineering student. *NOW AVAILABLE!*

Electrical and Electronic Devices, Circuits, and Materials: Technological Challenges and Solutions, edited by Suman Lata Tripathi, Parvej Ahmad Alvi, and Umashankar Subramaniam, ISBN: 9781119750369. Covering every aspect of the design and improvement needed for solid-state electronic devices and circuit and their reliability issues, this new volume also includes overall system design for all kinds of analog and digital applications and developments in power systems. *NOW AVAILABLE!*

Books in the series, "Rotating Machinery Fundamentals and Advances"

Condition Monitoring, Troubleshooting, and Reliability in Rotating Machinery, Edited by Robert X. Perez, ISBN: 9781119631545. This third volume in a broad collection of current rotating machinery topics, written by industry experts, is a must-have for rotating equipment engineers, maintenance personnel, students, and anyone else wanting to stay abreast with current rotating machinery concepts and technology.

DESIGN, MODELING AND RELIABILITY IN ROTATING MACHINERY, Edited by Robert X. Perez, ISBN: 9781119631682. This broad collection of current rotating machinery topics, written by industry experts, is a must-have for rotating equipment engineers, maintenance personnel, students, and anyone else wanting to stay abreast with current rotating machinery concepts and technology.

MAINTENANCE, RELIABILITY AND TROUBLESHOOTING IN ROTATING AND PROCESS MACHINERY, Edited by Robert X. Perez, ISBN: 9781119631644. For Rotating equipment professionals, maintenance personnel, project engineers, process engineers, mechanical engineering students, and anyone working around critical rotating equipment.

Troubleshooting Rotating Machinery: Including Centrifugal Pumps and Compressors, Reciprocating Pumps and Compressors, Fans, Steam Turbines, Electric Motors, and More, by Robert X. Perez and Andrew P. Conkey, ISBN: 9781119294139. A must-read for anyone whose job depends on operating or maintaining safe, reliable, and efficient machinery in a process facility, this volume shows the newcomer and veteran alike how you can start troubleshooting rotating equipment problems immediately by implementing this book's common-sense methodology. *NOW AVAILABLE!*

Operator's Guide to Process Compressors, by Robert X. Perez, ISBN: 9781119580614. The perfect primer for anyone responsible for operating or maintaining process gas compressors. *NOW AVAILABLE!*

Printed and bound by CPI Group (UK) Ltd, Croydon, CR0 4YY
17/09/2026
14964266-0001

The LINC Complex

Methods and Protocols

Edited by

Gregg G. Gundersen

Department of Pathology and Cell Biology, Columbia University, New York, NY, USA

Howard J. Worman

Department of Medicine and Department of Pathology and Cell Biology, Columbia University, New York, NY, USA

Humana Press

Editors
Gregg G. Gundersen
Department of Pathology and Cell Biology
Columbia University
New York, NY, USA

Howard J. Worman
Department of Medicine and Department of Pathology and Cell Biology
Columbia University
New York, NY, USA

ISSN 1064-3745 ISSN 1940-6029 (electronic)
Methods in Molecular Biology
ISBN 978-1-4939-8690-3 ISBN 978-1-4939-8691-0 (eBook)
https://doi.org/10.1007/978-1-4939-8691-0

Library of Congress Control Number: 2018950395

Printed on acid-free paper

This Humana Press imprint is published by the registered company Springer Science+Business Media, LLC part of Springer Nature.
The registered company address is: 233 Spring Street, New York, NY 10013, U.S.A.

Preface

LINCing Molecular Methodologies Across Disciplines

In 2006, Didier Hodzic, Brian Burke, and colleagues coined the term LINC (linker of nucleoskeleton and cytoskeleton) complex for the assembly of Klarsicht, ANC-1, and Syne homology (KASH) domain and Sad1 and UNC-84 (SUN) domain proteins that respectively span the inner and outer nuclear membranes, connecting the nucleoskeleton to the cytoskeleton [1]. That year, we predicted that research on this complex linking the nucleus and cytoplasm will undoubtedly have implications for our understanding of nuclear positioning, nuclear migration, and the pathogenesis of inherited diseases [2]. What we could not predict was the importance of the LINC complex for a myriad of other functions, including mechanotransduction, chromosome movements in meiosis, and DNA repair. Indeed, a growing number of biological scientists studying organisms from plants to people have focused on this fascinating molecular system. In this volume of *Methods in Molecular Biology*, we bring together leading scientists from diverse disciplines to describe research approaches and methodologies used to study the LINC complex and its cellular functions.

New York, NY, USA

Gregg G. Gundersen
Howard J. Worman

References

1. Crisp M, Liu Q, Roux K et al (2006) Coupling of the nucleus and cytoplasm: role of the LINC complex. J Cell Biol 172:41–53
2. Worman HJ, Gundersen GG (2006) Here come the SUNs: a nucleocytoskeletal missing link. Trends Cell Biol 16:67–69

Contents

Contributors

Angélique Millon-Frémillon • *Institute for Advanced Biosciences, Centre de recherche UGA—INSERM U1209—CNRS UMR 5309, Grenoble, France*

Isaac Angert • *School of Physics and Astronomy, University of Minnesota, Minneapolis, MN, USA*

Susumu Antoku • *Department of Pathology and Cell Biology, Columbia University, New York, NY, USA*

Paul T. Arsenovic • *Department of Biomedical Engineering, Virginia Commonwealth University, Richmond, VA, USA*

Alexander L. Auld • *Biology Department, Boston College, Chestnut Hill, MA, USA*

Ricardo Benavente • *Department of Cell and Developmental Biology, Biocenter, University of Würzburg, Würzburg, Germany*

James N. Brandt • *Department of Biology, Lewis and Clark College, Portland, OR, USA*

Brian Burke • *Institute of Medical Biology, Singapore, Singapore*

Natalie E. Cain • *Department of Molecular and Cellular Biology, University of California, Davis, CA, USA*

Ju Chen • *Department of Medicine, University of California, San Diego, La Jolla, CA, USA*

Christophe Guilluy • *Institute for Advanced Biosciences, Centre de recherche UGA, INSERM U1209, CNRS UMR 5309, Grenoble, France*

Mary Ann Collins • *Biology Department, Boston College, Chestnut Hill, MA, USA*

Daniel E. Conway • *Department of Biomedical Engineering, Virginia Commonwealth University, Richmond, VA, USA*

Victor E. Cruz • *Department of Biology, Massachusetts Institute of Technology, Cambridge, MA, USA*

Richard B. Dickinson • *Department of Chemical Engineering, University of Florida, Gainesville, FL, USA*

David E. Evans • *Faculty of Health and Life Sciences, Department of Biological and Medical Sciences, Oxford Brookes University, Oxford, UK*

Xi Fang • *Department of Medicine, University of California, San Diego, La Jolla, CA, USA*

Eric S. Folker • *Biology Department, Boston College, Chestnut Hill, MA, USA*

Heidi N. Fridolfsson • *Department of Molecular and Cellular Biology, University of California, Davis, CA, USA*

G.W. Gant Luxton • *Department of Genetics, Cell Biology, and Development, University of Minnesota, Minneapolis, MN, USA*

Edgar R. Gomes • *Instituto de Medicina Molecular, Faculdade de Medicina, Universidade de Lisboa, Lisboa, Portugal*

Katja Graumann • *Faculty of Health and Life Sciences, Department of Biological and Medical Sciences, Oxford Brookes University, Oxford, UK*

Gregg G. Gundersen • *Department of Pathology and Cell Biology, Columbia University, New York, NY, USA*

FREDERIK HELMPROBST • *Imaging Core Facility, Biocenter, University of Würzburg, Würzburg, Germany*
JARED HENNEN • *School of Physics and Astronomy, University of Minnesota, Minneapolis, MN, USA*
GREG J. HERMANN • *Department of Biology, Lewis and Clark College, Portland, OR, USA*
LESLIE A. HERRERA • *Department of Molecular and Cellular Biology, University of California, Davis, CA, USA*
MIKI HIEDA • *Graduate School of Medicine and Health Science, Osaka University, Osaka, Japan; Graduate School of Health Science, Ehime Prefectural University of Health Science, Ehime, Japan*
DIDIER HODZIC • *Department of Developmental Biology, Washington University School of Medicine, St Louis, MO, USA*
KWANG-HO HUR • *School of Physics and Astronomy, University of Minnesota, Minneapolis, MN, USA*
SUE L. JASPERSEN • *Stowers Institute for Medical Research, Kansas City, MO, USA; Department of Molecular and Integrative Physiology, University of Kansas Medical Center, Kansas City, KS, USA*
JULIEN AUREILLE • *Institute for Advanced Biosciences, Centre de recherche UGA—INSERM U1209—CNRS UMR 5309, Grenoble, France*
IAKOWOS KARAKESISOGLOU • *Department of Biosciences, University of Durham, Durham, UK*
JEREMY KEYS • *Meinig School of Biomedical Engineering and Weill Institute for Cell and Molecular Biology, Cornell University, Ithaca, NY, USA*
JAN LAMMERDING • *Meinig School of Biomedical Engineering and Weill Institute for Cell and Molecular Biology, Cornell University, Ithaca, NY, USA*
TANMAY P. LELE • *Department of Chemical Engineering, University of Florida, Gainesville, FL, USA*
TORREY R. MANDIGO • *Biology Department, Boston College, Chestnut Hill, MA, USA*
AYAKA MATSUMOTO • *Graduate School of Medicine and Health Science, Osaka University, Osaka, Japan*
NARIAKI MATSUURA • *Graduate School of Medicine and Health Science, Osaka University, Osaka, Japan; Osaka International Cancer Institute, Osaka, Japan*
IRIS MEIER • *Faculty of Health and Life Sciences, Department of Biological and Medical Sciences, Oxford Brookes University, Oxford, UK; Department of Molecular Genetics, The Ohio State University, Columbus, OH, USA*
CARMEN MROß • *Institute of Biochemistry I, Medical Faculty, University Hospital Cologne, Köln, Germany; Center for Molecular Medicine Cologne (CMMC) and Cologne Cluster of Excellence on Cellular Stress Responses in Aging-Associated Diseases (CECAD), University of Cologne, Köln, Germany*
JOACHIM D. MUELLER • *School of Physics and Astronomy, University of Minnesota, Minneapolis, MN, USA*
PETER L. NAGY • *MNG Laboratories, Atlanta, GA, USA*
SRUJANA NEELAM • *Department of Chemical Engineering, University of Florida, Gainesville, FL, USA*
NEJMA BELAADI • *l'institut du thorax, INSERM, CNRS, Université de Nantes, Nantes, France*

ANGELIKA A. NOEGEL • *Institute of Biochemistry I, Medical Faculty, University Hospital Cologne, Köln, Germany; Center for Molecular Medicine Cologne (CMMC) and Cologne Cluster of Excellence on Cellular Stress Responses in Aging-Associated Diseases (CECAD), University of Cologne, Köln, Germany*

MAFALDA R. PIMENTEL • *Instituto de Medicina Molecular, Faculdade de Medicina, Universidade de Lisboa, Lisboa, Portugal*

CHLOE POTTER • *Department of Developmental Biology, Washington University School of Medicine, St Louis, MO, USA*

AXEL POULET • *Laboratoire GReD, CNRS, INSERM, Université Clermont Auvergne, Clermont-Ferrand, France; Faculty of Health and Life Sciences, Department of Biological and Medical Sciences, Oxford Brookes University, Oxford, UK; Department of Biostatistics and Bioinformatics, Rollins School of Public Health, Emory University, Atlanta, GA, USA*

MÓNICA PRADILLO • *Departamento de Genética, Fisiología y Microbiología, Facultad de Biología, Universidad Complutense de Madrid, Madrid, Spain*

WILLIAM ROMAN • *Instituto de Medicina Molecular, Faculdade de Medicina, Universidade de Lisboa, Lisboa, Portugal*

THOMAS U. SCHWARTZ • *Department of Biology, Massachusetts Institute of Technology, Cambridge, MA, USA*

BRIAN D. SLAUGHTER • *Stowers Institute for Medical Research, Kansas City, MO, USA*

MARIE-CHRISTIN SPINDLER • *Department of Cell and Developmental Biology, Biocenter, University of Würzburg, Würzburg, Germany*

DANIEL A. STARR • *Department of Molecular and Cellular Biology, University of California, Davis, CA, USA*

CHRISTIAN STIGLOHER • *Imaging Core Facility, Biocenter, University of Würzburg, Würzburg, Germany*

MATTHEW J. STROUD • *Department of Medicine, University of California, San Diego, La Jolla, CA, USA; Cardiovascular Division, King's College London, British Heart Foundation Centre of Excellence, London, UK*

KENTARO TAMURA • *Department of Environmental and Life Sciences, School of Food and Nutritional Sciences, University of Shizuoka, Shizuoka, Japan*

CHRISTOPHE TATOUT • *CNRS, INSERM, Université Clermont Auvergne, Laboratoire GReD, Clermont-Ferrand, France*

VINCENT J. TOCCO • *Department of Chemical Engineering, University of Florida, Gainesville, FL, USA*

JAY R. UNRUH • *Stowers Institute for Medical Research, Kansas City, MO, USA*

EMMANUEL VANROBAYS • *CNRS, INSERM, Université Clermont Auvergne, Laboratoire GReD, Clermont-Ferrand, France*

JAVIER VARAS • *Departamento de Genética, Fisiología y Microbiología, Facultad de Biología, Universidad Complutense de Madrid, Madrid, Spain*

JENNIFER VEEVERS • *Department of Medicine, University of California, San Diego, La Jolla, CA, USA*

MAXIME VOISIN • *CNRS, INSERM, Université Clermont Auvergne, Laboratoire GReD, Clermont-Ferrand, France*

AARON WINDSOR • *Meinig School of Biomedical Engineering and Weill Institute for Cell and Molecular Biology, Cornell University, Ithaca, NY, USA*

HOWARD J. WORMAN • *Department of Medicine and Department of Pathology and Cell Biology, Columbia University, New York, NY, USA*
QIAO ZHANG • *Department of Chemical Engineering, University of Florida, Gainesville, FL, USA*
XIAO ZHOU • *Department of Molecular Genetics, The Ohio State University, Columbus, OH, USA; Department of Molecular and Cell Biology, University of California, Berkeley, Berkeley, CA, USA*
RUIJUN ZHU • *Department of Pathology and Cell Biology, Columbia University, New York, NY, USA*

Part I

Structural and Biochemical Analysis of LINC Complexes

Chapter 1

EM Tomography of Meiotic LINC Complexes

Marie-Christin Spindler, Frederik Helmprobst, Christian Stigloher, and Ricardo Benavente

Abstract

Electron microscope (EM) tomography is a powerful technique that enables the three-dimensional analysis of subcellular structures at high resolution. We have applied this method to the quantitative analysis of LINC complex distribution and interaction with the cytoskeleton in meiotic cells from male mice. In this chapter, we describe methods to generate and analyze the tomograms.

Key words LINC complex, Meiosis, Telomere, Nuclear envelope, Cytoskeleton, Microtubules, EM tomography

1 Introduction

LINC complexes of the nuclear envelope are essential in transducing cytoskeleton-derived forces. These forces are required for nuclear movement and positioning in the cell and are also involved in chromosome movement within the nucleus [1–3]. A particularly well-suited subject for the investigation of the interaction of LINC complexes with the cytoskeleton is the mouse meiotic cell [3, 4]. In contrast to somatic cells in which LINC complexes are distributed over the entire nuclear periphery, in meiotic cells they are restricted to the sites at which telomeres are attached to the nuclear envelope [5–9]. Furthermore, the composition of LINC complexes of mouse meiotic prophase cells has been characterized: the proteins SUN1 and SUN2 of the inner nuclear membrane and outer membrane protein KASH5 [6–9]. In meiotic cells, the interaction of LINC complexes with microtubules appears to be essential for telomere movements that lead to pairing, synapsis, and recombination of homologous chromosomes [3]. Early electron micrographs of murine spermatocytes show numerous filamentous structures at the attachment site of meiotic telomeres to the nuclear envelope [5]. Available immunolocalization data indicates that these filaments correspond to LINC complexes. Filaments at these

Gregg G. Gundersen and Howard J. Worman (eds.), *The LINC Complex: Methods and Protocols*, Methods in Molecular Biology, vol. 1840, https://doi.org/10.1007/978-1-4939-8691-0_1, © Springer Science+Business Media, LLC, part of Springer Nature 2018

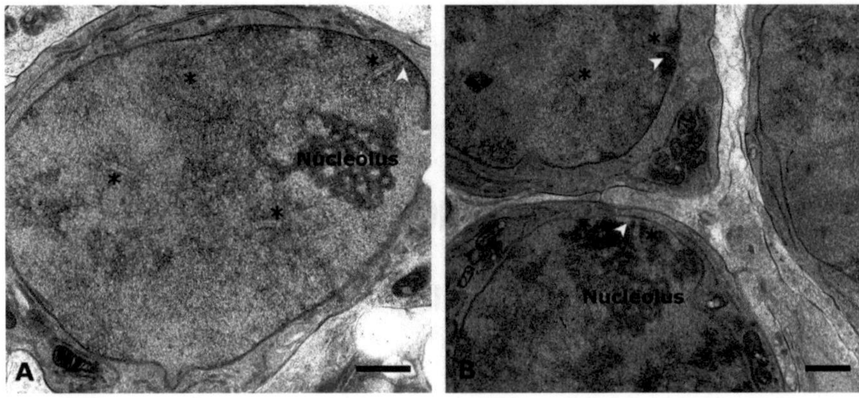

Fig. 1 Tissue preservation following chemical prefixation combined with high-pressure freezing. (**A, B**) Electron micrographs of murine pachytene spermatocytes with synaptonemal complexes (asterisks) in 14-day-old mice. White arrowheads denote attachment sites of synaptonemal complexes at the nuclear envelope. (**A**) Acquisition conducted at a magnification of 6000× with an Olympus Veleta 2k × 2k camera (Olympus Soft Imaging Solutions GmbH, Münster, Germany). (**B**) Synaptonemal complexes attached to the nuclear envelope (white arrowheads), the site of LINC complex accumulation. Acquisition at 5000× with an Olympus Veleta 2k × 2k camera. Tissue section thickness is 250 nm; scale bar, 1 μm (**A, B**)

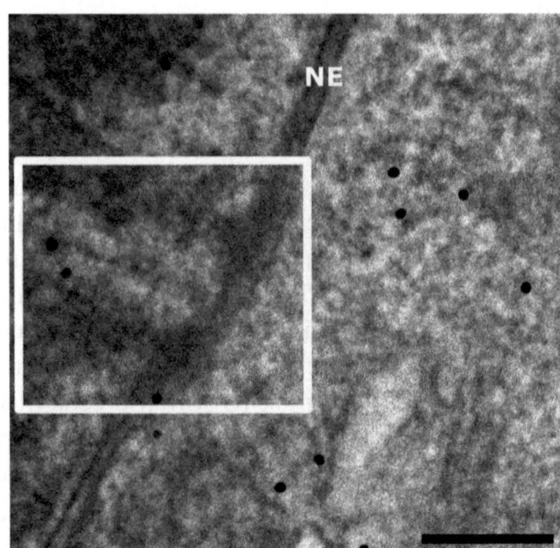

Fig. 2 Electron micrograph of a synaptonemal complex attached to the nuclear envelope (NE; white box) from a 250 nm testis section using a 4k × 4k camera at a magnification of 40,000×. Scale bar: 200 nm. Fiducial markers: 12 nm gold particles

sites can be resolved in 3D with transmission electron tomography. Here, we provide detailed protocols of our investigations of LINC complexes at the attachment sites of meiotic telomeres to the nuclear envelope (Figs. 1 and 2) using EM tomography [10]. This method can provide important three-dimensional information of LINC complex distribution and interaction with microtubules in the cytoplasm (Figs. 3 and 4).

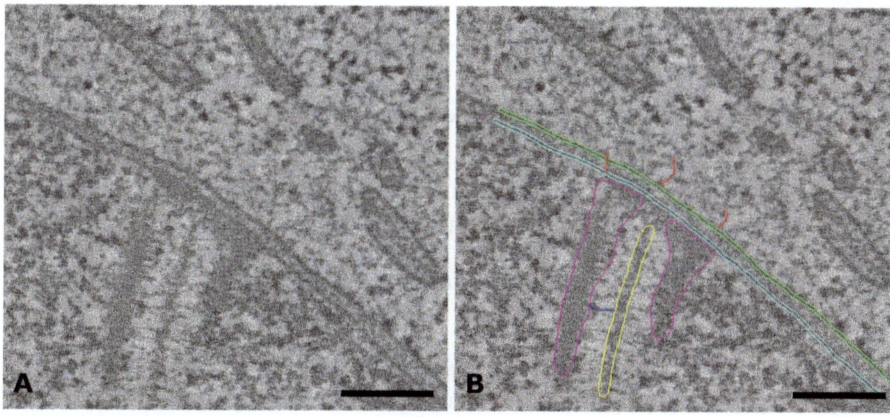

Fig. 3 Annotation of synaptonemal complex attachment including LINC complexes at the nuclear envelope in a dual-axis tomogram. Tomogram slice without (**A**) and with respective annotation (**B**). Annotation: inner nuclear membrane, cyan; outer nuclear membrane, green; lateral elements, magenta; central element, yellow; LINC complexes, red. Magnification, 40,000× using a 4k × 4k camera; scale bar, 200 nm

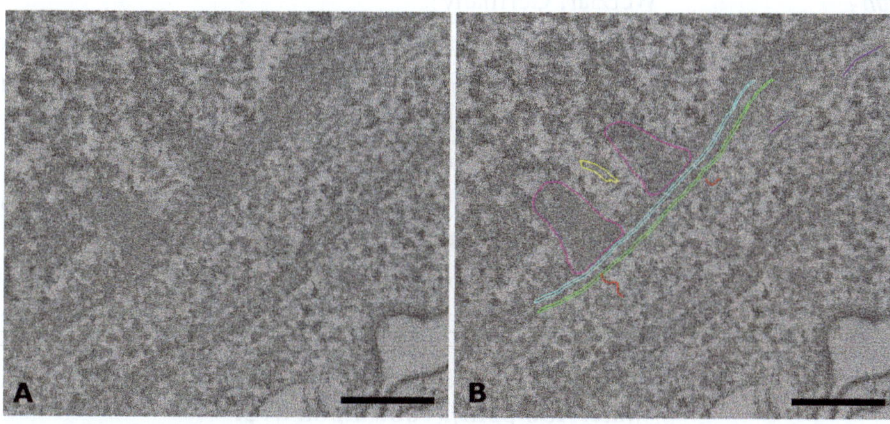

Fig. 4 Annotation of a synaptonemal attachment including LINC complexes at the nuclear envelope. Tomogram slice containing (longitudinal/transverse) microtubuli in proximity to the synaptonemal complex attachment site (**A**, **B**). Color coding according to Fig. 3B, microtubule annotated in purple (**B**)

2 Materials

2.1 Prefixation of Tissue

1. Fresh testis tissue from young 14-day-old mice (*see* **Note 1**). Prior to testis resection, animals are euthanized with CO_2 followed by cervical dislocation.
2. PBS (phosphate-buffered saline): 140 mM NaCl, 6.4 mM Na_2HPO_4, 1.4 mM KH_2PO_4, 2.6 mM KCl; adjust to pH 7.4 with NaOH.
3. Karnovsky solution (4% paraformaldehyde, 2.5% glutaraldehyde in 0.2 M cacodylate buffer): Dissolve 8% paraformaldehyde in 0.2 M cacodylate buffer at pH 7.4 in a glass beaker at

approximately 60 °C. Avoid boiling of the solution and cool to room temperature once dissolved. In parallel prepare an equal amount of 5% of glutaraldehyde in ddH_2O. Combine the cool paraformaldehyde and glutaraldehyde solution to reach the desired aldehyde concentrations.

4. 50 mM cacodylate buffer: 50 mM cacodylate, 50 mM KCl, 2.5 mM $MgCl_2$, pH 7.2.

2.2 High-Pressure Freezing

1. BSA solution: 10% BSA in PBS. Filter through a 0.45 μm filter and store at −20 °C until use.
2. Specimen carriers type A (200 μm recess) and type B as a lid (0 μm recess) coated with lecithin (*see* **Note 2**).
3. High-pressure freezer, e.g., EM HPM100 (Leica Microsystems, Wetzlar, Germany).
4. Liquid nitrogen.

2.3 Freeze Substitution and Embedding in Epon

1. Freeze substitution system, e.g., EM AFS2 (Leica Microsystem, Wetzlar, Germany).
2. Leica plastic capsules D 13 × H 18 mm with mesh bottoms.
3. Leica universal metal reagent bath container for plastic capsules covered with Teflon lids.
4. Anhydrous/dry acetone (water <50 ppm, *see* **Note 3**).
5. 0.5% glutaraldehyde and 0.1% tannic acid in anhydrous acetone (water <50 ppm, *see* **Note 3**).
6. 2% osmium tetroxide in anhydrous acetone (water <50 ppm, *see* **Note 3**).

2.4 Embedding in Epon

1. Epon series: 50% and 90% Epon 812 in anhydrous acetone (water <50 ppm), 100% Epon 812.
2. Equipment for embedding in Epon: snap-caps, plastic Pasteur pipettes, mounted needles, falcon tube with ethanol, jar with acetone, glass dish.
3. Gelatin capsules and capsule molds; alternatively, silicone rubber flat embedding molds.
4. Heating cabinet set to 60 °C.

2.5 Sectioning

1. Trimming: razor blades, binocular microscope, ultramicrotomy block.
2. Semithin sections: Histo Jumbo Diamond Knife (Diatome AG, Biel, Switzerland), poly-L-lysine 25 × 75 mm coverslips (Mentzel Glas, Braunschweig, Germany), mounted eyelash, syringe, methylene blue-azur II stain (*see* **Note 4**), wash bottle, light microscope, Formvar-coated slot, or mesh copper grids.

2.6 Contrasting and Carbon Coating

1. Equipment for contrasting: 15 × 15 cm glass plate, parafilm, glass vials with rolled rim, beaker, wash bottle, glass petri dish, paper box for light exclusion, microsurgery forceps, 1.5 mL LightSafe microcentrifuge tubes (Sigma-Aldrich, Darmstadt, Germany), 0.2 μm filter.
2. Chemicals and solutions for contrasting:
   (a) Contrasting in ethanolic conditions: 2.5% uranyl acetate in ethanol, ethanol, 50% ethanol in ddH_2O, Reynold's lead citrate (*see* **Note 5** [11]) diluted 1:1 in degassed ddH_2O, sodium hydroxide as pellets.
   (b) Contrasting in aqueous conditions: 2.5% uranyl acetate in ddH_2O, Reynold's lead citrate diluted 1:1 in degassed ddH_2O, degassed H_2O, sodium hydroxide pellets (*see* **Note 6**).
3. Equipment for carbon coating: carbon coater, e.g., Bal-Tec MED 010 planar-magnetron sputtering device (Balzers Union AG, Balzers, Liechtenstein), spectral carbon rods (Baltic-Präparation e.K., Niesgrau, Germany), filter paper, double-faced adhesive tape.
4. >12 nm gold-labeled secondary antibody as fiducial marker for precise tilt-series alignment.

2.7 Electron Tomography and Software

1. Transmission electron microscope (JEM-2100, JEOL, Munich, Germany) equipped with a high-resolution camera (e.g., TemCam F416, Tietz Video and Imaging Processing Systems, Gauting, Germany; *see* **Note 7**).
2. Tilt-series acquisition: SerialEM [12].
3. IMOD suite: ETomo for tomogram reconstruction and 3dmod for annotation (*see* **Note 8** [13]).

3 Methods

For testis tissue preservation, an adaption of the protocol by Dhanyasi and colleagues is followed [14].

3.1 Prefixation of Seminiferous Tubules for High-Pressure Freezing

1. Remove the tunica albuginea from freshly resected testes of 14-day-old mice in PBS using microsurgery forceps.
2. Transfer individual seminiferous tubules into fresh PBS. Keep the samples on ice continuously to preserve tissue integrity.
3. Transfer seminiferous tubules into Karnovsky fixative and incubate first for 30 min at room temperature, then another 60 min on ice.
4. Wash five times for 3 min each in 50 mM cacodylate buffer (can be stored several weeks at 4 °C).

3.2 High-Pressure Freezing

High-pressure freezing and freeze substitution is conducted based on previously described morphological protocols for *C. elegans* and *Danio rerio* [15–17].

1. Transfer prefixed seminiferous tubules from 50 mM cacodylate buffer to 10% BSA to immerse the tissue in freeze protectant.
2. Add 10 μL of 10% BSA in PBS to the bottom of 200-μm lecithin-coated specimen carriers type A (*see* **Note 9**).
3. Place prefixed seminiferous tubules within the platelets.
4. Overfill the recess with the freeze protectant (*see* **Note 10**).
5. Close carriers with lecithin-coated specimen carrier type B using the 0 μm recesses as lids.
6. Place the carrier sandwich into the high-pressure freezer and freeze samples at a cooling rate of >20,000 K/s and >2100 bar. Carriers are directly casted into liquid nitrogen to prevent samples from thawing (*see* **Note 11**).
7. Transfer the carriers into the freeze substitution machine (AFS) keeping the samples constantly at liquid nitrogen temperature (*see* **Note 11**).

3.3 Freeze Substitution

Use plastic Pasteur pipettes for the exchange of freeze substitution solutions. Equilibrate their temperature to −90 °C at the rim of the freeze substitution machine before bringing them in contact with the solutions.

1. Pre-cool freeze substitution solutions: 0.5% glutaraldehyde with 0.1% tannic acid in anhydrous acetone, anhydrous acetone, and 2% osmium tetroxide in anhydrous acetone to −70 °C. Store a second stock of anhydrous acetone at −20 °C.
2. Place universal metal reagent bath containers in the freeze substitution machine. Use all but one of the containers as specimen holders (*see* **Note 12**).
3. Cool the freeze substitution machine and the 0.5% glutaraldehyde with 0.1% tannic acid in anhydrous acetone to −90 °C. Distribute the solution among the universal metal reagent bath containers once cooled, and cover with Teflon lids.
4. Place the specimen carrier into the plastic capsules with mesh bottoms, and transfer them quickly into the AFS. Cover the containers with Teflon lids (*see* **Note 13**).
5. Incubate the samples for 96 h in 0.5% glutaraldehyde with 0.1% tannic acid in anhydrous acetone at −90 °C. Exchange for fresh solution once during this time.
6. Wash four times with anhydrous acetone over the course of 4–6 h at −90 °C.

7. Exchange anhydrous acetone for 2% osmium tetroxide in anhydrous acetone and incubate for 28 h.
8. Ramp the temperature from −90 °C up to −20 °C over the course of 14 h. Then incubate at a constant −20 °C for 16 h before steadily increasing the temperature up to 4 °C within a period of 4 h.
9. Wash four times with anhydrous acetone over the course of 2–3 h at 4 °C.
10. Remove the universal metal reagent bath container with the samples in anhydrous acetone from the AFS to start the embedding protocol for Epon.

3.4 Embedding in Epon

Infiltrate samples with an ascending Epon series.

1. Prepare a snap-on lid with Epon/anhydrous acetone (ratio 1:1) (*see* **Note 13**).
2. Quickly place plastic capsules with the sample carriers in the 50% epoxy solution. Avoid drying of the specimen.
3. Transfer the samples into fresh 50% Epon/anhydrous acetone: remove samples from the carriers by pipetting up and down with a plastic Pasteur pipette keeping them immersed in Epon/acetone constantly; if they remain attached to the carriers, carefully lift the samples with a mounted needle first before transferring them into a snap-on lid with fresh epoxy-acetone mixture using the plastic Pasteur pipette.
4. Clean individual AFS containers by placing them in a jar filled with acetone.
5. Incubate the samples for 5 h in the Epon/anhydrous acetone mixture (1:1 ratio) at room temperature.
6. Transfer specimens to a snap-on lid containing 90% Epon in anhydrous acetone to be infiltrated overnight at 4 °C; preferably in a cold room covered with a glass dish.
7. Transfer samples into pure Epon and infiltrate for 2–3 h at room temperature.
8. Exchange the epoxy resin twice and incubate for 2–3 h each.
9. Embed the samples either at the bottom of an Epon-filled capsule placed upright in a capsule mold or within silicone rubber flat embedding molds. Cure for at least 48 h at 60 °C.

3.5 Sectioning

1. Load the Epon embedded sample onto an ultramicrotomy block.
2. Trim away excess Epon with a razor blade until reaching the embedded tissue; observe cutting depth through a binocular microscope.
3. Fasten the block to the ultramicrotome.

4. Submerge a poly-L-lysine-coated coverslip within the boat of the Histo Jumbo Diamond Knife, and fill recess with ddH_2O using a syringe.
5. Cut semithin sections; detach them from the knife with a mounted eyelash, and direct them to the coverslip. Decrease the water level with a syringe, forcing the sections to adhere irreversibly to the coverslip. Then carefully remove the coverslip from the knife's boat.
6. Stain the semithin section with methylene blue for 1 min.
7. Remove the staining solution by rinsing the section with a wash bottle filled with ddH_2O. Let the coverslip dry for 1 min on a heating block without burning the section.
8. Search for a suitable tissue region using a transmission light microscope. Repeat **Steps 4–8** until a respective cross section of a seminiferous tubule is detected.
9. Cut 250 nm semithin sections and transfer them onto Formvar-coated slot or mesh grids. Thinner sections provide a better *z*-resolution for electron tomography. However, for the sections to contain entire telomere attachment sites, we choose 250 nm semithin sections.

3.6 Contrasting and Carbon Coating

1. Coat a glass plate with parafilm (*see* **Note 14**).
2. Contrasting with uranyl acetate: place 50 μL of uranyl acetate solution on the parafilm, and incubate the grids facing down with the sample side toward the drop of solution. In this step contrasting can be performed either in ethanolic or aqueous conditions. We achieve good results for both approaches (*see* **Note 15**).
   (a) Ethanolic conditions: incubate grids for 15 min with 2.5% uranyl acetate in ethanol.
   (b) Aqueous conditions: incubate grids for 10 min with 2% uranyl acetate in ddH_2O.
3. Washing: prepare three roll edge glasses with washing solutions; secure the grids with microsurgical forceps and dip them repeatedly into the first washing solution before moving to the next one. Proceed in the following order:
   (a) After ethanolic contrasting: 100% ethanol followed by ethanol/ddH_2O (1:1 ratio) followed by 100% ddH_2O.
   (b) After aqueous contrasting: immerse three times in fresh ddH_2O each.
4. Rinse grids with a wash bottle of ddH_2O. Avoid spraying the Formvar membrane directly as it is delicate and hence prone to tear.

5. Dry grids by gently swiping them over a piece of filter paper (*see* **Note 16**).
6. Contrasting with Reynold's lead citrate (diluted 1:1 in degassed ddH_2O): prepare fresh glass plate with parafilm. Place 50 μL of the diluted Reynold's lead citrate within a circle of sodium hydroxide pellets (*see* **Note 17**). Incubate samples previously contrasted with uranyl acetate in ethanol for 10 min and aqueously contrasted grids for 5 min.
7. Wash grids by repeated dipping in degassed ddH_2O; exchange degassed water three times.
8. Repeat **Steps 4** and **5**.
9. Carbon coat dry samples in a vacuum of 8×10^{-6} bar with a sputter coater to avoid charging of the grids due to long exposures to the electron beam during tilt-series acquisition (*see* **Note 18**).
10. Fiducial marker application: incubate grids for at least 1 min on each side with an unspecific secondary antibody coupled to at least 12 nm gold particles. First incubate one side, rinse with ddH_2O, and then proceed to the other side.

3.7 Electron Tomography, Tomogram Reconstruction, and Modeling

1. Introduce grid with a tomography sample holder to a transmission electron microscope equipped with a goniometer stage suited for high-tilt tomographic applications. We use the JEOL JEM-2100 at 200 kV which is connected to the TVIPS F416 4k × 4k camera.
2. Use SerialEM [12] to set following parameters for tilt-series acquisition: magnification of 40,000× to achieve good resolution of LINC complexes; recording angles from at least −65° to +65° in 1° increments.
3. Remove grid from the electron microscope and rotate it by 90° before reintroducing the grid to the TEM and recording a complementary tilt series (*see* **Note 19**).
4. Use ETomo of the IMOD suite [13] for tilt-series alignment and tomogram reconstruction based on a weighted back-projection algorithm.
5. Carry out manual reconstruction of the synaptonemal complex and its associated LINC complexes using 3dmod [13]:
   (a) Open the .rec or the .mrc file for reconstruction, and flip the *x*- and *y*-axis by rotating around *x* to get the *xy*-view in the main image display and model editing zap window.
   (b) Zoom in and out on the structure of interest using the "+" and "−" shortcuts.
   (c) Choose between the three possible object types, *closed*, *open*, and *scattered*: the *closed* option is suited for volumes

and surfaces such as the lateral and central element of the synaptonemal complex as well as nuclear membranes. The *open* object is best used for filamentous proteins like the transverse filaments and the LINC complexes (*see* **Note 20**). Use the shortcut "n" to start a new filament of the same object. The *open* object option can also be used to annotate longitudinal microtubules by placing a line in the middle of the tube, using the shortcut "v" to open the model view window and navigating to edit → objects → meshing → ticking the option *tube* and entering the diameter of a microtubules of 25 nm. For transverse microtubules a circular closed object with the same diameter is favored.

(d) In the main control window, the 3dmod information window navigate to special → drawing tools → drawing type and choose between *sculpt* in case of closed objects and *normal* for open objects.

(e) Trace an object in one plane, move to a section that is approximately 30–50 sections above or below the current plane, annotate the structure in this *z*-plane accordingly and interpolate using special → interpolator → linear (for linear interpolation), and choose the section difference plus one as the *z* bridge. Confirm by pressing enter. We advise that the last contour is chosen and used as the new first contour in the next interpolation step to avoid inconsistencies in the model.

4 Notes

1. The first wave of mouse spermatogenesis starts approximately 8–9 days postpartum and is characterized by the synchronous progression of all meiocytes within the testis. Consequently, in 14-day-old mice mid-pachytene spermatocytes are enriched. As we want to study the synaptonemal complex with associated LINC complexes and the full assembly of this tripartite structure is solely limited to pachytene spermatocytes, we take advantage of the overrepresentation of latter cells by using tissue of 14-day-old mice.
2. Leica specimen carriers Ø 3 mm × 0.5 mm gold-plated copper. Type A has circular indentations on both sides, one of Ø 0.1 mm and one of Ø 0.2 mm. Type B has an indentation on one side of Ø 0.3 mm and is flat on the other side. We use a carrier sandwich of the 0.2 mm indentation of a type A carrier in combination with the flat side of the type B carrier as a lid.
3. Freeze substitution prepares the samples for electron microscopy in high vacuum by dehydration. Cellular water is gradually substituted with the organic-solvent acetone in combination

with fixatives at low temperatures to avoid ice crystal formation. In order not to reintroduce water to the specimen, all freeze substitution solutions need to be anhydrous/dry.

4. Methylene blue-azur II stain: 1 g sodium tetraborate in 100 mL ddH_2O (1%) and heat whilst stirring until the sodium borate is completely dissolved, then add 1 g methylene blue (1%). Simultaneously dissolve 1 g azur II in 100 mL ddH_2O (1%) and stir the solution whilst heating. Combine the methylene blue and azur II stock 1:1 and filter the solution two times before use.
5. Reynold's lead citrate: dissolve 1.33 g lead nitrate and 1.76 g sodium citrate in 30 mL degassed water and shake strongly and repeatedly over the course of 30 min. Add 8 mL 1 M sodium hydroxide (0.4 g sodium hydroxide pellets dissolved in 10 mL degassed water) and make up to 50 mL with degassed water.
6. Centrifuge both the respective uranyl acetate solution and the Reynold's lead citrate at maximum speed for 5 min. Additionally, filter the uranyl acetate solutions through a 0.2 μm PTFE filter. Filtering the solution prevents contamination of section with spicular crystalline precipitation caused by the uranyl acetate.
7. CMOS cameras, such as the TemCam F416 (Tietz Video and Imaging Processing Systems, Gauting, Germany), combine high sensitivity and high resolution (4K) with fast readout times, fulfilling the requirements for the acquisition of LINC complexes at the attachment sites of meiotic telomeres to the nuclear envelope.
8. Excellent documentation on the individual IMOD programs including video tutorials can be found online [18].
9. Structure preservation during and after high-pressure freezing depends on strict air exclusion from the carrier sandwich to prevent ice crystal formation on the one hand and secure sample removal after high-pressure freezing on the other hand. Placing a small amount of freeze protectant/filler in the carrier before transferring the sample facilitates an air-tight seal. Lecithin coating of the carriers enables the removal of intact samples after high-pressure freezing.
10. Overfilling the chamber with freeze protectant/filler prevents ice crystal formation due to implosion during high-pressure freezing.
11. It is crucial to avoid thawing of the samples above -90 ± 2 °C to secure ultrastructure preservation.
12. Equilibrate the individual freeze substitution solutions for 10–20 min before use within the remaining container of the AFS.

13. Each of the universal metal reagent bath containers can carry up to four plastic capsules with different samples. To distinguish the samples, we carve a different number of notches in the individual capsules. We add the same notches to the snap-on lids used during embedding to assign the samples correctly.
14. The creation of a flat surface is of importance to ensure even incubation of the samples during contrasting, thus avoiding creasing of the parafilm.
15. Uranyl acetate is photosensitive; hence incubate grids in the dark, e.g., by covering the grids with a paper box.
16. Remove solution between the tweezers legs which might lead to accidental rehydration of the sample.
17. Lead citrate reacts with carbon dioxide to form lead carbonate. The latter precipitates in form of round, electron-dense hexagons contaminating the sections. To prevent precipitation, we not only degas the ddH_2O in which the lead citrate is diluted but also arrange sodium hydroxide pellets, which absorb CO_2, in a circular pattern around the lead citrate during incubation. We also advise that you hold your breath during placing the grids on the lead citrate before covering the reaction with a glass dish to further prevent carbon dioxide exchange.
18. Attach a piece of filter paper with double-faced adhesive tape next to the grids during coating; use the darkening of the filter paper as an indicator for sufficient coating. Carbon coating can be repeated until charging stops.
19. Using the JEOL EM-21311 high-tilt specimen retainer (JEOL, Munich, Germany), we acquire tilt series within a maximum range of 140° (from −70° to +70°). The structural information contained in the missing tilt angles (141–180°) results in a reduced 3D resolution after tomogram reconstruction. This so-called missing wedge effect can be partially compensated for by the recording of a second, perpendicular tilt axis.
20. We advise that criteria are chosen for continuous objects spanning multiple *z*-planes such as transverse filaments of synaptonemal complexes and LINC complexes. Our two main criteria have similarities in electron density and most importantly continuity.

Acknowledgments

We are highly indebted to Markus Engstler (Biocenter, Würzburg) for equipment available in our laboratories. R.B. is supported by grant Be 1168/8-1 of the DFG.

References

1. Ungricht R, Kutay U (2017) Mechanisms and functions of nuclear envelope remodeling. Nat Rev Mol Cell Biol 18:229–245
2. Chang W, Worman HJ, Gundersen GG (2015) Accessorizing and anchoring the LINC complex for multifunctionality. J Cell Biol 208:11–22
3. Link J, Jahn D, Alsheimer M (2015) Structural and functional adaptations of the mammalian nuclear envelope to meet the meiotic requirements. Nucleus 6:93–101
4. Link J, Benavente R, Alsheimer M (2016) Analysis of meiotic telomere behavior in the mouse. Methods Mol Biol 1411:195–208
5. Liebe B, Alsheimer M, Höög C et al (2004) Telomere attachment, meiotic chromosome condensation, pairing, and bouquet stage duration are modified in spermatocytes lacking axial elements. Mol Biol Cell 15:827–837
6. Schmitt J, Benavente R, Hodzic D et al (2007) Transmembrane protein Sun2 is involved in tethering mammalian meiotic telomeres to the nuclear envelope. Proc Natl Acad Sci U S A 104:7426–7431
7. Ding X, Xu R, Yu J et al (2007) SUN1 is required for telomere attachment to nuclear envelope and gametogenesis in mice. Dev Cell 12:863–872
8. Morimoto A, Shibuya H, Zhu X et al (2012) A conserved KASH domain protein associated with telomeres, SUN1 and dynactin during mammalian meiosis. J Cell Biol 198:165–172
9. Horn HF, Kim DI, Wright GD et al (2013) A mammalian KASH domain protein coupling meiotic chromosomes to the cytoskeleton. J Cell Biol 202:1023–1039
10. Spindler MC, Redolfi J, Helmprobst F et al (2018) Electron tomography of LINC complex association with microtubules at meiotic telomere attachment sites, submitted.
11. Reynolds ES (1963) The use of lead citrate at high pH as an electron-opaque stain in electron microscopy. J Cell Biol 17:208
12. Kremer JR, Mastronarde DN, McIntosh JR (1996) Computer visualization of three-dimensional image data using IMOD. J Struct Biol 116:71–76
13. Mastronarde DN (2005) Automated electron microscope tomography using robust prediction of specimen movements. J Struct Biol 152:36–51
14. Dhanyasi N, Segal D, Shimoni E et al (2015) Surface apposition and multiple cell contacts promote myoblast fusion in drosophila flight muscles. J Cell Biol 211:191–203
15. Weimer RM (2006) Preservation of *C. elegans* tissue via high-pressure freezing and freeze-substitution for ultrastructural analysis and immunocytochemistry. Methods Mol Biol 351:203–221
16. Stigloher C, Zhan H, Zhen M et al (2011) The presynaptic dense projection of the *Caenorhabditis elegans* cholinergic neuromuscular junction localizes synaptic vesicles at the active zone through SYD-2/Liprin and UNC-10/RIM-dependent interactions. J Neurosci 31:4388–4396
17. Helmprobst F, Frank M, Stigloher C (2015) Presynaptic architecture of the larval zebrafish neuromuscular junction. J Comp Neurol 523:1984–1997
18. http://bio3d.colorado.edu/imod/. Accessed 8 June 2017.

Chapter 2

Recombinant Purification of the Periplasmic Portion of the LINC Complex

Victor E. Cruz and Thomas U. Schwartz

Abstract

Recombinant expression of proteins and their complexes is the routine laboratory procedure to generate pure reagents for biochemical and structural studies. Here we present the standard procedure developed in our lab for the production of milligram quantities of stoichiometric SUN–KASH complexes. The protocol was specifically developed for the purification of the periplasmic portion of LINC complexes.

Key words LINC complex, Co-expression, Protein purification

1 Introduction

The nucleus of eukaryotic cells is separated from the cytoplasm through a double-layered membrane, the nuclear envelope (NE), composed of an inner nuclear membrane (INM) and an outer nuclear membrane (ONM). Sandwiched between INM and ONM lies a narrow element of the endoplasmic reticulum, the perinuclear space (PNS). INM and ONM are fused at distinct circular openings, which generate the conduits for macromolecular exchange through nuclear pore complexes (NPCs) [1]. Linker of nucleoskeleton and cytoskeleton (LINC) complexes physically connect the nucleus to its surrounding for mechanotransduction [2–7]. The core of a LINC complex consists of a SUN domain containing type II membrane protein spanning the INM and a tail-anchored KASH motif containing membrane protein spanning the ONM. SUN proteins typically have an N-terminal nucleoplasmic region, followed by a transmembrane helix, a coiled-coil element, and a C-terminal ~200 amino acid SUN domain [8]. In plants, the SUN domain can also occur in the middle of a protein (so-called mid-SUNs) [9, 10].

KASH proteins have an N-terminal cytoplasmic portion that often binds to a cytoskeletal element followed by a C-terminal KASH "domain" that consists of a single-pass transmembrane

Gregg G. Gundersen and Howard J. Worman (eds.), *The LINC Complex: Methods and Protocols*, Methods in Molecular Biology, vol. 1840, https://doi.org/10.1007/978-1-4939-8691-0_2, © Springer Science+Business Media, LLC, part of Springer Nature 2018

segment followed by a ~15–30 residue luminal peptide [8]. It is the luminal peptide that exclusively binds to SUN. The structures of the human SUN2–KASH1 and SUN2–KASH2 complexes have been solved [11, 12]. They show that the LINC complex is a heterohexamer, in which three KASH peptides independently bind a SUN trimer. The KASH peptides bind clefts generated at the interfaces of adjacent SUN protomers, thus explaining the need for SUN to trimerize to be binding-competent [13]. While there are multiple SUN and KASH homologs in most organisms, sequence analysis combined with structure modeling suggests that the principal binding mode is likely to be conserved.

Here we describe the purification of recombinant SUN–KASH complexes, focusing on the perinuclear core (Fig. 1). Similar procedures have been previously described [11, 13]; however, the method we present here is the most reproducible and universally applicable method that supersedes our earlier strategies. Using this procedure, we have co-purified various human SUN–KASH complexes and also the UNC84–UNC83 from *C. elegans* (*unpublished data*).

Our current method starts with co-expressing a 6×His–GB1 moiety [14] fused to the KASH peptide together with SUN that contains an N-terminal trimerizing-GCN4 tag [15]. In a first step, we co-purify the complex via immobilized Ni affinity chromatography. An excess of unbound GB1–KASH is then separated by size-exclusion chromatography, as the GB1 tag is only ~10 kDa in

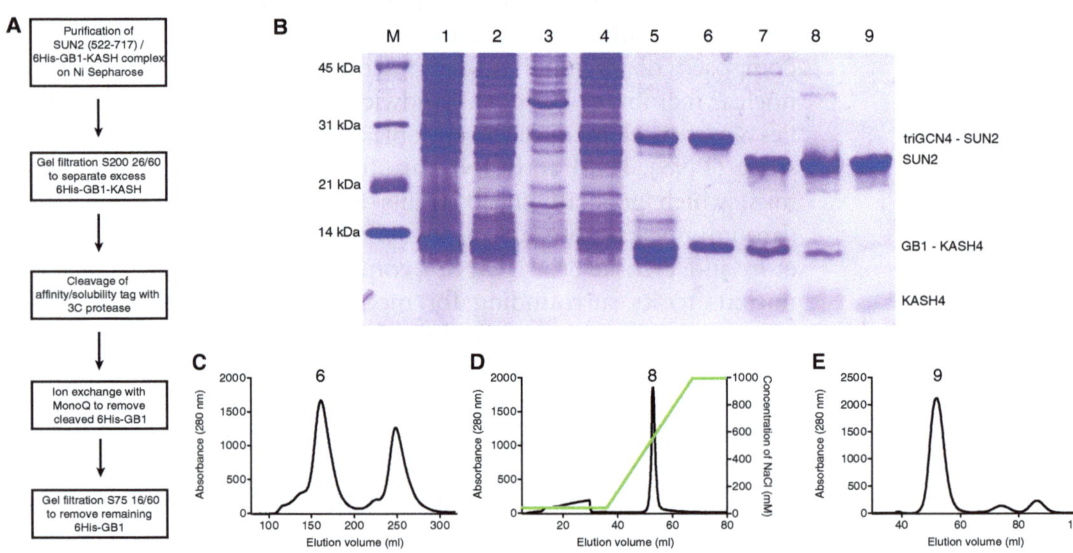

Fig. 1 *Purification of SUN–KASH complex.* (**A**) Flowchart of the SUN–KASH purification protocol. (**B**) SDS–PAGE analysis of critical fractions of the protein purification. Lanes correspond to 1, total lysate; 2, soluble fraction of lysate; 3, insoluble fraction of lysate; 4, unbound fraction from Ni-affinity chromatography; 5, Ni-affinity purified protein; 6, main peak from first size-exclusion step; 7, HRV 3C digest; 8, main peak from anion exchange step; and 9, main peak from second size-exclusion step. (**C**) Chromatogram of first size-exclusion step. (**D**) Chromatogram of anion exchange step. (**E**) Chromatogram of second size-exclusion step

size, compared to ~90 kDa for the heterohexameric SUN bound GB1–KASH complex. Then, we cleave the GB1 tag using human rhinovirus (HRV) 3C protease and separate it in a final size-exclusion chromatographic step. Some SUN–KASH complexes co-purify with chaperones or other contaminants that can be separated after 3C cleavage by anion exchange, followed by the final size-exclusion step (Fig. 1B).

The final yield of the purified complex is ~10 mg/L of bacterial culture and can be concentrated upward of 20 mg/mL if necessary. The SUN–KASH complexes can be flash frozen in liquid nitrogen and stored at −80 °C.

2 Materials

All solutions are prepared with ultrapure deionized water, filtered, and stored at 4 °C unless otherwise noted.

2.1 *E. coli* Manipulation and Growth

1. Bicistronic pET-DUET1 plasmid (Novagen). SUN2 (522–717) is cloned into the first cassette, and GB1–KASH is cloned into the second cassette (*see* **Note 1**).
2. Recombinant *E. coli* expression strain LOBSTR (DE3) (RIL) (Kerafast).
3. Luria–Bertani (LB) growth media.
4. 50 mL Erlenmeyer flasks for the starter culture and 2.8 L Erlenmeyer flasks for overnight expression.
5. Temperature-controlled shaking incubator.
6. 0.4 M Isopropyl β-D-1-thiogalactopyranoside (IPTG) solution (2000×).

2.2 *E. coli* Harvest and Lysis

1. Floor model centrifuge.
2. Lysis buffer: 50 mM potassium phosphate buffer (KPi), pH 8.0, 400 mM NaCl, 40 mM imidazole.
3. Phenylmethylsulfonyl fluoride (PMSF) (1000×): 100 mM in 100% ethanol.
4. TurboNuclease (Eton Biosciences).
5. LM20 Microfluidizer (Microfluidics) cell disruptor or equivalent.

2.3 Protein Purification

1. Ni Sepharose 6 Fast Flow beads (GE Healthcare).
2. 10 mL gravity flow columns (Thermo Scientific).
3. Elution buffer: 10 mM Tris–HCl, pH 7.4, 150 mM NaCl, 1 mM KCl, 250 mM imidazole pH 8.0. Final pH will be 8.0.
4. Vivaspin 20 mL 50 kDa molecular weight cutoff (Sartorius).

5. Fast protein liquid chromatography (FPLC) protein purification system.
6. Size-exclusion chromatography (SEC) buffer: 10 mM Tris–HCl, pH 7.4, 150 mM NaCl, 1 mM KCl, 0.2 mM EDTA.
7. HiLoad 26/60 Superdex 200 size-exclusion column (GE Healthcare).
8. Ion-exchange (IEX) buffer A: 10 mM Tris–HCl, pH 7.4, 1 mM KCl.
9. IEX buffer B: 10 mM Tris–HCl, pH 7.4, 1 mM KCl, 1 M NaCl.
10. Mono Q 5/50 GL column (GE Healthcare).
11. HiLoad 16/60 Superdex 75 size-exclusion column (GE Healthcare).
12. HRV 3C protease.

3 Methods

3.1 Protein Expression

1. Grow LOBSTR (DE3) (RIL) bacterial cells, transformed with the appropriate plasmid, in a shaker incubator overnight at 37 °C in 20 mL of LB as a starter culture (*see* **Note 2**).
2. The next morning inoculate 1 L of LB supplemented with 0.4% (w/v) glucose with the overnight starter culture. Grow culture at 37 °C while shaking at 220 rpm to an optical density at 600 nm (OD600 nm) of 0.6–0.8. Shift cells to 18 °C and induce protein expression by adding IPTG to a final concentration of 200 μM. Grow culture for another 12–16 h.
3. Centrifuge at 6000 × *g* for 6 min. Discard supernatant. The cell pellet can be frozen after this step and stored at −20 °C.

3.2 Cell Lysis and Nickel Affinity Chromatography

1. Resuspend cell pellet at 4 °C in lysis buffer. Use about 20 mL of buffer for every 200 mL of densely grown bacterial culture (OD600 nm ≈ 5 at time of harvest). Resuspend until the suspension is homogenous.
2. Lyse cell suspension by two passages through an LM20 Microfluidizer at 18 kpsi or equivalent. Upon lysis, add PMSF to a final concentration of 100 μM and 250 units of TurboNuclease. Mix contents by gently swirling the lysate. Collect a sample of the crude lysate for SDS–PAGE.
3. Centrifuge lysate at 9500 × *g* for 25 min. Collect the cleared lysate; take a small sample for SDS–PAGE analysis. Take sample of pellet, before discarding it (*see* **Note 3**).
4. Wash the Ni Sepharose 6 Fast Flow beads with lysis buffer. Use 1 mL of bead bed volume per 1000 OD of cells. Add washed

beads to cleared lysate and incubate in batch at 4 °C. Gently stir or rock the mixture during incubation.

5. Centrifuge the nickel bead suspension at 2000 × *g* for 2 min in 50 mL falcon tubes. Keep a small sample of the cleared supernatant as an unbound control. Resuspend the nickel beads in 5× bed volume of lysis buffer and centrifuge again with 5× the bed volume of nickel beads. Repeat this washing step twice. Resuspend the washed nickel beads and transfer to disposable 10 mL gravity flow columns. Drain column. Wash column with another five bed volumes of lysis buffer. Elute protein complex from the nickel beads with five bed volumes of elution buffer in one fraction. Collect a sample from the elution for SDS–PAGE analysis.

3.3 Separating SUN–KASH Complex from Excess GB1 and Proteolytic Cleavage of Tags

1. Concentrate the protein elution in a 20 mL Vivaspin column with a molecular weight cutoff of 50 kDa. Concentrate protein to a volume of 5 mL, and then load onto a Superdex S200 26/60 size-exclusion column (GE Healthcare) pre-equilibrated in SEC buffer.
2. The elution profile from the size-exclusion column will show three peaks: a void peak, the SUN–KASH complex peak, and excess GB1–KASH as the final peak (Fig. 1C). Pool the SUN–KASH complex peak and determine the protein concentration using a UV spectrophotometer. Proteolytically cleave the solubility tag by adding HRV 3C protease at a 1:200 (w/w) ratio and incubating at 4 °C overnight.
3. Determine purity of the protein complex by SDS–PAGE and stain using Coomassie Brilliant Blue. If the protein complex is pure and no additional protein bands are observed on the gel, skip to **step 4**. Otherwise, load the SUN–KASH complex (in SEC buffer) onto a Mono Q 5/50 anion exchange column. Elute using a 50–1000 mM NaCl in four column volumes. The SUN–KASH complex elutes at ~500 mM NaCl. This step has the added benefit of concentrating the protein for the final size-exclusion chromatography step (Fig. 1D).
4. If ion exchange step was unnecessary, then concentrate the protein in a 20 mL Vivaspin concentrator with a molecular weight cutoff of 50 kDa to a volume of 5 mL. Otherwise, collect peak fractions from the ion exchange purification. Inject protein complex onto a Superdex S75 16/60 column, pre-equilibrated in SEC buffer. The main peak should contain pure, homogeneous SUN2–KASH complex (Fig. 1E).

4 Notes

1. We have employed the general strategy, outlined here, to purify various human SUN–KASH complexes (including SUN2–KASH1/2/3/4/5, SUN2–LRMP, and SUN1–KASH1/2) and the UNC84–UNC83 complex from *C. elegans*. The tri-GCN4 tag helps to stabilize the apo-SUN–trimer, especially for constructs that only contain the SUN domains. Extended SUN constructs, containing portions of the coiled-coil domain, stably trimerize without the tri-GCN4 tag.
2. We routinely use LOBSTR (DE3) (RIL) bacterial cells for protein expression. They are particularly useful when the protein expression is poor. In such circumstances they help in suppressing co-purification of contaminating *E. coli* proteins [16].
3. To generate a protein sample of the insoluble fraction after cell lysis, we resuspend the pellet in 8 M urea by vortexing and then homogenize the sample using a dounce homogenizer. The equivalent of 50–100 μL bacterial culture of OD600 = 1 results in a balanced banding pattern following Coomassie Brilliant Blue-stained SDS–PAGE analysis using standard 15% polyacrylamide mini-gels.

Acknowledgments

This work was supported by NIH grant AR-065484 to T.U.S. and in part by the NIH Predoctoral Training Grant T32GM007287.

References

1. Knockenhauer KE, Schwartz TU (2016) The nuclear pore complex as a flexible and dynamic gate. Cell 164(6):1162–1171
2. Burke B, Roux KJ (2009) Nuclei take a position: managing nuclear location. Dev Cell 17(5):587–597
3. Starr DA, Fridolfsson HN (2010) Interactions between nuclei and the cytoskeleton are mediated by SUN–KASH nuclear-envelope bridges. Annu Rev Cell Dev Biol 26(1):421–444
4. Luxton GG, Starr DA (2014) KASHing up with the nucleus: novel functional roles of KASH proteins at the cytoplasmic surface of the nucleus. Curr Opin Cell Biol 28:69–75
5. Rothballer A, Kutay U (2013) The diverse functional LINCs of the nuclear envelope to the cytoskeleton and chromatin. Chromosoma 122(5):415–429
6. Chang W, Worman HJ, Gundersen GG (2015) Accessorizing and anchoring the LINC complex for multifunctionality. J Cell Biol 208(1):11–22
7. Meier I (2016) LINCing the eukaryotic tree of life-towards a broad evolutionary comparison of nucleocytoplasmic bridging complexes. J Cell Sci 129(19):3523–3531
8. Sosa BA, Kutay U, Schwartz TU (2013) Structural insights into LINC complexes. Curr Opin Struct Biol 23(2):285–291
9. Murphy SP, Simmons CR, Bass HW (2010) Structure and expression of the maize (*Zea mays* L.) SUN-domain protein gene family: evidence for the existence of two divergent classes of SUN proteins in plants. BMC Plant Biol 10:269
10. Graumann K, Vanrobays E, Tutois S, Probst AV, Evans DE, Tatout C (2014) Characterization of two distinct subfamilies of

SUN-domain proteins in Arabidopsis and their interactions with the novel KASH-domain protein AtTIK. J Exp Bot 65(22):6499–6512

11. Sosa BA, Rothballer A, Kutay U, Schwartz TU (2012) LINC complexes form by binding of three KASH peptides to domain interfaces of trimeric SUN proteins. Cell 149(5):1035–1047
12. Wang W, Shi Z, Jiao S, Chen C, Wang H, Liu G et al (2012) Structural insights into SUN–KASH complexes across the nuclear envelope. Cell Res 22(10):1440–1452
13. Demircioglu FE, Cruz VE, Schwartz TU (2016) Purification and structural analysis of SUN and KASH domain proteins. Methods Enzymol 569:63–78
14. Huth JR, Bewley CA, Jackson BM, Hinnebusch AG, Clore MG, Gronenborn AM (1997) Design of an expression system for detecting folded protein domains and mapping macromolecular interactions by NMR. Protein Sci 6:2359–2364
15. Ciani B, Bjelic S, Honnappa S, Jawhari H, Jaussi R, Payapilly A et al (2010) Molecular basis of coiled-coil oligomerization-state specificity. Proc Natl Acad Sci 107(46):19850–19855
16. Andersen KR, Leksa NC, Schwartz TU (2013) Optimized *E. coli* expression strain LOBSTR eliminates common contaminants from his-tag purification. Proteins Struct Funct Bioinform 81(11):1857–1861

Chapter 3

Analysis of High Molecular Weight Isoforms of Nesprin-1 and Nesprin-2 with Vertical Agarose Gel Electrophoresis

Chloe Potter and Didier Hodzic

Abstract

The biochemical characterization of proteins most often require their identification by immunoblotting. Whereas SDS-PAGE provides satisfactory results for most proteins, the identification of larger proteins requires alternative methods to ensure their separation and complete transfer onto nitrocellulose membranes. Here, we describe the application of vertical agarose gel electrophoresis to identify large isoforms of nesprin-1 and nesprin-2.

Key words Vertical agarose gel electrophoresis, VAGE, Nesprin-1 giant, Nesprin-2 giant, Nesprins, SUN proteins, LINC complexes

1 Introduction

SYNE1 and *SYNE2* (*Sy*naptic *n*uclear *e*nvelope 1 and 2) are notoriously large genes that encode multiple isoforms of nesprin-1 and nesprin-2 (*n*uclear *e*nvelope *sp*ect*rin* repeats 1 and 2), a family of spectrin repeat-containing proteins that dock to the outer nuclear membrane of the nuclear envelope through an evolutionary-conserved C-terminal KASH domain [1]. The KASH domain of nesprins interacts with the SUN domain of SUN proteins, a family of transmembrane proteins of the inner nuclear membrane. The direct interaction of nesprins with SUN proteins forms the so-called linkers of the nucleoskeleton and the cytoskeleton (LINC) complexes that span the whole nuclear envelope [2, 3]. Studies in various genetic models have now clearly established that LINC complexes mediate nuclear positioning during organogenesis [4, 5].

Understanding the biology of nesprins is stressed by the identification of an increasing array of human pathologies associated with mutations of these genes [6, 7]. However, the size of large isoforms encoded by *SYNE1* and *SYNE2* such as nesprin-1 giant (~1 MDa), nesprin1β (~350 kDa), and nesprin-2 giant (~800 kDa) significantly hampers both the separation and the reproducible

Gregg G. Gundersen and Howard J. Worman (eds.), *The LINC Complex: Methods and Protocols*, Methods in Molecular Biology, vol. 1840, https://doi.org/10.1007/978-1-4939-8691-0_3, © Springer Science+Business Media, LLC, part of Springer Nature 2018

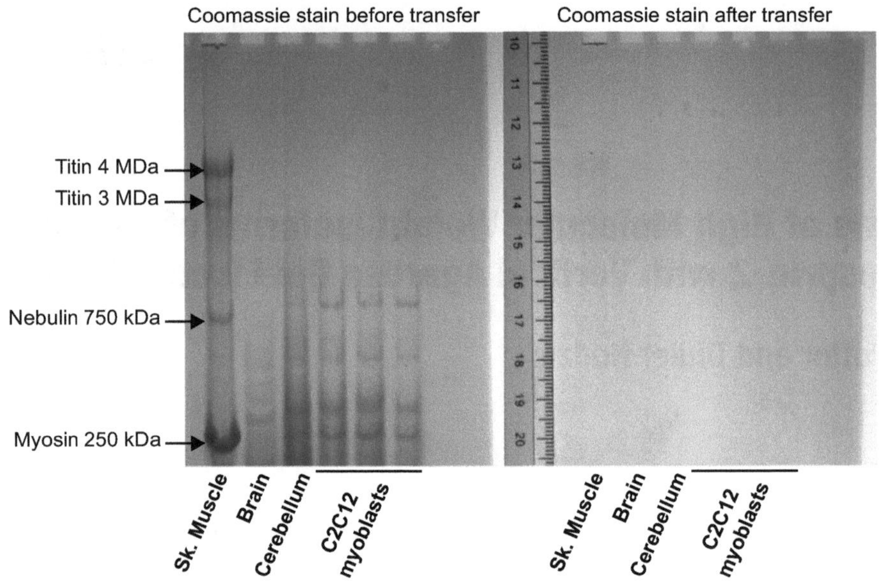

Fig. 1 VAGE allows for the efficient separation and transfer of high molecular weight proteins. A single agarose gel was loaded in duplicate with the indicated samples and cut in two halves. A half gel was stained with Coomassie blue before transfer (left), and the other was used for protein transfer and then stained with Coomassie blue (right). Note the efficient transfer of high molecular weight proteins. Labels on the left point to two isoforms of titin, nebulin, and myosin that are used as molecular weight markers after Ponceau S staining of nitrocellulose membranes

transfer of these proteins to nitrocellulose membrane using classical SDS-PAGE. To palliate this drawback, we now routinely analyze large isoforms of nesprins using vertical gel agarose electrophoresis (VAGE), a method that was initially developed for the analysis of large structural proteins of the skeletal muscle [8]. Here, we provide a detailed experimental description of this method applied to the identification of large isoforms of nesprin-1 and nesprin-2 extracted from mouse tissues. This protocol allows for the efficient separation, transfer (Fig. 1), and identification of large isoforms of nesprins (Figs. 2 and 3).

2 Materials

2.1 Mouse Tissue Lysates

1. Freshly isolated mouse tissue. Mice should be euthanized by CO_2 inhalation and dissected immediately to isolate tissues of interest. Appropriate animal protocols approved by the institution are necessary before proceeding with isolation of mouse tissues.
2. Bullet Blender (Next Advance).
3. Zirconium beads (Next Advance, Cat#ZrOB05-RNA).

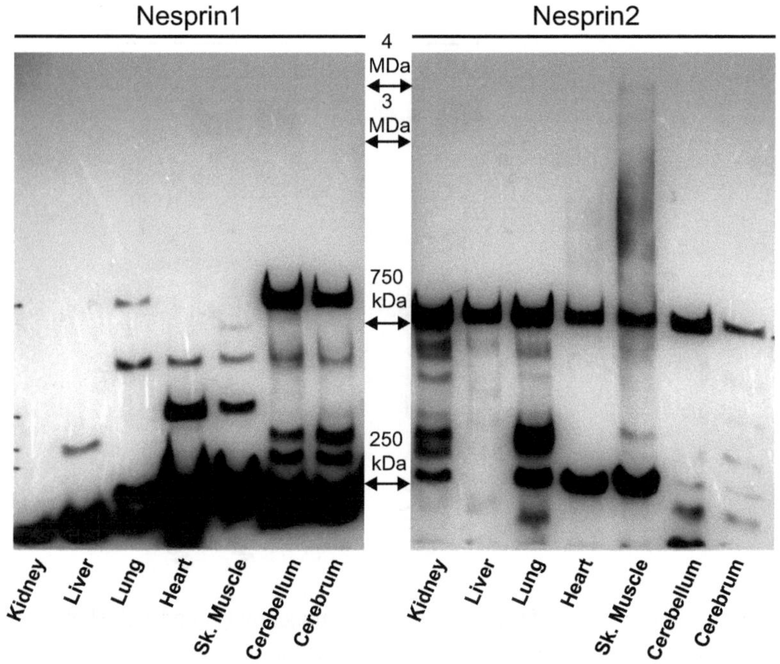

Fig. 2 Detection of high molecular weight isoforms of nesprin-1 and nesprin-2 in mouse tissues. A single agarose gel was loaded in duplicate with indicated samples, transferred onto nitrocellulose membrane that was subsequently cut into two pieces after Ponceau S staining. The left side was blotted with nesprin-1 antibody and the right side with nesprin-2 antibody. The respective epitopes of nesprin-1 and nesprin-2 antibodies are shared by the C-terminal region of all nesprin-1 and nesprin-2 isoforms. Note that nesprin-1 giant, whose molecular weight was estimated at 980 kDa, is expressed in the cerebellum and the brain and to a lower extent in the lungs. By contrast, nesprin-2 giant, whose molecular weight was estimated at 800 kDa, is expressed in all tissues. Note that additional immunoreactive bands of lower molecular weights are also detected. These bands may correspond to smaller nesprin isoforms (*see* Fig. 3) or to degradation products of larger isoforms

4. VAGE buffer: 8 M urea, 2 M thiourea, 3% SDS, 75 mM dithiothreitol, 0.05 M, 0.04% bromophenol blue, 50 mM Tris–HCl pH 6.8.

2.2 Vertical Agarose Gel Electrophoresis

1. Protean II slab cell system (Biorad).
2. SeaKem Gold Agarose (Lonza, Cat#50152).
3. Frosted inner glass plates (Biorad, Cat#1651825, 16 × 20 cm).
4. Outer glass plates (Biorad, Cat#1651822, 18.3 × 20 cm).
5. Power supply (PowerPac HV, Biorad).
6. 5× running buffer: 0.25 M Tris–Base, 1.92 M glycine, 0.5% SDS (no need to pH).
7. β-mercaptoethanol.

2.3 Transfer of Agarose Gel to Nitrocellulose

1. Trans-Blot cell (Biorad).
2. Nitrocellulose membrane (Protran, Cat#10600015, GE Healthcare).

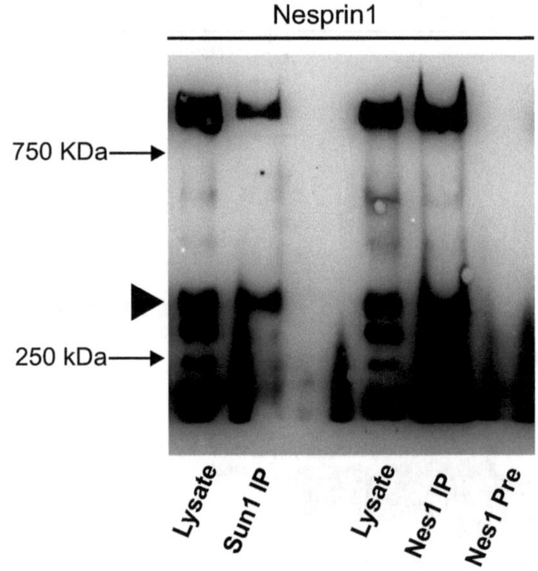

Fig. 3 VAGE analysis of large isoforms of nesprin-1 that coimmunoprecipitated with SUN1 in cerebellar tissues. Cerebellar lysates were immunoprecipitated with antibodies either against SUN1 (left), a known binding partners of nesprin-1 at the nuclear envelope, or against nesprin-1 (right). Note the coimmunoprecipitation of nesprin-1 giant and of a smaller nesprin-1 isoform (arrowhead) with SUN1. Both proteins are specifically immunoprecipitated by nesprin-1 antibodies. *IP* immunoprecipitation, *Pre* immunoprecipitation performed with the corresponding preimmune serum

3. Transfer buffer: 20% methanol, 40 mM glycine, 0.05% SDS, 50 mM Tris–HCl pH 8.3.
4. Ponceau S: 0.1% Ponceau S, 5% glacial acetic acid.
5. TBST: 150 mM NaCl, 0.1% Tween 20, 10 mM Tris–HCl, pH 7.8.
6. Blocking solution: 5% blotting-grade blocker (Biorad, Cat#170-6404) in TBST.
7. Coomassie blue (SimplyBlue Safe Stain, Cat#LC6060, Thermo Fisher).
8. West Pico signal (Thermo Fisher, Cat#1868123 and #1862124).
9. Nesprin-1, nesprin-2, and Sun1 antibodies [9, 10].
10. Horseradish peroxidase-conjugated donkey anti-rabbit secondary antibody.
11. X-ray film.

2.4 Immunoprecipitation

1. RIPA buffer: 150 mM NaCl, 0.5% sodium deoxycholate, 0.1% SDS, 1% Triton X-100, 50 mM Tris–HCl, pH 7.6 supplemented with protease inhibitors (Roche, Cat#11697498001).
2. Protein A/G beads (Pierce, Cat#20421).

3 Methods

3.1 Preparation of Mouse Tissue Lysates

1. To 1.5 mL tube, add ~0.1 g of zirconium oxide beads and weigh each tube.
2. Euthanize mice by CO_2 inhalation using a controlled flow chamber.
3. Dissect out the tissues of interest (as well as skeletal muscle whose lysates are used as molecular weight markers; *see* Subheading 3.3, **Step 6,** and Fig. 1), and mince with a razor blade.
4. Add ~100 mg of minced tissues to 1.5 mL tube with zirconium beads, and weigh again to know the exact weight of wet tissue.
5. Add freshly prepared VAGE buffer to obtain a final concentration of 100 mg of wet tissue/mL of VAGE buffer.
6. Beat all tubes in a Bullet Blender for 4 min at speed 7.
7. Boil samples for 5 min and centrifuge at 16,000 × *g* for 10 min.
8. Collect the supernatants and use immediately or store in small aliquots at −80 °C for future use.

3.2 Agarose Gel Preparation and Electrophoresis

1. Assemble a frosted inner glass plate together with spacers and an outer glass plate using sandwich clamps provided with the Protean II slab cell system.
2. Mount the assembly on the casting stand (*see* **Note 1**).
3. Place the casting stand as well as a 35 mL syringe barrel in an oven set at 70 °C for equilibration.
4. For a 1% agarose gel, weigh 0.5 g SeaKem Gold Agarose in a 250 mL conical flask. In a cylinder, mix 15 mL of 100% glycerol with 10 mL of 5× running buffer and 25 mL of water, add to the agarose powder, swirl, and cover with aluminum foil before heating at 95 °C in a water bath until the solution is completely clear.
5. Let the agarose solution equilibrate at 70 °C in the oven next to the casting stand.
6. With the tip of the syringe barrel touching the upper part of the outer glass plate, slowly pour the agarose solution through the barrel until it reaches the top of the inner plate and insert a 15-well comb.
7. Turn off the oven and leave open until the system reaches room temperature.
8. Place the cast on a benchtop for 30 min and then at 4 °C for 30 min to polymerize the agarose completely (*see* **Note 2**).
9. Remove the comb just before loading the samples and discard any excess agar with a razor blade (*see* **Note 3**).

10. Remove the gel from the casting stand and secure it to the running cell (*see* **Note 4**).
11. Fill the lower chamber with 3 L of 1× running buffer and the upper chamber with 600 mL of running buffer supplemented with 400 μL of β-mercaptoethanol (*see* **Note 5**).
12. Load 40 μL of 5× dilution of each tissue sample in VAGE buffer, i.e., the equivalent of 800 μg of wet tissue/well (*see* **Note 6**). Make sure to load at least one well with skeletal muscle lysates that will be used as a molecular weight marker (*see* Subheading 3.3, **Step 6**).
13. Move the running cell to a cold room, and run the gel at 60 V for at least 16 h using an appropriate power supply.

3.3 Gel Transfer and Nitrocellulose Membrane Staining and Immunoblotting

1. Disassemble the glass plates from the running cell, unscrew the clamps, delicately remove the spacers, lift one glass plate to access the gel, and briefly rinse in transfer buffer.
2. Assemble gel, nitrocellulose membrane, filter papers, and foam pads in a gel holder cassette, and insert in the Trans-Blot cell so that the nitrocellulose is between the gel and the +electrode.
3. Fill the Trans-Blot cell tank with transfer buffer, and transfer for 2 h 20 min at 70 V at 4 °C.
4. Open the gel holder cassette; remove the foam pad and the filter paper to access the gel. Mark the bottom of each well on the membrane with a pencil; discard the gel (*see* **Note** 7).
5. Remove the nitrocellulose membrane. With protein side up, immerse in Ponceau S for 1 min and rinse with MilliQ water until protein bands are clearly visible.
6. Using a pencil, mark the strong stained bands that correspond to titin (4 MDa), titin (3 MDa), nebulin (750 kDa), and myosin (250 kDa) that are clearly visible in skeletal muscle lysates (Fig. 1). These marks are used as molecular weight markers (*see* **Note 8**).
7. Incubate the membrane in 25 mL of blocking solution for 1 h at room temperature. Replace with blocking solution supplemented with 1:1000 dilutions of nesprin-1 and nesprin-2 antibodies. Incubate overnight at 4 °C.
8. Wash the membrane three times 10 min in TBST.
9. Incubate the membrane in 25 mL of blocking solution supplemented with 5 μL of HRP-conjugated goat anti-rabbit secondary antibodies for 1 h at room temperature.
10. Wash the membrane three times in TBST.

11. Cover the membrane with 3 mL stable peroxide solution and 3 mL of luminol/enhancer solution from the West Pico Signal kit.
12. Expose the membrane to X-ray films for 30 s or 2 min or in a gel imaging system (GBox, Syngene).
13. Typical immunoblotting results are shown in Fig. 2 for immunoblotting of mouse tissue lysates or of immunoprecipitations thereof (*see* **Note 9**).

3.4 Immunoprecipitation

1. Dissect out the tissues of interest and mince with a razor blade.
2. Add ~100 mg of minced tissues to 1.5 mL tube with zirconium beads and weigh again to know the exact weight of wet tissue.
3. Add freshly prepared RIPA buffer to obtain a final concentration of 100 mg of wet tissue/mL of RIPA buffer.
4. Beat the samples in a Bullet Blender for 4 min at speed 7.
5. Centrifuge at 12,000 × *g* for 10 min at 4 °C and collect the supernatant.
6. Preclear 500 μL of lysate for 2 h with 50 μL of Protein A/G beads slurry.
7. Centrifuge at 5000 × *g* for 1 min.
8. Collect the supernatant and add 20 μL of fresh Protein A/G slurry and 3 μg of immunoprecipitating antibody.
9. Rotate overnight at 4 °C.
10. Wash beads three times with RIPA buffer.
11. After the last wash, remove the supernatant and add 200 μL of VAGE buffer.
12. Boil for 5 min.

4 Notes

1. After assembling the glass plates on the cast stand, check for potential leaks by pouring water between both glass plates. Most leaks usually occur because the basis of the spacers is not perfectly flush with the bottom of the glass plates.
2. Polymerized gels can be unmounted from the cast stand and stored in Saran wrap (to avoid desiccation) at 4 °C for up to 2 days.
3. To avoid damaging the wells while removing the comb, slightly bend the comb first toward you in order to let air bubbles enter the well. Then, slowly pull the comb upward from the left side and then from right side multiple times.

4. After securing the gel casts to the running unit, it is preferable to fill the upper chamber with 600 mL of water to check for potential leaks.
5. Before loading the samples on the gels, flush all wells to remove the glycerol at the bottom of each well and load 10 μL of VAGE buffer both to visualize the bottom of each well and to remove potential agarose debris that may prevent the sample from reaching the bottom of the well.
6. We find that a 5× dilution of lysates purified from 100 mg/mL of wet tissue gives optimal signal to noise ratios. More concentrated lysates lead to very strong signal that produce band smearing in the immunoblot especially for nesprin-1.
7. It is preferable to stain the gel upon completion of protein transfer to nitrocellulose with Coomassie blue to ensure that transfer proceeded efficiently. No proteins should be visible on the stained gel (Fig. 1).
8. Ensure to load one well with undiluted skeletal muscle lysate to be able to distinguish bands of both isoforms of titin, nebulin, and myosin (Fig. 1) to unequivocally mark these proteins on the nitrocellulose membrane after Ponceau S staining. We usually run that concentrated sample of skeletal muscle in the middle of the gel (masked in Fig. 2) and cut the Ponceau S-stained membrane along that sample in order to blot the left and right side of the membrane with different antibodies. Half membranes are then carefully realigned vertically before exposure so that molecular weight differences between nesprin-1 and nesprin-2 isoforms can be clearly emphasized.
9. As shown in Fig. 2, 1% agarose gels allow to separate proteins with molecular weights between 3 MDa and 250 kDa over more than 6 cm, thereby allowing for a relatively precise estimation of molecular weight of nesprin-1 isoforms. Using titin isoforms, nebulin, and myosin as reference, the molecular weights of nesprin-1 giant and nesprin-2 giant were estimated at 980 kDa and 800 kDa, respectively. These sizes are in agreement with the theoretical molecular weight of the corresponding primary sequences.

Acknowledgment

This work was funded by an NIH grant from the National Eye Institute (#EY022632 to D.H.) and an unrestricted grant to the Department of Ophthalmology and Visual Sciences at Washington University from *Research to Prevent Blindness*.

References

1. Warren DT, Zhang Q, Weissberg PL, Shanahan CM (2005) Nesprins: intracellular scaffolds that maintain cell architecture and coordinate cell function? Expert Rev Mol Med 7(11):1–15. https://doi.org/10.1017/S1462399405009294
2. Crisp M, Liu Q, Roux K, Rattner JB, Shanahan C, Burke B, Stahl PD, Hodzic D (2006) Coupling of the nucleus and cytoplasm: role of the LINC complex. J Cell Biol 172(1):41–53. https://doi.org/10.1083/jcb.200509124
3. Chang W, Worman HJ, Gundersen GG (2015) Accessorizing and anchoring the LINC complex for multifunctionality. J Cell Biol 208(1):11–22. https://doi.org/10.1083/jcb.201409047
4. Bone CR, Starr DA (2016) Nuclear migration events throughout development. J Cell Sci 129(10):1951–1961. https://doi.org/10.1242/jcs.179788
5. Razafsky D, Hodzic D (2015) Nuclear envelope: positioning nuclei and organizing synapses. Curr Opin Cell Biol 34:84–93. https://doi.org/10.1016/j.ceb.2015.06.001
6. Cartwright S, Karakesisoglou I (2014) Nesprins in health and disease. Semin Cell Dev Biol 29:169–179. https://doi.org/10.1016/j.semcdb.2013.12.010
7. Synofzik M, Smets K, Mallaret M, Di Bella D, Gallenmuller C, Baets J, Schulze M, Magri S, Sarto E, Mustafa M, Deconinck T, Haack T, Zuchner S, Gonzalez M, Timmann D, Stendel C, Klopstock T, Durr A, Tranchant C, Sturm M, Hamza W, Nanetti L, Mariotti C, Koenig M, Schols L, Schule R, de Jonghe P, Anheim M, Taroni F, Bauer P (2016) SYNE1 ataxia is a common recessive ataxia with major non-cerebellar features: a large multi-centre study. Brain 139(Pt 5):1378–1393. https://doi.org/10.1093/brain/aww079
8. Warren CM, Krzesinski PR, Greaser ML (2003) Vertical agarose gel electrophoresis and electroblotting of high-molecular-weight proteins. Electrophoresis 24(11):1695–1702. https://doi.org/10.1002/elps.200305392
9. Razafsky DS, Ward CL, Kolb T, Hodzic D (2013) Developmental regulation of linkers of the nucleoskeleton to the cytoskeleton during mouse postnatal retinogenesis. Nucleus 4(5):399–409. https://doi.org/10.4161/nucl.26244
10. Khatau SB, Bloom RJ, Bajpai S, Razafsky D, Zang S, Giri A, Wu PH, Marchand J, Celedon A, Hale CM, Sun SX, Hodzic D, Wirtz D (2012) The distinct roles of the nucleus and nucleus-cytoskeleton connections in three-dimensional cell migration. Sci Rep 2:488. https://doi.org/10.1038/srep00488

Chapter 4

Analysis of Nesprin-2 Interaction with Its Binding Partners and Actin

Susumu Antoku and Gregg G. Gundersen

Abstract

Nuclei are connected to the actin cytoskeleton for controlling its position in the cell and for mechanochemical signaling. Nesprin-2G is one of the major outer nuclear membrane proteins that links the nucleus to the actin cytoskeleton. In addition to its paired calponin homology (CH) domains, nesprin-2G interacts with actin filaments by binding the actin-bundling proteins FHOD1 and fascin. We describe methods to measure the interaction of nesprin-2G with actin filaments using an actin co-sedimentation assay and with its binding partner FHOD1 using a GST pull-down method.

Key words LINC complex, Calponin homology domains, Spectrin repeats, Actin filaments, FHOD1, Fascin

1 Introduction

Nesprin-2G is a member of the KASH family of proteins in vertebrates that together with the SUN proteins comprise the linker of nucleoskeleton and cytoskeleton (LINC) complex [1, 2]. Nesprin-2G is one of two giant vertebrate KASH proteins (nesprin-1G being the other) that has paired CH domains that allow it to directly interact with actin filaments [3, 4]. Unlike nesprin-1G, nesprin-2G additionally interacts with actin filaments by binding the actin-bundling proteins FHOD1 and fascin [5]. Nesprin-2G directly binds both these proteins through specific spectrin repeats (SRs): SRs11–12 near the N-terminus in the case of FHOD1 [6, 7] and SRs51–53 near the C-terminus in the case of fascin [8]. The multipoint attachment of nesprin-2G to actin filaments is necessary to assemble and reinforce transmembrane actin-associated nuclear (TAN) lines that move the nucleus rearward in fibroblasts and myoblast polarizing for migration [4, 6–11] and after centrifugal displacement [12]. Here, we describe biochemical methods to analyze the interaction of nesprin-2G with actin and FHOD1.

Gregg G. Gundersen and Howard J. Worman (eds.), *The LINC Complex: Methods and Protocols*, Methods in Molecular Biology, vol. 1840, https://doi.org/10.1007/978-1-4939-8691-0_4,

There are a number of assays suitable for probing the interaction of nesprin-2G with actin and its binding partners. Because of their ease, co-immunoprecipitation and GST-pulldown from cellular lysates are the first method of choice and can provide valuable initial evidence supporting an interaction. Both were used in the analysis of nesprin-2G's interaction with FHOD1 and fascin [6, 8]. To ascertain direct binding, in vitro binding assays with purified proteins need to be performed. For these assays, either native proteins purified from cells or tissues or recombinant proteins purified after expression in bacteria or cultured cells need to be prepared. A variety of methods are available for detecting binding in these assays including pull-down and Western blot assays with tagged protein, surface plasmon resonance binding analysis, and fluorescence anisotropy with fluorescent derivatives of the proteins.

A specialized form of binding assay is available for proteins that bind to filamentous actin (F-actin). Because F-actin is substantially larger that soluble actin subunits, it can be pelleted at high centrifugal force. Actin-binding proteins that associate with F-actin are brought down with the F-actin, whereas those that do not bind remain in the supernatant. By quantification of protein bands on SDS-polyacrylamide gels, this method can be used to determine the K_d of the interaction between the actin-binding protein and actin. Here, we will describe methods to examine the interaction of fragments of nesprin-2G with F-actin and FHOD1.

2 Materials

2.1 Recombinant Fragment of Nesprin-2G

1. BL21 (DE3) bacteria (NEB, #2527I) transformed with pGEX 6P-1 plasmid (GE Healthcare, #28-9546-48) encoding GST-HA-nesprin-2G CH (7–891 a.a. of mouse nesprin-2G) and GST-nesprin-2G SR11–12 (1414–1635 a.a. of mouse nesprin-2G).
2. LB media.
3. 100 mg/mL ampicillin. Store at −20 °C.
4. 1 M isopropyl β-D-1-thiogalactopyranoside (IPTG). Store at −20 °C.
5. Bacteria lysis buffer: 1× phosphate-buffered saline (PBS), pH 7.4, 50 mM EDTA, 1% (v/v) Triton-X, 10% (v/v) glycerol, and 1 mM dithiothreitol (DTT). Store at 4 °C.
6. 100 mM phenylmethylsulfonyl fluoride (PMSF) in 100% ethanol. Store at −20 °C.
7. Ultrasonic cell disruptor for sonication of bacteria.
8. TNE buffer: 50 mM Tris–HCl, pH 7.5, 150 mM NaCl, 10 mM EDTA. Store at 4 °C.

9. Cleavage buffer: 50 mM Tris–HCl pH 7.0, 150 mM NaCl, 1 mM EDTA, and 0.1% (v/v) Triton-X. Store at 4 °C.
10. Glutathione-Sepharose beads (GE Healthcare, #17-0756-01). Store at 4 °C.
11. HRV 3C protease (Accelagen, #H0101S). Store at −20 °C.
12. Coomassie Plus protein reagent (Thermo Fisher, #23236). Store at room temperature.
13. Elution buffer: 1× PBS, 10% glycerol, and 1 mM DTT. Store at 4 °C.
14. PD-10 column (GE Healthcare, #17-0851-01).
15. 10K MW cutoff centrifugal concentrator tube (MilliporeSigma, #UFC801024).

2.2 GST Pull-Down Assay

1. GST-nesprin-2G SR11–12 immobilized beads.
2. pMYC-C4 FHOD1 GBD-DID (1–339 a.a. of human FHOD1). pMYC-C4 vector is derived from pEGFP-C1 vector (Clontech) by replacing the EGFP sequence with the MYC tag.
3. 293T cells (ATCC, #CRL-3216).
4. Culture medium: DMEM with 10% fetal bovine serum and 10 mM Na-HEPES, pH 7.4.
5. Transfection reagent: 55 mM Na-HEPES, 270 mM NaCl, and 1.5 mM Na_2HPO_4; pH 7.05. Store at −20 °C.
6. 2 M $CaCl_2$. Store at −20 °C.
7. Lysis buffer: 25 mM Tris–HCl, pH 7.4, 150 mM NaCl, 5 mM EDTA, 1% Triton X-100, 10 mM sodium pyrophosphate, 10 mM β-glycerophosphate, 10 mM NaF, 1 mM Na_3VO_4, and 10% glycerol. Store at 4 °C.
8. 100× Halt Protease and Phosphatase Inhibitor Cocktail (Thermo Scientific, #1861281).
9. SDS sample buffer (5×): 312.5 mM Tris–HCl, pH 6.8, 10% (w/v) SDS, 50% (v/v) glycerol, 8% (w/v) DTT, and 0.02% (w/v) bromophenol blue. Store at −20 °C.
10. SDS/lysis sample buffer: 1× SDS sample buffer in lysis buffer. Store at −20 °C.

2.3 Actin Co-Sedimentation Assay

1. G-buffer (5×): 25 mM Tris–HCl, pH 8.0, 1 mM $CaCl_2$, 1 mM ATP, and 2.5 mM DTT. Store at −20 °C.
2. F-buffer (10×): 500 mM KCl, 20 mM $MgCl_2$, and 10 mM ATP, pH 7.4. Store at −20 °C.
3. F-actin washing buffer: 1× F-buffer and 0.8× G-buffer. Store at −20 °C.

4. Phalloidin: Prepare 2.5 mM phalloidin by dissolving 1 mg of phalloidin (Setareh Biotech, #6901) in 0.5 mL of water. Aliquot in small volumes and store in −20 °C.
5. Actin: Dissolve 1 mg of rabbit skeletal muscle actin (Cytoskeleton, Inc., #AKL99) in 0.1 mL of water to yield 238 μM actin. Aliquot in small volumes and store at −80 °C.
6. HA-nesprin-2G CH protein. Store at −80 °C.
7. SDS sample buffer (5×). Store at −20 °C.
8. SDS/actin sample buffer: 1× SDS sample buffer in F-actin washing buffer. Store at −20 °C.
9. Polycarbonate thick wall ultracentrifugation tubes (Beckman Coulter, #343775).
10. TLA-100 rotor and TL-100 centrifuge (Beckman Coutler).

2.4 SDS Polyacrylamide Gel Electrophoresis (PAGE) and Western Blotting

1. Running buffer (10×): 1.92 M glycine, 0.25 M Tris–HCl, and 10% (w/v) SDS, pH 8.3. Store at room temperature.
2. SDS polyacrylamide gel.
3. Molecular weight standard markers.
4. Gel apparatus and power supply.
5. Coomassie Brilliant Blue reagent: 0.25% (w/v) Coomassie Brilliant Blue R-250, 40% methanol, and 10% glacial acetic acid. Keep at room temperature.
6. Destaining solution: 20% methanol and 10% glacial acetic acid. Store at room temperature.
7. Transfer buffer: 25 mM Tris–HCl, 192 mM glycine, 0.02% (w/v) SDS, 20% (v/v) methanol. Store at 4 °C.
8. TBST: 50 mM Tris–HCl pH 7.4, 150 mM NaCl, 0.05% Tween-20. Store at room temperature.
9. Blocking buffer: 2% nonfat milk in TBST. Store at 4 °C.
10. Primary antibody.
11. Secondary antibody conjugated with Alexa Flour 680 or 800 (Jackson ImmunoResearch Laboratories, Inc.).
12. Blot and gel imaging system, Odyssey CLx (LI-CORE).

3 Methods

3.1 Preparation of Recombinant Nesprin-2G CH and SR11–12 Domains and FHOD1

1. Shake BL21(DE3) bacteria transformed with GST-HA-nesprin-2G CH and GST-nesprin-2G SR11–12 plasmids in 12 mL of LB with 100 μg/mL ampicillin at 37 °C overnight at 200–250 rpm.
2. Next day, inoculate 10 mL of overnight culture into 1 L LB with 100 μg/mL ampicillin. Shake the culture at 37 °C until

OD600 reaches 0.6. Cool down the culture on the ice. Add 1 mL of 1 M IPTG and shake the culture overnight at 16 °C at 200–250 rpm.

3. Next day, spin down the culture at 15,000 × *g* at 4 °C for 10 min.
4. Add 6 mL of lysis buffer, and resuspend the bacteria pellet. Keep lysates on ice at all times.
5. Add 60 μL of 100 mM PMSF, and sonicate the lysate on ice for several cycles without heating up lysates.
6. Spin down crude lysate at 13,000 × *g* for 10 min. Transfer the soluble fraction to a new tube.
7. Mix 400 μL of glutathione-Sepharose beads into the lysate and rotate the mixture at 4 °C for 2 h to overnight.
8. Wash the beads four times with 10 mL of TNE buffer. At this step, protein concentration can be measured with Coomassie Plus protein reagent. At this stage, beads can be stored at 4 °C or frozen at −80 °C for longtime storage.
9. For cleaving GST, the beads are washed with 10 mL cleavage buffer twice and resuspended in a total volume of 800 μL. Apply 1 μg of HRV 3C protease per 200 μg of protein, and rotate the mixture at 4 °C overnight.
10. Elute the cleaved protein with cleavage buffer three times to yield a total elution volume of 2.5 mL.
11. Equilibrate PD-10 column with total volume of 25 mL of elution buffer.
12. Apply the 2.5 mL of cleaved protein solution into PD-10 column and collect 500 μL fractions. Check the fractions with Coomassie Plus protein reagent to identify protein-containing fractions.
13. Combine the protein fractions and concentrate with 10K MW cutoff centrifugal concentrator tube.
14. Aliquot the purified protein and snap-freeze in liquid nitrogen. Store at −80 °C.

3.2 GST Pull-Down Assay

1. Plate 5.0×10^6 293T cells on a 100 mm plate in a total volume of 11 mL of culture medium.
2. Next day, mix 15 μg of pMYC-C4 FHOD1-GBD-DID and 125 μL of 2 M $CaCl_2$ solution in total volume of 500 μL water in a plastic tube. While vortexing the tube, add dropwise 2× transfection buffer to DNA mixture. The solution now should look a little hazy (*see* **Note 1**).
3. Add dropwise the DNA mixture to the 100 mm plate containing the 293T cells while swirling the plate.

4. After 6 h, replace the medium with fresh culture medium.
5. After 2 days, put the plate on ice, and wash the plate with ice-cold PBS. Add 1 mL of lysis buffer with 1× Halt protease and phosphatase cocktail. Scrape the plate and transfer the lysates to a microfuge tube. Keep the tube on ice for 30 min.
6. Spin down the crude lysate at 13,000 × *g* for 10 min. Transfer the soluble fraction to a new tube.
7. Mix 100 μg of GST-nesprin-2G SR11–12 immobilized beads with soluble fraction of lysate and rotate the mixture at 4 °C overnight (*see* **Notes 2** and **3**).
8. Wash the beads four times with 1 mL of lysis buffer. Carefully aspirate all of the solution (*see* **Note 4**).
9. Add 50 μL of 1.5× SDS/lysis sample buffer to the beads. Boil the samples for 5 min before running on SDS–PAGE.

3.3 Actin Co-sedimentation Assay

1. Pre-cool TLA-100 rotor and TL-100 ultracentrifuge to 4 °C (*see* **Note 5**).
2. Prepare 10 μM actin in 1× G-buffer in a tube and incubate on ice for at least 1 h to overnight (*see* **Note 6**).
3. After incubation, transfer the G-actin solution to ultracentrifugation tubes (maximum volume of 200 μL). Put the tubes into a pre-cooled TLA-100 rotor. Label the tubes so that the position of pellet can be identified after the spin. Centrifuge at 436,000 × *g* for 20 min at 4 °C.
4. After the spin, carefully avoid touching any pelleted material in the tube and transfer the supernatant containing G-actin solution to a new tube.
5. Add 10× F-buffer and ddH_2O to the G-actin solution to make 8 μM actin and 1× F-buffer. Keep the tube at room temperature (24 °C) for at least 2 h.
6. After a 2 h incubation, add phalloidin to 8 μM final concentration to the F-actin solution.
7. Next, dilute HA-nesprin-2G CH into 1× F-actin buffer in an ultracentrifugation tube in the range of 200 nM to 2 μM and incubate on ice for 30 min (*see* **Note 7**).
8. After the incubation, centrifuge at 436,000 × *g* for 20 min at 4 °C. Carefully transfer the supernatant to a new tube.
9. Mix HA-nesprin-2G CH and F-actin at 1:1 volume ratio so that the final F-actin concentration is 4 μM and HA-nesprin-2G CH is in the range of 100 nM to 1 μM.
10. Incubate at room temperature for 1 h.
11. After the incubation, centrifuge the tubes at 436,000 × *g* for 20 min at 4 °C. Carefully transfer the supernatant to a tube containing 1/5 volume of 5× SDS sample buffer. Boil the

samples for 5 min for running on SDS–PAGE. This is the supernatant fraction of the sample (*see* **Note 8**).

12. Without disturbing the pellet fraction, carefully add 200 μL of pre-cooled F-actin washing buffer to the ultracentrifugation tube.
13. Centrifuge at 436,000 × *g* for 10 min at 4 °C.
14. Carefully remove the supernatant and discard. Add 62.5 μL of 90 °C 1× SDS sample buffer to the tube. Pipet up and down to solubilize the pellet and transfer to a new tube. Boil the samples for 5 min for running on SDS–PAGE. This is the pellet fraction of the sample (*see* **Note 9**).

3.4 SDS–PAGE and Western Blotting

1. Assemble SDS–polyacrylamide gel in a gel apparatus, and fill the apparatus with 1× running buffer.
2. Load the samples and molecular weight markers and run the gel.
3. When the dye front has reached the bottom of the gel, turn off the power.
4. To directly stain the gel to detect bound proteins, put the gel into Coomassie Brilliant Blue staining reagent for 1 h.
5. Discard Coomassie Brilliant Blue staining reagent and incubate the stained gel with destaining reagent until the gel is destained well. After the color intensity of destaining reagent and the gel becomes similar, it is good to change the destaining reagent to destain more.
6. After destaining is optimal, discard the destaining reagent, and incubate the gel with water.
7. Scan the gel to quantify the protein bands. A typical F-actin pelleting assay for nesprin-2G CH is shown in Fig. 1.

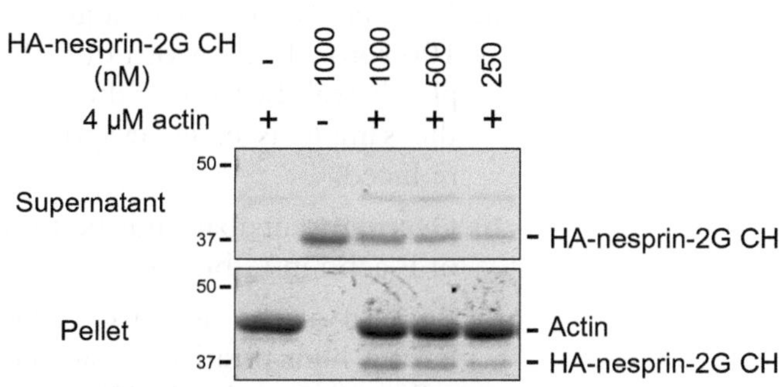

Fig. 1 Interaction of nesprin-2G CH with F-actin. Supernatant and pellet fraction samples from an actin pelleting assay with HA-nesprin-2G CH were run on 10% SDS–PAGE and separated. Proteins on the gels were visualized by Coomassie staining. The numbers on the left side of the gels indicate molecular weight in kDa

8. For Western blotting, put nitrocellulose paper on the gel, and place it into the gel cassette. Assemble the cassette into a transfer apparatus, and fill it with 1× transfer buffer. Turn on the power supply.
9. After transfer is done, wash the nitrocellulose with TBST once.
10. Put the nitrocellulose into blocking buffer, and incubate on a shaker at room temperature for 1 h.
11. Incubate the nitrocellulose with primary antibody in blocking buffer on a shaker for overnight at 4 °C.
12. Wash the nitrocellulose with TBST for 5 min for five times.
13. Incubate the nitrocellulose with secondary antibody on a shaker for at least 1 h at room temperature.
14. Wash the membrane with TBST for 5 min for five times.
15. Scan the Western blots with an Odyssey scanner (or similar scanner) to quantify the signals.

4 Notes

1. Alternative transfection protocols, such as liposome-mediated DNA delivery can be used instead of this calcium phosphate transfection method.
2. For distinguishing the interaction from background non-specific interaction, use GST-immobilized beads as a negative control.
3. Previously, we showed that in addition to nesprin-2G SR11–12, SR10–13 interacts with FHOD1 GBD-DID [6], so SR10–13 can also be used for pulling down FHOD1.
4. To reduce non-specific interaction, the salt concentration of the washes can be increased.
5. It is important to have a good O-ring on the lid of the TLA-100 rotor. If it is worn out, the inside of rotor will not completely seal. Because of the vacuum during ultracentrifugation, the sample is easily evaporated, and the sample volume is reduced.
6. Overnight incubation is better for complete depolymerization of F-actin in G-buffer.
7. Because of the negative surface charge of F-actin, proteins having high positive surface charge may interact electrostatically with F-actin. As His tag has a high positive charge, it is advisable to avoid His-tagged proteins for interaction studies with F-actin. Alternatively, the His tag can be removed from the protein after purification.

8. If the amount of protein is too low to be detected, the volume of the reaction should be increased. The supernatant fraction can be concentrated by a centrifuge vacuum concentrator. This will allow the loading of protein on SDS–PAGE to be increased enhancing the possibility that the proteins will be detected.
9. For testing the interaction of a protein that strongly bundles F-actin, the pellet fraction tube with sample buffer should be boiled. Otherwise, the pellet does not go into the solution.

Acknowledgments

The methods described in this paper were supported by NIH grants R01GM099481 and R01 AR068636 to G.G.G.

References

1. Gundersen GG, Worman HJ (2013) Nuclear positioning. Cell 152(6):1376–1389
2. Starr DA, Fridolfsson HN (2010) Interactions between nuclei and the cytoskeleton are mediated by SUN-KASH nuclear-envelope bridges. Annu Rev Cell Dev Biol 26:421–444
3. Zhen YY et al (2002) NUANCE, a giant protein connecting the nucleus and actin cytoskeleton. J Cell Sci 115(Pt 15):3207–3222
4. Luxton GW et al (2010) Linear arrays of nuclear envelope proteins harness retrograde actin flow for nuclear movement. Science 329(5994):956–959
5. Chang W, Worman HJ, Gundersen GG (2015) Accessorizing and anchoring the LINC complex for multifunctionality. J Cell Biol 208(1):11–22
6. Kutscheidt S et al (2014) FHOD1 interaction with nesprin-2G mediates TAN line formation and nuclear movement. Nat Cell Biol 16(7):708–715
7. Antoku S et al (2015) Reinforcing the LINC complex connection to actin filaments: the role of FHOD1 in TAN line formation and nuclear movement. Cell Cycle 14(14):2200–2205
8. Jayo A et al (2016) Fascin regulates nuclear movement and deformation in migrating cells. Dev Cell 38(4):371–383
9. Chang W et al (2015) Linker of nucleoskeleton and cytoskeleton (LINC) complex-mediated actin-dependent nuclear positioning orients centrosomes in migrating myoblasts. Nucleus 6(1):77–88
10. Folker ES et al (2011) Lamin a variants that cause striated muscle disease are defective in anchoring transmembrane actin-associated nuclear lines for nuclear movement. Proc Natl Acad Sci U S A 108(1):131–136
11. Luxton GW et al (2011) TAN lines: a novel nuclear envelope structure involved in nuclear positioning. Nucleus 2(3):173–181
12. Zhu R, Antoku S, Gundersen GG (2017) Centrifugal displacement of nuclei reveals multiple LINC complex mechanisms for homeostatic nuclear positioning. Curr Biol 27(20):3097–3110 e5.

Chapter 5

Interactions of Nesprin-4-Containing LINC Complexes in Outer Hair Cells Explored by BioID

Brian Burke

Abstract

As components of diverse tissues and organs, metazoan cells have to display a wide variety of specialized functions. Implementation of such functions invariably entails the establishment of tissue-specific cellular architecture (Bone and Starr, J Cell Sci 129:1951–1961, 2016). In animal cells, the nucleus is typically the largest organelle and in many respects acts as a landmark for multiple subcellular structures. For instance, in epithelial cells, the nucleus is frequently positioned close to the basal membrane via association with the cytoskeleton. Clearly such associations must be mediated by protein components of the outer nuclear membrane. One such protein is Nesprin-4, a member of the KASH domain family that is expressed in a variety of epithelial cells, including sensory outer hair cells of the inner ear. In this chapter, I describe a proximity-based biotinylation technique, BioID, that can be applied to Nesprin-4 to map its interactions at the nuclear periphery.

Key words Nesprin-4, LINC complex, BioID, Biotinylation, Interactome, Proteomics

1 Introduction

Association of cytoskeletal elements with the nucleus is mediated in large part by LINC (linker of the nucleoskeleton and cytoskeleton) complexes [1, 2]. These are composed of SUN domain protein trimers that reside within the inner nuclear membrane (INM) [3, 4]. While their nucleoplasmic N-terminal domains associate with nuclear components, their luminal C-terminal domains function as tethers for KASH domain proteins that are residents of the outer nuclear membrane (ONM) [5–8]. It is the cytoplasmic domains of KASH proteins that provide a connection to the cytoskeleton [1, 2, 9].

In the inner ear, outer hair cells (OHCs) form a specialized sensory epithelium within the organ of Corti [10]. These cells are characterized by the presence of sensory stereocilia on the apical plasma membrane and a basally situated nucleus. OHCs appear to express a single class of LINC complex composed of the SUN

Gregg G. Gundersen and Howard J. Worman (eds.), *The LINC Complex: Methods and Protocols*, Methods in Molecular Biology, vol. 1840, https://doi.org/10.1007/978-1-4939-8691-0_5,

domain protein, Sun1, and the KASH domain protein Nesprin-4 [11, 12]. In mice loss of either of these proteins leads to mislocalization of the nucleus from basal to apical regions of the cell. This is associated with erosion of OHC viability resulting in progressive deafness. In humans, mutations that result in loss or truncation of Nesprin-4 cause early-onset hearing loss (OMIM DFNB76) [11]. Nesprin-4 acts as an adaptor for Kinesin-1 at the nuclear surface [12]. It is likely that it is the microtubule plus-end motor activity of Kinesin-1 that is responsible for the Nesprin-4-dependent basal positioning of nuclei in OHCs.

Further exploration of the molecular interactions of Nesprin-4 in situ has been confounded by the fact that the mouse inner ear typically contains only a few thousand OHCs. Moreover, LINC complexes themselves are quite insoluble [5]. Consequently, it is simply impractical to use techniques such as co-immunoprecipitation to characterize the Nesprin-4 interactome in OHCs. Even the original identification of Kinesin-1 as a Nesprin-4-binding partner was carried out in tissue culture cells expressing a soluble truncated version of Nesprin-4 that lacks a transmembrane and luminal domain [12].

To circumvent problems associated with interactome analysis of insoluble protein complexes, Roux and colleagues developed the BioID technique [13, 14]. This method takes advantage of a promiscuous *E. coli* biotin protein ligase (BirA, referred to here as BioID1) that may be fused to a protein of interest, in this case Nesprin-4. The BioID1-Nesprin-4 fusion protein is stably expressed in a suitable cell type where it should localize appropriately to the nuclear envelope. This may be monitored by immunofluorescence microscopy. Upon addition of biotin to the culture medium, the BioID1 moiety will catalyze the biotinylation of proteins within a 10–15 nm radius of the BioID1-Nesprin-4 fusion protein [15]. These proteins would then represent candidate interactors of Nesprin-4. Biotinylated proteins are recovered on streptavidin-coated magnetic beads and identified by mass spectrometry. Obviously, each protein identified in this way should be validated as a Nesprin-4 interactor by alternative methods such as pulldown of recombinant proteins (or fragments thereof) following either transfection of tissue culture cells or in in vitro transcription translation mixes.

The choice of cells for these experiments will ultimately be determined by the investigator. Currently, there are few options for culturing cells of the inner ear. However, Rivolta and Holley [16] have described several cell lines such as UB/OC-1 that share some properties with cochlear hair cells and which could be suitable for a Nesprin-4 BioID screen. A complementary approach would be to employ epithelial cells that express Nesprin-4 but which are derived from an alternative tissue source. The expectation is that these might express at least some Nesprin-4-associated proteins that would also be present in OHCs. This could be confirmed later by, for instance, immunohistochemical examination of OHCs. We have previously shown that mouse HC11 mammary epithelial cells [17] express Nesprin-4 upon differentiation [12]. These might therefore represent a suitable option for a Nesprin-4 interactome analysis by BioID.

2 Materials

1. 1 mM Biotin 20× stock (working concentration, 50 μM): Dissolve 12.2 mg of Biotin (Sigma, B4501) in 50 mL of serum-free tissue culture medium (e.g., Dulbecco's modified Eagle's medium). Sterilize by passage through a 0.22 μm filter unit (Millex). Dispense into sterile 15 mL Falcon tubes in 5 mL aliquots. Store at 4 °C.
2. Buffer A: For 20 mL, 100 mg sodium deoxycholate, 16.2 mL water, 600 μL 5 M NaCl, 2 mL 10% w/v NP-40 (1% final), 200 μL 10% SDS (0.1% final), 1 mL 1 M Tris–HCl pH 7.4 (50 mM final). Check that the pH is 7.4. Before use, add 800 μL of protease inhibitor 25×.
3. Buffer B: For 30 mL, 28.2 mL water, 600 μL of Tris–HCl 1 M pH 7.4 (20 mM Tris–HCl pH 7.4 final). Before use, add 1.2 mL of protease inhibitor 25×.
4. Buffer C: SDS-PAGE sample buffer.
5. Buffer D: Prepare in fume hood. For 200 mL: 39 mL of water, 1 mL of LC-MS grade acetic acid 100% (0.5% final), 160 mL of LC-MS grade acetonitrile 100% (80% final). Stable at room temperature in the dark.
6. Buffer E: For 1 L, ~995 mL water, 5 mL of LC-MS grade acetic acid 100% (0.5% final). Stable at room temperature.
7. Buffer ED1: Prepare in fume hood, use double (nitrile) gloves. Add 2,2,2-trifluoroethanol (TFE) last. For 100 mL, 50 mL of triethylammonium bicarbonate (TEAB), 100 mM (50 mM final), 50 mL of TFE 100% (50% final). Store at room temperature.
8. C18 cartridges: Sep-pak C18, Waters (cat# WAT051910). Store at room temperature. If packs are opened, seal for storage.
9. Chloroacetamide (CAA) 550 mM: Prepare in fume hood. Dissolve 102.8 mg of CAA in 2 mL of water, in the dark, aliquot by 100 μL, and store at −20 °C. Can be thawed only once.

10. 1 M dithiothreitol (DTT): Dissolve 1.55 g of DTT (stored at 4 °C) in 10 mL of water, aliquot, and store at −20 °C. Store at −20 °C in 100 μL aliquots.
11. 3% formaldehyde from paraformaldehyde: Heat about 80 mL of Ca- and Mg-free phosphate-buffered saline (PBS) to 80 °C and add 3 g paraformaldehyde with continuous stirring until the solution is clear. Add 100 μL each of 100 mM $CaCl_2$ and $MgCl_2$ with stirring while the solution is still warm. Allow to cool and clear by passage through a 0.45 μm filter. Freeze in aliquots (~12 mL) at −20 °C.
12. HEPES 50 mM pH 8.5: Make sure that pH is 8.5.
13. Magnetic tube holder: For example, Invitrogen DynaMag™-2 Magnet, cat# 123-21D.
14. LysC 0.5 μg/mL: Reconstitute 1 vial of Lys-C Protease (WAKO, cat# 129-02541 10 AU per vial, order by two vials) with 4.4 mL of water (0.5 μg/μL final). Aliquot by 100 μL and store at −20 °C. Can be thawed up to two times.
15. Lysis/extraction buffer: For 50 mL, 38 mL water, 2.5 mL of 1 M Tris–HCl pH 7.4 (50 mM final), 5 mL of 5 M NaCl (500 mM final), 2 mL of 10% SDS (0.4% final), 2.5 mL 0.1 M EDTA pH 7.4 (5 mM final). Before use, add DTT to 1 mM and 1× protease inhibitor.
16. Protease inhibitor 25×: Dissolve completely in 2 mL to generate a 25× solution (Roche cOmplete EDTA-free Protease Inhibitor Tablet 05056489001).
17. Rotary mixer: Biosan or equivalent.
18. Sonicator: Sonifier-250, Branson, or equivalent.
19. Streptavidin Dynabead: Dynabead, Invitrogen myOne Streptavidin C1.
20. Tris–HCl(2-carboxyethyl) phosphine hydrochloride (TCEP) solution: 0.5 M (Sigma 64654-7, in glass ampules). Store at RT for 2 years unopened, otherwise store at −20 °C. Water soluble.
21. TEAB 100 mM: For 50 mL Falcon tubes (avoid glass containers), 45 mL of water, 5 mL of TEAB 1 M (pH 8.5). Store at room temperature.
22. TFA 10%: Prepare in fume hood. For 50 mL to 45 mL of water, add 5 mL of TFA 100% (10% final). Stable at room temperature.
23. TFE 100%: Creates toxic fumes, use nitrile gloves. Store at room temperature.
24. Tandem Mass Tag (TMT) labels: TMT10plex Isobaric Label Reagent Set, 3 × 0.8 mg, cat# 90111 or TMTsixplex Label Reagent Set, 2 × 0.8 mg, cat# 90062 (Thermo Fisher

Scientific/Pierce). The 10-plex set can be used in the following combinations for a 6-plex analysis (do not exceed 6-plex to ensure MS run time is kept to a minimum): A (126, 127N, 128C, 129N, 130C, 131) or B (126, 127C, 128N, 129C, 130N, 131).

25. Triton X-100, 20% (w/v): For 100 mL, 20 g Triton X-100, to 100 mL with deionized water. Stir overnight to dissolve. Triton X-100 can be easily weighed in a glass beaker on a top-loading balance. It is much easier than measuring volume since Triton X-100 is extremely viscous. Filter sterilize using a 0.22 μm vacuum filtration unit. Store at 4 °C.
26. Trypsin 1 mg/mL. per tube: resuspend 20 μg of sequencing grade modified trypsin (stored lyophilized at −20 °C) in 20 μL of resuspension buffer (or 50 mM acetic acid) and store at −80 °C if necessary (can only be thawed once for each frozen aliquot).

3 Methods

The first and most important aspect of the BioID protocol is the design and construction of the BioID fusion construct [14]. In the case of KASH domain proteins such as Nesprin-4, the BioID1 enzyme must be placed at the N-terminus. The reasons for this are twofold. Firstly, it is the N-terminal domain that is exposed to the cytoplasm and which confers Nesprin-4 with its unique properties. Secondly, the C-terminus of Nesprin-4, and indeed of all KASH domain proteins, is essential for association with INM SUN proteins and hence for localization to the ONM [5, 18]. Consequently, this region of the molecule cannot be modified in any way. A basic BioID1 plasmid, pcDNA3.1 MycBioID, that can be used as starting point for the BioID1-Nesprin-4 fusion construct can be obtained from Addgene.org (plasmid #35700). This BioID1 cDNA features a 5′ sequence encoding a Myc epitope tag, which will simplify monitoring protein expression [13]. While the pcDNA3.1 vector can be employed directly for fusion protein expression following stable transfection, we prefer to use the lentiviral vector, pTRIPZ which incorporates a Tet-on promoter [19]. This allows us to induce fusion protein expression by addition of doxycycline (DOX) to the cell culture medium (at a concentration of about 1 μg/mL) [20]. pTRIPZ also simplifies the derivation of stable cell lines with selection in puromycin (1 μg/mL).

It cannot be overemphasized that transient transfection should not be employed for BioID analysis. The reason for this is the extreme heterogeneity in protein expression. Overexpression of the BioID fusion protein may result in mis-targeting of a significant fraction. This in turn will almost certainly produce spurious association and interaction data. While the use of an inducible expression system is not essential, in our experience it provides us with the means to fine-tune expression levels.

We normally use up to 5 × 10^7 cells per BioID sample depending upon the nature and abundance of the BioID fusion protein. This translates to roughly 1–5 10 cm tissue culture Petri dishes or 1–2 15 cm dishes. Ultimately, the sample size should be determined by the investigator, based upon the quality of preliminary data.

A key determinant of success in any BioID experiment is that the behavior of the BioID fusion protein should mimic in every way that of the endogenous protein. With Nesprin-4, we would want to ensure that Myc-BioID-Nesprin-4 localizes to the NE and at the same time recruits Kinesin-1 [12]. These properties of the BioID-Nesprin-4 fusion protein should be monitored by immunofluorescence microscopy prior to embarking on the BioID screen. Our immunofluorescence microscopy protocol is included below.

3.1 Cell Culture

Cells should be seeded at such a density that they will be just confluent at the time they are to be harvested. For a single 10 cm tissue culture Petri dish and depending upon the cell type, this would typically correspond to roughly 10^6 cells. Seed all replicates and controls at the same cell density. Appropriate experimental and control combinations such as with and without DOX, with and without biotin, should be prepared in parallel. Timing and concentration of DOX addition to induce fusion protein expression must be determined empirically. It is essential to ensure that the level of expression of the BioID fusion protein never exceeds that of the endogenous protein (Nesprin-4 in this case). If necessary, determine a DOX dose-response curve. When the induced cells are approximately 80% confluent, replace the growth medium with cell culture medium supplemented with 50 μM Biotin, and incubate for a further 6–24 h before proceeding with the protein extraction.

3.2 Cell Lysis and Protein Extraction

In most situations, cells can be lysed directly in the tissue culture dish. However, in some situations it may be preferable if the cells are first trypsinized and recovered by centrifugation prior to lysis (for instance, if the cells are grown in multilayered flasks). Determine cell density where possible to ensure that there are equal numbers of cells across all conditions at extraction time. Avoid cross contamination of proteins by using disposable labware.

1. Wash the cells 2–3× with 10–20 mL PBS to remove any free biotin since this may interfere with subsequent pulldowns.
2. Remove residual PBS with a vacuum aspirator.

3. Lyse the cells at room temperature with 600 μL (for a 10 cm dish) of lysis buffer (containing protease inhibitors and 1 mM DTT). The solution will become viscous with the release of DNA. Scrape any residual cells off of the dish and transfer to a 15 mL Falcon tube.
4. Add 240 μL of 20% Triton X-100 and transfer to ice.
5. Add 2.16 mL of 50 mM Tris–HCl pH 7.4.
6. Sonicate the lysate (using a Branson Sonifier or equivalent). Use a low-power setting and short duty cycle. Apply 10–20 pulses or until the solution is no longer viscous. Alternatively, multiple passes through a 27G needle fitted to a 3 mL syringe may be employed here.
7. Distribute the extract in to 3 × 1.5 mL Eppendorf tubes.
8. Centrifuge the lysate for 10 min at 16,000 × *g* at 4 °C in a pre-cooled table-top microcentrifuge.
9. Transfer the supernatant carefully into fresh Eppendorf tubes.
10. As a matter of routine, save 40 μL of the lysate in a 1.5 mL tube, and add 40 μL of buffer C (SDS-PAGE buffer) for Western blot analysis later. Store at −20 °C.

3.3 Recovery of Biotinylated Proteins

3.3.1 Preparation of Streptavidin Beads

1. Resuspend the beads by vortexing the shipping vial for 30 s.
2. Pipette 100 μL of Dynabeads into 3 × 1.5 mL Eppendorf tubes. Unless using unusually large quantities of cell extract, 100 μL should be sufficient for most purposes.
3. Add 1 mL of PBS and shake to mix.
4. Transfer the Eppendorf tubes to the magnetic tube rack for 1 min.
5. Aspirate the PBS.
6. Repeat the last three steps 1–2 times.

3.3.2 Biotin Capture

1. Add the protein extracts to the washed Dynabeads in the Eppendorf tubes.
2. Incubate overnight (16 h) at 4 °C on a rotary mixer at medium speed (or 2 h minimum at room temperature). The next morning, incubate for a further 30 min at room temperature.
3. Place samples on the magnetic tube holder for 2 min and discard supernatant.
4. Add 1 mL of buffer A. Mix quickly to ensure that all beads are resuspended.
5. Incubate 10 min on the rotary mixer.
6. Place samples on the magnet tube rack for 2 min, discard supernatant.
7. Wash once more for 10 min with 1 mL buffer A.

8. Wash for 2 × 10 min with 1 mL buffer B. Prior to recovering the washed beads, remove 40 μL of the bead suspension for Western blot analysis later.
9. Place samples on the magnet tube rack for 2 min, aspirate supernatant.

At this point the beads and adsorbed biotinylated proteins may be consolidated in a single tube, frozen in liquid nitrogen, and submitted for analysis to a mass spectrometry/proteomics facility. Some investigators may wish to continue processing the samples with on-bead proteolytic digestion with subsequent TMT labelling.

3.4 On-Bead Proteolytic Digestion

3.4.1 Digest

Work in the fume hood and use doubled nitrile gloves. Prepare buffer ED1. All reagent volumes (TCEP, CAA, LysC) are based on 100 μL of beads.

1. Set shaking heat block to 55 °C.
2. Resuspend the beads in 50 μL of buffer ED1 per 100 μL of beads.
3. Add 2 μL TCEP 0.5 M (20 mM final) and mix well.
4. Centrifuge briefly and incubate at 55 °C for 20 min with shaking.
5. Add 5.2 μL of CAA 550 mM (1/10 dilution, 55 mM final), and incubate 30 min at room temperature in the dark.
6. Add 542.8 μL of 100 mM TEAB (brings sample to 600 μL).
7. Add 20 μL of LysC 0.5 μg/μL (10 μg per reaction) and incubate 3–4 h at 37 °C with shaking.
8. Add 130 μL of 100 mM TEAB (brings sample to 750 μL).
9. Add 10 μL of trypsin 1 μg/μL (10 μg per reaction) and incubate overnight at 37 °C with shaking.
10. Transfer the tubes to the magnetic tube holder and transfer the supernatants to fresh Eppendorf tubes.
11. Centrifuge for 5 min at 16,000 × *g* at RT, transfer the supernatant to fresh Eppendorf tubes.
12. Add 84.4 μL of TFA 10% (1/10 dilution, 1% final).

3.4.2 C18 Cartridges Preparation

Condition C18 cartridges just before the procedure. Always use the same flow direction, and pass samples *slowly* through the cartridge using a syringe to ensure maximum binding/recovery. Work under hood or air intake. Prepare 50 mL Falcon tubes with buffer D and E.

1. Label the required number of C18 cartridges.
2. Aspirate 2 mL of buffer D with a 3 mL syringe.

3. Connect the cartridge and slowly pass the liquid through. Use a 15 mL Falcon tube as waste collector.
4. Disconnect cartridge.
5. Equilibrate the cartridge with 2 mL of buffer E using a fresh syringe following the same procedure.
6. Use the same syringe to force air through the cartridge 2–3 times (disconnect and reconnect the syringe) to completely eliminate the remaining buffer E.
7. Use cartridges conditioned in this way within 30 min.

3.4.3 Desalting with Conditioned C18 Cartridges

1. Add 4 mL of buffer E in a labeled 15 mL falcon tube, and top up with the digested sample.
2. Apply sample (in buffer E) to the cartridge with a 10 mL syringe: remove plunger, and connect the syringe barrel to the cartridge; pipette in sample, insert plunger, and pass the sample slowly through the cartridge and collect the effluent in the original 15 mL tube; disconnect cartridge.
3. Reapply the effluent from **step 2** and then repeat.
4. Using a new 10 mL syringe, wash the cartridge with 4 mL of buffer E using the same procedure.
5. Disconnect cartridge, and circulate air 2–3× through the cartridge with the syringe to eliminate completely the remaining buffer E.
6. Elute sample in to a 1.5 mL Eppendorf tube using a 1 mL syringe containing 1 mL of Buffer D
7. For maximum recovery, repeat the elution 2× using the sample eluate from **step 6** instead of buffer D.
8. Complete the recovery by circulating air in the cartridge (as above) and collecting the remaining eluate in the same tube. Samples stored frozen if necessary.
9. Concentrate sample using speedvac (at 45 °C or room temperature) or freeze drying.
10. Submit samples directly for mass spectrometry. Alternatively, investigators may wish to insert a TMT labelling step at this point [21]. This option may be employed where multiple samples, both experimental and control, are analyzed simultaneously. Multiplexing in this way reduces total machine time while improving the reliability of quantitative comparisons between parallel samples.

3.4.4 Isobaric TMT Labelling

1. Resuspend the dried, purified, and desalted samples in 25 μL of HEPES 50 mM pH 8.5.
2. Add precisely 10 μL of appropriate TMT labels (*see* Subheading 2) to each sample.

3. Tap to mix and briefly spin down samples.
4. Incubate 1 h at RT.
5. Add 50 μL of 1 M Tris–HCl to quench the labelling, and incubate for 15 min at room temperature.
6. Add 9.4 μL of 10% TFA (1/10 dilution, 1% final).
7. Desalt using conditioned C18 cartridges exactly as described above.

3.5 Immunofluorescence Microscopy

This procedure [22] should be employed to monitor both fusion protein expression using an anti-myc primary antibody (e.g., Hybridoma clone 9E10, ATCC.org; Abcam #ab32) as well as biotinylation of proximal proteins using Alexa-dye conjugated streptavidin.

1. Sterilize round or square 0.17 mm thick (#1.5) tissue culture compatible glass coverslips (such as Thermo Fisher Scientific Gold Seal #12-519-21A). This is most conveniently accomplished by autoclaving in a glass Petri dish containing layers of Watman #1 filter paper circles.
2. Place a coverslip into each well of a six-well tissue culture plate. Distribute cells in appropriate growth medium to yield a 50–80% confluent monolayer after about 2 days. As necessary, timing and cell density can be adjusted to allow for BioID fusion protein induction and biotinylation in the presence of 50 μm biotin.
3. Remove cells from incubator, and wash the monolayers twice at room temperature in PBS. After the final wash, aspirate the PBS and replace in each well with 2 mL 3% formaldehyde.
4. Incubate for 10–20 min at room temperature.
5. Wash monolayers in each well in PBS.
6. Replace PBS with PBS containing 50 mm NH_4Cl. Leave at room temperature for 10 min. This step will neutralize any residual formaldehyde.
7. Wash each well with 2 × 2 mL PBS.
8. Replace PBS with 0.2% (w/v) Triton X-100 in PBS. Leave for 5 min at room temperature.
9. Wash each well with 2 × 2 mL PBS.
10. Replace PBS with 0.2% (w/v) fish skin gelatin in PBS (FSG-PBS).
11. Dilute primary antibodies or Alexa 568-streptavidin in FSG-PBS.
12. Spread a piece of parafilm (~10 × 15 cm) on the bench. Onto the parafilm pipette 50 μL drops of diluted primary antibody.

13. Remove each coverslip from the six-well plate using a pair of fine forceps. Remove residual FSG-PBS by gently touching the edge of the coverslip to a piece of filter paper.
14. Place each coverslip *cell-side down* onto an antibody droplet. Incubate for 20–30 min at room temperature. Evaporation can be minimized by placing the lid of the six-well plate over the coverslips.
15. Return the coverslips to the six-well plate.
16. Wash each well with 3 × 2 mL FSG-PBS.
17. Dilute Alexa dye-conjugated secondary antibodies in FSG-PBS.
18. Perform incubations with appropriate secondary antibodies exactly as described for the primary antibodies.
19. Return the coverslips to the six-well plate (keeping track of the "cell side" of each coverslip).
20. Wash each well with 3 × 2 mL FSG-PBS.
21. Wash each well with 2 × 2 mL PBS.
22. Carefully rinse off the back of each coverslip with deionized water. This will eliminate the appearance of any salt crystals.
23. Mount the coverslip cell-side down on a drop of mounting medium (e.g., ProLong™ Gold, Thermo Fisher P36931) placed on a standard microscope slide.
24. Drain off excess mounting medium with a piece of filter paper, and allow to harden at room temperature.
25. Observe and record images using a suitable fluorescence microscope system (confocal or widefield) equipped with a high numerical aperture oil immersion lens.

Finally, once fusion protein expression and biotinylation have been established by immunofluorescence microscopy, the full spectrum of biotinylated proteins may be documented by Western blot analysis. This would involve probing blots with HRP-streptavidin as previously described [13, 14]. Where appropriate antibodies are available, candidate interactors may also be followed in this way.

References

1. Bone CR, Starr DA (2016) Nuclear migration events throughout development. J Cell Sci 129:1951–1961
2. Gundersen GG, Worman HJ (2013) Nuclear positioning. Cell 152:1376–1389
3. Sosa BA, Rothballer A, Kutay U et al (2012) LINC complexes form by binding of three KASH peptides to domain interfaces of trimeric SUN proteins. Cell 149:1035–1047
4. Zhou Z, Du X, Cai Z et al (2012) Structure of Sad1-UNC84 homology (SUN) domain defines features of molecular bridge in nuclear envelope. J Biol Chem 287:5317–5326
5. Crisp M, Liu Q, Roux K et al (2006) Coupling of the nucleus and cytoplasm: role of the LINC complex. J Cell Biol 172:41–53
6. Haque F, Lloyd DJ, Smallwood DT et al (2006) SUN1 interacts with nuclear lamin A

and cytoplasmic nesprins to provide a physical connection between the nuclear lamina and the cytoskeleton. Mol Cell Biol 26:3738–3371

7. Hasan S, Guttinger S, Muhlhausser P et al (2006) Nuclear envelope localization of human UNC84A does not require nuclear lamins. FEBS Lett 580:1263–1268
8. Padmakumar VC, Libotte T, Lu W et al (2005) The inner nuclear membrane protein Sun1 mediates the anchorage of Nesprin-2 to the nuclear envelope. J Cell Sci 118:3419–3430
9. Burke B, Roux KJ (2009) Nuclei take a position: managing nuclear location. Dev Cell 17:587–597
10. Hudspeth AJ (2014) Integrating the active process of hair cells with cochlear function. Nat Rev Neurosci 15:600–614
11. Horn HF, Brownstein Z, Lenz DR et al (2013) The LINC complex is essential for hearing. J Clin Invest 123:740–750
12. Roux KJ, Crisp ML, Liu Q et al (2009) Nesprin-4 is an outer nuclear membrane protein that can induce kinesin-mediated cell polarization. Proc Natl Acad Sci U S A 106:2194–2199
13. Roux KJ, Kim DI, Raida M et al (2012) A promiscuous biotin ligase fusion protein identifies proximal and interacting proteins in mammalian cells. J Cell Biol 196:801–810
14. Roux KJ, Kim DI, Burke B (2013) BioID: a screen for protein-protein interactions. Curr Protoc Protein Sci 74:Unit 19.23
15. Kim DI, Birendra KC, Zhu W et al (2014) Probing nuclear pore complex architecture with proximity-dependent biotinylation. Proc Natl Acad Sci U S A 111:E2453–E2461
16. Rivolta MN, Holley MC (2002) Cell lines in inner ear research. J Neurobiol 53:306–318
17. Ball RK, Friis RR, Schoenenberger CA et al (1988) Prolactin regulation of beta-casein gene expression and of a cytosolic 120-kd protein in a cloned mouse mammary epithelial cell line. EMBO J 7:2089–2095
18. Sosa BA, Kutay U, Schwartz TU (2013) Structural insights into LINC complexes. Curr Opin Struct Biol 23:285–291
19. Dreesen O, Chojnowski A, Ong PF et al (2013) Lamin B1 fluctuations have differential effects on cellular proliferation and senescence. J Cell Biol 200:605–617
20. Chojnowski A, Ong PF, Wong ES et al (2015) Progerin reduces LAP2alpha-telomere association in Hutchinson-Gilford progeria. Elife 4:e07759
21. Thompson A, Schafer J, Kuhn K et al (2003) Tandem mass tags: a novel quantification strategy for comparative analysis of complex protein mixtures by MS/MS. Anal Chem 75:1895–1904
22. Ash JF, Louvard D, Singer SJ (1977) Antibody-induced linkages of plasma membrane proteins to intracellular actomyosin-containing filaments in cultured fibroblasts. Proc Natl Acad Sci U S A 74:5584–5588

Part II

Mechanical Aspects of the LINC Complex

Chapter 6

Using Nesprin Tension Sensors to Measure Force on the LINC Complex

Paul T. Arsenovic and Daniel E. Conway

Abstract

Mechanotransduction, or the process by which mechanical forces regulate cellular functions, is increasingly studied in a variety of different physiological and pathological contexts. Although these forces are most often studied at cell-matrix and cell-cell adhesions, recent work has shown that the nuclear LINC complex is also subject to mechanical forces. Here we describe how to use a FRET-based biosensor, known as TSmod, in the LINC complex protein nesprin-2G. This approach allows for measurement of LINC complex forces in living cells with spatial-temporal resolution.

Key words Mechanobiology, FRET tension biosensors, Nuclear LINC complex

1 Introduction

Evidence for mechanical forces at cell-matrix adhesions has existed for over 30 years [1]. More recently mechanical forces have been directly measured across cell-cell adhesions [2, 3]. The model of cellular tensegrity predicts that cell-matrix and cell-cell forces are readily transferred across the cytoskeleton and applied to intracellular structures such as the nucleus [4]. Over 20 years ago, Ingber and coworkers showed that changes in actomyosin forces altered nuclear shape [5]. Subsequent experiments showed that externally applied forces to the perimeter of the cell also altered nuclear shape [6–8], suggesting that the cytoskeleton is "hardwired" into the nuclear membrane [9]. More recent work in the field of cell biology has identified a group of nuclear membrane-associated proteins, known as the LINC (linker of nucleoskeleton and cytoskeleton) complex, that mediate the connection of the cytosolic cytoskeleton to the nuclear membrane [9]. The LINC complex is formed by nesprin and SUN proteins that cross both the inner and outer nuclear membranes to mechanically tether the cytosolic cytoskeleton to the nuclear lamina (Fig. 1). The LINC complex is evolutionarily conserved across virtually all eukaryotes [9],

Gregg G. Gundersen and Howard J. Worman (eds.), *The LINC Complex: Methods and Protocols*, Methods in Molecular Biology, vol. 1840, https://doi.org/10.1007/978-1-4939-8691-0_6, © Springer Science+Business Media, LLC, part of Springer Nature 2018

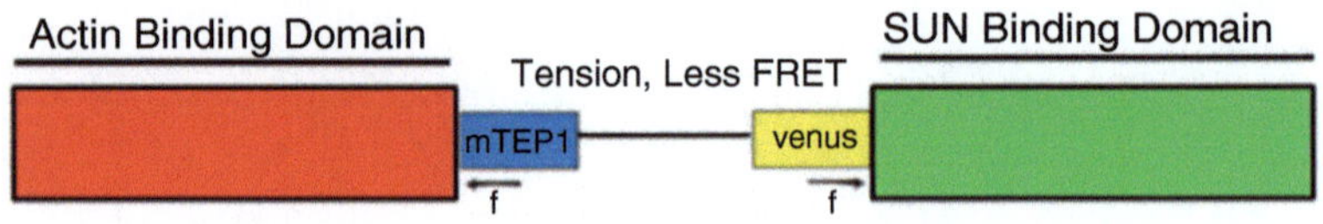

Fig. 1 Diagram of nesprin-TS. TSmod, the FRET-based force sensing module is flanked by actin and sun binding domains to capture actin-based tensile forces exerted on sun proteins embedded in the nuclear membrane

suggesting that the mechanical linkage of the cytoskeleton to the nucleus may be essential to cell function and homeostasis. One potential function of the LINC complex is the transfer of forces from the cytoplasmic cytoskeleton onto the nucleus.

The nucleus itself has been proposed to be a mechanosensor, where nuclear forces could regulate cellular functions, including nuclear transport, DNA structure, and gene expression [10]. However, studies to examine the role of nuclear forces have been hindered by a lack of techniques capable of directly measuring nuclear forces in living cells. Current knowledge of nuclear mechanical forces is based on experimental measurements of nuclear geometry and positioning, knockdown of LINC complex proteins, and overexpression of dominant negative proteins that disrupt the LINC complex. Disruption of the LINC complex perturbs cell migration, nuclear shape, and position [11, 12]. However, these experiments do not establish if forces directly regulate these functions—the LINC complex may also serve to regulate biochemical signaling pathways correlated with force [9].

Recently, a genetically encoded, calibrated Förster resonance energy transfer (FRET)-based tension biosensor (known as TSmod) was developed [13]. This sensor consists of a pair of fluorescent proteins capable of FRET separated by an elastic peptide (flagelliform). Application of force to the sensor results in increased elongation of the elastic peptide (strain), resulting in decreased FRET. The FRET-force relationship of the sensor was previously calibrated and found to have a dynamic range of 1–5 pN [13]. This sensor module was recently shown to respond to compressive loading, with higher FRET than in the unloaded state [14]. We and others have inserted the sensor into a number of proteins at cell-cell [15–18] and cell-matrix [13, 19] adhesions, indicating that the dynamic range of the sensor is well suited to study forces on proteins. In addition, the length of the flagelliform elastic linker in the TSmod sensor can be adjusted to shift the FRET-force dynamic range [20], which may be important for proteins subjected to higher or lower levels of force. Recently a new elastic peptide, known as HP35 (consisting of the villin headpiece), has been used to measure higher forces (7–10 pN) [19].

To extend this force measurement technique to the nuclear LINC complex, we recently developed a nesprin force sensor (known as nesprin-TS) in which TSmod was inserted into mini-nesprin-2G [21]. This sensor captures the mechanical forces exerted between the actin cytoskeleton and the nuclear envelope where nesprin-2G binds (*see* **Notes 12–14** for additional considerations regarding how force is applied to the nucleus and limitations of the mini-nesprin-2G sensor). This sensor behaved similarly to nesprin-2G, in that it was able to rescue nuclear positioning in nesprin-2G-depleted cells [21]. Furthermore, the sensor exhibited FRET changes in response to changes in actomyosin contractility [21]. Nesprin-2G was found to be under constitutive mechanical tension in adherent fibroblasts. Mechanical forces on nesprin-2G were found to be spatially different (apical versus basal sides of the nucleus) and were increased in elongated cells [21]. Thus, TSmod-based biosensors are well suited to study LINC complex forces in living cells.

In this protocol, we provide a detailed methodology for using the nesprin-2G force sensor, which details how to express the nesprin tension sensor (nesprin-TS) in mammalian cells and how to acquire and analyze FRET images of these cells expressing nesprin-TS. Although this protocol discusses expressing nesprin-TS in fibroblasts, it is easily adaptable to other mammalian cells, including both cell lines and primary cells. In our hands nesprin-2G is under tensile force in a variety of different cells (fibroblasts, epithelial, endothelial), but these forces vary depending on the cell type (Arsenovic, unpublished) (*see* **Note 10** concerning alternate FRET force senseors). Furthermore, this protocol as it relates to the use of the nesprin-TS can be readily adapted to other FRET-based force biosensors that have been developed for other proteins.

2 Materials

2.1 Growth and Amplification of Biosensor Constructs

1. DNA plasmids for nesprin-TS and the force-insensitive control nesprin-HL are available from Addgene (plasmids 68127 and 68128). Plasmids encoding mTFP1 (54613), venus (27793), and TSmod (26021) are can also be obtained from Addgene.
2. Standard lysogeny broth (LB).
3. LB plates and medium with ampicillin or kanamycin as appropriate.
4. 1 L bacterial flask.
5. Temperature-controlled orbital shaker.
6. Midi-prep DNA isolation kit.

2.2 Cell Culture and Transfection

1. NIH3T3 fibroblasts (ATCC CRL-1658).
2. DMEM (Thermo Fisher 11995).
3. Bovine calf serum (Thermo Fisher 16170).
4. Lipofectamine 2000 (Thermo Fisher 11668).
5. #1.5 coverglass bottom dishes (Cellvis D35-20-1.5-N).
6. Bovine fibronectin (Alfa Aesar J65696).
7. Dulbecco's phosphate buffered saline (Thermo 14190-144).

2.3 Imaging

1. Live Cell Imaging Solution (Thermo Fisher A14291DJ).
2. Temperature-controlled inverted confocal with 458 and 514 nm laser lines (Zeiss LSM 710 with spectral META detector).

3 Methods

3.1 Plasmid Amplification

1. The use of DNA-based biosensors requires *E. coli* amplification and DNA isolation. DNA obtained directly from Addgene arrives as a live bacterial stab and can be directly streaked onto an LB plate (with correct antibiotic). DNA obtained in water or Tris–HCl must first be transformed into a standard competent *E. coli* strains such as DH5-α.
2. After overnight growth at 37 °C, a single colony is selected from a plate and grown in 2 mL LB medium (with antibiotic) for 8 h with orbital shaking at 37 °C. After 8 h 1–2 mL is added to a large culture flask containing 150 mL LB medium (with antibiotic) and then grown overnight with orbital shaking at 37 °C. This larger flask is sometimes referred to as a midi-prep.
3. Spin down bacteria (5000 × *g* for 30 min), and isolate the DNA using a standard midi-prep DNA isolation kit, following manufacturer instructions.

3.2 Cell Transfection

Carry out all cell culture experiments in a BSL-2 equipped tissue culture facility, and follow all institutional guidelines for working with recombinant DNA.

1. NIH3T3 fibroblasts are cultured in DMEM with 10% bovine serum, using standard cell culture growth and passaging methods.
2. Nesprin-TS, nesprin-HL, and fluorescent protein control plasmids can readily be transfected into NIH3T3 fibroblasts using Lipofectamine 2000 per manufacturer instructions.

3. For all imaging experiments, cells are grown on #1.5 coverglass bottom dishes (Cellvis D35-20-1.5-N). Dishes are first coated with a layer of fibronectin at a concentration of 20 μg/mL (in PBS) for 20 min at room temperature, which is then removed immediately before cell seeding.
4. After 4–6 h of incubation with transfection reagents, cells are trypsinized, centrifuged, and resuspended on glass bottom dishes in fresh medium.
5. Cells are allowed to attach overnight and can be imaged the next day.
6. Viral methods of transfection (*see* **Note 8**) stable cell lines (*see* **Note 9**) can also be used.

3.3 Imaging

1. Prior to imaging phenol red containing culture medium is replaced with HEPES-buffered clear Live Cell Imaging Solution with 10% calf serum. The Live Cell Imaging Solution has improved clarity and signal to noise ratio and provides for proper pH at ambient atmosphere. Additionally, cells should be maintained in a 37 °C enclosure on the microscope, if possible. Maintaining cells at a constant physiological temperature prevents temperature-dependent focal drift and improves cell viability in longer experiments.
2. Image cells using a laser scanning confocal (Zeiss LSM 710 with spectral META detector). However, confocal imaging is not required for the sensors. Standard epifluorescence can be used with filters to separately excite mTFP1 (458 nm) and yellow (515 nm) with appropriate band pass filters to separately resolve blue and yellow emission; however this protocol will detail the use of spectral imaging-based FRET (*see* **Note 1**), not epifluorescence standard filter set-based FRET.
3. Collect spectral images (*see* **Note 1**) at 458 nm excitation using a 40× water or 63× oil objective. The 40× objective allows for acquisition of a larger group of cells, whereas the 63× objective (with additional optical zoom) can provide additional spatial information about the distribution of FRET across a single nucleus.
4. Select cells for imaging that are expressing the nesprin-TS at a high enough level to be readily visualized and imaged, but avoid imaging cells in which expression levels are too high (example images of appropriate and inappropriate cells are shown in Fig. 2A, B; also *see* **Note 2**). Additionally, cells that looked stressed or abnormal should be excluded from analysis.
5. To account for cell-cell variations in FRET, image at least 10–20 cells per condition (*see* **Note 3**).

6. Include a known high or low force control to confirm that the sensor and analysis of FRET is working. Examples include the force-insensitive nesprin-HL (headless) sensor, as well as myosin agonists and antagonists (*see* **Notes 4–6**).

3.4 Image Analysis

1. Use open-source ImageJ software (http://fiji.sc/) or similar software to open and further process FRET images.
2. Manually mask images such that the only FRET signal is from the nuclear envelope (Fig. 2). Without masking, FRET signal from other regions of the cell may represent a large fraction of pixels and significantly affect the mean FRET for the image. Delineating the nuclear envelope is easiest for cells with lower expression of nesprin-TS (*see* **Note 2**).
3. Further process images by performing a background correction (also known as background subtraction) by subtracting the average pixel value of nonfluorescent cells. Untransfected

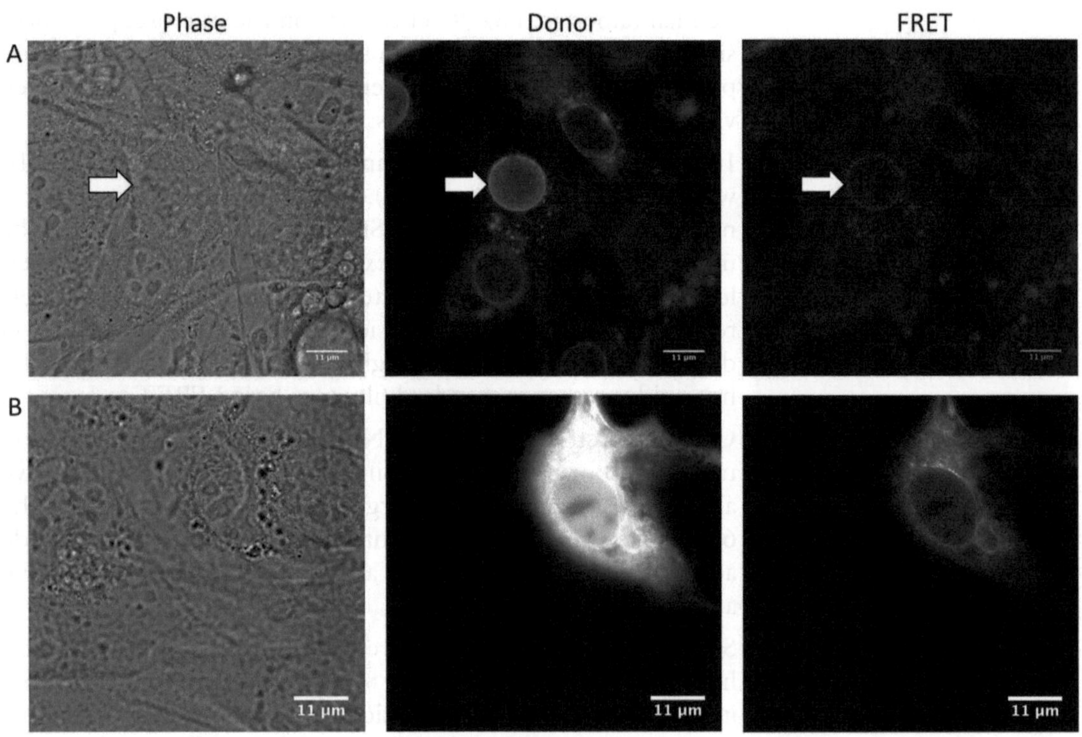

Fig. 2 Phase-contrast and fluorescent images of nesprin-TS expressing NIH3T3 fibroblasts. Row **A**: Phase-contrast, donor, and FRET channel fluorescent images of an optimally expressing cell. The best nucleus for analysis is indicated by the white arrow. Note that while other nuclei are discernable in the donor channel image, the intensity is too low in the FRET channel for accurate analysis. Row **B**: Phase-contrast, donor, and FRET fluorescent channel images of a cell with high expression. The high level of expression of nesprin-TS in both the ER and nucleoplasm makes it difficult to discern the boundaries of the nuclear envelope. These cells are excluded from analysis

cells imaged under the same power and gain are used to calculate the average background pixel value.

4. The masked and background FRET images need to be further processed to allow FRET images to be compared between conditions. Ratiometric FRET offers the most simple and straightforward method to measure relative FRET differences between conditions. To calculate the FRET ratio, background-corrected pixel values from the FRET channel are normalized to the background-corrected donor channel by dividing the unmixed acceptor image (FRET channel) by the unmixed donor image. Normalization to the acceptor-only signal (excited by the 514 nm acceptor laser) is also possible, but this will be less sensitive to changes in FRET with the benefit of reduced noise in the ratio image. Ratiometric FRET does not provide an absolute measure of FRET, but provides a relative measure of FRET changes between conditions. Because it can be influenced by background subtraction values, we do not recommend comparing FRET ratios obtained from experiments collected from different microscope sessions (*see* **Note 7**). Furthermore, it is essential to hold all laser power and detector gain settings constant across all samples. Repeated imaging of the same region may result in photobleaching effects (*see* **Note 11**).

4 Notes

1. FRET signal (acceptor emission obtained with donor excitation) can be contaminated with both donor and acceptor signals unrelated to FRET when using standard band pass filters. The FRET signal must be corrected to remove this bleed-through, and these corrections are beyond the scope of this protocol. Alternatively, spectral unmixing allows for real-time separation of donor and acceptor signal, eliminating the additional step of removing donor bleed-through. If a spectral detector is used, a single excitation wavelength (458 nm) may be selected to illuminate the sample and capture the entire fluorescent emission (420–720 nm) simultaneously. We have published a detailed methods paper in *Journal of Visualized Experiments* (JOVE) which details the use of the Zeiss 710 for spectral unmixing of the FRET signal from nesprin-TS [22].

2. In high-expressing cells, nesprin-TS frequently localizes inside the nucleus, in addition to the nuclear envelope. In these cells it is difficult to discern the fluorescence from the nuclear envelope from that inside of the nucleus, and therefore we exclude these cells from FRET analysis. Cells with lower levels of expression often give better contrast where it is easier to

identify and mask the nuclear envelope. However this can be a trade-off as images acquired from lower intensity result in FRET with a lower signal to noise ratio.

3. Many times FRET differences are minimal between conditions. Furthermore, we have observed large cell-to-cell variations in FRET within the same condition. We therefore recommend the acquisition of many images (10–20 per condition) to more accurately determine if two conditions are significantly different in nesprin-2G forces. Paired comparisons of the same cell before and after treatment can be extremely useful for minimizing biological variability in baseline nesprin-2G force.
4. In addition to nesprin-HL, we frequently employ the use of ROCK and MLCK inhibitors (Y-27632 and ML7, respectively) to inhibit myosin. We have also used actin-destabilizing agents, such as latrunculin A, to reduce force (Fig. 3). Blebbistatin has yellow fluorescence and may exhibit phototoxicity at the wavelengths used to acquire nesprin-TS [23]. In our experience the yellow color of blebbistatin prevents its use with any mTFP1-venus TSmod sensor. The use of chemical inhibitors can be a quick way to determine whether measurable nesprin-2G tensile forces exist in your cell of interest. The use of these inhibitors has the added advantage of enabling before and after treatment imaging of the same cell, allowing for paired comparisons to be made (reducing cell-to-cell variability; *see* Fig. 3).
5. The largest variations in nesprin-2G forces we have observed were between cells grown on micropatterned 20 μm lines as compared to cells grown on non-patterned surfaces [24]. This may also be a useful positive control to confirm the responsiveness of the nesprin-TS to measure differences in force on the nucleus.

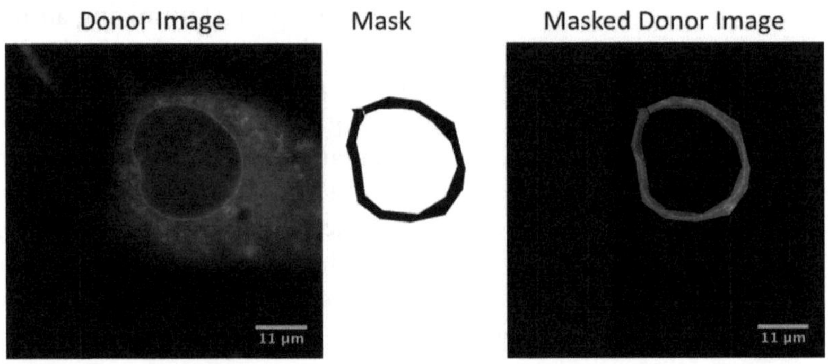

Fig. 3 Demonstration of image masking. Images may be manually masked using the polygon selection tool in ImageJ or a paintbrush tool of fixed width. A masked or binary image (values of 0 or 1) is created from the user selection. Donor image and FRET image (not shown) are each multiplied by the mask to create masked donor and FRET images of the nuclear envelope

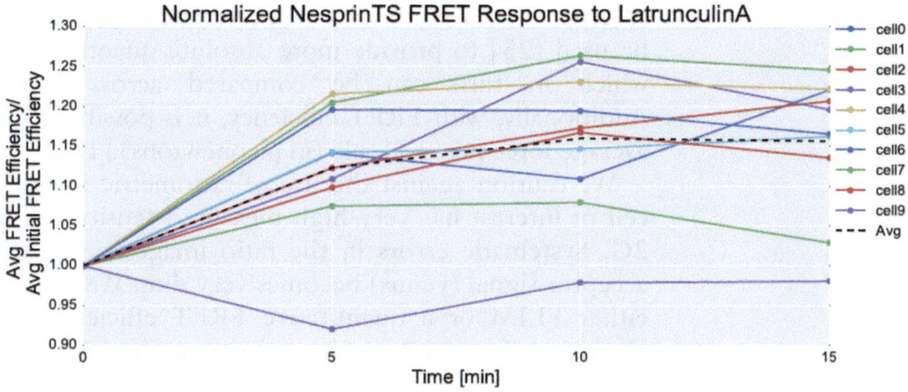

Fig. 4 Response of nesprin-TS to disruption of actin filaments. Exposure of NIH3T3 fibroblasts to 10 μM latrunculin A results in a sustained increase in nesprin-TS FRET, indicating decreased force. The same cells were tracked before and during latrunculin treatment, which allowed for normalization to the initial FRET efficiency

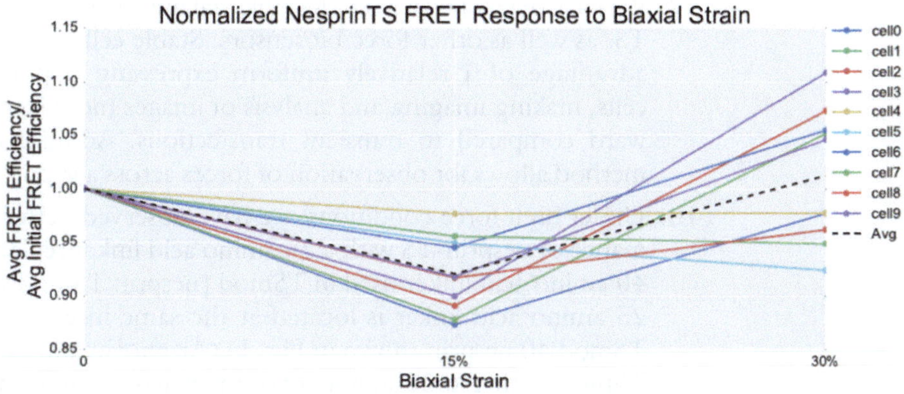

Fig. 5 Response of nesprin-TS to biaxial strain. Exposure of NIH3T3 fibroblasts to 15% biaxial strain results in decreased FRET for nesprin-TS, indicating increased force. However higher levels of stretch result in a varied response with some cells increasing and decreasing in force. The same cells were tracked before and during stretch

6. Although we have shown calyculin A and other activators of myosin can increase nesprin-2G force [24], these compounds, when used for too long or at too high of a dose, can cause cell detachment and rounding, eventually leading to a reduction in nuclear forces at longer time points. Recent work by our lab has shown that biaxial stretch can also be used to increase nesprin-2G force (Fig. 4). However too large of a deformation can also reduce nesprin-2G force in some cells, presumably due to dissociation or rupture of nesprin-2G linkage to cytoskeletal proteins (Fig. 5).

7. Alternate FRET methods which calculate FRET efficiency can be used [25] to provide more absolute quantities of FRET, which in turn can be compared across experiments. Additionally, with FRET efficiency, it is possible to estimate average force per molecule (in piconewtons) [13].

   We caution against the use of ratiometric imaging if the cell of interest has very high measured tension on nesprin-2G. Systematic errors in the ratio images occur when the acceptor signal (venus) becomes very dim. We suggest using either FLIM or a quantitative FRET efficiency unmixing algorithm if you suspect the sensor is under high loads [25].

8. We have developed an adenovirus for nesprin-TS, which enables more uniform and longer-term expression of the sensor in a variety of cells, including primary cells. Lentiviral versions (particularly tetracycline inducible versions which could be used to control expression levels) may also provide a robust way to express sensors.

9. We have developed a MDCK stable cell line expressing nesprin-TS, as well as other force biosensors. Stable cell lines offer the advantage of a relatively uniform expressing population of cells, making imaging and analysis of images more straightforward compared to transient transfections. Additionally this method allows for observation of forces across a group of cells.

10. Under high force conditions, we have observed better performance of nesprin-TS with a 25 amino acid linker replacing the 40 amino acid linker used in TSmod (nesprin TSmod with the 25 amino acid linker is located at the same insert site as the TSmod 40 nesprin shown in Fig. 1) (Arsenovic, unpublished). TSmod with the 25 amino acid linker has a higher dynamic range of FRET, but is less responsive at lower forces [20]. Although we have not used the newer HP35 elastic peptide [19] as the tension responsive element in nesprin tension sensors, it is possible that it may be more responsive under conditions when nesprin-2G forces are high.

11. When making repeated measurements of the same cell, it is important to control for photobleaching effects that may influence FRET. This can be done by imaging a control sample expressing nesprin-TS without treatment and comparing the change in FRET after repeated measurements.

12. Our sensor is based on the design of mini-nesprin-2G, a much smaller version of the giant form of nesprin-2G. This was chosen for ease of cloning and expression. However, this sensor lacks a number of important binding domains, such as FHOD1 [26], and does not contain binding sites for the microtubule motor proteins kinesin and dynein [27–29]. Thus, the sensor should be viewed only as a tool to study actin-based forces on

the LINC complex, using a minimal form of nesprin-2G. Although we showed that this truncated form of nesprin retains the ability to rescue reward nuclear movement [24], it may be missing binding sites for key regulatory molecules that could influence forces. In addition, microtubule and intermediate-based connections to the LINC complex (through nesprin-2 and other nesprin isoforms) may also contribute tensile and compressive forces to the nucleus. It will be interesting to determine forces across other nesprin isoforms with similar biosensors for these proteins.

13. The mechanical model for how forces are transmitted into the nucleus remains unclear. While we hypothesize that forces applied to nesprin are transmitted across SUN1 and SUN2 and into the nucleus, this cannot be concluded from our observation that nesprin-2G is under tensile load. Additional force sensors for SUN1/2 or proteins inside the nucleus are needed to better understand force transmission.
14. Forces applied to the nucleus may not be exclusively through the LINC complex. Wirtz and colleagues have suggested that an actin cap may apply compressive forces onto the nucleus [30], which may or may not be LINC complex dependent. In addition Lele and colleagues have suggested that LINC-independent compressive forces from the plasma membrane may also be applied to the nucleus during cell spreading [31]. Results obtained from LINC complex biosensors may not capture all forces applied onto the nucleus. This raises the possibility that experiments which affect cell spread area (e.g., hydrogels of varying elastic moduli, experiments in which cell confluence is varied) may be altering both LINC-dependent and LINC-independent forces on cell nuclei.

Acknowledgments

This work was supported by NIH grant R35GM119617 and NSF CAREER CMMI1653299.

References

1. Harris AK, Wild P, Stopak D (1980) Silicone rubber substrata: a new wrinkle in the study of cell locomotion. Science 208:177–179
2. Liu Z, Sniadecki NJ, Chen CS (2010) Mechanical forces in endothelial cells during firm adhesion and early transmigration of human monocytes. Cell Mol Bioeng 3:50–59
3. Maruthamuthu V, Sabass B, Schwarz US, Gardel ML (2011) Cell-ECM traction force modulates endogenous tension at cell-cell contacts. Proc Natl Acad Sci U S A 108:4708–4713
4. Ingber DE (2003) Tensegrity I. Cell structure and hierarchical systems biology. J Cell Sci 116:1157–1173
5. Sims JR, Karp S, Ingber DE (1992) Altering the cellular mechanical force balance results in integrated changes in cell, cytoskeletal and

nuclear shape. J Cell Sci 103(Pt 4):1215–1222

6. Maniotis AJ, Chen CS, Ingber DE (1997) Demonstration of mechanical connections between integrins, cytoskeletal filaments, and nucleoplasm that stabilize nuclear structure. Proc Natl Acad Sci U S A 94:849–854
7. Hu S, Eberhard L, Chen J, Love JC, Butler JP, Fredberg JJ, Whitesides GM, Wang N (2004) Mechanical anisotropy of adherent cells probed by a three-dimensional magnetic twisting device. Am J Physiol Cell Physiol 287:C1184–C1191
8. Hu S, Chen J, Fabry B, Numaguchi Y, Gouldstone A, Ingber DE, Fredberg JJ, Butler JP, Wang N (2003) Intracellular stress tomography reveals stress focusing and structural anisotropy in cytoskeleton of living cells. Am J Physiol Cell Physiol 285:C1082–C1090
9. Alam S, Lovett DB, Dickinson RB, Roux KJ, Lele TP (2014) Nuclear forces and cell mechanosensing. Prog Mol Biol Transl Sci 126:205–215
10. Wang N, Tytell JD, Ingber DE (2009) Mechanotransduction at a distance: mechanically coupling the extracellular matrix with the nucleus. Nat Rev Mol Cell Biol 10:75–82
11. Lombardi ML, Jaalouk DE, Shanahan CM, Burke B, Roux KJ, Lammerding J (2011) The interaction between nesprins and sun proteins at the nuclear envelope is critical for force transmission between the nucleus and cytoskeleton. J Biol Chem 286:26743–26753
12. Luxton GWG, Gomes ER, Folker ES, Vintinner E, Gundersen GG (2010) Linear arrays of nuclear envelope proteins harness retrograde actin flow for nuclear movement. Science 329:956–959
13. Grashoff C, Hoffman BD, Brenner MD, Zhou R, Parsons M, Yang MT, McLean MA, Sligar SG, Chen CS, Ha T, Schwartz MA (2010) Measuring mechanical tension across vinculin reveals regulation of focal adhesion dynamics. Nature 466:263–266
14. Rothenberg KE, Neibart SS, LaCroix AS, Hoffman BD (2015) Controlling cell geometry affects the spatial distribution of load across Vinculin. Cell Mol Bioeng 8:364–382
15. Conway DE, Breckenridge MT, Hinde E, Gratton E, Chen CS, Schwartz MA (2013) Fluid shear stress on endothelial cells modulates mechanical tension across VE-cadherin and PECAM-1. Curr Biol 23:1024–1030
16. Kuriyama S, Theveneau E, Benedetto A, Parsons M, Tanaka M, Charras G, Kabla A, Mayor R (2014) In vivo collective cell migration requires an LPAR2-dependent increase in tissue fluidity. J Cell Biol 206:113–127
17. Borghi N, Sorokina M, Shcherbakova OG, Weis WI, Pruitt BL, Nelson WJ, Dunn AR (2012) E-cadherin is under constitutive actomyosin-generated tension that is increased at cell-cell contacts upon externally applied stretch. Proc Natl Acad Sci U S A 109:12568–12573
18. Cai D, Chen S-C, Prasad M, He L, Wang X, Choesmel-Cadamuro V, Sawyer JK, Danuser G, Montell DJ (2014) Mechanical feedback through E-cadherin promotes direction sensing during collective cell migration. Cell 157:1146–1159
19. Austen K, Ringer P, Mehlich A, Chrostek-Grashoff A, Kluger C, Klingner C, Sabass B, Zent R, Rief M, Grashoff C (2015) Extracellular rigidity sensing by talin isoform-specific mechanical linkages. Nat Cell Biol 17:1597–1606
20. Brenner MD, Zhou R, Conway DE, Lanzano L, Gratton E, Schwartz MA, Ha T (2016) Spider silk peptide is a compact, linear nanospring ideal for intracellular tension sensing. Nano Lett 16(3):2096–2102
21. Ostlund C, Folker ES, Choi JC, Gomes ER, Gundersen GG, Worman HJ (2009) Dynamics and molecular interactions of linker of nucleoskeleton and cytoskeleton (LINC) complex proteins. J Cell Sci 122:4099–4108
22. Arsenovic PT, Bathula K, Conway DE (2017) A protocol for using Förster resonance energy transfer (FRET)-force biosensors to measure mechanical forces across the nuclear LINC complex. J Vis Exp. https://doi.org/10.3791/54902
23. Kolega J (2004) Phototoxicity and photoinactivation of blebbistatin in UV and visible light. Biochem Biophys Res Commun 320:1020–1025
24. Arsenovic PT, Ramachandran I, Bathula K, Zhu R, Narang JD, Noll NA, Lemmon CA, Gundersen GG, Conway DE (2016) Nesprin-2G, a component of the nuclear LINC complex, is subject to myosin-dependent tension. Biophys J 110:34–43
25. Arsenovic PT, Mayer CR, Conway DE (2017) SensorFRET: a standardless approach to measuring pixel-based spectral bleed-through and FRET efficiency using spectral imaging. Sci Rep 7:15609
26. Kutscheidt S, Zhu R, Antoku S, Luxton GWG, Stagljar I, Fackler OT, Gundersen GG (2014) FHOD1 interaction with nesprin-2G mediates TAN line formation and nuclear movement. Nat Cell Biol 16:708–715
27. Chang W, Worman HJ, Gundersen GG (2015) Accessorizing and anchoring the LINC com-

plex for multifunctionality. J Cell Biol 208:11–22

28. Wilson MH, Holzbaur ELF (2015) Nesprins anchor kinesin-1 motors to the nucleus to drive nuclear distribution in muscle cells. Development 142:218–228

29. Zhu R, Antoku S, Gundersen GG (2017) Centrifugal displacement of nuclei reveals multiple LINC complex mechanisms for homeostatic nuclear positioning. Curr Biol 27:3097–3110.e5

30. Khatau SB, Hale CM, Stewart-Hutchinson PJ, Patel MS, Stewart CL, Searson PC, Hodzic D, Wirtz D (2009) A perinuclear actin cap regulates nuclear shape. Proc Natl Acad Sci U S A 106:19017–19022

31. Li Y, Lovett D, Zhang Q, Neelam S, Kuchibhotla RA, Zhu R, Gundersen GG, Lele TP, Dickinson RB (2015) Moving cell boundaries drive nuclear shaping during cell spreading. Biophys J 109:670–686

Chapter 7

Analyzing Mechanotransduction Through the LINC Complex in Isolated Nuclei

Nejma Belaadi, Angélique Millon-Frémillon, Julien Aureille, and Christophe Guilluy

Abstract

The mechanical properties of the cellular microenvironment can impact many aspects of cell behavior, including molecular processes in the nucleus. Recent studies indicate that the LINC complex and its associated nuclear envelope transmit and transduce mechanical stress into biochemical pathways that ultimately regulate nuclear structure or gene expression. Here we describe a method to apply tensional forces to the LINC complex of isolated nuclei. Using magnetic beads and magnets, this technique can be used to explore the biochemical pathways that are activated in response to tension applied to the surface of isolated nuclei.

Key words Mechanotransduction, Lamina, Nucleus, Mechanical tension, Nesprin

1 Introduction

Almost three centuries after the pioneering observations of the nucleus using light microscopy [1], electron microscopy images revealed that the nucleus is not isolated within the cell, but is connected to cytoskeletal filaments instead [2, 3]. The proteins that connect the cytoskeleton to the nuclear envelope constitute the linker of the nucleoskeleton and cytoskeleton (LINC) complex [4, 5]. This complex is composed of SUN (Sad1 and UNC-84) proteins anchored in the inner nuclear membrane and nesprin (nuclear envelope spectrin-repeat-containing proteins) anchored in the outer nuclear membrane [4, 6]. SUN proteins interact with nesprins within the perinuclear space through their highly conserved SUN (Sad1 UNC-84 homology) domain and KASH (Klarsicht, ANC-1, and Syne homology) domain, respectively [4].

The LINC complex constitutes a network of transmembrane proteins that transmit force between the cytoskeleton and the nuclear envelope [7, 8]. Using various techniques to apply mechanical stress to nuclei, recent work showed that mechanotransduction

Gregg G. Gundersen and Howard J. Worman (eds.), *The LINC Complex: Methods and Protocols*, Methods in Molecular Biology, vol. 1840, https://doi.org/10.1007/978-1-4939-8691-0_7, © Springer Science+Business Media, LLC, part of Springer Nature 2018

mechanisms exist in the nucleus and transduce mechanical stress into signaling pathways that ultimately regulate nuclear structure or gene expression [9–12]. Interestingly, these recent advances revealed the LINC complex behaves like cell surface adhesion and undergoes remodeling in response to mechanical tension [8]. Here we describe a method that we designed to apply tensional forces to isolated nuclei using magnetic beads and magnets. This method is adapted from techniques that have been used to analyze the cellular response to forces applied to cell surface adhesion receptors [13–16].

In this protocol, tensional forces are applied to the LINC complex of isolated nuclei in order to mimic mechanical stress transmission from the cytoskeleton to the nucleus. This experimental system can be used to study either the mechanical response of isolated nuclei using magnetic tweezers or the force-dependent nuclear signaling pathways by various biochemical approaches. Another advantage of this method is the isolation of LINC complexes through a straightforward ligand affinity purification procedure. Here we detail the methods to isolate nuclei from HeLa cells and apply forces to nesprin-2 using magnetic beads. Additionally, we describe how to purify the LINC complex and analyze its protein composition following force application.

2 Materials

All solutions were prepared using ultrapure water, and common chemicals were from Sigma-Aldrich and Thermo Fisher Scientific.

2.1 Magnetic Bead Preparation

1. Magnetic beads: 2.8 μm diameter M-280 tosyl-activated (Invitrogen, cat. no. 142.03).
2. Dynamag magnet separator (Thermo Fisher Scientific, cat. no. 12321D).
3. Buffer A: 0.1 M sodium phosphate buffer, pH 7.4.
4. Buffer B: Combine 0.01 M sodium phosphate, pH 7.4, 0.0137 M NaCl, and 0.5% (w/v) delipidated BSA.
5. Nesprin-2 antibody (Abcam).
6. Phosphate-buffered saline (PBS): 137 mM NaCl, 2.7 mM KCl, 10 mM Na_2HPO_4, 1.8 mM KH_2PO_4, pH 7.4.

2.2 Isolation of Nuclei

1. HeLa cells (ATCC, cat. no. CCL-2.1).
2. Medium: DMEM plus 10% fetal bovine serum.
3. Cell scraper.
4. Dounce homogenizer (7 mL capacity; Bellco, cat. no. 1984-10007).

5. Hypotonic buffer: 10 mM HEPES, pH 7.9, 1 mM KCl, 1.5 mM $MgCl_2$, 0.5 mM dithiothreitol, and protease inhibitors (freshly added before use).
6. Buffer C: 20 mM HEPES, pH 7.8, 25 mM KCl, 5 mM $MgCl_2$, 0.25 M sucrose, and 1 mM ATP (freshly added before use).

2.3 Force Application with Magnetic Beads for Biochemical Measurements

1. Poly-lysine-coated dishes (35 mm). To coat the dish, incubate with poly-L-lysine solution (Sigma, cat. no. P4707) 1 h in a 37 °C incubator, then remove the solution by vacuum aspiration, and allow surface to dry.
2. Isolated nuclei (as in Subheading 2.2).
3. Magnetic beads coated with anti-nesprin-2 or other nesprin antibodies (as in Subheading 2.1).
4. The magnets used to stimulate isolated nuclei with 30–40 pN force are the 1.25-in.-diameter × 0.25-in.-thick nickel-plated neodymium (grade N52) magnet (K&J Magnetics).
5. Buffer D: 20 mM HEPES, pH 7.8, 25 mM KCl, 5 mM $MgCl_2$, and 1 mM ATP (freshly added before use).
6. Lysis buffer: 20 mM Tris–HCl at pH 7.6, 150 mM NaCl, 1% NP-40, 2 mM $MgCl_2$, and protease inhibitors.
7. 2× SDS sample buffer: 0.125 M Tris–HCl (pH 6.8), 4% (wt/vol) SDS, 0.005% (wt/vol) bromophenol blue, 20% (vol/vol) β-mercaptoethanol (β-mercaptoethanol should be freshly added before use), 10% (vol/vol) glycerol.

3 Methods

3.1 Magnetic Bead Conjugation

1. Wash 90 μL of 2.8 μm tosyl-activated Dynabeads in 1 mL of buffer A in 1.5 mL microcentrifuge tube, and use the Dynamag magnet separator to collect the beads and aspirate the medium.
2. Mix 20 μg of anti-nesprin-2 antibody (Abcam) with appropriate volume of buffer B to bring the total volume to 200 μL and mix by pipetting (other antibodies targeting different isoforms of nesprin can be used here).
3. Combine antibody with beads, and incubate for 24 h at 37 °C on a rotor to allow covalent linkage of antibody to the bead.
4. Collect the beads using the magnetic separator (Dynamag).
5. Wash beads three times with 1 mL PBS using the magnetic separator to collect the beads.
6. Resuspend the beads in 1 mL PBS to give a concentration of 6×10^8 beads/mL. Store beads at 4°C for up to 1 month (*see* **Note 1**).

3.2 Nuclei Isolation

1. Use near-confluent HeLa cells plated on 150 mm dishes and serum starved with DMEM lacking serum for 16 h.
2. Aspirate the medium and wash two times with 10 mL PBS at 37 °C.
3. Aspirate the PBS.
4. Add ice-cold hypotonic buffer (6 mL), and detach the cell bodies using a cell scraper.
5. Homogenize the samples using 30 strokes of a Dounce homogenizer.
6. Transfer the sample to a 15 mL conical tube on ice. Incubate for 5 min on ice.
7. Centrifuge at 700 × *g* for 5 min at 4 °C.
8. Resuspend the pellet in hyponic buffer and centrifuge at 700 × *g* for 5 min at 4 °C.
9. Resuspend the nuclear pellet in buffer C (10 mL) and store on ice (*see* **Note 2**).
10. Following their isolation, the nuclei can be plated and stimulated in order to perform biochemical assay (as described below in Subheading 3.3) or force experiments using magnetic tweezers (*see* **Note 3**).

3.3 Application of Tension to the LINC Complex Using Magnetic Beads and Permanent Magnets for Biochemical Analysis

1. Plate 10^6 nuclei on poly-L-lysine-coated dishes (35 mm) for 30 min at 37 °C in 1 mL of buffer C.
2. Incubate nuclei with nesprin antibody-coated magnetic beads (2×10^6 beads per dish) at 37 °C for 20 min.
3. Wash twice with buffer C, and add 1 mL of buffer D for 15 min at 37 °C.
4. Stimulate with force by suspending the permanent magnet at 4.5 mm over nuclei (permanent magnet can be placed on the lid of the dish) for the appropriate amount of time (*see* **Notes 4** and **5**).
5. Aspirate buffer D.
6. Add 300 μL of lysis buffer and use the cell scraper to detach the nuclei.
7. Collect the lysate in a 1.5 mL microcentrifuge tube and incubate on ice for 5 min.
8. Homogenize by pipetting and collect 50 μL for total nuclear fraction.
9. Collect the beads using the magnetic separator.
10. Wash three times using lysis buffer.
11. Collect the beads using the magnetic separator, and add 2× SDS sample buffer (50 μL) to constitute the bead fraction (*see* **Note 6**).

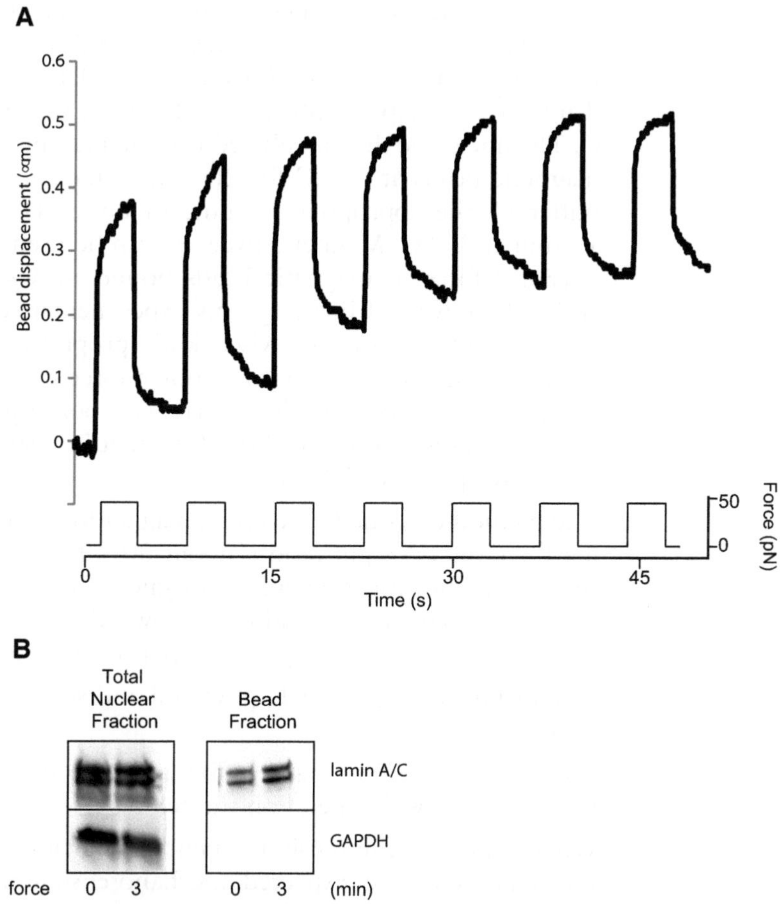

Fig. 1 Isolated nuclei respond to force applied to the LINC complex. (**A**) Typical displacement of a 2.8 μm bead coated with anti-nesprin-2 antibody bound to an isolated nucleus during force pulse application. Stiffening is indicated by decreased displacement during later pulses. (**B**) Nuclei isolated from HeLa cells were incubated with anti-nesprin-2-coated magnetic beads. After stimulation with a permanent magnet for 3 min, the nuclei were lysed, and the protein complexes associated with the beads (bead complex) were isolated from the lysate using a magnetic separation stand, and both fractions were solubilized in SDS buffer and analyzed by Western blotting

12. Analyze the total nuclear fraction and bead fraction using Western blot (Fig. 1B) or mass spectrometry (*see* **Note 7**).

4 Notes

1. Beads can aggregate after conjugation. If vortexing is not sufficient to homogenize, beads separation can be achieved either by pipetting or by using sonication for a short period (20 s).
2. Depending on cell type, the nuclear fraction can be contaminated by cytosolic elements (tubulin), whose presence can be assessed using Western blot. If this is the case, the **steps 5–7** can be repeated.

3. For mechanical force experiments, the format of the slide or coverslip and volume of buffer depend on the magnetic tweezers set-up. Typically 10,000 nuclei are plated on a 12 mm diameter coverslip for 30 min at 37 °C in 0.5 mL of buffer C. The nuclei are then incubated with nesprin antibody-coated magnetic beads at 37 °C for 20 min. After two washes with buffer C, the appropriate volume of buffer D is added for 15 min at 37 °C. Magnetic tweezers can be used to apply pN tensional force to magnetic beads bound to nesprins. When coupled to live cell imaging, this experimental system allows the measurement of local viscoelastic properties by tracking the bead displacement due to a known force generated by the magnetic field from an electromagnetic pole tip. Figure 1A shows a representative bead displacement in response to constant force pulses of 50 pN.
4. The method described here uses a magnet for which the resulting force on a 2.8 μm bead is about 30–40 pN (calculated by measuring the displacement of magnetic beads in undiluted glycerol, a Newtonian liquid with known viscosity); however, it is possible to use larger magnets to apply larger amount of force.
5. The magnets are extremely powerful and should be manipulated carefully.
6. The volume of SDS sample buffer added to the beads is only indicative and will depend on the cell type.
7. While exploring the tension-dependent signaling pathways, if no changes in the expected mechanosensing processes are detected, different possibilities can be considered. The duration of the stimulation with the magnet may not be sufficient and can be modulated (typical stimulation range from 1 to 10 min). The amount of force may not be sufficient and can be increased by using a thicker magnet (grade N52) or larger beads. To ensure that the LINC complex purification was efficient, Western blot can be used to probe for SUN proteins and/or nuclear envelope components. Additionally, analyzing lamin A/C recruitment to the LINC complex in response to force (using Western blot in Subheading 3.2) can be useful to determine if the force stimulation was sufficient.

5 Conclusion

Recent work shows that the nucleus may act as a mechanosensitive organelle, whose nucleoskeleton can dynamically remodel in response to mechanical stress. This dynamic reorganization hinges on nuclear mechanotransduction mechanisms, which are only beginning to be elucidated [8]. Various methods have been developed to manipulate and measure mechanical tension at the molecular level, such as atomic force microscopy and molecular "tension sensors" [17]. Here we describe a method using magnetic beads and magnets to stimulate isolated nuclei with pN tensional forces and explore the force-dependent nuclear biochemical pathways.

This experimental system can be coupled with magnetic tweezers to explore the nuclear mechanical response to force of single isolated nucleus. Alternatively, a large numbers of isolated nuclei can be stimulated using a permanent magnet to yield sufficient material for biochemical or molecular biology assays. Additionally, this straightforward technique can allow LINC complex purification following force application. The purified LINC complex can then be used for various biochemical approaches, such as Western blot or kinase assays to investigate kinase activity in response to mechanical tension. This method is a simple way to explore the nuclear molecular mechanisms that are activated in response to mechanical tension and may help understanding how the mechanical properties of the cellular microenvironment regulate cell behavior.

Acknowledgments

We thank Keith Burridge for his continuous support. C.G. is supported by grants from the Agence Nationale de la Recherche (ANR-13-JSV1-0008) and from European Research Council (ERC) under European Union's Horizon 2020 research and innovation program (ERC Starting Grant n°639300).

References

1. Pederson T (2011) The nucleus introduced. Cold Spring Harb Perspect Biol 3:a000521. https://doi.org/10.1101/cshperspect.a000521
2. Capco DG, Wan KM, Penman S (1982) The nuclear matrix: three-dimensional architecture and protein composition. Cell 29:847–858
3. Lehto VP, Virtanen I, Kurki P (1978) Intermediate filaments anchor the nuclei in nuclear monolayers of cultured human fibroblasts. Nature 272:175–177
4. Crisp M, Liu Q, Roux K, Rattner JB, Shanahan C, Burke B, Stahl PD, Hodzic D (2006) Coupling of the nucleus and cytoplasm: role of the LINC complex. J Cell Biol 172:41–53. https://doi.org/10.1083/jcb.200509124
5. Chang W, Worman HJ, Gundersen GG (2015) Accessorizing and anchoring the LINC complex for multifunctionality. J Cell Biol 208:11–22. https://doi.org/10.1083/jcb.201409047
6. Gundersen GG, Worman HJ (2013) Nuclear positioning. Cell 152:1376–1389. https://doi.org/10.1016/j.cell.2013.02.031
7. Maniotis AJ, Chen CS, Ingber DE (1997) Demonstration of mechanical connections between integrins, cytoskeletal filaments, and nucleoplasm that stabilize nuclear structure. Proc Natl Acad Sci U S A 94:849–854
8. Aureille J, Belaadi N, Guilluy C (2017) Mechanotransduction via the nuclear envelope: a distant reflection of the cell surface. Curr Opin Cell Biol 44:59–67. https://doi.org/10.1016/j.ceb.2016.10.003
9. Guilluy C, Osborne LD, Van Landeghem L, Sharek L, Superfine R, Garcia-Mata R, Burridge K (2014) Isolated nuclei adapt to force and reveal a mechanotransduction path-

way in the nucleus. Nat Cell Biol 16:376–381. https://doi.org/10.1038/ncb2927

10. Swift J, Ivanovska IL, Buxboim A, Harada T, Dingal PCDP, Pinter J, Pajerowski JD, Spinler KR, Shin J-W, Tewari M, Rehfeldt F, Speicher DW, Discher DE (2013) Nuclear lamin-A scales with tissue stiffness and enhances matrix-directed differentiation. Science 341:1240104. https://doi.org/10.1126/science.1240104
11. Enyedi B, Jelcic M, Niethammer P (2016) The cell nucleus serves as a mechanotransducer of tissue damage-induced inflammation. Cell 165:1160–1170. https://doi.org/10.1016/j.cell.2016.04.016
12. Kumar A, Mazzanti M, Mistrik M, Kosar M, Beznoussenko GV, Mironov AA, Garrè M, Parazzoli D, Shivashankar GV, Scita G, Bartek J, Foiani M (2014) ATR mediates a checkpoint at the nuclear envelope in response to mechanical stress. Cell 158:633–646. https://doi.org/10.1016/j.cell.2014.05.046
13. Marjoram RJ, Guilluy C, Burridge K (2016) Using magnets and magnetic beads to dissect signaling pathways activated by mechanical tension applied to cells. Methods San Diego Calif 94:19–26. https://doi.org/10.1016/j.ymeth.2015.09.025
14. Zhao X-H, Laschinger C, Arora P, Szászi K, Kapus A, McCulloch CA (2007) Force activates smooth muscle alpha-actin promoter activity through the Rho signaling pathway. J Cell Sci 120:1801–1809. https://doi.org/10.1242/jcs.001586
15. Guilluy C, Swaminathan V, Garcia-Mata R, O'Brien ET, Superfine R, Burridge K (2011) The Rho GEFs LARG and GEF-H1 regulate the mechanical response to force on integrins. Nat Cell Biol 13:722–727. https://doi.org/10.1038/ncb2254
16. Millon-Frémillon A, Aureille J, Guilluy C (2017) Analyzing cell surface adhesion remodeling in response to mechanical tension using magnetic beads. J Vis Exp. https://doi.org/10.3791/55330
17. Arsenovic PT, Ramachandran I, Bathula K, Zhu R, Narang JD, Noll NA, Lemmon CA, Gundersen GG, Conway DE (2016) Nesprin-2G, a component of the nuclear LINC complex, is subject to myosin-dependent tension. Biophys J 110:34–43. https://doi.org/10.1016/j.bpj.2015.11.014

Chapter 8

Direct Force Probe for Nuclear Mechanics

Vincent J. Tocco, Srujana Neelam, Qiao Zhang, Richard B. Dickinson, and Tanmay P. Lele

Abstract

We describe a recently reported method for directly applying a known, nanonewton-scale force to the nucleus in a living, intact cell. First, a suction seal is applied on the nuclear surface using a micropipette. Then, the micropipette is translated away from the nucleus. The nucleus deforms and translates with the moving micropipette and then eventually detaches from the micropipette and recovers (roughly) its original shape and position. At the point of detachment, the resisting force (from the deformed nucleus and connected cytoskeleton) balances the suction force. Because the suction force is precisely known and reproducibly applied, this method therefore allows comparisons of nuclear response across disruptions to the cytoskeleton, nucleus, or cell. This method is useful for quantifying nuclear elastic properties in its native, integrated environment.

Key words Nuclear forces, Nuclear shaping, Nuclear positioning, Nuclear mechanics

1 Introduction

A variety of forces act on the nucleus, including mechanical forces generated by molecular motors [1–6], frictional forces in the cytoskeleton [7], and osmotic forces dependent on the concentrations of macromolecules in the nucleus and cytoplasm [8, 9]. These forces, which can be propagated to the nucleus through the linker of nucleus and cytoskeleton (LINC) complex [10–12], act to shape and position the nucleus in the cell. Abnormal nuclear shaping and positioning in the cell characterize multiple human diseases [13–16]. It is therefore important to understand how forces on the nucleus and the mechanical properties of the nucleus become abnormal in human diseases.

Because the nucleus is integrated with the different cytoplasmic structures through the LINC complex, it becomes necessary to probe nuclear properties while it is mechanically integrated with the cell. This is challenging because most methods such as micropipette aspiration [17–20] require either isolated nuclei or nuclei

Gregg G. Gundersen and Howard J. Worman (eds.), *The LINC Complex: Methods and Protocols*, Methods in Molecular Biology, vol. 1840, https://doi.org/10.1007/978-1-4939-8691-0_8, © Springer Science+Business Media, LLC, part of Springer Nature 2018

in suspended cells. Under such conditions, the surrounding cytoskeletal network (which exerts stress on the nucleus in a spread, adherent cell) is either removed completely or greatly reorganized relative to its normal state. Atomic force microscopy [20] or applying force to cells adherent to substrates [21] can be used to probe the nucleus. In these methods, force is not applied *directly* to the nucleus, but rather is transmitted through the intervening cytoskeletal network. Therefore, the magnitude of the force applied to the nuclear surface is unknown. This makes it difficult to compare nuclear response across different perturbations to the nucleus.

We have recently described a direct force probe, which applies mechanical force directly to the nucleus in an adherent cell [22]. The method is easy to implement with the commonly used Eppendorf microinjection system. A fine micropipette (0.5 μm tip diameter) is suction-sealed against the nuclear surface and moved away from the nucleus (Fig. 1A). Motion of the pipette tip deforms and displaces the nucleus. Eventually, the nucleus detaches from

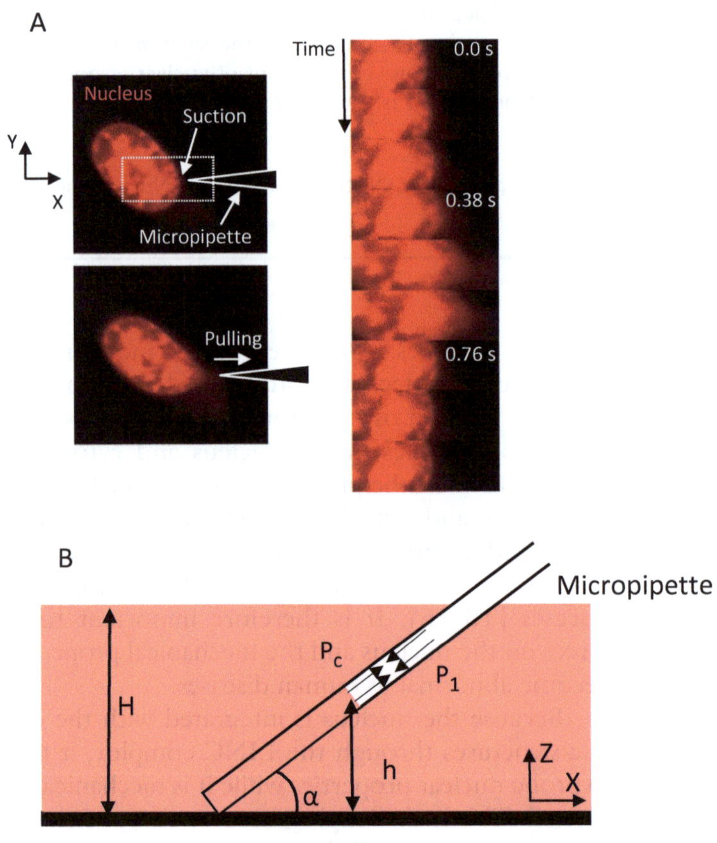

Fig. 1 Schematic of direct force probe experiments. (**A**) A micropipette is suction-sealed against the nuclear surface of an NIH3T3 fibroblast and moved, deforming the nucleus until it detaches. The region bounded by the white box is shown in a kymograph, displaying the nuclear deformation over time. (**B**) Free-body schematic diagram of the forces in the micropipette

the pipette and recovers the original shape and position when the restoring forces exerted by cellular structures and the nucleus itself equal the suction force. We have reported that the extent of nuclear deformation is proportional to the force. One can then perturb the cytoskeleton, chromatin the LINC complex, and determine the contribution of these components to the nuclear deformation response *while it is integrated* with the cell. Here we describe a systematic protocol for using this method to probe the nucleus in adherent cells mechanically.

2 Materials

2.1 Preparation of Cells

1. Adherent cell line, such as NIH3T3, MEF, etc.
2. Cell media (composition dependent on cell line).
3. 35 mm glass-bottomed dishes (World Precision Instruments).
4. Phosphate-buffered saline.
5. Fibronectin.
6. Trypsin (0.25% w/v).
7. SYTO10/SYTO59 nucleic acid dye (or equivalent; Thermo Fisher Scientific).
8. 0.3 mg/mL PLL-g-PEG solution (SuSoS, Dübendorf, Switzerland).
9. Plasmid or siRNA (with appropriate transfection reagents) or small-molecule inhibitor for perturbing components of the LINC complex, nucleus, or cytoskeleton.

2.2 Equipment for Probing the Nucleus

1. Femtotips Microinjection capillary tips (0.5 μm-inner diameter; Thermo Fisher Scientific).
2. InjectMan® NI2 Micromanipulation system (Eppendorf, Hamburg, Germany; discontinued, current equivalent model is InjectMan® 4).
3. FemtoJet® Microinjection system (Eppendorf).
4. Nikon TE-2000 Eclipse Inverted Microscope (or equivalent inverted epifluorescent or confocal microscope), equipped with a temperature-controlled environmental chamber (to maintain cells at 37 °C and 5% CO_2), and a CCD camera, and connected to a computer with Nikon Elements software.
5. Filter cubes corresponding to fluorescent proteins/probes used in the experiment.

2.3 Data Analysis

1. Quantitative image-processing software, for example, MATLAB (The MathWorks Inc., Natick, MA) or FIJI (open-source, NIH).

3 Methods

3.1 Preparation of Cells and Microscope Setup

1. The FemtoJet® Microinjection system installs to the arm of the microscope. Install in accordance with manufacturer's recommendations.
2. Coat sterile 35 mm glass-bottomed dishes with 5 μg/mL fibronectin in PBS for 1 h at room temperature or overnight at 4 °C.
3. Wash the dishes multiple times with PBS.
4. Trypsinize and plate cells on the dishes (roughly 3×10^5; corresponding to 25% confluency in 35 mm dish). Allow cells to spread overnight.
5. (Optional) Transfect cells with a plasmid or siRNA to perturb the nucleus, cytoskeleton, or LINC complex (e.g., overexpression of GFP-KASH4 to disrupt the LINC complex). Allow the appropriate length of time for protein expression, knockdown, or drug treatment.
6. (Optional) Treat cells with a small-molecule inhibitor to perturb the nucleus, cytoskeleton, or LINC complex (e.g., nocodazole to inhibit microtubule polymerization). Allow the appropriate length of time for inhibition.
7. Wash cells with PBS and change media at least once prior to imaging. Add SYTO dye at manufacturer-recommended concentration (*see* **Note 1**) 15 min prior to the experiment to image the nucleus.
8. Choose the appropriate microscope objective. For most experiments, we used a Plan Fluor oil immersion 40×/1.3 NA objective. Apply a small drop of immersion oil, as needed.
9. Clamp the glass-bottomed dish tightly into the dish holder and load onto the stage. Maintain cells at 37 °C and 5% CO_2 at all times while on microscope stage.
10. In order to immerse the micropipette into the culture media (as specified below), we had to modify our closed environmental chamber that supplied the cells with humidified CO_2. We did so by removing the hard plastic cover and replacing it with plastic wrap (the transparent wrap maintains the CO_2 concentration inside the chamber).
11. Using DIC microscopy, find the appropriate focal plane to visualize cells. Correct for uneven illumination in the field of view by performing both Kohler illumination (DIC) and shading correction (fluorescence channels).

3.2 Conversion of Micropipette Pressure to Force

A major advantage of this method is that it applies a precisely known suction pressure reproducibly to the nucleus. We begin by initially imposing an external (gauge) pressure P_1 using the InjectMan® system to force out liquid from the pipette using a

Table 1
P_1 is approximately equal to P_c; other pressures are negligible; therefore, the suction pressure is equal to P_1 to a good approximation

P_1 (hPa)	h (mm)	P_h (hPa)	P_c (hPa)	Force (nN)
12.72	3.53	0.347	12.58	0.2516
50	0.88	0.086	50.1	1.002
100	0.44	0.043	100.1	2.002
200	0.21	0.021	200.2	4.004
300	0.14	0.014	300.2	6.004
400	0.11	0.01	400.2	8.004
500	0.08	0.008	500.2	10.004
600	0.06	0.006	600.2	12.004

microinjection system. At equilibrium, the pressure P_1 plus the fluid head in the micropipette (P_h) balances the capillary pressure (P_c) p the fluid head in the culture dish (P_H) (Fig. 1B). Table 1 shows the magnitude of these pressures; P_1 and P_c are approximately equal and opposite for a broad range of values of P_1 (Table 1). We then insert the micropipette tip into the cell, touch the micropipette to the nuclear surface, and set P_1 to zero (gauge pressure) by opening the pipette tubing to atmospheric pressure. A suction pressure ($P_c \approx P_1$) now exists on the nuclear surface. The force on the nuclear surface is calculated by multiplying the pressure by the area of the micropipette tip (0.20 μm^2).

In the table, we first chose a value for P_1, which can be set with the Eppendorf system. Next, we calculated the height of the fluid inside the micropipette, h, from the pressure balance $P_1 = P_c + P_H - P_h$. $P_h = \rho g h$ is the pressure head in the pipette; ρ is the density of the culture media (approximately 1000 g/L), and g is the gravitational constant (9.8 m/s^2). The pressure head of the fluid in the dish (P_H; not shown in the table) is constant for the range of P_1 and ~ 0.2 hPa in magnitude. $P_c = 4\gamma \cos\theta/d$ is the capillary pressure; γ is the surface tension (72.8 dyn/cm), θ is the wetting angle of water on glass (30°), and d is the diameter of the micropipette, which decreases as the tube tapers toward the tip (accounted for in calculation by approximating the slope of micropipette). Force is calculated by multiplying P_c by the area of the micropipette tip (0.20 μm^2).

3.3 Applying Force to the Nucleus

Before the micropipette can be used to probe nuclei, it must be positioned in the field of view (*see* **Note 2**).

1. (Optional) To prevent the micropipette from adhering to the cells, immerse the tip in 0.3 mg/mL PLL-g-PEG solution for 1 h at room temperature (*see* **Note 3**).

2. Load the micropipette (0.5 μm diameter tip) into the capillary holder.
   (a) CAUTION: Fine micropipettes are very sharp; use care when handling to avoid personal injury.
3. Peel back the plastic wrap on the environmental chamber (as described above) to allow entry of the micropipette.
4. Lower the micropipette by twisting the joystick that controls the manipulator clockwise until the tip immerses in media. Position the tip as close to the center of the objective as possible (use "coarse" or "fine" mode on the micromanipulator for large or small micropipette movements, respectively).
5. Adjust the focus to a plane above the cells, switch from "DIC" to the "phase" optical configuration on the NIS-Elements software, and remove the polarizer from the light path. This configuration makes the shadow of the micropipette easier to see.
6. Move the micropipette back and forth (along *y*-axis), watching the screen carefully for a shadow.
7. Locate the tip of the micropipette using fine movements on the micromanipulator.
8. Bring the cells back into focus and lower the micropipette until the shape of the pipette is visible in the focal plane of the cells. Return to the DIC optical configuration and return the polarizer into the optical path.
9. The tip should always be clear of any cell debris and cleaned (*see* **Note 4**) regularly.
10. Set the desired pressure P_1 (in hPa) on the microinjector. Allow the height of the fluid in the micropipette to equilibrate (Fig. 1B).
11. Insert the tip of the micropipette into the cell, as close to the edge of the nucleus as possible (*see* **Note 5**).
12. Touch the tip of the micropipette to the nuclear surface (*see* **Note 6**), and then seal it to the surface by disconnecting the outlet tube of the microinjector to open the end of the capillary to the atmosphere.
13. Set up an .avi acquisition in Nikon Elements software. Set the desired frames per second and then begin recording.
14. Switch to fluorescent microscopy by adding the appropriate excitation and emission filters to the light path.
15. Translate the micropipette away from the nucleus by steadily moving the joystick to the right (laterally) (*see* **Note 7**).
16. Continue moving the tip and recording until the nucleus eventually detaches from the micropipette. Work quickly to avoid photobleaching.
17. End the .avi acquisition and save the raw data file for later analysis.

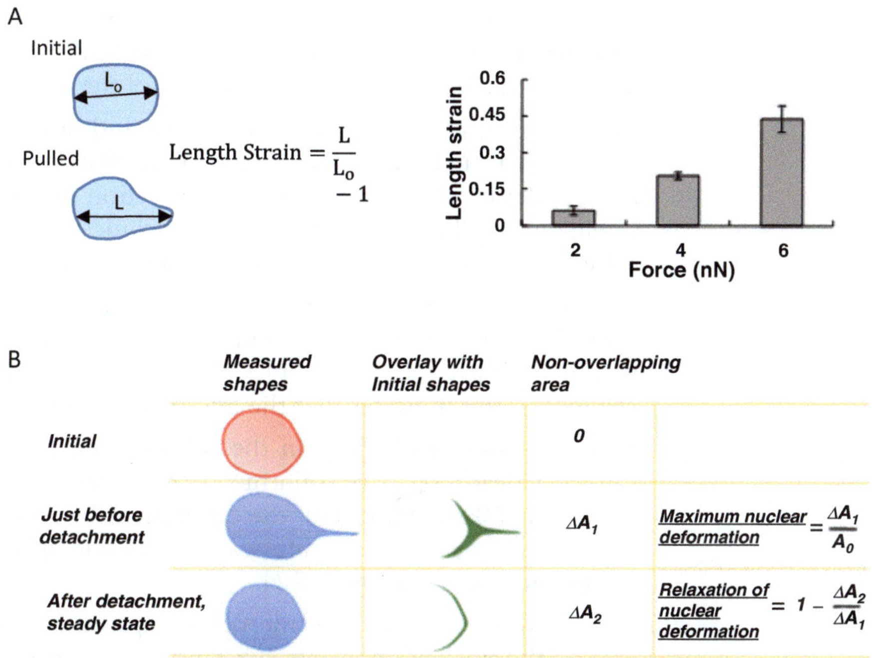

Fig. 2 Useful nuclear properties measurable with the direct force probe. (**A**) Length strain schematic and calculation are shown with average length strain as a function of suction force. Data is mean ± SEM, and all are statistically different at $p < 0.05$. $n = 6, 7, 14$ for 2 nN, 4 nN, and 6 nN, respectively (plot reproduced with permission from PNAS [22]). (**B**) Schematic explaining how to measure the maximum nuclear deformation and the relaxation of maximum deformation by measuring non-overlapping areas of the nucleus (Reproduced with permission from PNAS [22])

3.4 Data Analysis

We use a custom MATLAB script to analyze images, but any other image-processing software (e.g., FIJI) will work as well. The extent of deformation is quantifiable in multiple ways from images of the initial nuclear shape and the deformed shape: length strain (Fig. 2A) or maximum nuclear deformation (Fig. 2B). The time scale of relaxation is calculated by fitting an exponential decay to a plot of length strain as a function of time.

4 Notes

1. We used the DNA minor groove-binding SYTO dyes because we found that they did not affect nuclear mechanical properties when compared to unlabeled nuclei. Labeling the nucleus with GFP-histone H1 made the nucleus stiffer in fibroblasts [22]. Other small-molecule dyes (e.g., Hoechst 33342) may damage DNA or otherwise alter nuclear properties in live cells [23]. Whenever possible, use fluorophores with long wavelengths (i.e., red-shifted) to minimize the deleterious effects of phototoxicity. Also, minimize exposure time and excitation light power.

2. It is very important to avoid lowering the tip of the micropipette too far before the tip is located in the field of view, as the fragile tip can easily break on the surface of the culture dish.
3. The blocking step can prevent cell debris and dead cells from adhering to the body of the micropipette. We did not detect nuclear deformation when there is no suction pressure in the capillary (not shown); therefore, we conclude that there is no adhesion between the nucleus and the micropipette tip.
4. To check whether the pipette is clogged, apply the maximum possible pressure with the microinjector system (ensure the tip is away from live cells). If bubbles emerge from the tip into the cell media, this indicates that the micropipette is clear.
5. If the micropipette is held in the cell for long times, it may cause cell death. We have confirmed that the cell remains viable within the 30 s time scale of the experiment by adding ethidium homodimer-2, which diffuses into cell and appears red if the membrane is damaged [22].
6. The relatively small area of the tip (compared to the surface area of the nucleus) prevents substantial flow of water across nuclear pores into the capillary, which would decrease the actual pressure applied. The formula below (derived in [22]) illustrates the higher resistance to flow in the micropipette compared to through nuclear pores:

$$\frac{(P_0 - P_m)}{(P_0 - P_s)} = \frac{1}{1 + \varphi / \varphi_p}$$

In the above equation, P_0 is the hydrostatic pressure inside the nucleus, P_m is the pressure at the outer membrane surface, P_s is the applied suction pressure, and φ and φ_p are the filtration coefficients in nuclear pores and in the micropipette, respectively. Using common literature values, we estimated the ratio in the denominator to be extremely small ($\varphi/\varphi_p < 10^{-5}$). Therefore, there is a negligible pressure drop across the nuclear membrane if a 0.5 μm-diameter micropipette is used, despite the fact that fluid flows through nuclear pores (for the complete derivation, see the supplemental information of [22]).
7. The pulling rate can be controlled automatically with a computer, but we found it simpler to perform the pulling manually with the micromanipulator joystick. To test whether variations in the manual loading rate could affect the nuclear response, we calculated the loading rate from several experiments by tracking the deformation of the leading nuclear edge over time (loading rate is the maximum slope of the cubic fit; Fig. 3B). We found no correlation between nuclear mechanical response and loading rate (Fig. 3C).

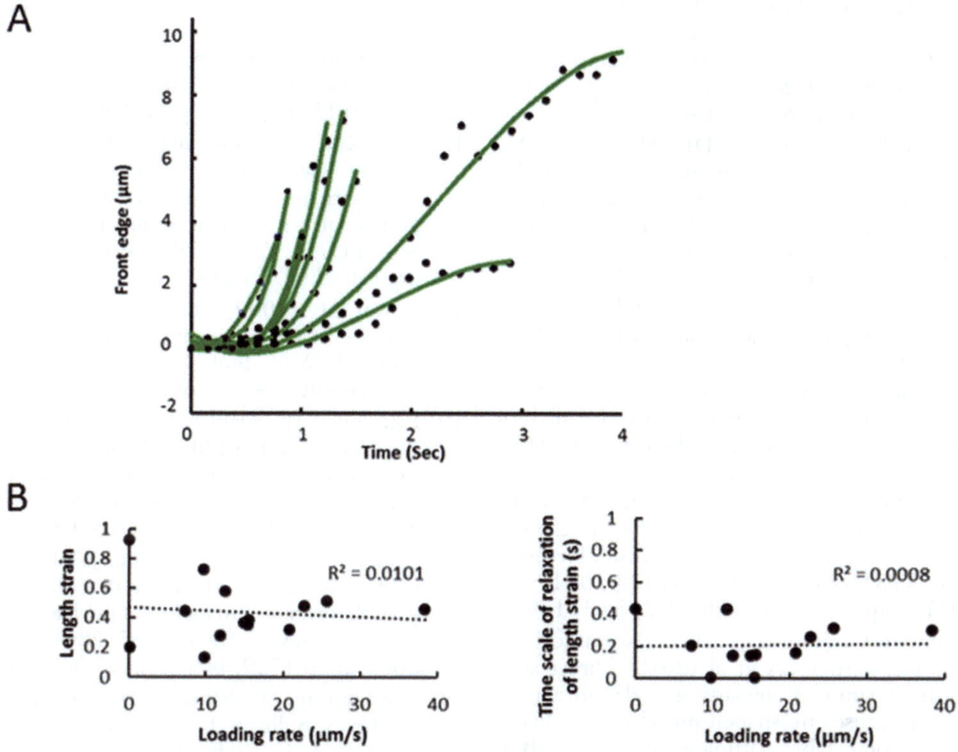

Fig. 3 Deformation is independent of loading rate. (**A**) Front edge nuclear displacements are plotted as functions of time. Data is plotted until detachment of the nucleus from the micropipette in eight different cells and fit with a cubic polynomial. The slope at maximum displacement is the loading rate of the nucleus. (**B**) Both the length strain of the moved nucleus just before detachment and the dynamics of length strain relaxation were uncorrelated with the loading rate (Reproduced with permission from PNAS [22])

References

1. Wu J, Lee KC, Dickinson RB, Lele TP (2011) How dynein and microtubules rotate the nucleus. J Cell Physiol 226(10):2666–2674. https://doi.org/10.1002/jcp.22616
2. Starr DA (2011) Watching nuclei move: insights into how kinesin-1 and dynein function together. Bioarchitecture 1(1):9–13. https://doi.org/10.4161/bioa.1.1.14629
3. Wu J, Kent IA, Shekhar N, Chancellor TJ, Mendonca A, Dickinson RB et al (2014) Actomyosin pulls to advance the nucleus in a migrating tissue cell. Biophys J 106(1):7–15. https://doi.org/10.1016/j.bpj.2013.11.4489
4. Chancellor TJ, Lee J, Thodeti CK, Lele T (2010) Actomyosin tension exerted on the nucleus through nesprin-1 connections influences endothelial cell adhesion, migration, and cyclic strain-induced reorientation. Biophys J 99(1):115–123. https://doi.org/10.1016/j.bpj.2010.04.011
5. Arsenovic PT, Ramachandran I, Bathula K, Zhu R, Narang JD, Noll NA et al (2016) Nesprin-2G, a component of the nuclear LINC complex, is subject to myosin-dependent tension. Biophys J 110(1):34–43. https://doi.org/10.1016/j.bpj.2015.11.014
6. Denais CM, Gilbert RM, Isermann P, McGregor AL, te Lindert M, Weigelin B et al (2016) Nuclear envelope rupture and repair during cancer cell migration. Science 352(6283):353–358. https://doi.org/10.1126/science.aad7297
7. Li Y, Lovett D, Zhang Q, Neelam S, Kuchibhotla RA, Zhu R et al (2015) Moving cell boundaries drive nuclear shaping during cell spreading. Biophys J 109(4):670–686. https://doi.org/10.1016/j.bpj.2015.07.006
8. Finan JD, Guilak F (2010) The effects of osmotic stress on the structure and function of the cell nucleus. J Cell Biochem 109(3):460–467. https://doi.org/10.1002/jcb.22437

9. Finan JD, Leddy HA, Guilak F (2011) Osmotic stress alters chromatin condensation and nucleocytoplasmic transport. Biochem Biophys Res Commun 408(2):230–235. https://doi.org/10.1016/j.bbrc.2011.03.131
10. Lombardi ML, Jaalouk DE, Shanahan CM, Burke B, Roux KJ, Lammerding J (2011) The interaction between nesprins and sun proteins at the nuclear envelope is critical for force transmission between the nucleus and cytoskeleton. J Biol Chem 286(30):26743–26753. https://doi.org/10.1074/jbc.M111.233700
11. Cain NE, Tapley EC, McDonald KL, Cain BM, Starr DA (2014) The SUN protein UNC-84 is required only in force-bearing cells to maintain nuclear envelope architecture. J Cell Biol 206(2):163–172. https://doi.org/10.1083/jcb.201405081
12. Alam SG, Lovett D, Kim DI, Roux K, Dickinson RB, Lele TP (2015) The nucleus is an intracellular propagator of tensile forces in NIH 3T3 fibroblasts. J Cell Sci 128(10):1901–1911. https://doi.org/10.1242/jcs.161703
13. Coffinier C, Jung HJ, Li Z, Nobumori C, Yun UJ, Farber EA et al (2010) Direct synthesis of lamin A, bypassing prelamin a processing, causes misshapen nuclei in fibroblasts but no detectable pathology in mice. J Biol Chem 285(27):20818–20826. https://doi.org/10.1074/jbc.M110.128835
14. Sullivan T, Escalante-Alcalde D, Bhatt H, Anver M, Bhat N, Nagashima K et al (1999) Loss of A-type lamin expression compromises nuclear envelope integrity leading to muscular dystrophy. J Cell Biol 147(5):913–920
15. Roux KJ, Burke B (2007) Nuclear envelope defects in muscular dystrophy. Biochim Biophys Acta 1772(2):118–127. https://doi.org/10.1016/j.bbadis.2006.06.001
16. Meinke P, Mattioli E, Haque F, Antoku S, Columbaro M, Straatman KR et al (2014) Muscular dystrophy-associated SUN1 and SUN2 variants disrupt nuclear-cytoskeletal connections and myonuclear organization. PLoS Genet 10(9):e1004605. https://doi.org/10.1371/journal.pgen.1004605
17. Dahl KN, Kahn SM, Wilson KL, Discher DE (2004) The nuclear envelope lamina network has elasticity and a compressibility limit suggestive of a molecular shock absorber. J Cell Sci 117(Pt 20):4779–4786. https://doi.org/10.1242/jcs.01357
18. Rowat AC (2009) Physical properties of the nucleus studied by micropipette aspiration. Methods Mol Biol 464:3–12. https://doi.org/10.1007/978-1-60327-461-6_1
19. Pajerowski JD, Dahl KN, Zhong FL, Sammak PJ, Discher DE (2007) Physical plasticity of the nucleus in stem cell differentiation. Proc Natl Acad Sci U S A 104(40):15619–15624. https://doi.org/10.1073/pnas.0702576104
20. Dahl KN, Engler AJ, Pajerowski JD, Discher DE (2005) Power-law rheology of isolated nuclei with deformation mapping of nuclear substructures. Biophys J 89(4):2855–2864. https://doi.org/10.1529/biophysj.105.062554
21. Lammerding J, Schulze PC, Takahashi T, Kozlov S, Sullivan T, Kamm RD et al (2004) Lamin A/C deficiency causes defective nuclear mechanics and mechanotransduction. J Clin Invest 113(3):370–378. https://doi.org/10.1172/JCI19670
22. Neelam S, Chancellor TJ, Li Y, Nickerson JA, Roux KJ, Dickinson RB et al (2015) Direct force probe reveals the mechanics of nuclear homeostasis in the mammalian cell. Proc Natl Acad Sci U S A 112(18):5720–5725. https://doi.org/10.1073/pnas.1502111112
23. Durand RE, Olive PL (1982) Cytotoxicity, mutagenicity and DNA damage by Hoechst 33342. J Histochem Cytochem 30(2):111–116

Chapter 9

Centrifugal Displacement of Nuclei in Adherent Cells to Study LINC Complex-Dependent Mechanisms of Homeostatic Nuclear Positioning

Ruijun Zhu and Gregg G. Gundersen

Abstract

The positioning of the nucleus is critical for key cellular processes including division, migration, and differentiation. Traditional approaches to understanding the functions and mechanisms of nuclear positioning have relied upon cellular systems in which nuclei move in response to stimuli or developmental programs and use molecular or pharmacological perturbations of nuclear and cytoskeletal elements. Here, we describe a complimentary approach to perturbing nuclear position in adherent cells using centrifugal force and how this may be used to understand LINC complex mechanisms of homeostatic nuclear positioning.

Key words LINC complex, KASH protein, Nesprin, SUN protein, Nuclear positioning

1 Introduction

Eukaryotic cells specifically position their nucleus in many cellular contexts, from division to migration to tissue homeostasis [1]. Studies of actively moving nuclei during these processes have revealed an important role for nuclear membrane KASH (Klarsicht, ANC-1, Syne homology) proteins and SUN (Sad1, UNC-84 homology) proteins, which together comprise the LINC (linker of nucleoskeleton and cytoskeleton) complex [2–4]. The LINC complex spans the inner and outer nuclear membrane and through the KASH proteins (also referred to as nesprins in vertebrates) allows the nucleus to attach to actin filaments, microtubules, and intermediate filaments. Nuclear positioning can be altered by disruption of the LINC complex, proteins associated with LINC complex, or the cytoskeleton (and its regulators), and this is accompanied by defects in cell polarization, migration, and/or division [5–18]. To date, virtually all studies of the mechanism of nuclear positioning and its possible role have relied on molecular perturbations such as

Gregg G. Gundersen and Howard J. Worman (eds.), *The LINC Complex: Methods and Protocols*, Methods in Molecular Biology, vol. 1840, https://doi.org/10.1007/978-1-4939-8691-0_9,

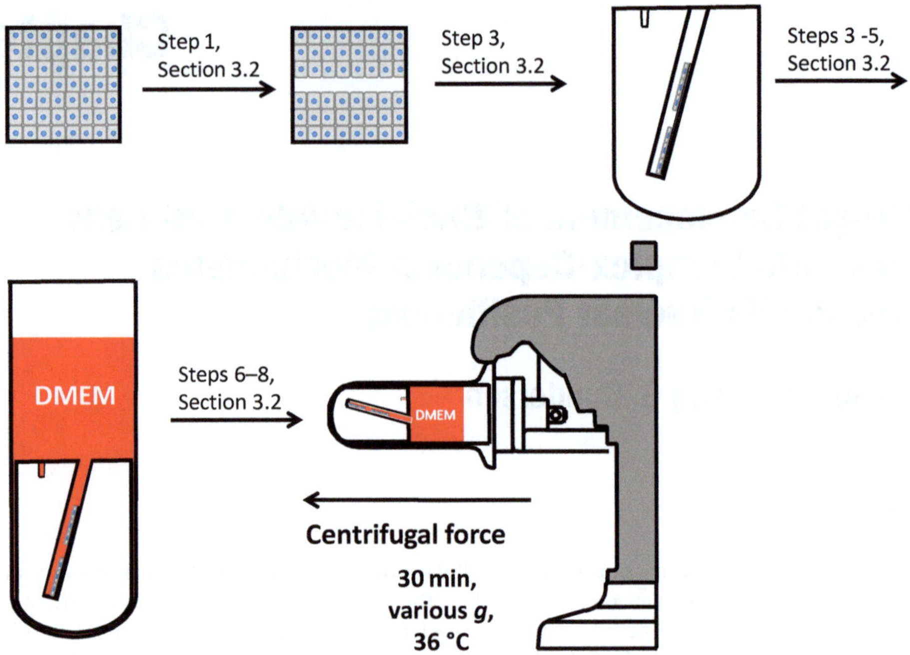

Fig. 1 Schematic of the centrifugation method to displace nuclei illustrating how a wounded monolayer is oriented within the adaptor and centrifuge tube. The rotor diagram was adapted from Beckman booklet PN L5-TB-069PE. Redrawn from ref. [26]

knockdown, expression of disease variants or dominant negative inhibitors, or treatment with small molecule inhibitors. Because of the difficulty in ruling out possible roles of the disrupted or inhibited proteins in processes other than positioning the nucleus, it has been difficult to determine whether nuclear positioning per se has a direct role in the cellular behaviors under study.

Several approaches have been developed to exert physical force on the nucleus to alter its position in cells, including applying suction to the nucleus with a microneedle [19] or hydrodynamic drag to cells (and their nuclei) using air bubbles under shear flow [20]. Although each method has its advantages, they are primarily restricted to analysis of single cells, can involve tedious micromanipulation, and may present technical challenges in determining the amount of force applied.

To develop a more efficient method to move the nucleus, we decide to use centrifugal force. Centrifugation has been used to enucleate cultured cells since the 1970s [21–23]. It has also been used to displace nuclei in yeast to study the relationship between the nucleus and the cell division plane [24]. We modified a method used to enucleate adherent mammalian cells [25], so that it could be used instead to displace the nucleus (Fig. 1) [26]. We omitted cytoskeletal drugs used in the enucleation procedure and

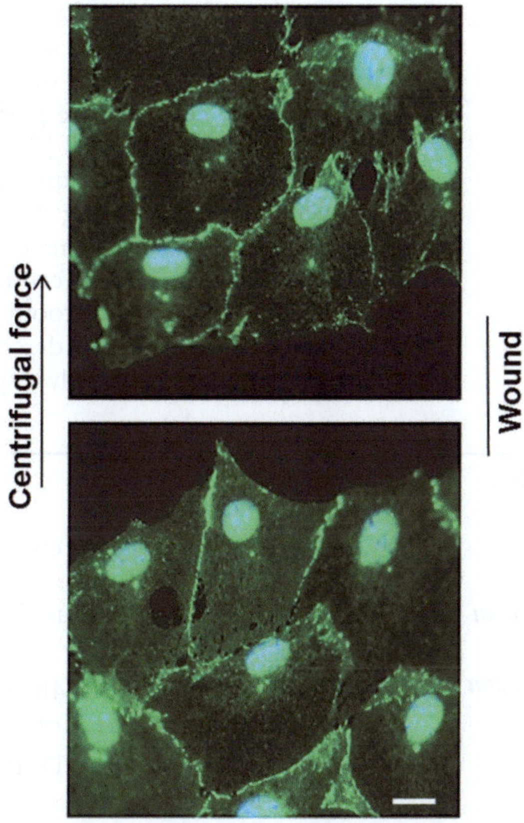

Fig. 2 Centrifugal force displaces nuclei in wounded monolayers of NIH3T3 fibroblasts. Monolayers were centrifuged at 5000 × *g* for 30 min and then fixed and stained for nuclei (DAPI) shown in blue and cell-cell junctions (β-catenin) and centrosomes (pericentrin) shown in green. Cells from both sides of the wound are shown. The nuclei are displaced toward the direction of centrifugal force indicated by the arrow. Note that the nuclei are toward the rear of the cell on one side of the wound and toward the front on the other (based on the direction of migration into the wound). Bar: 10 μm

additionally serum-starved cells to reduce levels of filamentous actin. With this method, nuclei can be displaced by centrifugal force in a wide variety of adherent cultured cells, including mouse NIH3T3 fibroblasts and C2C12 myoblasts and human HeLa carcinoma cells and HT1080 fibrosarcoma cells [26]. Additionally, if wounded monolayers of cells are centrifuged with the force orthogonal to the wound, nuclei are displaced toward the rear of the cells on one side of the wound and toward the leading edge on the other (Fig. 2). The displacement of the nucleus can be controlled by varying the centrifugal force (or time), and at least at 5000 × *g*, displacement of other organelles is minimal [26].

It is important to note that the nuclei displaced by centrifugation do not remain at their new location, but instead return to their original position over the course of 1–2 h. Thus, centrifugation revealed mechanisms of homeostatic nuclear positioning, which were found to depend on different LINC complexes and cytoskeletal elements depending on the confluency of the cells, and in wound edge cells, on the position of the nucleus relative to the front-back axis of the cell [26]. As pretreating the cells with lysophosphatidic acid, a serum-derived factor that triggers nuclear movement via activating Cdc42-dependent pathways [27], or expressing dominant negative LINC complex proteins, altered the extent of displacement by centrifugation [26], this method can also be used as an analytical tool to study the mechanisms of nuclear anchoring. Here we will describe the detailed methods for using centrifugation to artificially displace nuclei in adherent cells.

2 Materials

All solutions should be prepared with ddH_2O.

2.1 Preparation of Cells for Centrifugation

1. Adherent tissue cultured cells (e.g., NIH3T3 fibroblasts or other adherent cells).
2. Growth medium: DMEM, 10 mM Na-HEPES, pH 7.4, and 10% calf serum (*see* **Note 1**).
3. Serum-free medium: DMEM, 10 mM Na-HEPES; 0.1% fatty acid-free BSA.
4. 1 M Na-HEPES, pH 7.4 stock: 1 M HEPES (4-(2-hydroxyethyl)piperazine-1-ethanesulfonic acid) is dissolved in ddH_2O and the pH adjusted to 7.4 with NaOH, sterile filtered, and stored at 4 °C. The 1 M Na-HEPES stock is diluted 1:100 in DMEM to give a final concentration of 10 mM.
5. Conditioned medium: Serum-free medium harvested from 100 mm dishes of cells (*see* Subheading 3.1 below).
6. Acid washed coverslips: 22 mm^2 or 10 × 12 mm No. 1.5 glass coverslips are placed in porcelain coverslip racks (Thomas Scientific); washed in 1 N HCl for 10 min; rinsed for 10 min with running water, followed by two rinses each in ddH_2O and 95% ethanol; and then dried in a covered beaker and stored sterile (*see* **Note 2**).
7. Forceps for handling glass coverslips.
8. Tissue culture plates: 6-well plate for 22 mm^2 coverslips; 12-well plate for 10 × 12 mm coverslips.

2.2 Centrifugation of Cells

1. Centrifuge tube adaptors: 22 mm (fitted for 22 mm^2 coverslips) or 10 mm (fitted for 10 × 12 mm coverslips) adaptors made of polysulfone (Fig. 3).
2. Centrifuge tubes: ultraclear centrifuge tubes (Beckman No. 344058) for 22 mm adaptor or polyallomer centrifuge tubes (Beckman No. 326819) for 10 mm adaptor.

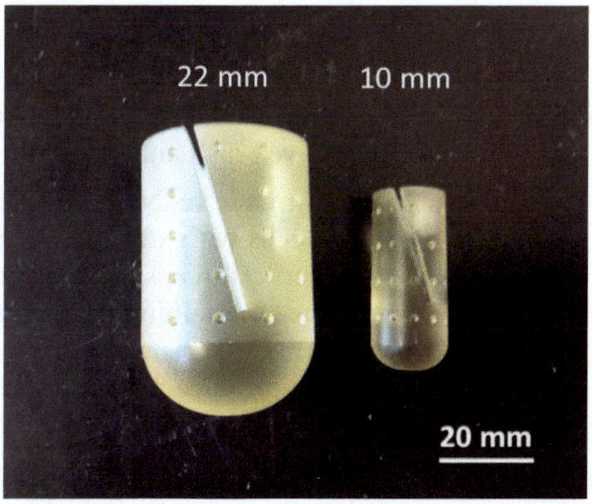

Fig. 3 Image of adaptors for holding coverslips during centrifugation. The large adaptor holds 22 mm^2 coverslips; the small adaptor holds 10 × 12 mm coverslips. The original design was by Vladimir Rodionov, University of Connecticut

3. Swinging bucket rotor: We use Beckman model SW28 (for 22 mm adaptor) or SW55 (for 10 mm adaptor) rotors.
4. Ultracentrifuge.
5. Forceps.

2.3 Fixation and Immunofluorescence Staining

1. Phosphate-buffered saline (PBS): 137 mM NaCl, 2.7 mM KCl, 4.3 mM Na_2HPO_4, 1.47 mM KH_2PO_4. Adjust pH to 7.4.
2. 4% paraformaldehyde (PFA) fixation buffer: 32% PFA aqueous solution (EM Grade, Purified, Electron Microscopy Sciences No. 15714) is diluted with PBS to a final concentration of 4%.
3. Methanol.
4. Blocking buffer: PBS, 0.3% Triton-100, 5% BSA.
5. 1 μg/mL 4′,6-diamidino-2-phenylindole (DAPI).
6. Rhodamine phalloidin (A12379, Thermo Fisher Scientific).
7. Mounting media: Fluoromount-G (Southern Biotech. Cat # 0100-01).
8. Glass microscope slides.
9. Imaging chamber: Chamber for replaceable 22 mm^2 square coverslips (Bioscience Tools, Cat # CSC-22x22).
10. Widefield epifluorescence microscope equipped with a digital camera and imaging software.

3 Methods

3.1 Preparation of Cells

1. Passage NIH3T3 fibroblasts onto an acid-washed coverslips in a 6-well dish and grow until ~80% confluent.
2. Serum starve NIH3T3 fibroblasts. Wash the coverslips with the cells with sterile PBS and then remove the coverslips with a sterile forceps, and dip them sequentially into 4 × 60 mm dishes containing 10 mL serum-free medium before putting them into a new 6-well dish containing 2 mL serum-free medium in each well. Incubate cells in serum-free medium for 48–72 h in a 37 °C incubator.
3. Prepare conditioned medium. While the cells on coverslips are being serum starved, starve one or more 100 mm dish(es) of near confluent NIH3T3 fibroblasts by changing the medium from serum containing to serum-free medium. Wash cells with PBS once followed by three washes with serum-free medium. Add 8 mL of serum-free medium to each dish. After 48 h, collect serum-free medium from the dish(es) and filter through a 0.22 μm sterile filter to remove debris in the medium. Conditioned medium can be stored for several weeks at 4 °C.

3.2 Centrifugation of Cells

1. Cells for centrifugation can be sparse or confluent. For wounded monolayers, gently scrape one or more wounds across the monolayer with a sterile 200 μL pipette tip (Fig. 1). The wounds should be aligned parallel with the coverslip edge. Allow wounded monolayers to recover for at least 1 h in a 37 °C incubator.
2. Prewarm the ultracentrifuge and the rotor to 36 °C (*see* **Note 3**).
3. Assemble the centrifuge tube with the coverslip (Fig. 1). Place the coverslip in the adaptor with the cell side facing up. For wounded monolayers, orient the wound so that it will be orthogonal to the centrifugal force (*see* Fig. 1). Make sure the coverslip is well seated in the adaptor by gently pressing the coverslip against the adaptor. This will help prevent the coverslip from floating. Before centrifuging, it is useful to break one corner of the coverslip that is closer to the opening of the adaptor to mark the centripetal side.
4. Fill the centrifugal tube with conditioned medium (*see* **Notes 4** and **5**).
5. Use forceps to hold the adaptor from the cleft as well as the hole drilled in the top surface to accommodate a screw (the dotted box shown in Fig. 1), and put the adaptor into the centrifugal tube. When the adaptor is at the bottom of the tube, make sure the cell side of the coverslip is not touching the adaptor.
6. Assemble the centrifuge tube into the bucket (Fig. 1). Make sure the adaptor is aligned such that the cells face upward

during centrifugation; if using wounded monolayers, align the adaptor and coverslip so that centrifugal force will be orthogonal to the wound (*see* **Note 6**).

7. Balance centrifuge tubes by adding conditioned medium as necessary.
8. Centrifuge the samples at desired speed (e.g., 5000 × *g*) for 30 min at 36 °C (*see* **Notes 7** and **8**).
9. After the centrifugation is complete, remove the adaptor with the coverslip from the centrifuge tube using a pair of forceps. Remove the coverslip from the adaptor for either fixation or live recording (*see* **Note 9**).

3.3 Analysis of Nuclear Position by Immunofluorescence

1. Place the centrifuged coverslip into a 35 mm dish containing ~2 mL PBS.
2. For PFA fixation: Aspirate the PBS from the dish and add 1 mL 4% PFA. Fix the sample for 10 min at room temperature.
3. For methanol fixation: Plunge the coverslip into methanol pre-chilled to −20 °C and fix for 5 min. The coverslip can either be placed into a porcelain coverslip rack in a beaker containing the pre-chilled methanol (−20 °C) or any glass dish or plate that will hold a 22 mm coverslip. Fix for 5 min (*see* **Note 10**).
4. Wash the PFA or methanol-fixed coverslip three times with PBS (5 min each rinse) at room temperature.
5. Incubate the coverslip with blocking buffer in room temperature for 1 h.
6. Wash the coverslip briefly and then incubate the coverslip with the primary antibody diluted in blocking buffer for 1 h at room temperature.
7. Wash the coverslip three times with PBS, 5 min each time at room temperature.
8. Incubate the coverslip with secondary antibody diluted in blocking buffer containing 1:1000 DAPI (to stain the nucleus) and/or 1:200 rhodamine phalloidin (to label F-actin to show cell contour if cell periphery is not depicted by primary antibody) for 1 h at room temperature.
9. Wash the coverslip three times with PBS, 5 min each time at room temperature.
10. Mount the coverslip by placing a drop of mounting media onto a clean glass slide. Seal the coverslip with nail polish after the mounting media is solidified to avoid unnecessary movement of the coverslip. Store slides at 4 °C in the dark.
11. Take fluorescence images of the cells with a widefield epifluorescence microscope, and construct an overlay image showing the nucleus (DAPI stain) and the cell contours (actin) using imaging software.

12. Analyze position of the nucleus by using image analysis software to calculate the centroid of the cells and the centroid of the nucleus. For wounded monolayers the position of the nucleus can also be plotted with respect to the front and the back of the cells (*see* **Note 11**).

4 Notes

1. This media is optimal for growth of NIH3T3 fibroblasts. Other cell lines may require different media or serum for optimal growth.
2. The 22 mm^2 coverslips are used for the large (22 mm) centrifuge adaptor. For the smaller (10 mm) centrifuge adaptor, 22 × 60 mm No. 1.5 coverslips are cut with a diamond pen into 10 × 12 mm coverslips.
3. Depending on the specific model of the ultracentrifuge, it usually takes about 1–2 h for the machine and the rotor to equilibrate to 36 °C. Previous studies of enucleation found that when the temperature is below 25 °C, enucleation efficiency decreased dramatically [28], suggesting that the appropriate temperature is important for centrifugal force to displace the nucleus.
4. Conditioned media is used when starved cells are used. If serum is not deprived from the cell culture, Subheading 3.1, **steps 2** and **3** can be omitted.
5. 10 mL conditioned media is put into 22 mm adaptor tube, while 2 mL conditioned media is put into 10 mm adaptor tube in order to make sure the coverslip and adaptor are fully immersed in the media during centrifugation.
6. The centrifuge tubes may need to be trimmed with a pair of scissor to avoid friction-induced rotation by the bucket lid when closing and opening the bucket.
7. The centrifugal force and time of centrifugation can be adjusted to obtain different degrees of nuclear displacement. With NIH3T3 fibroblasts, forces between 1000 and 20,000 × *g* for 30 min yielded a linear increase in nuclear displacement. We also found that other cell types may require higher *g* forces [26].
8. The ultracentrifuge is set to 36 °C to avoid overheating the sample. Note that the chamber temperature will increase during de-vacuum.
9. Centrifuged cells can also be used for real-time live cell imaging. Place the centrifuged coverslip in a 35 mm dish filled with prewarmed PBS (37 °C), and then assemble it into an imaging

slide or dish for live cell recording. During the assembly of imaging device for live cell imaging, remember to mark the device so that the orientation of the centrifugal force relative to the cells is known.

10. The choice between PFA and methanol fixation depends on the proteins being stained by immunofluorescence. For example, PFA fixation must be used for staining F-actin with rhodamine phalloidin, whereas methanol fixation is optimal for staining microtubules with antibodies.

11. It is useful to develop an automated method to measure the position of nuclei in cells. We use in-house developed software called Cell Plot that is available online at http://www.columbia.edu/~wc2383/software.html. This software automatically recognizes the nucleus by the DAPI signal. The cell contour is manually drawn and then used to calculate the cell centroid. The position of the nucleus determined by the DAPI signal is then used to calculate the distance relative to the cell centroid. For cells at the wound edge, the position of the nucleus is reported as the distance along the front-back axis of the cell. For all measurements, the position of the nucleus is calculated as a value relative to the cell size to allow cells of different size to be compared and included in the analysis [29].

References

1. Gundersen GG, Worman HJ (2013) Nuclear positioning. Cell 152(6):1376–1389
2. Chang W, Worman HJ, Gundersen GG (2015) Accessorizing and anchoring the LINC complex for multifunctionality. J Cell Biol 208(1):11–22
3. Crisp M et al (2006) Coupling of the nucleus and cytoplasm: role of the LINC complex. J Cell Biol 172(1):41–53
4. Haque F et al (2006) SUN1 interacts with nuclear lamin A and cytoplasmic nesprins to provide a physical connection between the nuclear lamina and the cytoskeleton. Mol Cell Biol 26(10):3738–3751
5. Adames NR, Cooper JA (2000) Microtubule interactions with the cell cortex causing nuclear movements in *Saccharomyces cerevisiae*. J Cell Biol 149(4):863–874
6. Borrego-Pinto J et al (2012) Samp1 is a component of TAN lines and is required for nuclear movement. J Cell Sci 125(Pt 5):1099–1105
7. Chang W et al (2013) Emerin organizes actin flow for nuclear movement and centrosome orientation in migrating fibroblasts. Mol Biol Cell 24(24):3869–3880
8. Folker ES et al (2011) Lamin a variants that cause striated muscle disease are defective in anchoring transmembrane actin-associated nuclear lines for nuclear movement. Proc Natl Acad Sci 108(1):131–136
9. Kutscheidt S et al (2014) FHOD1 interaction with nesprin-2G mediates TAN line formation and nuclear movement. Nat Cell Biol 16(7):708–715
10. Luxton GW et al (2010) Linear arrays of nuclear envelope proteins harness retrograde actin flow for nuclear movement. Science 329(5994):956–959
11. Fridolfsson HN et al (2010) UNC-83 coordinates kinesin-1 and dynein activities at the nuclear envelope during nuclear migration. Dev Biol 338(2):237–250
12. Fridolfsson HN, Starr DA (2010) Kinesin-1 and dynein at the nuclear envelope mediate the bidirectional migrations of nuclei. J Cell Biol 191(1):115–128
13. Lee KK et al (2002) Lamin-dependent localization of UNC-84, a protein required for nuclear migration in *Caenorhabditis elegans*. Mol Biol Cell 13(3):892–901

14. Meyerzon M et al (2009) UNC-83 is a nuclear-specific cargo adaptor for kinesin-1-mediated nuclear migration. Development 136(16): 2725–2733
15. Starr DA, Han M (2002) Role of ANC-1 in tethering nuclei to the actin cytoskeleton. Science 298(5592):406–409
16. Roux KJ et al (2009) Nesprin 4 is an outer nuclear membrane protein that can induce kinesin-mediated cell polarization. Proc Natl Acad Sci 106(7):2194–2199
17. Hu DJ et al (2013) Dynein recruitment to nuclear pores activates apical nuclear migration and mitotic entry in brain progenitor cells. Cell 154(6):1300–1313
18. Jayo A et al (2016) Fascin regulates nuclear movement and deformation in migrating cells. Dev Cell 38(4):371–383
19. Neelam S et al (2015) Direct force probe reveals the mechanics of nuclear homeostasis in the mammalian cell. Proc Natl Acad Sci 112(18):5720–5725
20. Tkachenko E et al (2013) The nucleus of endothelial cell as a sensor of blood flow direction. Biol Open 2(10):1007–1012
21. Veomett G et al (1974) Reconstruction of mammalian cells from nuclear and cytoplasmic components separated by treatment with cytochalasin B. Proc Natl Acad Sci U S A 71(5):1999–2002
22. Prescott DM, Kirkpatrick JB (1973) Mass enucleation of cultured animal cells. Methods Cell Biol 7:189–202
23. Poste G (1973) Anucleate mammalian cells: applications in cell biology and virology. Methods Cell Biol 7:211–249
24. Daga RR, Chang F (2005) Dynamic positioning of the fission yeast cell division plane. Proc Natl Acad Sci U S A 102(23):8228–8232
25. Rodionov V et al (2001) Digital fluorescence microscopy of cell cytoplasts with and without the centrosome. Methods Cell Biol 67: 43–51
26. Zhu R, Antoku S, Gundersen GG (2017) Centrifugal displacement of nuclei reveals multiple LINC complex mechanisms for homeostatic nuclear positioning. Curr Biol 27(20):3097–3110
27. Gomes ER, Jani S, Gundersen GG (2005) Nuclear movement regulated by Cdc42, MRCK, myosin, and actin flow establishes MTOC polarization in migrating cells. Cell 121(3):451–463
28. Wright WE (1973) The production of mass populations of anucleate cytoplasms. Methods Cell Biol 7:203–210
29. Chang W, Antoku S, Gundersen GG (2016) Wound-healing assays to study mechanisms of nuclear movement in fibroblasts and myoblasts. Methods Mol Biol 1411:255–267

Chapter 10

Assembly and Use of a Microfluidic Device to Study Cell Migration in Confined Environments

Jeremy Keys, Aaron Windsor, and Jan Lammerding

Abstract

Cells migrating in tissues must often pass through physical barriers in their surroundings in the form of fibrous extracellular matrix or other cells. To improve our understanding of how cells move in such confined microenvironments, we have designed a microfluidic device in which cells migrate through a series of three-dimensional polydimethylsiloxane (PDMS) constrictions with precisely controlled geometries that mimic physiological pore sizes. The migration device offers an experimental platform that combines a well-defined three-dimensional (3D) environment with a setup well suited for imaging confined cell migration at high spatial and temporal resolution. In this protocol, we describe the fabrication and use of these devices using standard soft lithography techniques and light microscopy. Analysis of live-cell time-lapse series of cells with fluorescently labeled nuclear and/or cytoskeletal structures migrating in the devices can reveal new insights into the molecular processes required for confined migration, including the role of the linker of nucleoskeleton and cytoskeleton (LINC) complex, which has been implicated in 3D migration.

Key words Cell migration, Confined environments, Microfluidics, Nucleus, LINC complex, Microscopy, Live-cell imaging

1 Introduction

Cell migration represents a crucial step in a variety of biological processes, including cancer metastasis, inflammation, and wound healing. In the in vivo tissue environment, interstitial spaces, extracellular matrix networks, and other cells constitute a confined environment for migrating cells. Dense fibrous tissue matrix and layers of endothelial cells form narrow constrictions, which measure 0.1–30 μm in diameter [1, 2]. To move through such tight spaces, cells must produce significant intracellular forces to compress their nucleus, as it is typically the largest and stiffest organelle [3–5]. Prior work has suggested that the linker of nucleus and cytoskeleton (LINC) complex facilitates the transmission of forces from contractile actomyosin fibers to the nucleus during confined migration [6–10]. Consequently, depletion of LINC complex proteins such as Nesprin-2 and Nesprin-3 reduces migration speed through

Gregg G. Gundersen and Howard J. Worman (eds.), *The LINC Complex: Methods and Protocols*, Methods in Molecular Biology, vol. 1840, https://doi.org/10.1007/978-1-4939-8691-0_10,

confining environments and impairs the cell's ability to deform its nucleus [6–9]. Since the dimensionality of the cell environment modulates migratory behavior, the study of cell migration in vitro requires model systems that faithfully reproduce the 3D confinements of biological tissue (Fig. 1A, B) [11]. Additionally, to elucidate the dynamic nuclear and cytoskeletal processes that mediate nuclear translocation in confined spaces, these model systems must also enable the detailed observation of cells by time-lapse microscopy by confocal or wide-field microscopy.

Many tools have been developed for the study of confined migration in vitro [12, 13]. One popular group of devices are transwell invasion assays, such as the Boyden chamber. However, these systems have a limited ability to directly observe cells during migration, as cells in these chambers migrate perpendicular to the imaging plane. Another approach for studying confined migration are extracellular matrix scaffolds (e.g., collagen or Matrigel). These scaffolds are generated through random self-assembly of matrix fibers, which produces an environment that closely mimics biological tissue but provides only limited control over the size of individual pores. In recent years, microfluidic devices have been developed using soft

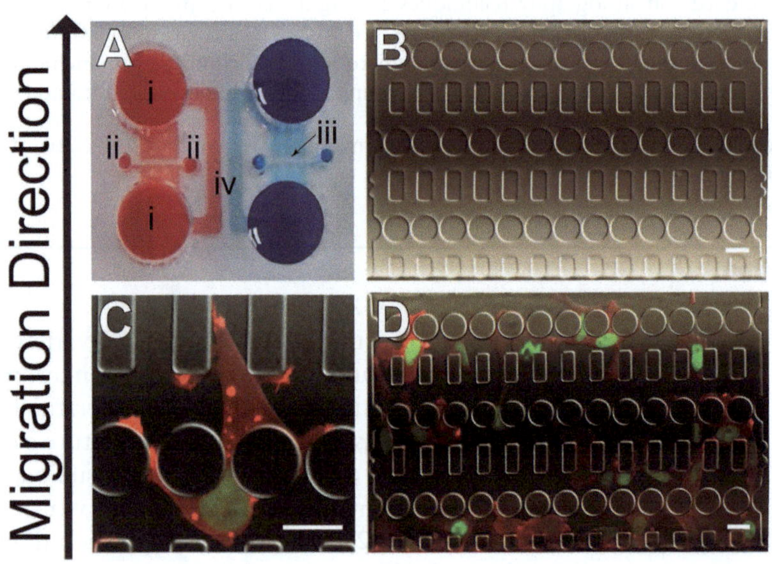

Fig. 1 Overview of the migration device. (**A**) Top-down view of migration device, with food coloring added to enhance details. Cell culture media is added into the media reservoirs (i). Cells are seeded into the devices through seeding ports (ii) that lead to the 5-μm-tall area with the constrictions (iii). Larger bypass channels (iv) allow rapid equilibration of media reservoir levels to prevent flow through constrictions, facilitating formation of a chemotactic gradient by diffusion if serum or growth factors are added to one of the reservoirs. (**B**) Image of the 5-μm-tall area of the device with 2-μm-wide constrictions without cells present. Scale bar 25 μm. (**C**, **D**) Images of constriction area containing MDA-MB-231 breast cancer cells expressing an mCherry-actin chromobody and an mNeonGreen-H2B histone label. Panel (**C**) depicts a cell as it starts to pass through a constriction. Scale bar 25 μm

lithography techniques to reproduce the physical confinements of the in vivo tissue environment [12, 14, 15]. These devices feature channels of defined geometries through which the cells must migrate, either spontaneously or following a chemotactic gradient. These microfluidic platforms are quite versatile as they provide the user with flexibility in the design and layout of constrictions.

In this protocol, we describe the design and use of a polydimethylsiloxane (PDMS) microfluidic device with a series of tight constrictions located between two larger chambers (Fig. 1B, C) [14]. Cells are seeded in one of the chambers and then migrate through a section 5 μm in height containing constrictions between 1 and 2 μm in width, formed by 30 μm wide circular pillars (Fig. 1). Additionally, there is a set of 15-μm-wide constrictions, which cells can migrate through without deforming their nucleus and which serve as an important control to assess effects independent of nuclear confinement. Unlike other microfluidic devices, in which cells often move through long, continuously confining channels, the confinement in the device described here is limited to a very short segment, i.e., cells move through a single, tight constriction, enter a less confined region, and then encounter the next constriction. This design seeks to reproduce the varied, discontinuous sequence of pores and confinement which cells are subjected to during in vivo migration, such as matrix fibers or endothelial cell layers [1, 16]. These devices have already aided in our initial studies of dynamic processes that occur during confined migration, such as the rupture of the cell nucleus and the essential role of perinuclear myosin IIB in moving the nucleus through narrow constrictions [7, 17].

To produce these devices, SU-8 microfluidic features are formed onto a silicon wafer through photolithography. Next, a PDMS replica is cast from the SU-8 features; the PDMS is cut into individual devices, bound to a glass slide, functionalized with extracellular matrix (ECM) solution, and seeded with cells. Following a period of incubation to allow cells to enter the constrictions, analysis of cell migration can be performed by live-cell imaging or standard immunofluorescence techniques. This protocol will outline the procedures necessary for both producing these devices and using them for the study of confined migration (Fig. 2). The protocol assumes basic familiarity with SU-8 and PDMS soft lithography. For users new to soft lithography, we recommend Qin et al. as a good starting point [18].

2 Materials

2.1 Photolithography

1. CZ silicon wafer, 4 in. diameter, type N, 525 μm thick, <1-0-0> orientation (Silicon Quest International).
2. Chrome photomask on a quartz substrate, 5″ × 5″ × 0.090″ (Telic, Valencia, CA).

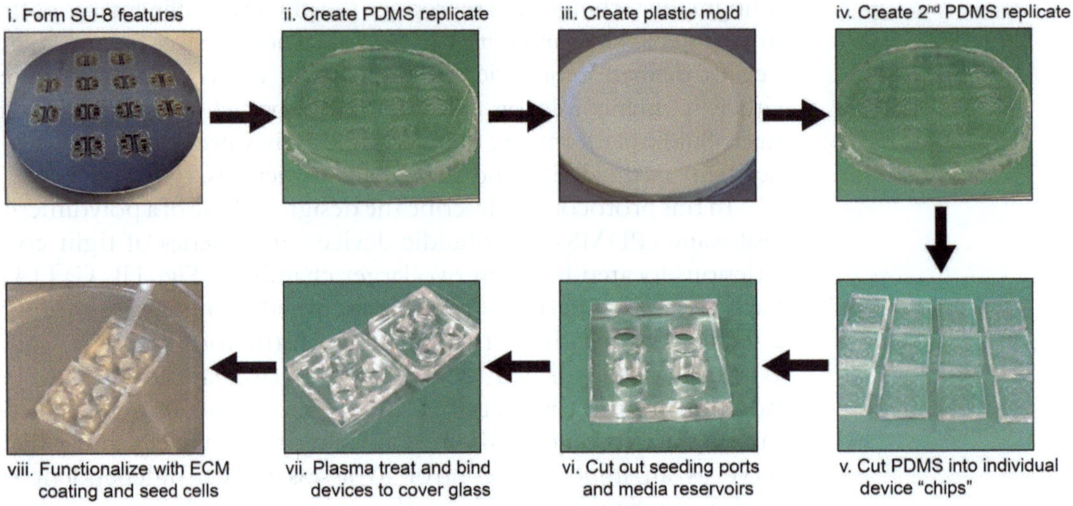

Fig. 2 Overview of migration device fabrication. (i) SU-8 microfluidic features are developed onto a silicon wafer through photolithography. (ii) A PDMS replicate of these features is formed using a two-part elastomer curing-base mixture. (iii) Optionally (but recommended), a secondary plastic mold can be cast from the initial PDMS replicate, which will be used to form additional PDMS devices. (iv) Using a similar process to step *ii*, a PDMS replicate is formed from the secondary plastic mold. (v) The PDMS replicate is cut into 12 device "chips." Each "chip" contains two independent microfluidic devices, each with the features shown in Fig. 1. (vi) Media reservoirs (labeled i in Fig. 1A) and seeding ports (labeled ii in Fig. 1A) are cut out using biopsy punches. (vii) Devices are cleaned with IPA and DI water, treated using a plasma cleaner, and then covalently bound to glass slides. (viii) Devices are functionalized with extracellular matrix proteins prior to adding cells through the seeding ports (labeled ii in Fig. 1A) and filling reservoirs with cell media. At this point, devices may be incubated until ready for analysis using live-cell imaging or immunofluorescence

3. Heidelberg DWL 2000 mask writer (Heidelberg Instruments, Heidelberg, Germany).
4. Long-pass filter for near-UV light (PL-360LP from Omega Optical or equivalent).
5. AutoCAD software (Autodesk, Mill Valley, CA) or equivalent.
6. Mask aligner system (ABM, San Jose, CA).
7. Molecular vapor deposition system (SPTS Technologies, Newport, RI).
8. Kapton polyimide film (DuPont, Wilmington, DE).
9. Oven suitable for temperatures up to 150 °C.
10. CEE Model 100 spin coater, or equivalent (Brewer Sciences, Rolla, MO).
11. SU-8 2005 photoresist (MicroChem, Newton, MA).
12. SU-8100 photoresist (MicroChem, Newton, MA).
13. SU-8 developer (MicroChem, Newton, MA).
14. Semiconductor grade acetone.

15. Cleanroom swab (Texwipe TX761 Alpha Swab with long handle, or equivalent).
16. CMOS grade isopropyl alcohol (IPA).
17. Deionized (DI) water.
18. >95% (1*H*,1*H*,2*H*,2*H*-perfluorooctyl)trichlorosilane (FOTS) (Gelest Inc., Morrisville, PA).

2.2 Casting of Migration Devices in PDMS

1. 150 mm petri dish.
2. Sylgard 184 silicone elastomer base and curing agent (Dow Corning, Midland, MI).
3. Stirring rod for mixing elastomer base and curing components.
4. Vacuum pump (Oerlikon Trivac D2.5E).
5. Vacuum desiccator (Catalog# Fisher 08-594-16C).
6. Oven suitable for temperatures of 65 °C (e.g., VWR Gravity Convection Oven, Catalog# 414005-108, or equivalent).

2.3 Mounting and Seeding of Devices

1. Biopsy punches (1.2 and 5 mm).
2. 24 mm × 60 mm Number 1.5 Micro Cover Glass (VWR, Radnor, PA, Catalog# 48393-251) stored overnight in 0.2 M hydrochloric acid.
3. Isopropyl alcohol (IPA).
4. DI water.
5. 70% ethanol.
6. Oxygen plasma cleaner (Harrick Plasma, Catalog# PDC-001).
7. Type I collagen (50 μg/mL in 0.02 M glacial acetic acid) or fibronectin (5 μg/mL in PBS) solution.
8. Dulbecco's phosphate-buffered saline (DPBS), no calcium, no magnesium (Thermo Fisher, Waltham, MA, Catalog# 14200-075 or equivalent).
9. Cells of interest (*see* **Note 1**).
10. Cell culture media appropriate for cells of interest.
11. Pipettes and tips for loading devices (20 and 200 μL pipettes work well).

2.4 Microscopy and Analysis

1. Inverted fluorescence microscope. Microscope should have objective with 20× magnification and fluorescence excitation/emission filters for GFP and/or other fluorophores of interest.
2. Stage-top incubation chamber for microscope to maintain temperature at 37 °C. Humidity control is optional, as microfluidic devices can be sealed.
3. Microscope-mounted CCD or CMOS camera for image acquisition.

4. Image acquisition software, such as ZEN BLUE (Zeiss), Micromanager, or others.
5. ImageJ, FIJI, MATLAB, or other software for image analysis.

3 Methods

3.1 Photolithography

All photolithography steps should be performed in a dedicated clean-room facility under standard clean-room conditions, with protective equipment and sufficient ventilation underneath a fume hood. The development of SU-8 microfluidic features onto a silicon wafer is described below:

1. Generate a design for the mask using CAD software and the downloaded CAD files. The device design and geometry are described in detail in Davidson et al. [14]. The CAD files for the device can be downloaded at http://lammerding.wicmb.cornell.edu/.
2. Using a Heidelberg DWL 2000 Mask Writer, develop the chrome photomask based on the CAD design layout. This photomask will be used to expose the design features in SU-8 photoresist onto the silicon wafer using near-ultraviolet (UV) light.
3. Clean silicon wafer using standard metal oxide semiconductor (MOS) cleaning procedures (*see* **Note 2**), and bake overnight at 90 °C to dehydrate.
4. Coat wafer with ~2 mL of SU-8 2005 (MicroChem), and spin down using a spin coater (CEE Model 100 spin coater, or equivalent) at 3000 RPM for 30 s (*see* **Note 3**) to obtain a 5-μm-thick layer, which will be used for the design of the first layer with the 5-μm-tall features.
5. Allow the SU-8 to relax for 10 min at room temperature (RT). Then remove 5 mm of excess SU-8 from the edge of the wafer perimeter using a clean-room swab soaked in acetone.
6. Bake the wafer on a hot plate from RT to 65 °C at a rate of 2.0 °C/min, hold at 65 °C for 10 min, then remove from plate, and allow to cool back to RT (*see* **Note 4**).
7. Expose the wafer to near-UV light at 365 nm using an ABM contact aligner with a long-pass filter for 40 s (*see* **Note 5**).
8. An hour after exposure, bake the wafer on a hot plate from RT to 95 °C at a rate of 2.0 °C/min, hold at 95 °C for 1 min, and then leave to cool back to RT.
9. Place wafer in SU-8 developer overnight to remove unexposed SU-8.

10. Rinse wafer with fresh SU-8 developer, then rinse with isopropyl alcohol and deionized water two times each to clean.
11. Bake wafer in an oven from RT to 150 °C for 20 min. Then shut off oven, and allow wafer to cool to RT. This step is a "hard-bake" step, which solidifies the first SU-8 layer and prevents unwanted merging with the second layer.
12. Dehydrate wafer overnight by baking at 90 °C.
13. Cover the alignment marks on the wafer for the 5 μm layer using Kapton tape (*see* **Note 6**).
14. For a thickness of 200 μm, coat wafer with ~2 mL SU-8100, and spin at 1500 RPM for 60 s (*see* **Note 7**).
15. Allow the SU-8 to relax for 10 min, and then remove 5 mm of excess SU-8 from the edge of the wafer perimeter using a clean-room swab soaked in acetone.
16. Bake the wafer on a hot plate with a lid (any cover is sufficient; we use a Pyrex petri dish cover) from RT to 55 °C at a rate of 2.0 °C/min for 14 h, then increase to 60 °C at a rate of 2.0 °C/min for 14 h, and then leave to cool back to RT.
17. Score the edges of the Kapton tape with a razor and gently remove from the wafer.
18. Expose the wafer on the contact aligner with a long-pass filter for 1 min, six times with 1 min of rest in between (*see* **Note 8**).
19. Twenty minutes after exposure, bake the wafer on a hot plate with a lid. Increase the temperature from RT to 95 °C at a rate of 1.5 °C/min, hold at 95 °C for 1 min, and then leave to cool back to RT.
20. Leave wafers in SU-8 developer overnight.
21. Rinse wafer with fresh SU-8 developer, followed by a wash with isopropyl alcohol and deionized water, two times each.
22. Bake wafers from RT to 60 °C for 2 h in an oven on an aluminum plate to remove moisture, and leave to cool in the oven to RT.
23. Coat wafers with FOTS using a molecular vapor deposition (MVD) tool.
24. Wafers can now be removed from the clean room and used for casting into PDMS (Fig. 2ii, iv).

3.2 Casting of Migration Devices in PDMS

From this point on, all components should be handled with nitrile gloves to minimize the risk of contamination of devices. All steps may be performed on a lab bench, until devices are assembled, sprayed with ethanol, and moved into a cell culture hood.

1. Place the silicon wafer with SU-8 features facing upward into a 150 mm petri dish (or use secondary plastic mold; *see* **Note 9**).

2. Add PDMS base and curing agents into a standard plastic cup at a 10:1 ratio (typically 50 g of base and 5 g of curing agent is sufficient for a set of 12 device chips), and stir vigorously for 5 min to fully combine. The stirring of these components will cause many air bubbles to form in the mixture, and these must be removed through degassing prior to curing of the PDMS.
3. Place PDMS mixture into a vacuum desiccator at 30 psi for 20 min to eliminate bubbles and accelerate degassing of the polymer.
4. Pour PDMS mixture over wafer (or secondary device mold), and allow 5–10 min to set at room temperature.
5. Using a very light stream of pressurized air, blow directly over the surface of the PDMS in order to eliminate all remaining bubbles. After this step, PDMS should be completely clear.
6. Preheat an oven to 65 °C, and then bake PDMS mixture in oven for at least 2 h (*see* **Note 10**).
7. Remove mold from oven and allow PDMS to cool to room temperature (*see* **Note 11**).
8. Using a razor, cut around edges of PDMS mold, ensuring that there is sufficient space in between the cut-edge and device features.
9. Carefully peel PDMS out of the mold, ensuring that PDMS does not tear during removal. The molded PDMS will contain 12 device "chips," each of which contains 2 migration devices (Fig. 1A).
10. Using a razor, cut PDMS into 12 device chip segments (Fig. 2v).
11. Place PDMS devices onto clean packing tape, feature-side down, to protect devices from dust during storage.
12. Store devices at room temperature, or proceed immediately to mounting onto glass slides for use with cells.

3.3 Mounting and Seeding of Devices

1. Prior to mounting, store glass slides overnight in 0.2 M hydrochloric acid (*see* **Note 12**).
2. Remove a device chip from packing tape and immediately prepare for cleaning.
3. Cut out holes for seeding ports and media reservoir into PDMS using biopsy punches (Fig. 1A, i and ii). Location of holes is marked on PDMS as part of the mask design. Media reservoirs are cut out using a 5 mm punch, while the seeding inlets should be cut out with a 1.2 mm punch (*see* **Note 13**).
4. Hold device with forceps, and rinse with isopropyl alcohol (IPA), followed by deionized water. Repeat once (*see* **Note 14**).

5. Using pressurized air, dry device thoroughly, and then place inside of plasma cleaner, feature-side up.
6. Repeat washing and drying steps on cover glass slides, and place inside of plasma cleaner alongside device (*see* **Note 15**).
7. Close plasma cleaner and turn on pump and power switches. Wait a few minutes to allow the plasma cleaner to warm up, and then turn on the RF level (which regulates the intensity of plasma within the chamber) to high.
8. Plasma treat the PDMS devices and glass cover slides for 5 min, adjusting the air intake in order to keep the plasma active (a bright pink color) throughout the treatment procedure (*see* **Note 16**).
9. Turn off the plasma cleaner and release the air pressure very slowly.
10. Remove the treated glass slide and PDMS device from the plasma cleaner, and place the device onto the glass slide, feature-side down (*see* **Note 17**).
11. Using your thumb, press the device down firmly onto the glass slide. Press around the device to ensure that the whole device is firmly bonded to the glass cover slide (*see* **Note 18**).
12. To improve adhesion of the PDMS to the glass, place the bonded device onto a hot plate at 95 °C for 5 min (*see* **Note 19**).
13. Remove the device from the hot plate, and allow the device to cool for a few minutes, before spraying the outside of the device completely with ethanol and moving the device into a cell culture hood.
14. Fill the media reservoirs with ethanol, and allow the device to incubate for 10 min at RT for sterilization.
15. Remove ethanol from the device, and rinse the media reservoirs three times with PBS for 5 min each. Each device will hold between 150 and 200 μL of fluid.
16. At this stage, the inside surfaces of the device can be functionalized with various biologically relevant coating, depending on the cell line to be used and the experimental goals. We use fibronectin or collagen coatings for most cell lines. To functionalize the inside of the device, fill devices with protein solution through seeding ports (Fig. 1A, ii), and allow the coatings several hours to set to the device surface (*see* **Note 20**, Table 1).
17. Remove coating solution, and rinse the inside of the device three times using 180 μL of cell culture media applied to one of the reservoirs, allowing 5 min of incubation between each rinse.
18. Prepare cells for seeding into devices (*see* **Note 21**, Table 1).
19. Completely aspirate all media from devices, ensuring that bypass channels and device features are clear.

Table 1
Optimal cell seeding densities and concentrations of ECM coating to prepare migration devices

Cell line	Cells seeded per device	ECM coating	ECM incubation
HT1080	80,000 cells seeded minimum 2 h before imaging	50 μg/mL corning 354,236 rat tail collagen type I diluted in 0.02 M acetic acid	4 °C overnight or longer
MDA-MB-231 on collagen	50,000 seeded minimum 2 h before imaging	50 μg/mL corning 354,236 rat tail collagen type I diluted in 0.02 M acetic acid	4 °C overnight or longer
MDA-MB-231 on fibronectin	30,000 seeded 24 h before imaging	5 μg/mL Millipore FC010 human plasma fibronectin diluted in PBS	4 °C overnight or longer, or 4 h at 37 °C
Human fibroblasts	30,000 seeded 24 h before imaging	3 μg/mL Millipore FC010 human plasma fibronectin diluted in PBS	4 °C overnight

20. Pipette 6 μL of cell suspension into the seeding port on the same side of the device as the bypass channel (Fig. 1A). Seed cells into the right port on left device and left port on right device (*see* **Note 22**).
21. Check seeding of the cells underneath a bright-field microscope. The cells should be distributed evenly across the front of the device constriction channels (Fig. 3A). If the cells are biased toward one end of the constrictions (Fig. 3B), aspirate cells from the device, and repeat seeding process.
22. Slowly add 180 μL of cell culture media into the media reservoir at the end of the device opposite from where cells were seeded (Fig. 1A, into top reservoirs). When media is added to one reservoir, it will flow through the constrictions and bypass channels to fill the reservoir on the opposite side. By adding media on the reservoir on the end of the device opposite the seeding channels, this prevents the inflow of media from pulling cells off of the glass and potentially pushing them prematurely into the constrictions.
23. Check cell seeding under microscope after this step to ensure that addition of media did not move cells from their original, uniformly seeded position.
24. Place the device with cells into a 37 °C cell culture incubator, and incubate until ready to image.

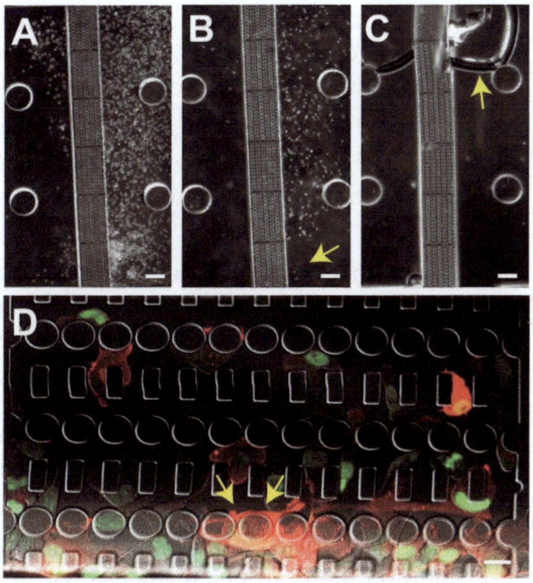

Fig. 3 Recognizing potential issues with device loading and bonding. (**A**) Properly loaded cells with even distribution across the device. Scale bar 200 μm. (**B**) Uneven loading of cells in front of constrictions; bottom constriction section has no cells at entrance (arrow). Scale bar 200 μm. (**C**) Air bubbles in cell media formed within the device, blocking entry into the constriction channels. Scale bar 200 μm. (**D**) Cells migrating underneath constriction pillars (arrows), indicating insufficient bonding of PDMS pillars to glass. Scale bar 25 μm

3.4 Time-Lapse Imaging and Analysis of Cell Migration

Time-lapse imaging of cells requires a microscope with an incubation chamber to maintain optimal cell culture conditions (i.e., adequate temperature, CO_2 concentration, humidity) throughout the imaging process (*see* **Note 23**). When using fluorescence microscopy, the experiments may require some troubleshooting to determine suitable excitation intensity, imaging intervals, exposure times, and imaging duration to avoid phototoxicity caused by repeated imaging throughout the experiment. Here, we will briefly outline our analysis protocol for studying nuclear transit in confined migration, after a time series of migrating cells within these devices has been collected. Alternatively, cells can also be fixed within migration devices and processed with standard immunofluorescence staining techniques for further analysis of nuclear and cytoskeletal elements involved in nuclear translocation (*see* **Note 24**):

1. Place the migration device in the microscope incubation chamber, and bring the cells into focus. We find that a single region of constrictions is best visualized under 20× magnification to measure transit times, but higher magnification may be required to capture subcellular dynamics.
2. Let the microscope with mounted migration device thermally equilibrate for 15–30 min to avoid drift of focus.
3. Set up image acquisition software to capture region of interest within the device at regular time intervals (minimum of one frame

every 10 min, more frequent imaging may be necessary for faster migrating cells). Each device has six regions to capture, representing areas with different constriction sizes (three with 1 × 5 μm^2 constrictions, two with 2 × 5 μm^2 constrictions, and one with 15 × 5 μm^2 constrictions), and each chip contains two devices.

4. Acquire time-lapse image series of all regions of interest overnight (*see* **Note 25**).
5. Using ImageJ, or an equivalent image analysis software, define the nuclear perimeter of each cell throughout the time series of images. If using cells modified to express fluorescently labeled proteins within the nucleus, you may define the nuclear perimeter using an intensity threshold (*see* Fig. 4). For cells expressing fluorescent nuclear markers, such as histone H2B-GFP, we have developed a MATLAB automated image analysis program for cell tracking and transit time analysis [19].
6. For each cell attempting to pass through a constriction, define the time point at which the cell has "committed" to enter the constriction. This can be done by noting when the front of the nucleus crosses an imaginary line parallel to the center of the constriction (*see* Fig. 4, **Note 26**). Similarly, define the time point at which the nucleus has exited the constriction, either when the rear of the nucleus crosses a second imaginary line toward the rear of the constriction (successful pass; *see* Fig. 4) or when the nucleus backs out of the constriction, i.e., the front of

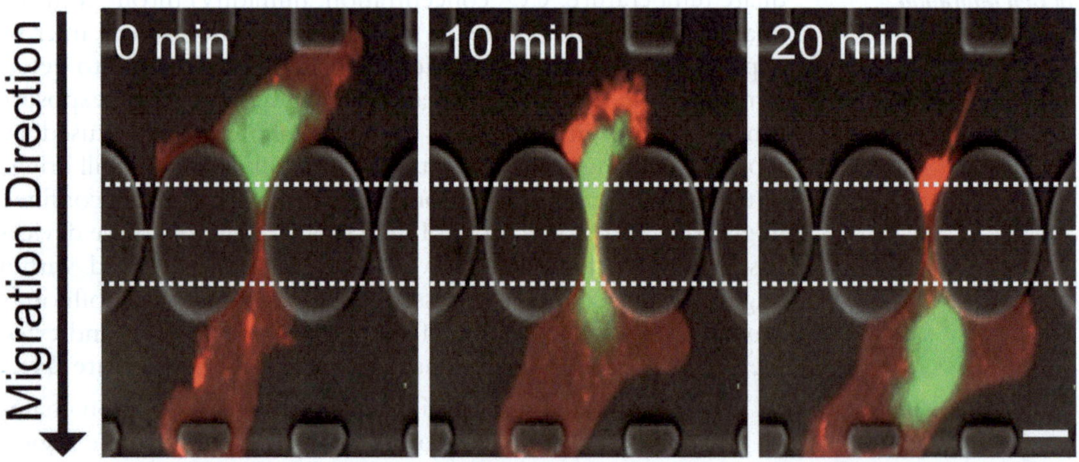

Fig. 4 Analysis of nuclear transit time through constriction. Nuclear transit time is defined as the time for a cell to completely translocate its nucleus through a single constriction. The most robust metric for this measurement is the time from when the nucleus "commits" to enter a constriction (i.e., crossing an imaginary dashed line, located 7 μm outside the constriction center, top dashed line) and begins to deform, to when it has fully passed the constriction (i.e., the trailing edge of the nucleus has passed an imaginary line 7 μm past the constriction center, bottom dashed line). This analysis can be performed manually or using automated particle tracking programs. The example shows an MDA-MB-231 breast cancer cell expressing an mCherry-actin chromobody and an mNeonGreen-H2B histone label. Scale bar 10 μm

the nucleus is no longer inside the region between the imaginary lines (unsuccessful attempt). For successful passes, the "transit time" is defined as the time between the entry and exit point.

7. Repeat this process for cells in the 15-μm-wide channels. This measurement is crucial when comparing different cell lines or treatment conditions, which may affect the overall motility of cells, regardless of nuclear confinement (*see* **Note 27**).
8. Normalize the transit time of cells moving through the 1- and 2-μm-wide constrictions to the average transit time of cells under the same condition (genotype, treatment) moving through the 15-μm-wide channels. This is considered the "normalized transit time" and describes the effect of the nuclear confinement on the migration efficiency.
9. Compare normalized transit times between different constriction sizes (e.g., 1 μm vs. 2-μm-wide constrictions), genotype, and treatment (e.g., LINC complex disruption vs. mock control) for analysis of migration times throughout the device.

4 Notes

1. Most migratory cell lines should work well in these devices. We have had success with HT1080 fibrosarcoma cells, mouse embryonic fibroblasts, differentiated HL60 neutrophil-like cells, MDA-MB-231 metastatic breast cancer cells, and a variety of other invasive breast cancer cell lines.
2. MOS cleaning is a 10 min base dip in one part NH_4OH, one part H_2O_2, and six parts water for 10 min, DI water rinse, and an acid dip in one part HCl, one part H_2O_2, and six parts water for another 10 min with a final DI water rinse.
3. It is important to slowly ramp up and ramp down from the 3000 RPM top speed to ensure even spreading of the SU-8. The spin protocol we have found to be successful is ramped up to 500 RPM at 100 RPM/s for 10 s, increaseed to 3000 RPM at 300 RPM/s for 30 s, and then ramped down to 100 RPM at 100 RPM/s for 30 s before stopping.
4. This preexposure baking step removes excess solvents, improves the photoresistive profile of the SU-8, and prevents adhesion to the contact mask.
5. ABM Contact Aligner dose with long-pass filter: 8.8 mW/cm^2 at 365 nm wavelength and 12.9 mW/cm^2 at 405 nm wavelength. Long-pass filter is a PL-360LP from Omega Optical or equivalent.
6. Tape may be forgone by using a clean-room wipe with acetone to remove the freshly spun SU-8 from the alignment marks.

7. Same as in **Note 2**: ramp up to 500 RPM at 100 RPM/s for 10 s, increase to 1500 RPM at 300 RPM/s for 30 s, and ramp down to 100 RPM at 100 RPM/s for 15 s before stopping.
8. It may be possible to use shorter exposure times, but 1 min exposure is typically sufficient to achieve good SU-8 feature sizes. The UV lamp intensity can decrease over its lifetime, and therefore the optimal exposure time for this step may vary. Over- or underexposure will affect the dimensions of the SU-8 features and of the final PDMS device. Thus, constriction dimensions should be validated for new devices. This can be done by imaging the SU-8 features, or by filling the assembled PDMS devices with fluorescently labeled dextran or other fluorescent solution and acquiring confocal image stacks of the constriction channels.
9. In order to protect the silicon wafer and delicate SU-8 features from damage during regular device production, we recommend that the first set of PDMS cast migration devices is used to produce a secondary plastic mold. This plastic mold should then be used for subsequent reproduction of migration devices. Detailed notes on producing a secondary plastic mold can be found in Desai et al. [20].
10. Ensure that the mold sits completely flat in the oven to prevent PDMS from tilting during curing. Tilted devices can still be used; however the devices will vary in thickness and therefore will not be able to hold the same volume of cell media.
11. When using a secondary plastic mold to make devices, the PDMS can be immediately removed following baking. When using the silicon wafer as your device mold, be sure to allow the device to cool completely before proceeding to avoid damaging SU-8 features.
12. 0.2 M hydrochloric acid (HCl) solution should be replaced on a weekly basis. Over time, HCl will evaporate, and devices will not properly adhere to glass slides.
13. During this step, be very careful with the placement of the punch-outs. The 1.2 mm punch sits between the bypass channel and the constrictions; interference with either will likely affect your results. It is also advised that you punch through the PDMS with the feature-side up to prevent poorly located cutouts.
14. When rinsing device, hold device upward, above your hands, and pour IPA and water downward onto device. Otherwise, runoff of material from gloves may fall onto the device and prevent adhesion between PDMS and the glass slide.
15. During drying of cover glass with pressurized air, hold glass firmly, and slowly increase air pressure, blowing parallel to the length of the glass to prevent bending and breaking of the glass.

16. This will activate the surface of the PDMS, allowing better adhesion between the device features and the glass slide. Pay close attention, and adjust air intake to keep plasma bright pink throughout cleaning, as we have found that poor device adhesion often results from poor plasma cleaning technique.
17. At this stage, ensure that you do not touch the surface of the glass slide nor the device-side surface of the PDMS. Additionally, check orientation of the device constrictions, and place very carefully, as you will not be able to adjust the orientation after it contacts the glass. Orientation of devices is entirely up to the user, but if mounting multiple devices on a single glass slide, analysis is generally easier if all cells are migrating in the same direction. Generally, we align the devices so that the cells will migrate "upward" (i.e., in the *y*-direction) while observing under a microscope.
18. This step may take some trial and error; if the PDMS device is not pressed sufficiently firmly against the glass, some device features may not fully adhere to the glass. In this case, cells can migrate underneath the PDMS pillars (Fig. 3D). If the PDMS device is pressed too hard against the glass, the features may collapse, preventing cells from migrating through the device. We have found that gently pressing around the device perimeter, evenly rolling one finger across the whole device with very light pressure, works best.
19. At this step, the device can be inspected underneath a bright-field microscope to see if the PDMS features are fully adhered. There will be a notably different color coming through regions of the device that are not bonded to the glass.
20. When coating, check the inside of the device for pockets of air bubbles, which may form during addition of protein solutions into the device. These air bubbles can form around constrictions, which can prevent the protein from coating these features. To eliminate the bubbles, vigorously pipette protein solution through the device. Optimal seeding densities of cells and concentrations of ECM solutions that we have determined are found in Table 1.
21. The exact number of cells to be seeded will be dependent on the specific cell line and may require optimization depending on what cell density is desired for the experiments. We have found that seeding 30,000 MDA-MB-231 cells per device works well for live-imaging experiments. In order to concentrate this number of cells into 6 μL of cell suspension, cells should be suspended in media at a concentration of 5 million cells per mL.
22. Cells must be seeded slowly and steadily with the pipette. Pushing the cells into the device too quickly may lead to the

cells spreading unevenly across the device (Fig. 3B). Additionally, when seeding cells, do not dispense the pipette past the first point of resistance; injection of air behind the cells results in the formation of bubbles within the device and can hinder cell migration (Fig. 3C).

23. If necessary, HEPES or other buffers may be added to cell culture media to maintain pH balance, particularly for longer duration studies. To prevent evaporation of cell media, devices should be sealed with cover glass laid across the media reservoirs before moving to microscope. In experiments that run longer than 12 h, it may be necessary to change cell culture media on a regular basis (typically every 24 h).

24. Fixation and staining of cells can be performed using standard immunofluorescence protocols; however it is recommended that longer times are used for each incubation and washing step to ensure that reagents have sufficient time to distribute completely throughout the microfluidic device. For immunofluorescence studies, cells should be seeded at lower densities, as large numbers of cells can block up the entry to the constriction channels, preventing staining reagents from reaching some cells within the devices. Alternatively, experiments can be performed with devices without covalently attaching the PDMS to the glass slides (i.e., without plasma treatment). In this case, the PDMS can be removed after fixing the cells inside the devices, and staining can be performed on cells adhering to the glass slide. The fixation step may require some optimization to identify conditions that promote cells to preferentially adhere to the glass cover slide and not the PDMS features.

25. Total imaging time is up to the user and depends on the cell line being used. For most cell lines, imaging overnight (12–14 h) is sufficient to yield a good number of cells passing through constrictions (10–30 cells per region of interest). For longer studies, it may be necessary to periodically replace cell media to keep cells healthy.

26. We have found that the most consistent measurement for a "point of commitment," i.e., when a cell is attempting to pass through a constriction, is an imaginary line drawn 7 μm away from the centerline of the constrictions, toward the entry side (Fig. 4). Another imaginary line can be drawn on the opposite side center line to define when the nucleus has fully passed through the constriction. It is also necessary to check image sequences for signs of potential issues with the constrictions, which would result in exclusion of the affected cells. For example, if it appears that some part of the nucleus or the cell body passes underneath the device pillars during migration, instead of in between the pillars that form the constriction, this indi-

cates that the PDMS pillars were not sufficiently bonded to the glass and the nucleus is not fully confined.

27. As in **Note 26**, it is also important to establish criteria for exclusion of particular cells migrating through the 15 μm channels. For example, when determining the migration speed or nuclear transit time of cells in the 15-μm-wide channels, it may be necessary to exclude cells that spontaneously switch their migration direction, as this would affect the results.

Acknowledgments

This work was supported by awards from the National Institutes of Health [R01 HL082792 and U54 CA210184], the Department of Defense Breast Cancer Research Program [Breakthrough Award BC150580], and the National Science Foundation [CAREER Award CBET-1254846].

References

1. Weigelin B, Bakker GJ, Friedl P (2012) Intravital third harmonic generation microscopy of collective melanoma cell invasion. Intravital 1(1):32–43
2. Doerschuk CM, Beyers N, Coxson HO, Wiggs B, Hogg JC (1993) Comparison of neutrophil and capillary diameters and their relation to neutrophil sequestration in the lung. J Appl Physiol 74:3040–3045
3. McGregor AL, Hsia C, Lammerding J (2016) Squish and squeeze—the nucleus as a physical barrier during migration in confined environments. Curr Opin Cell Biol 40:32–40
4. Caille N, Thoumine O, Tardy Y, Meister JJ (2002) Contribution of the nucleus to the mechanical properties of endothelial cells. J Biomech 35(2):177–187
5. Davidson PM, Lammerding J (2014) Broken nuclei-lamins, nuclear mechanics, and disease. Trends Cell Biol 24(4):247–256
6. Petrie R, Koo H, Yamada K (2014) Generation of compartmentalized pressure by a nuclear piston governs cell motility in a 3D matrix. Science 342(6200):1062–1065
7. Thomas DG, Yenepalli A, Denais CM, Rape A, Beach JR, Wang Y-L, Schiemann WP, Baskaran H, Lammerding J, Egelhoff TT (2015) Non-muscle myosin IIB is critical for nuclear translocation during 3D invasion. J Cell Biol 210:583–594
8. Jayo A, Malboubi M, Antoku S, Chang W, Ortiz-Zapater E, Groen C, Pfisterer K, Tootle T, Charras G, Gundersen G, Parsons M (2016) Fascin regulates nuclear movement and deformation in migrating cells. Dev Cell 38(4):371–383
9. Khatau S, Bloom R, Bajpai S, Razafsky D, Zang S, Giri A, Wu P, Marchand J, Celedon A, Hale C, Sun S, Hodzic D, Wirtz D (2012) The distinct roles of the nucleus and nucleus-cytoskeleton connections in three-dimensional cell migration. Sci Rep 2:488
10. Bone C, Chang Y, Cain N, Murphy S, Starr D (2016) Nuclei migrate through constricted spaces using microtubule motors and actin networks in *C. elegans* hypodermal cells. Development 143(22):4193–4202
11. Doyle AD, Petrie RJ, Kutys ML, Yamada KM (2013) Dimensions in cell migration. Curr Opin Cell Biol 25(5):642–649
12. Lautenschlager F, Piel M (2012) Microfabricated devices for cell biology: all for one and one for all. Curr Opin Cell Biol 25(1):116–124
13. Kramer N, Walzl A, Unger C, Rosner M, Krupitza G, Hengstschlager M, Dolznig H (2013) In vitro cell migration and invasion assays. Mutat Res 752(1):10–24
14. Davidson PM, Sliz J, Isermann P, Denais CM, Lammerding J (2015) Design of a microfluidic device to quantify dynamic intranuclear deformation during cell migration through confining environments. Integr Biol 7:1534–1546
15. Tong Z, Balzer EM, Dallas MR, Hung WC, Stebe KJ, Konstantopolous K (2012) Chemotaxis of cell populations through con-

fined spaces at single-cell resolution. PLoS One 7:1–10

16. Wolf K, te Lindert M, Krause M, Alexander S, te Riet J, Willis LA, Hoffman RM, Figdor CG, Weiss SJ, Friedl P (2013) Physical limits of cell migration: control ECM space and nuclear deformation and tuning by proteolysis and traction force. J Cell Biol 201:1069–1084
17. Denais CM, Gilbert RM, Isermann P, McGregor AL, te Lindert M, Weigelin B, Davidson PM, Friedl P, Wolf K, Lammerding J (2016) Nuclear envelope rupture and repair during cancer cell migration. Science 352(6283):353–358
18. Qin D, Xia Y, Whitesides GM (2010) Soft lithography for micro- and nanoscale patterning. Nat Protoc 5(3):491–503
19. Elacqua JJ, McGregor AL, Lammerding J (2018) Automated analysis of cell migration and nuclear envelope rupture in confined environments. PLoS One 13(4):e0195664
20. Desai SP, Freeman DM, Voldman J (2009) Plastic masters-rigid templates for soft lithography. Lab Chip 9:1631–1637

Part III

Analysis of the LINC Complex in Model Systems and Development

Chapter 11

Investigating LINC Complex Protein Homo-oligomerization in the Nuclear Envelopes of Living Cells Using Fluorescence Fluctuation Spectroscopy

Jared Hennen, Isaac Angert, Kwang-Ho Hur, G. W. Gant Luxton, and Joachim D. Mueller

Abstract

Linkers of nucleoskeleton and cytoskeleton (LINC) complexes are conserved nuclear envelope (NE) spanning molecular bridges which mechanically integrate the nucleus with the cytoskeleton and mediate force transmission into the nucleoplasm. Despite their critical roles in fundamental cellular processes such as meiotic chromosome and nuclear positioning, the mechanism of LINC complex assembly in cells remains unclear. To begin to address this deficit, we recently developed z-scan fluorescence fluctuation spectroscopy (FFS) and brightness analysis as a method for quantifying the oligomeric states of fluorescent protein-tagged NE proteins including nesprins and SUN proteins. Since the homo-oligomerization of SUN2 is critical for its ability to interact with nesprins within the perinuclear space, the knowledge obtained through quantitative brightness experiments reveals important insights into the in vivo mechanisms of LINC complex assembly. Here we describe the procedure we use to determine the brightness of proteins in the NE of living cells. In addition to the measurement procedure, we discuss the instrumentation requirements and present the results of applying this procedure to measure the brightness of nesprin-2 and SUN2.

Key words Brightness, FFS, KASH, LINC complex, Nesprin, Nuclear envelope, Perinuclear space, SUN protein

Abbreviations

b	Normalized brightness
KASH	*Klarsicht*, *ANC-1*, SYNE homology
LINC	Linker of nucleoskeleton and cytoskeleton
N	Number concentration
NE	Nuclear envelope
SUN	*Sad1/UNC-84*

Gregg G. Gundersen and Howard J. Worman (eds.), *The LINC Complex: Methods and Protocols*, Methods in Molecular Biology, vol. 1840, https://doi.org/10.1007/978-1-4939-8691-0_11,

1 Introduction

LINC complexes span the NE and physically couple chromatin and the nucleoskeleton with the cytoskeleton [1]. These conserved mechanosensitive molecular bridges are essential for the transmission of mechanical stimuli from the extracellular matrix through the cytoskeleton and into the nuclear interior [2, 3]. Consequently, LINC complex-mediated mechanotransmission is required for several fundamental cellular processes such as meiotic chromosome pairing, mechanotransduction, and nuclear positioning [4, 5]. Further highlighting their central importance in cellular mechanobiology is a growing list of genetic mutations in LINC complex proteins that are associated with a myriad of human diseases including cancer, hearing loss, and muscular dystrophy [4, 6].

LINC complexes consist of the outer and inner nuclear membrane *Klarsicht/ANC-1/SYNE* homology (KASH) and *Sad1/UNC-84* (SUN) proteins, respectively [7]. KASH proteins are identified by their conserved C-terminal KASH domain, which contains a transmembrane domain followed by the luminal ~10–32 residue KASH peptide [8]. The divergent spectrin repeat-containing N-termini of KASH proteins project away from the nucleus into the cytoplasm where they interact with the cytoskeleton [4]. SUN proteins directly interact with KASH peptides within the perinuclear space via their conserved and eponymous C-terminal SUN domain [9, 10]. Within the nucleoplasm, the divergent N-termini of SUN proteins engage A-type lamins, chromatin, and other inner nuclear membrane proteins [11]. Mammals encode six KASH proteins (nesprins-1 to nesprins-4, lymphocyte-restricted membrane protein (LRMP), and KASH5) and five SUN proteins (SUN1to SUN5) [5].

Several recent in vitro studies have provided invaluable structural insights into the SUN-KASH interaction and the ability of LINC complexes to transmit mechanical forces. Specifically, SUN2 homo-trimerization was demonstrated to be a prerequisite for KASH-binding, and an intermolecular disulfide bond was shown to exist between conserved cysteine residues in the SUN domain and KASH peptides of nesprin-1 and nesprin-2 [12–14]. Notwithstanding these significant advances, key questions remain. For example, do SUN2 homo-trimers exist in the NE of living cells? Do all SUN proteins homo-trimerize in order to interact with KASH proteins? How is the SUN-KASH interaction regulated?

To begin to address these questions, we recently extended the application of FFS to quantify protein-protein interactions in the NE in living cells [15–17]. The procedure outlined here can be used to measure the number concentration and brightness of LINC complex components, as well as other NE proteins, within their native environment. The brightness λ recovered from an FFS measurement represents the mean photon count rate of a fluorescently labeled protein complex and is proportional to its average oligomeric state [18]. To highlight this we determine the normalized brightness b where a monomeric protein would have $b = 1$, while a trimeric protein would have $b = 3$. Here, we demonstrate how this approach has been used to determine the oligomeric state of the KASH peptide of nesprin-2 and the luminal domain of SUN2 in the NE of living U2OS cells.

2 Materials

2.1 FFS Instrumentation

FFS experiments can be performed in cells using a research-grade confocal or multiphoton inverted microscope that is equipped with sensitive photon-counting detectors (quantum efficiency >40% in the wavelength region of interest). While many commercial microscope systems are suitable for FFS experiments, we will briefly describe our home-built two-photon (2P) microscope that we use for collecting FFS data from cells (*see* **Note 1**). This description of our custom 2P microscope should provide a useful resource for those investigators considering employing FFS in their research. A schematic of our instrumental setup is provided in Fig. 1:

1. Zeiss Axiovert 200 microscope (Carl Zeiss AG, Jena, Germany) which is mounted on a research-grade optical table (RS2000, Newport, Irvine, CA) supported by pneumatic isolators (S-2000A, Newport, Irvine, CA) for vibration isolation.
2. Mode-locked Ti/sapphire laser (Tsunami, Spectra Physics, Mountain View, CA) for 2P excitation. We prefer to use excitation wavelengths in the range of 900–1000 nm for measuring EGFP in living cells; here we chose a wavelength of 1000 nm.
3. Beam expander to achieve overfilling of the back aperture of the objective (*see* **Note 2**).

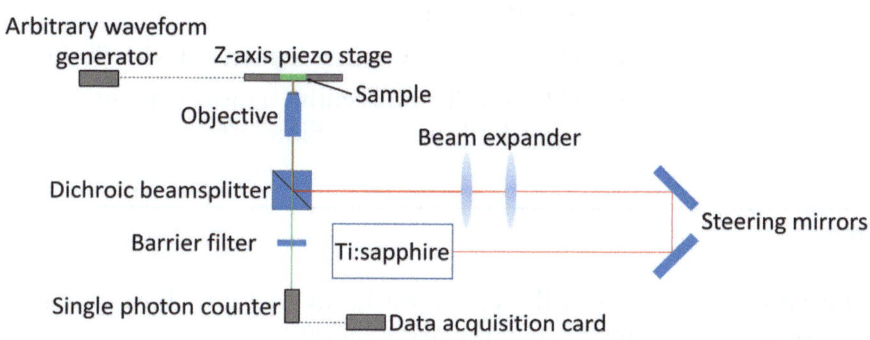

Fig. 1 The FFS experimental setup used in this work

4. Two steering mirrors (10Q20UF.35S, Newport, Irvine, CA) to align the laser beam into the beam expander (*see* **Note 3**).
5. Multiphoton short-pass dichroic beam splitter (675DCSXR, Chroma Technology, Bellows Falls, VT) to separate the excitation light from the emitted fluorescence.
6. Short-pass barrier filter (FF01-750/SP-25, Semrock, Rochester, NY) to remove any residual scattered laser light.
7. Zeiss 63× C-Apochromat water immersion objective with NA = 1.2 (*see* **Note 4**) to focus the excitation light into the sample.
8. PZ-2000 XYZ series automated stage with a piezo *z*-axis top plate (ASI, Eugene, OR) fitted on the microscope stage for performing z-scan measurements in cells.
9. 30 MHz 33522A Function/Arbitrary waveform generator (Agilent Technologies, Santa Clara, CA) to drive the *z*-axis piezo controller via an external analog input.
10. SPCM-AQ-141 single-photon counting module (Perkin-Elmer, Dumberry, Quebec) which is mounted on an XYZ linear translation stage (562-XYZ, Newport, Irvine, CA) for alignment and produces transistor-transistor logic (TTL) pulses that are recorded by a Flex04-12D data acquisition card (correlator.com, Bridgewater, NJ) and stored in computer memory for subsequent analysis.

2.2 Samples and Microscope Slide

1. 24-well glass-bottom slide with #1.5H cover glass (In Vitro Scientific, Sunnyvale, CA) (*see* **Note 5**).
2. The 24-well should contain the following in separate wells but on the same 24-well plate (*see* **Note 6**):
   (a) Texas Red at 200 nM or other standard fluorescent dye solution.
   (b) U2OS cells transiently transfected with EGFP (*see* **Note 7**).
   (c) U2OS cells transiently transfected with tandem dimeric EGFP ($EGFP_2$).
   (d) U2OS cells transiently transfected with SS-EGFP.
   (e) U2OS cells transiently transfected with other EGFP-tagged NE proteins of interest.

3 Methods

3.1 Optimize Collar Position of Objective

1. Focus the excitation beam ~40 μm above the glass-solution interface in the dye solution.
2. Record photon counts for ~60 s at a sampling rate of 100 kHz for dye measurements.

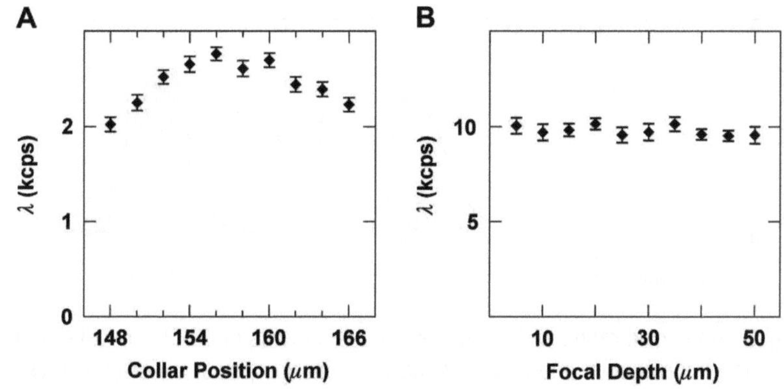

Fig. 2 Calibration and verification of objective correction collar position. (**A**) Brightness of Texas Red vs. water-immersion objective correction collar position. (**B**) Brightness of Texas Red vs. focal depth measured with a calibrated correction collar

3. Analyze the photon counts using a standard algorithm [19] to obtain the brightness λ in units of counts per second.
4. Systematically adjust the collar, and measure the brightness of the dye to determine the collar position that maximizes the brightness (Fig. 2A). The resulting resolution in collar position is typically ±2 μm.
5. If the measured brightness of the dye solution is independent of focal depth (Fig. 2B), the position of the correction collar has been successfully verified.

3.2 Select an Appropriate Excitation Power

1. Identify an EGFP-expressing cell that has a uniform distribution of cytoplasmic fluorescence using epifluorescence.
2. Use bright-field illumination to identify an area of the cytoplasm in the cell identified in **step 1** that lacks obvious large organelles.
3. Aim and focus the laser in the cytoplasmic area identified in **step 2** so as to maximize the intensity of the detected fluorescence signal.
4. Collect photon counts for ~10 s to determine the mean intensity.
5. Adjust the laser power and repeat **step 4**.
6. Plot fluorescence intensity vs. the squared power. There should be a linear increase at low powers with deviation from linearity at higher powers due to photobleaching and excitation saturation (Fig. 3A) (*see* **Note 8**).
7. Brightness measurements must be taken in the regime where intensity scales linearly with the squared power (solid red line,

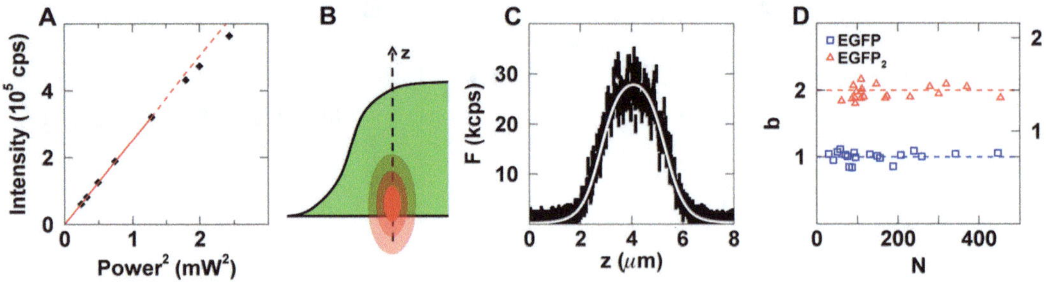

Fig. 3 Power calibration and brightness measurements of EGFP and $EGFP_2$ in the cytoplasm of living cells. (**A**) Average fluorescence intensity vs. squared excitation power for EGFP expressed in U2OS cells. The solid red line is a linear fit to the data below 1.5 mW^2, which is extended beyond the linear regime with the dashed red line. (**B**) Illustration of a z-scan through the cytoplasm of a cell expressing EGFP. (**C**) Intensity trace (black) and fit (white) of a z-scan through the cytoplasm of an EGFP expressing cell. (**D**) Plot of *b* vs. *N* for EGFP and $EGFP_2$ expressed in U2OS cells. The blue dashed line denotes the reference brightness λ_{EGFP}, while the red line denotes the value expected for a dimer

Fig. 3A). Typically, we conservatively set the excitation power to ~1/5th of the maximum power of the linear regime for brightness experiments.

8. The limiting excitation power should be experimentally determined for every fluorescently labeled protein to verify that the chosen power is still in the linear power regime. Once the excitation power is chosen, it must be kept constant throughout all subsequent experiments to ensure a fixed brightness value of EGFP.

3.3 Measure the Reference Brightness of EGFP in Living Cells

1. Choose a cell and then position and focus the laser as in Subheading 3.2, **steps 1–3**.
2. Collect FFS data from the focused beam at this cytoplasmic location for ~60 s with a 20 kHz acquisition rate.
3. Perform a z-scan (Fig. 3B) through the cell at this cytoplasmic location at a rate of ~5 μm/s with a peak-to-peak amplitude of ~20 μm:
   (a) The z-scan intensity data (Fig. 3C) along with the data collected in **step 4** are analyzed as previously described to determine the brightness λ and the number concentration N of the sample [20].
   (b) N represents the average number of EGFP monomers within the point-spread function (PSF) and is calculated by dividing the mean intensity $\langle F \rangle$ by the reference brightness, $N = \langle F \rangle / \lambda_{EGFP}$. If the PSF volume is known, the number concentration can be converted to a molar concentration as described elsewhere [17].

4. Repeat **steps 1–3** for a minimum of ten cells with varied expression levels of EGFP:
   (a) The resulting brightness values should be concentration independent to within a relative standard deviation ≤10% (Fig. 3D).
   (b) The mean value defines the reference brightness λ_{EGFP}.
5. Repeat **steps 1–4** for cells expressing $EGFP_2$:
   (a) The resulting brightness should be twice the average brightness of EGFP (Fig. 3D).
   (b) To emphasize the relationship between brightness and oligomeric state, we define the normalized brightness *b* as $b = \lambda/\lambda_{EGFP}$ [21]. Measurements of $EGFP_2$ should have an average *b* within ~10% of $b = 2$ (*see* **Note 9**).

3.4 Brightness Measurement of EGFP in the NE

The procedure we use for measuring brightness in the NE is similar to the one used to measure brightness in the cytoplasm. However, we have found the NE to be much more challenging experimentally. Consequently, brightness measurements in the NE require additional precautions as we describe below. The following experimental procedure demonstrates how to measure the brightness of EGFP that is targeted to the endoplasmic reticulum lumen and the contiguous perinuclear space of the NE by fusing the signal sequence (SS) of the luminal protein torsinA to the N-terminus of EGFP (SS-EGFP) [17, 22]. Measurements of EGFP in the NE serve as a control, and a result of $b = 1$ establishes the fidelity of brightness analysis in the NE:

1. Use epifluorescence to identify an SS-EGFP-expressing cell that displays a clear and distinct ring of fluorescence around the nucleus (Fig. 4A).
2. Switch to bright-field illumination, and identify a location within the center of the nucleus from the cell identified in **step 1** that is devoid of visible structures (i.e., nucleoli) (Fig. 4A) (*see* **Note 10**).
3. Perform a z-scan at the location identified in **step 2** (Fig. 4B).
4. Fit the z-scan intensity data generated in **step 3** using the procedure described in Smith et al. [23]:
   (a) The fit identifies the fluorescence contributions from the ventral and dorsal NEs as well as the background signal originating from sources outside the NE (Fig. 4C).
   (b) Analysis further determines the intensity fraction of the fluorescence contributions from each NE. Only cells with an NE intensity fraction ≥0.9 are selected for brightness measurements. This criterion ensures that contributions from the background signal are at most 10%, which represents a

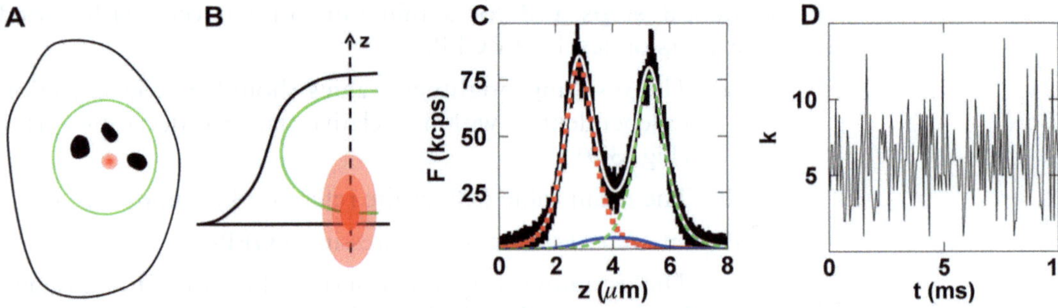

Fig. 4 Performing brightness measurements in the NE of living cells. **(A)** Illustration of a cell viewed via bright-field illumination. The NE is highlighted in green and nucleoli are shown in black. The red focus denotes a typical measurement position in the example cell. **(B)** Illustration of a z-scan with the excitation volume moving axially through the cell passing through the ventral and dorsal NE. **(C)** Intensity trace (black) of a z-scan through the nucleus of a SS-EGFP expressing cell with fit (solid white line). The fit is deconvolved to identify the contributions from the ventral NE (red squares), nucleoplasm (solid blue line), and dorsal NE (dashed green line). This analysis returned intensity fractions of 0.96 and 0.95 for the ventral and dorsal NE, respectively. **(D)** A 10-ms window of photon count data from a stationary measurement of SS-EGFP in the NE at full time resolution. The fluctuations encode information about the concentration and stoichiometry of labeled proteins

negligible amount for brightness analysis [17]. The remaining steps are performed only with cells that satisfy this criterion.

5. Focus the PSF on the ventral NE by maximizing the collected fluorescence intensity.
6. Collect ~60 s of data with a 20 kHz data acquisition rate (Fig. 4D):
   (a) Plot the intensity averaged over 1 s vs. time (Fig. 5).
   (b) Data showing no significant change in the average intensity (Fig. 5A) can be used for further analysis. Data containing a peak-to-peak difference exceeding 20% of the mean in the averaged intensity (Fig. 5B) should be discarded and retaken (*see* **Note 11**).
   (c) If it is difficult to obtain data without sudden changes in intensity, we have found that moving to a different location within the cell tends to result in stable data.
7. Determine the brightness of SS-EGFP from the data taken in **step 6** using the analysis procedure described in Hennen et al. [17].
8. Repeat **steps 5–7** in the dorsal NE.
9. Collect a final z-scan:
   (a) The intensity traces of the initial and final z-scan are compared to ensure that no detectable changes occurred during the measurement process. Focus drift of the instrument or motion of the cell would lead to a mismatch between

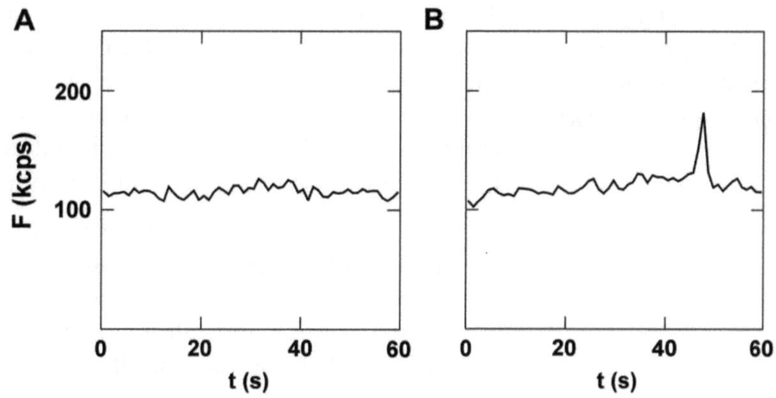

Fig. 5 Intensity traces of SS-EGFP vs. time in the NE. Intensity data were averaged over 1 s intervals. (**A**) Example of a stable intensity trace with no significant changes in the average intensity. (**B**) Example of an intensity trace containing a spike in the average intensity. This trace is not suitable for further analysis

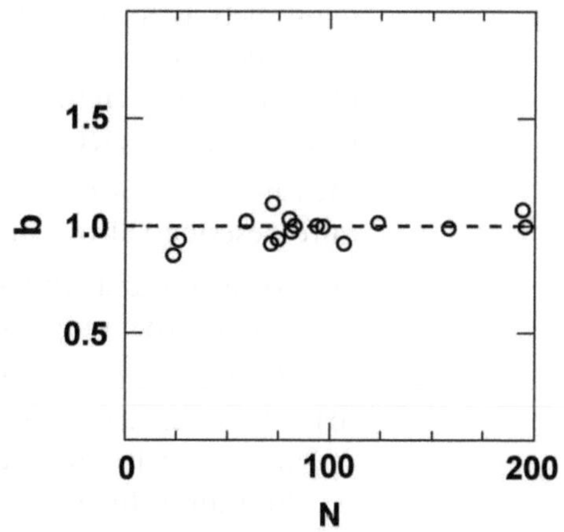

Fig. 6 Brightness of SS-EGFP in the NE. Plot of *b* vs. *N* of SS-EGFP in the NE

the z-scan intensity profiles. If a mismatch is detected, the data are discarded.

10. Repeat **steps 1–9** for at least ten cells;
    (a) Select cells that vary in expression level to obtain the brightness over a wide range of concentrations. A plot of the normalized brightness of SS-EGFP vs. number concentration is shown in Fig. 6. The experimental brightness is independent of concentration and has an average value of 0.99 ± 0.06. This result is consistent with SS-EGFP being a monomer, as expected. It is important to check that SS-EGFP can be accurately measured as a monomer before attempting to measure functional proteins in the NE.

3.5 Measuring the Oligomerization of LINC Complex Constituents in the NE

The procedure described above in Subheading 3.4 can also be used to measure the brightness of EGFP-tagged proteins in the NE; therefore, in the following section, we will demonstrate its application to the study of the molecular mechanisms underlying LINC complex assembly. To do this, we will describe how to measure the brightness of EGFP-tagged mouse nesprin-2 and SUN2 in the NE:

1. The technique described here requires the proteins being measured to diffuse in order to observe intensity fluctuations. Since both EGFP-tagged mini-nesprin-2G, a functional nesprin-2G construct [24], and full-length SUN2 are highly immobile in the NE [25], they are rapidly photobleached by 2P excitation and therefore unsuitable for standard quantitative brightness experiments. For this reason, we limit our analyses here to the luminal domain of each protein, which is targeted to the perinuclear space of the NE by an N-terminal SS-EGFP fusion (SS-EGFP-KASH2 and SS-EGFP-SUN2$^{261–731}$) [15, 17].
2. Perform the same procedure described in Subheading 3.2 for each protein being measured. While mobile, functional proteins such as SS-EGFP-SUN2$^{261–731}$ may have lower mobility than the calibration proteins discussed in previous sections [17]. Proper excitation power should be determined prior to measuring the reference brightness.
3. Measure the b of the proteins of interest following the procedure described above in Subheading 3.4:
   (a) Measurements of SS-EGFP-KASH2 showed no change in b over the range of measured N, with an average of $b = 1.05 \pm 0.1$ consistent with a monomer (Fig. 7A).
   (b) Measurements of SS-EGFP-SUN2$^{261–731}$ showed b increased in an N-dependent manner and approached a limiting value of $b = 3$ (Fig. 7B).
4. The brightness curve for SS-EGFP-SUN2$^{261–731}$ shown in Fig. 7B represents a binding titration. These measurements are performed in the presence of endogenous non-labeled SUN2, which may complicate the interpretation of this curve (*see* **Note 12**). Nevertheless, the effect of endogenous binding competition becomes negligible once the concentration of the exogenous labeled protein significantly exceeds that of the endogenous protein. In this regime, the b curve should asymptotically approach a limiting value which corresponds to the limiting oligomeric state of the protein. Consequently, the saturating brightness value of $b = 3$ reached at high concentrations of SS-EGFP-SUN2$^{261–731}$ reflects the limiting oligomeric state of the protein.
5. Repeat brightness experiments over several days on new cell preparations to ensure that the results obtained are both robust

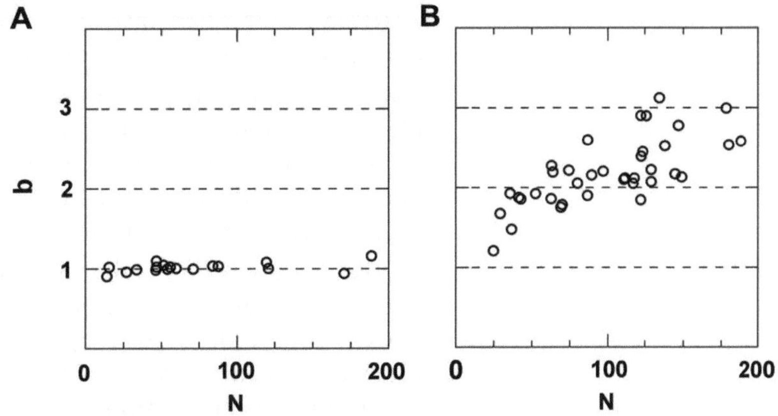

Fig. 7 *b* vs. *N* for LINC complex proteins in the NE. Plots for (**A**) SS-EGFP-KASH2 and (**B**) SS-EGFP-SUN2$^{261-731}$. Previously published in *Biophysical Journal* [17]

and reproducible. In addition to ensuring the reproducibility of the results, this generates a sufficient number of data points so that a reasonably dense *b* curve can be constructed. A single experiment will typically consist of *b* measurements performed in 15 cells after accounting for the calibrations, reference measurements, and analysis criteria described above. The data shown in Fig. 7A, B represent two and six separate experiments, respectively.

4 Notes

1. We prefer 2P excitation to one-photon confocal microscopy for FFS because it reduces the levels of out-of-focus photobleaching and phototoxicity experienced in living cells [26, 27]. In addition, the absence of the confocal pinhole in a 2P microscope results in a less complicated optical system, which facilitates instrument alignment and simplifies ongoing maintenance. Moreover, 2P excitation generates an excitation volume, or PSF, that can be accurately described by heuristic analytical functions such as the modified Gaussian-Lorentzian model [20]. Such knowledge regarding the nature of the PSF is essential for the successful execution of the quantitative FFS method described here.
2. A typical value for the excitation beam overfill factor is $2\omega_0/D{\sim}2$, where ω_0 and D are the radial beam waist and the pupil diameter of the objective, respectively. Before initiating a quantitative FFS experiment, it is essential that the excitation beam be properly aligned such that it passes through the center of the objective and the detector. Stable experimental setups will require minimal beam alignment adjustments from

day to day. Established, instrument-specific procedures should be followed to ensure that the beam is correctly aligned.

3. While inexpensive dielectric mirrors offer excellent reflectivity, only metallic mirrors or dielectric mirrors specially designed for femtosecond pulsed lasers should be used. This choice will minimize the reduction of the peak power of the laser due to group velocity dispersion [28].
4. The choice of objective is of utmost importance when designing an FFS system, as the signal-to-noise ratio of FFS experiments is highly dependent upon the efficient collection of fluorescence from individual molecules. Since the 2P excitation efficiency is proportional to the fourth power of the NA of the objective, FFS experiments performed in cells require the use of high NA objectives (NA > 1) [26]. We find C-Apochromat water-immersion objectives with a correction collar to compensate for variations in coverslip thickness to be the most suitable for collecting FFS data from cells. We discourage the use of oil-immersion objects because they introduce spherical aberrations, which complicate the interpretation of FFS experiments [29].
5. While the nominal thickness of standard #1.5 cover glass is 0.17 mm, actual thicknesses may vary between 0.16 and 0.19 mm. These deviations in thickness are a major source of spherical aberrations for high NA objectives. Such aberrations alter the size and shape of the focused excitation beam as a function of focal depth and lead to biased FFS data [30]. Therefore, the use of a high NA water-immersion objective equipped with a correction collar designed to compensate for the actual cover glass thickness is imperative for quantitative FFS experiments. Although we use a micrometer to directly measure the thickness of our cover glasses, we find that in practice the optimal collar position for FFS differs from the values printed on the objective collar by a constant offset, which is unique to each objective. Whenever possible, we recommend the use of #1.5H cover glass, which has a tighter tolerance (0.170 ± 0.005 mm) than regular #1.5 slides. The consistency in cover glass thickness minimizes the range of collar positions that must be tested when changing slides.
6. All of the experiments must be done in the same 24-well glass bottom slide in order to take advantage of the procedure described in Subheading 3.1. This procedure must be repeated for every new 24-well glass bottom slide used.
7. For the experiments described here, we used the human osteosarcoma U2OS cell line grown in Dulbecco's modified Eagle's medium supplemented with 10% fetal bovine serum. The cells are plated in wells and transiently transfected with the relevant

cDNA constructs. Finally, the cell growth media is exchanged for phosphate-buffered saline immediately before the start of measurement.

8. Quantitative *b* measurements in cells necessitate the selection of an appropriate excitation power to minimize photobleaching and excitation saturation [31]. The presence of either photobleaching or excitation saturation biases the interpretation of FFS data, which further complicates the analysis [32, 33].
9. Measurements of $EGFP_2$ are required in order to validate that the system is behaving as expected. Brightness values significantly above or below $b = 2$ are indicative of issues that complicate the interpretation of FFS measurements of the proteins of interest.
10. We avoid performing measurements near the edge of the nucleus as these areas of the NE may have significant curvature.
11. All fluorescence fluctuation methods implicitly assume a stationary signal that remains constant throughout the measurement period [34]. While the intensity is typically stable for measurements performed in the cytoplasm, we have found a higher fraction of measurements in the NE with unstable intensities. Data containing large changes in the average intensity are not conducive to our analysis.
12. An approximate measure of the effective dissociation coefficient is given by *N* where the brightness is halfway between its minimum and maximum value. However, since the data were taken in the presence of endogenous SUN2, binding competition between endogenous protein and SS-EGFP-$SUN2^{261-731}$ could lead to complexes containing a mixture of labeled and unlabeled protein, which reduce the observed *b*. The net effect of this competition is a shift of the brightness curve to higher *N*. Thus, the estimated dissociation coefficient provides an upper limit of the true value. Depleting the endogenous protein using methods such as RNA interference or genetic knockout will remove endogenous competition, thereby allowing measurement of the true binding affinity. While we recently demonstrated that the presence of endogenous SUN proteins had a negligible impact on the reported *b* of constructs encoding the EGFP-tagged luminal domains of SUN1 or SUN2 [15], we recommend testing the effect of unlabeled endogenous protein on all FFS-generated *b* measurements.

Acknowledgments

This work was financially supported in part by the NIH (J.D.M. (GM064589), and C.A.S. (AR007612)), and the Dystonia Medical Research Foundation (G.W.G.L. and J.D.M.).

References

1. Starr DA, Fridolfsson HN (2010) Interactions between nuclei and the cytoskeleton are mediated by SUN-KASH nuclear-envelope bridges. Annu Rev Cell Dev Biol 26:421–444
2. Alam S, Lovett DB, Dickinson RB et al (2014) Nuclear forces and cell mechanosensing. Prog Mol Biol Transl Sci 126:205–215
3. Kaminski A, Fedorchak GR, Lammerding J (2014) The cellular mastermind(?)—mechanotransduction and the nucleus. Prog Mol Biol Transl Sci 126:157–203
4. Luxton GWG, Starr DA (2014) KASHing up with the nucleus: novel functional roles of KASH proteins at the cytoplasmic surface of the nucleus. Curr Opin Cell Biol 28:69–75
5. Meinke P, Schirmer EC (2015) LINC'ing form and function at the nuclear envelope. FEBS Lett 589:2514–2521
6. Horn HF (2014) LINC complex proteins in development and disease. Curr Top Dev Biol 109:287–321
7. Crisp M, Liu Q, Roux K et al (2006) Coupling of the nucleus and cytoplasm: role of the LINC complex. J Cell Biol 172:41–53
8. Starr DA, Han M (2002) Role of ANC-1 in tethering nuclei to the actin cytoskeleton. Science 298:406–409
9. Malone CJ, Fixsen WD, Horvitz HR et al (1999) UNC-84 localizes to the nuclear envelope and is required for nuclear migration and anchoring during *C. elegans* development. Dev Camb Engl 126:3171–3181
10. Kim DI, Birendra KC, Roux KJ (2015) Making the LINC: SUN and KASH protein interactions. Biol Chem 396:295–310
11. Chang W, Worman HJ, Gundersen GG (2015) Accessorizing and anchoring the LINC complex for multifunctionality. J Cell Biol 208:11–22
12. Sosa BA, Rothballer A, Kutay U et al (2012) LINC complexes form by binding of three KASH peptides to domain interfaces of trimeric SUN proteins. Cell 149:1035–1047
13. Zhou Z, Du X, Cai Z et al (2012) Structure of Sad1-UNC84 homology (SUN) domain defines features of molecular bridge in nuclear envelope. J Biol Chem 287:5317–5326
14. Wang W, Shi Z, Jiao S et al (2012) Structural insights into SUN-KASH complexes across the nuclear envelope. Cell Res 22:1440–1452
15. Hennen J, Saunders CA, Mueller JD, Luxton GWG (2018) Fluorescence fluctuation spectroscopy reveals differential SUN protein oligomerization in living cells. Mol Biol Cell 29:1003–1011. PMID 29514929
16. Slaughter BD, Li R (2010) Toward quantitative "in vivo biochemistry" with fluorescence fluctuation spectroscopy. Mol Biol Cell 21:4306–4311
17. Hennen J, Hur K-H, Saunders CA et al Quantitative brightness analysis of protein oligomerization in the nuclear envelope. Biophys J (in press) Biophys J 113:138–147
18. Chen Y, Wei L-N, Müller JD (2003) Probing protein oligomerization in living cells with fluorescence fluctuation spectroscopy. Proc Natl Acad Sci U S A 100:15492–15497
19. Sanchez-Andres A, Chen Y, Müller JD (2005) Molecular brightness determined from a generalized form of Mandel's Q-parameter. Biophys J 89:3531–3547
20. Macdonald PJ, Chen Y, Wang X et al (2010) Brightness analysis by Z-scan fluorescence fluctuation spectroscopy for the study of protein interactions within living cells. Biophys J 99:979–988
21. Chen Y, Müller JD (2007) Determining the stoichiometry of protein heterocomplexes in living cells with fluorescence fluctuation spectroscopy. Proc Natl Acad Sci U S A 104:3147–3152
22. Saunders CA, Harris NJ, Willey PT et al (2017) TorsinA controls TAN line assembly and the retrograde flow of dorsal perinuclear actin cables during rearward nuclear movement. J Cell Biol 216:657–674
23. Smith EM, Hennen J, Chen Y et al (2015) Z-scan fluorescence profile deconvolution of cytosolic and membrane-associated protein populations. Anal Biochem 480:11–20
24. Luxton GWG, Gomes ER, Folker ES et al (2010) Linear arrays of nuclear envelope proteins harness retrograde actin flow for nuclear movement. Science 329:956–959
25. Ostlund C, Folker ES, Choi JC et al (2009) Dynamics and molecular interactions of linker of nucleoskeleton and cytoskeleton (LINC) complex proteins. J Cell Sci 122:4099–4108
26. Denk W, Strickler JH, Webb WW (1990) Two-photon laser scanning fluorescence microscopy. Science 248:73–76
27. Squirrell JM, Wokosin DL, White JG et al (1999) Long-term two-photon fluorescence imaging of mammalian embryos without compromising viability. Nat Biotechnol 17:763–767

28. Guild JB, Xu C, Webb WW (1997) Measurement of group delay dispersion of high numerical aperture objective lenses using two-photon excited fluorescence. Appl Opt 36: 397–401
29. Gibson SF, Lanni F (1992) Experimental test of an analytical model of aberration in an oil-immersion objective lens used in three-dimensional light microscopy. J Opt Soc Am A 9:154–166
30. Nasse MJ, Woehl JC (2010) Realistic modeling of the illumination point spread function in confocal scanning optical microscopy. J Opt Soc Am A Opt Image Sci Vis 27:295–302
31. Patterson GH, Piston DW (2000) Photobleaching in two-photon excitation microscopy. Biophys J 78:2159–2162
32. Delon A, Usson Y, Derouard J et al (2004) Photobleaching, mobility, and compartmentalisation: inferences in fluorescence correlation spectroscopy. J Fluoresc 14:255–267
33. Berland K, Shen G (2003) Excitation saturation in two-photon fluorescence correlation spectroscopy. Appl Opt 42:5566–5576
34. Hur K-H, Macdonald PJ, Berk S et al (2014) Quantitative measurement of brightness from living cells in the presence of photodepletion. PLoS One 9:e97440

Chapter 12

Functional Analysis of the Yeast LINC Complex Using Fluctuation Spectroscopy and Super-Resolution Imaging

Jay R. Unruh, Brian D. Slaughter, and Sue L. Jaspersen

Abstract

The *Saccharomyces cerevisiae* and *Schizosaccharomyces pombe* genomes encode a single SUN domain-containing protein, Mps3 and Sad1, respectively. Both localize to the yeast centrosome (known as the spindle pole body, SPB) and are essential for bipolar spindle formation. In addition, Mps3 and Sad1 play roles in chromosome organization in both mitotic and meiotic cells that are independent of their SPB function. To dissect the function of Mps3 at the nuclear envelope (NE) and SPB, we employed cell imaging methods such as scanning fluorescence cross-correlation spectroscopy (SFCCS) and single particle averaging with structured illumination microscopy (SPA-SIM) to determine the strength, nature, and location of protein-protein interactions in vivo. We describe how these same techniques can also be used in fission yeast to analyze Sad1, providing evidence of their applicability to other NE proteins and systems.

Key words Mps3/Sad1, Mps2/Kms1/Kms2, Line scanning, Fluorescence cross-correlation spectroscopy, Single particle averaging, Fluorescence resonance energy transfer, Super-resolution microscopy, Fluctuation spectroscopy

1 Introduction

In contrast to higher eukaryotes and plants, the genomes of fungi such as *Saccharomyces cerevisiae* and *Schizosaccharomyces pombe* encode for a single SUN domain-containing protein, Mps3 and Sad1, respectively. Both proteins were originally identified as membrane components of the spindle pole body (SPB), the yeast centrosome-equivalent organelle, that is embedded in the nuclear envelope (NE) during all or part of cell division [1–3]. At the SPB, the SUN domain of Mps3 interacts with the short C-terminal region of the membrane protein Mps2 in the lumenal space [4], an interaction reminiscent of the LINC complex, although Mps2 bears little homology to other KASH domain-containing proteins. Genetic and biochemical data suggests that Sad1 interacts with the KASH protein Kms2 at the SPB [5], although this interaction has not been directly verified in vivo at the SPB.

Gregg G. Gundersen and Howard J. Worman (eds.), *The LINC Complex: Methods and Protocols*, Methods in Molecular Biology, vol. 1840, https://doi.org/10.1007/978-1-4939-8691-0_12,

In addition to roles at the SPB, both Mps3 and Sad1 are important for positioning chromosomes within nuclei. First shown in budding yeast, Mps3 and other SUN proteins are important for the repair of certain types of double-stranded DNA breaks, presumably by sequestering them at the NE [6, 7]. Both Mps3 and Sad1 also tether other chromosomal regions at the nuclear periphery during mitosis, including extra TFIIIC sites, telomeres, and centromeres [8–11]. The role of SUN proteins in DNA damage repair and chromosome organization is not restricted to fungi, as shown by emerging data in other eukaryotes (reviewed in ref. 12). During meiosis in fission yeast, telomeric binding proteins recruit Sad1 to chromosome ends to form the meiotic bouquet. Chromosome movement within the nucleus is stimulated by dynein, which interacts with the outer nuclear membrane (ONM) KASH protein Kms1 which in turn transmits force to the nucleus via Sad1 [13]. Although aspects of the bouquet differ between organisms (e.g., budding yeast uses actin to stimulate movement through a Csm4 (KASH)-Mps3-Ndj1 (telomere) linkage [14–17]), the role of the LINC complex seems to be important for meiotic chromosome dynamics.

Here, we describe methods used to dissect the function of Mps3 at the SPB and the NE. As with any protein, it is necessary to understand protein-protein interactions that occur at both locations as well as to determine how Mps3 is distributed between the SPB and NE. Affinity purification/proteomics, in vitro binding assays, and yeast two-hybrid analysis lack the spatial resolution needed to demonstrate if the interaction under investigation occurs at the SPB, NE, or both. These conventional methods also do not report on temporal changes in protein binding that might occur during the cell cycle or as cells switch from a mitotic to meiotic program. Although fluorescence resonance energy transfer (FRET) in theory could be used to query interactions with SUN proteins, the constraints needed for FRET (~10 nm between donor and acceptor fluorophores) are difficult to satisfy given the topology of Mps3 in the membrane. The low abundance of Mps3, particularly at the NE, combined with its rapid diffusion on the membrane further complicates FRET analysis (*see* ref. 18).

We employ two complimentary methods to query putative Mps3 binding partners at the NE and SPB in yeast cells using fluorescently labeled proteins (FPs): scanning fluorescence cross-correlation spectroscopy (SFCCS) and single particle averaging with structured illumination microscopy (SPA-SIM). In SFCCS, the intensity fluctuations of two fluorescently labeled proteins are assayed over time at a defined location within the cell, such as the NE or even the inner nuclear membrane (INM). Using correlation analysis, it is possible to determine if the two proteins under study transit through the focal volume together, which would suggest they are part of a protein complex. Using SFCCS, we showed that a novel interaction between Mps3 and Ndc1 (a shared component of SPBs and nuclear pore complexes (NPCs)) [19] occurs on the NE at sites distinct from the SPB and NPCs [18]. Based on the correlation analysis, we could extract quantitative information about the Mps3-Ndc1 complex, including the fraction of Mps3 and Ndc1 within the mobile NE complex, their relative affinity, and when/where binding occurs. Two limitations of SFCCS are that it is only able to query homo- and heterotypic binding interactions on mobile (not immobile) molecules and it does not demonstrate direct binding between proteins. Thus, the utility of SFCCS to interrogate large structures such as the SPB (an ~1 GDa complex) is limited. Therefore, we developed SPA-SIM to study the spatial organization of proteins within *S. cerevisiae* and *S. pombe* SPBs, including the position and organization of Mps3 and Sad1 [20, 21]. Although SIM has the poorest *x–y* resolution of all the super-resolution techniques, we have found it to be most compatible with yeast since we can examine cells containing commonly used FPs directly. Because the construction of yeast strains containing endogenously expressed FP fusions is straightforward, this eliminates the need for extensive sample preparation and concerns over antibody labeling effects, which are required for other types of super-resolution imaging. Using computational and imaging processing methods derived in part from single particle averaging in electron microscopy (EM), we can align dual color SIM images to provide relative spatial information at a resolution between 10 and 30 nm, effectively bridging a gap between FRET and traditional SIM imaging [20, 21]. Below, we discuss how SPA-SIM can be used to provide an overall map of large protein complexes in budding yeast and serve as a guide for targeted protein-protein interaction studies using FRET and yeast two-hybrid, in vitro binding, and genetic analyses. We also discuss how FRET and SPA-SIM can be extended to studies of the LINC complex in fission yeast with only minor modifications.

2 Materials

One of the strengths of yeast as a system for studying LINC complex function is the ability to analyze endogenously expressed wild-type and mutant proteins in both wild-type and mutant backgrounds. Efficient homologous recombination in yeast means that genes can be fused to the coding sequence for FPs. Because the fusion is made in the genomic context as the sole copy of Mps3 in the cell, the tagged gene is subject to native regulation, so gene function, localization, and regulation can be studied without lingering concerns over artifacts associated with expression levels, functionality, or competing untagged protein. The most important

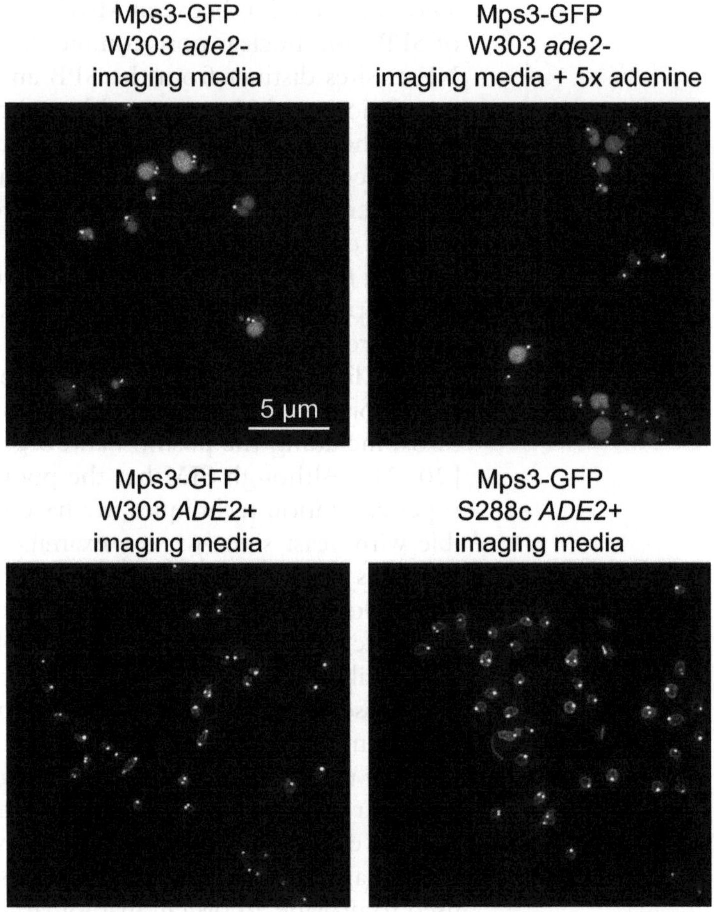

Fig. 1 Effect of strain background on Mps3-GFP. Confocal images showing the distribution of Mps3-GFP to the NE and SPB in the W303 and BY strain backgrounds. All strains are tagged with an identical *MPS3-GFP-HIS3MX* fusion at the endogenous locus and were grown to mid-log phase in imaging media. The top images are W303 strains containing the *ade2-1* mutation. One was grown in media supplemented with five times the normal amount of adenine. On the bottom, the lesion in the W303 strain was repaired so that the strain was *ADE2*. The BY strain was also *ADE2*. Bar, 5 μm

steps in setting up SFCCS and SPA-SIM experiments involve the design, the construction and growth of yeast strains (Fig. 1), and the selection of the appropriate imaging system.

2.1 Yeast Strains and Media for Cell Culture

1. *S. cerevisiae* strains: Methods for creating N- or C-terminal fusions to FPs have been previously described [22]. We create FP fusions in diploid strains to mitigate possible growth issues associated with loss of gene function due to the protein tags. Once positive transformants are obtained, strains can be sporulated, dissected, and analyzed to obtain isogenic double and single FP strains and the wild-type control using conventional

methods of yeast analysis (*see* **Notes 1** and **2**). These strains should be frozen as a glycerol stock by growing cells overnight in liquid culture to an OD_{600} of ~1.0 and then mixing 50:50 with 30% sterile glycerol. Frozen glycerol stocks can be kept indefinitely at −80 °C.

2. Media for YPD plates for *S. cerevisiae*: For 1 L, which makes ~40 plates, mix 10 g bacto-yeast extract, 20 g bacto-peptone, 20 g bacto-agar, and 950 mL ddH_2O. Autoclave. Add 50 mL 40% (w/v) sterile glucose after autoclaving. Pour into sterile 100 mm petri plates once liquid has cooled to ~55 °C, and let harden at room temperature. Imaging media: For 1 L, 6.7 g yeast nitrogen base with ammonium sulfate without amino acids, 5 g casamino acids, 16.6 mg uracil, and 950 mL ddH_2O. After autoclaving, add 4 mL 4 mg/mL adenine, 2 mL 4 mg/mL tryptophan, and 50 mL 40% (w/v) sterile glucose. Keep media in closed cabinet or wrapped in foil as light exposure over long periods will result in the formation of a precipitate that increases autofluorescence (*see* **Notes 1** and **3**). Amino acid stocks of adenine and tryptophan should be filter sterilized and stored in foil-wrapped containers at 4 °C.

3. Adaptions for *S. pombe*: Fission yeast strains containing endogenously tagged FPs can also be constructed using PCR-based methods as described elsewhere [23–25]. We typically generate clones directly in haploid yeast since it is difficult to propagate *S. pombe* for long periods in the diploid form. Once we have obtained the correct transformant, we generate doubly tagged strains through crosses. In doing this, we obtain isogenic single and doubly tagged strains as well as the untagged control. A glycerol stock of these cells is created by concentrating a 5 mL liquid culture grown overnight to an OD_{600} of ~1.0 in YES to 1 mL and then mixing it 50:50 with 50% sterile glycerol. Frozen glycerol stocks can be kept indefinitely at −80 °C. Although fission yeast can grow in the same media that support growth of budding yeast, we find that cell growth is increased and autofluorescence is decreased by the use of media optimized for *S. pombe*. YES plates: for 1 L which makes ~40 plates, mix 10 g bacto-yeast extract; 20 g bacto-agar; 30 g glucose; 0.225 g of each adenine, histidine, leucine, lysine, and uracil; and 1000 mL ddH_2O. Autoclave. Pour onto sterile 100 mm petri plates once liquid has cooled to ~55 °C. Let harden at room temperature. YES media: for 1 L, as above except omit bacto-agar. EMM5S media: for 1 L, 3 g potassium hydrogen phthalate; 2.2 g sodium phosphate dibasic; 20 g glucose; 5 g ammonium chloride; 0.225 g of each adenine, histidine, leucine, lysine, and uracil; 20 mL salt stock (52.5 g magnesium chloride, 0.735 g calcium chloride, 50 g potassium chloride, 2 g sodium sulfate, and 1000 mL ddH_2O; filter sterilize

and store at 4 °C); 1 mL vitamin stock (1 g pantothenic acid, 10 g nicotinic acid, 10 g inositol, 0.01 g biotin, and 1000 mL ddH_2O; filter sterilize and store at 4 °C); 100 μL mineral stock (5 g boric acid, 4 g manganese sulfate, 4 g zinc sulfate, 1 g iron chloride, 0.4 g molybdic acid, 1 g potassium iodide, 0.4 g copper sulfate, 10 g citric acid, and 1000 mL ddH_2O; filter sterilize and store at 4 °C); and 970 mL ddH_2O. Autoclave media. Keep in closed cabinet or wrapped in foil as light exposure over long periods will result in the formation of a precipitate that increases autofluorescence (*see* **Notes 3** and **4**).

2.2 Microscope and Supplies for SFCS and SFCCS

No sample preparation is required for SFCS or SFCCS since these methods require living yeast cells.

1. 22 × 22 mm number 1.5 coverslips: Fluorescence intensity is affected by the thickness of the coverslip. Because SFCS and SFCCS are quantitative methods, it is important to verify that all coverslips measure within 0.001 mm of each other using a caliper tool. Both coverslips and glass slides should be cleaned using 70% ethanol and allowed to dry on a lint-free surface such as lens paper (*see* **Notes 5** and **6**).
2. Microscope: Most commercial and homebuilt confocal microscopes can be used for SFCS and SFCCS. Microscope requirements for SFCS and SFCCS are:
   (a) *Confocality.* As with any fluorescence fluctuation methodology, information is obtained as molecules diffuse in and out of a focal volume. The ability to detect fluctuations from individual molecules requires a small volume—on the order of femtoliters. In addition, for quantification of fluctuation data, it is necessary to have a well-defined shape and size of focal volume, which are best achieved through a pinhole, which comes with a confocal (but not a wide-field) system.
   (b) *Multicolor.* While SFCS data can be achieved with a single laser and single detector, SFCCS data is two channel by definition. Excitation of two FPs with a single two-photon laser is possible in principle; however, we find it much simpler to obtain SFCCS data with two, single-photon laser lines. Standard argon or krypton lasers have more than enough power for SFCCS. As we will detail below, in SFCCS it is highly recommended to scan in alternating excitation (also called multi-track) mode to avoid spectral cross talk and to collect blue and yellow, for example, emission onto two identical detectors. Most, if not all, commercial confocal microscopes have these features.
   (c) *Sensitivity.* The principle of SFCS and SFCCS does not necessitate single-photon counting detectors. However,

we find that with endogenously expressed FPs in both budding and fission yeast, expression levels are sufficiently low that data is improved and photobleaching limited, with the use of lower laser powers (~10 μW at the sample) and single-photon counting detectors. Microscopes employing avalanche photodiode detectors (APDs) or gallium arsenide phosphide (GaAsP) detectors are sufficiently sensitive. The importance of eliminating photobleaching cannot be overstated. Even small amounts of photobleaching can make data analysis very difficult.

(d) *Scanning*. SFCS and SFCCS require the ability to select a line, or circle, and repeatedly scan this region. Most commercial microscopes have this function.

(e) *Speed*. Rapid scanning is not required, provided SFCS is being used for examining membrane diffusion, as will be discussed in this article. In fact, the line scanning itself allows us to examine slow diffusion while limiting photobleaching. The system needs to be able to image lines covering 256–1024 pixels, with individual pixel dwell times on the order of 3–10 μs.

(f) *Software*. Most commercial acquisition software packages are not equipped to align, calculate correlation, and fit correlation curves for SFCS data. This is in contrast to single-point FCS and FCCS software packages. It is best to acquire the line scanning data as an image, and then extract the intensity trace, and fit it with custom-built software. One option is to use ImageJ (NIH). Our custom ImageJ plug-ins are available at http://research.stowers.org/imagejplugins. The easiest way to download the software is by following the Stowers Fiji update site (https://imagej.net/Following_an_update_site).

2.3 Microscope and Materials for SPA-SIM

1. Four percent paraformaldehyde solution: 20 mL 16% paraformaldehyde (EM grade, methanol free), 2.7 g sucrose, and 60 mL ddH_2O. To remove particulates that may interfere with imaging, pass through a 0.2 μm filter. Store at 4 °C in a container shielded from light (*see* **Note** 7).
2. PBS: 8 g sodium chloride, 0.2 g potassium chloride, 1.42 g disodium phosphate, 0.24 g monopotassium phosphate, and 1000 mL ddH_2O. Adjust the pH to 7.4. Filter to remove particulates.
3. Dako fluorescence mounting media: (Agilent Technologies, S302389-2) (*see* **Note 8**).
4. 22 × 22 mm number 1.5 coverslips: SIM is less dramatically affected by coverslip thickness than SFCCS. Nevertheless, it is important to use number 1.5 coverslips. Both coverslips

and glass slides should be cleaned using 70% ethanol and allowed to dry on a lint-free surface such as lens paper (*see* **Notes 5** and **6**).

5. Microscope: The choice of a microscope for SIM is like choices for other kinds of high-resolution and high-sensitivity microscopes.
   (a) *Resolution.* Clearly, it is important to have the resolution necessary to resolve heterogeneity in the protein complex under study. SIM achieves ~100 nm resolution with typical GFP-labeled structures [26]. Switching to blue fluorophores can increase the resolution, though often with the trade-off of lower signal and faster photobleaching (*see* **Note 2**). We use an OMX BLAZE (GE Healthcare, Issaquah, WA) microscope for our SIM protocol. Note that for the specimens studied here, any microscope capable of resolving the major sub-components of the yeast SPB should be sufficient. This includes other commercially available SIM systems, as well as the recently developed Airyscan microscope (Carl Zeiss, Jena, Germany) and lattice light sheet [27] modalities, which have slightly reduced resolution relative to 3D SIM. Axial (z) resolution is less important in our work in that we can simply acquire more images that are oriented in a similar manner in the lateral (x–y) plane. However, the difference in axial resolution compared to lateral is still a crucial component to consider when realigning SPBs. Finally, it is important that cells are maintained less than 10 μm from the coverslip surface.
   (b) *Multicolor.* The alignment methodologies described here require a fiducial marker for alignment. The logical choice for this is a well-defined structural component labeled with a different color. Many super-resolution methodologies require unique combinations of labels and/or immunofluorescence for high-sensitivity multicolor labeling.
   (c) *High sensitivity.* Most super-resolution methodologies require more signal than a high-sensitivity wide-field or confocal microscope. This is, at the minimum, simply a result of needing more photons per unit resolved area given the higher resolution [28]. Obviously, sensitivity is going to be a function of the fluorophores used (*see* **Note 2**). The chosen modality should employ low noise but high-sensitivity detectors. For the SIM modality, the sCMOS cameras seem to work best.
   (d) *High speed.* There are two major reasons for high speed acquisition. The first is sample drift. Although we generally fix and mount cells prior to imaging, we must be cautious to preserve fluorophore integrity, so fixation is

not always ideal, and occasionally there is a slight amount of drift. Keeping the acquisition per channel under 30 s helps with this issue. Secondly, a large sample size is desirable for single particle averaging as described here. The faster samples can be acquired, the greater the accuracy of reconstruction and the throughput of characterization for a large multi-subunit structure.

(e) *Software*. The software shipped with most microscopes is sufficient for good reconstruction. It is nice to have the flexibility to change reconstruction frequency filters (e.g., the Wiener filter for SIM). Given enough signal, the Wiener filter can be decreased to produce noisier data with slightly higher resolution. Particle discovery, realignment, averaging, and analysis must be done with custom ImageJ plug-ins available at http://research.stowers.org/imagej-plugins. The easiest way to download the software is by following the Stowers Fiji update site (https://imagej.net/Following_an_update_site).

3 Methods

3.1 Yeast Growth and Culture

1. Use a sterile toothpick to streak for single colonies from *S. cerevisiae* glycerol stocks on a YPD plate. Grow cells for 2–3 days at 23–30 °C until isolated single colonies are visible. Use a single colony to inoculate a liquid culture for imaging experiments. Plates containing yeast can be stored at 4 °C for up to 10 days; however, long-term storage is not recommended as it is associated with poor growth, increased photobleaching, and autofluorescence.
2. Grow budding yeast cells for 12–18 h at 23–30 °C in imaging media until the OD_{600} is between 0.5 and 0.8. If the OD_{600} is above 0.8, dilute cells back to an OD_{600} of 0.3 and grow out 3–5 h. If OD_{600} is below 0.1, continue growing cells until culture reaches an OD_{600} of at least 0.3.
3. When performing experiments in *S. pombe*, use a sterile toothpick to streak for single colonies from the glycerol stock on a YES plate. Grow cells for 2–3 days at 30 °C until isolated single colonies are visible. Use a single colony to inoculate a liquid culture for imaging experiments. Plates containing fission yeast can be stored at 23 °C for up to 7 days; however, long-term storage is not recommended as it is associated with poor growth, increased photobleaching, and autofluorescence.
4. Grow fission yeast cells for ~24 h at 25 °C in YES with back dilutions to ensure that the cell OD_{600} remains between 0.5 and 0.8. Pellet cells, resuspend in an equivalent volume of EMM5S, and grow for ~4 h at 25–30 °C before imaging.

3.2 SFCS and SFCCS in Budding Yeast

SFCS and SFCCS are extensions of single-point fluctuation spectroscopy, the theory and methods of which have been described in detail previously [29–31]. As members of the fluctuation spectroscopy family, they require a measure of intensity fluctuations over time. Single-point FCS and FCCS methods have been used to study protein dynamics throughout the cell. These single-point fluctuation methods hold the focal volume fixed at one location; however, such an approach does not suffice on membranes such as the NE. This is because membrane diffusion is slow relative to cytosolic diffusion, and continuous excitation at a single point will bleach molecules prior to them diffusing through the focal volume, leading to artifacts in the correlation decay. Furthermore, even with the most rigid immobilization, drift can occur, and membranes may shift due to their natural movements within cells during the time of measurements (typically 3–5 min). A shift in the membrane will cause a large fluctuation in the intensity profile, which again will lead to artifacts in the correlation decay. These two challenges are solved, for the most part, by scanning a line (Fig. 2A).

In SFCS, the diffusion rate and number of mobile molecules can be determined for any fluorescently labeled protein. Because endogenous loci can be fused to FPs with relative ease in yeast, this method is easily applied to the NE pool of Mps3 and other NE proteins. SFCCS is a powerful and surprisingly simple method to study protein complexes in their most native form. It is an extension of SFCS in that the co-diffusion of two fluorescently labeled molecules is investigated; SFCCS is best thought of as a combination of three things, two of which are collected—the autocorrelation of two distinct individual channels—and a third thing that is only calculated, the cross-correlation between the individual channels. Below, we describe how to perform SFCS on an individual fluorescently labeled protein. We then discuss how to extend this method to SFCCS. A detailed data analysis tutorial and sample data can be found here: http://research.stowers.org/imagejplugins/sfccs.html.

1. Yeast sample preparation. Fluctuation methods only work on live cells. For SFCS, only one protein needs to be fused to a FP; for SFCCS, two proteins need to be fused to compatible FPs (*see* **Note 2**).
   (a) After cultures have reached the desired OD_{600}, pellet 1–2 mL of cells by spinning for 1 min at 12,000 × *g* in an Eppendorf centrifuge.
   (b) Aspirate off media and resuspend cells in 100 μL of imaging media or PBS by vortexing briefly (*see* **Note 10**).
   (c) Place 2.5 μL cells onto a clean 22 × 22 mm number 1.5 coverslip.

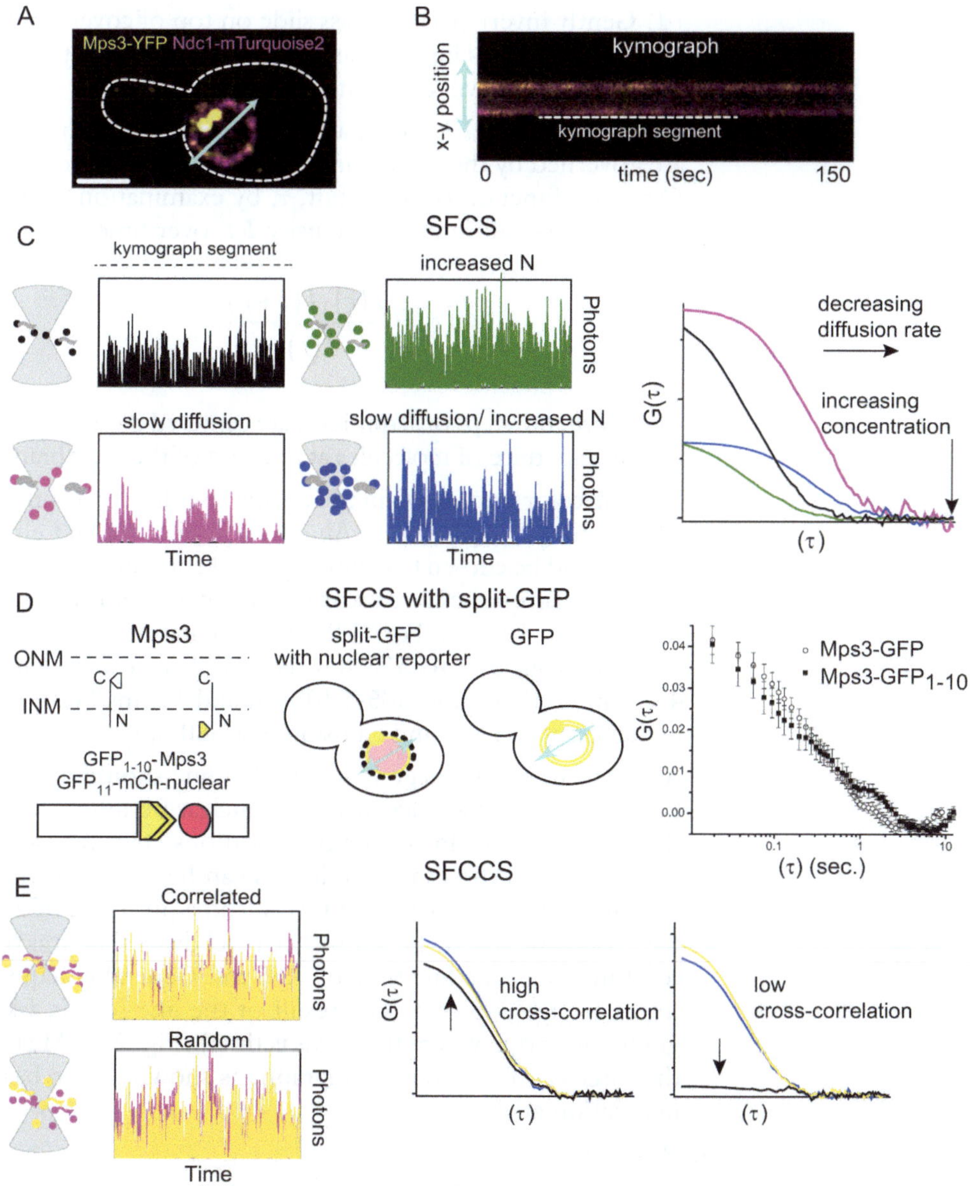

Fig. 2 SFCS and SFCCS of Mps3. (**A**) Image of budding yeast cell containing Mps3-YFP (yellow) and Ndc1-mTurquoise2 (magenta). The cell outline is shown with dashed lines. The line profile for SFCS and SFCCS is shown. Bar, 2 μm. (**B**) Line scans can be visualized as a kymograph, which shows the fluctuations of molecules as they traverse the NE at the two points chosen in (**A**). For fluctuation analysis, the intensity along a segment of the kymograph is measured. (**C**) Schematic demonstrating the principle of SFCS. Simulated data and subsequent correlation curves for diffusing molecules through the focal volume of samples with increased concentration, decreased diffusion rate, or both. The amplitude of the correlation is inversely proportional to the number of diffusing molecules (*N*), and the slope (τ_D) is related to the diffusion rate. (**D**) SFCS of Mps3-GFP and Mps3-GFP$_{1-10}$ in a strain containing the nuclear marker Pus1-mCherry-GFP$_{11}$. (**E**) Principle of SFCCS is illustrated for randomly diffusing magenta and yellow particles and co-diffusing particles using simulated data. The cross-correlation curve is shown in black

(d) Gently invert cleaned glass slide on top of coverslip, pressing out excess liquid. Care must be taken to not press too hard, as this will cause the cells to burst.

2. *Image acquisition.* As fluctuation methods, SFCS and SFFCS are governed by the basic principle of determining correlation, $G(\tau)$, as a function of time shift, τ, by examination of fluctuation traces of fluorescence intensity $I(t)$ over time.

$$G(\tau) = \frac{I(t) \cdot I(t+\tau)}{I(t)^2}. \tag{1}$$

Therefore, the steps of acquisition are centered around obtaining an intensity trace of molecules at a region of the membrane.

For one color acquisition in SFCS, any laser line that sufficiently excites the FP can be used. An emission dichroic and emission filter should be chosen that allows good throughput for the given FP. For mTurquoise2, we use excitation at 440 nm and emission from 470 to 495 nm. For YFP we use excitation at 514 nm and emission from 530 to 575 nm. For GFP we use excitation at 488 nm and emission 505–540 nm, and for mCherry we use excitation at 561 nm and emission past 580 nm.

After an overview image is acquired, a line should be selected that is drawn perpendicular to the nuclear periphery, crossing the nucleus either in one or two locations (Fig. 2A; *see* **Note 11**). Although the exact conditions can be varied, in the past we have used a line with 512 pixels, and a pixel size of 22 nm, for a total line size of 11.3 μm. The pixel dwell time was 6.4 μs, resulting in a total line time of 7.65 ms. Generally, acquisition traces are for 3–5 min. The result of the line scan is a kymograph (an image where the *y*-axis is time) (Fig. 2B). Measuring the intensity across the kymograph, as shown, is the basis for fluctuation analysis.

3. *Analysis.*

(a) *Line selection.* The first step in analysis is to determine what region on the kymograph to use to generate an intensity trace. Raw data (we will call kymograph 1) is first binned by a factor of 2 in time (y) for a total time per line of 15.3 ms and by 4 in x for a total pixel size of 88 nm (Plugins>Stowers>Jay_Unruh>Image_Tools>bin image jru v1). Binning is employed to reduce noise in the subsequent analysis. This second binned kymograph (kymograph 2) will be used to generate the intensity trace; however, it is necessary to further bin kymograph 2 in time in order to best see what region should be selected for analysis (a highly binned image best shows the times the membrane shifted, which are avoided during analysis). The data is binned by another 20× in time (y) to generate kymograph 3. A region is selected on kymograph 3 using the simple segmented line tool and mapped back to kymograph 2 using Plugins>Stowers>Jay_Unruh>Carpet Tools>get line traj jru v1.

(b) *Detrending.* Even with line scanning with low laser power and sensitive detectors, photobleaching can be a problem. In our hands, it is best to detrend the intensity traces by dividing the data into segments and subtracting a linear fit from each part. This and subsequent steps will be performed with the plug-in, Plugins>Stowers>Jay_Unruh>Trajectory Tools>analysis auto corr v2. The program will ask how many segments the data should be divided into for detrending and what shape of focal volume should be used. We recommend 3–4 segments for detrending. Each selected segment should be at least 20 s.

(c) *Correlation and fitting.* With our corrected intensity trace, we calculate $G(\tau)$ using Eq. (1) and then fit it to determine a transit time (τ_D) and concentration (N) using Eq. (2). The correlation fitting function for SFCS is unique from normal FCS because of the restricted motion of the molecules along the membrane in a vertical (x–z) direction [32]:

$$G(\tau)=\frac{1}{N}\cdot\left(1+\frac{\tau}{\tau_D}\right)^{-1/2}\cdot\left(1+\frac{\tau}{\tau_D S^2}\right)^{-1/2}. \quad (2)$$

Here τ_D is the transit time of the molecule through the focal volume, S is the structure parameter of the focal volume $S = z_0/w_0$, and w_0 and z_0 are the radial and axial "waists" of the focal volume, which double the Gaussian standard deviation in each dimension [32]. S is typically assumed to be five for a confocal microscope [33]. Importantly, N represents the average number of diffusing entities in the focal volume over the length of the measurement, proportional to concentration (Fig. 2C) (*see* **Note 12**).

(d) Once τ_D is extracted from the fit, the diffusion coefficient can be calculated as follows:

$$D=\frac{w_0^{\,2}}{4\cdot\tau_D}. \quad (3)$$

(e) Note that SFCS using GFP at the NE cannot distinguish between protein located at the INM or the ONM since the separation between these two NE compartments falls below the resolution limit of all confocal systems and most super-resolution methods. One way to compare mobility or concentration of proteins at the INM and ONM is to use split GFP, as recently demonstrated for Asi1 and other proteins [34] and shown in Fig. 2D for Mps3 (*see* **Note 13**).

4. *SFCCS.* A major challenge and limitation for single-point FCCS are cross talk between channels. However, membrane diffusion, assayed with SFCCS, is often sufficiently slow such that alternating excitation (multi-track) mode is both possible and highly recommended. In some cases, the use of multi-track mode nearly eliminates the possibility of cross talk. We have had the best success using the mTurquoise2 (a vastly improved isoform of CFP; [35]) and YFP for SFCCS experiments on the yeast NE [18, 34]. We excite mTurquoise2 with a 440 or 458 nm line and YFP with a 514 nm line, in alternating excitation mode. Importantly, mTurquoise2 is collected with a narrow band-pass (BP) filter, such as BP 470–495 nm. Even though the 440 or 458 nm line will excite YFP, little or no emission from YFP will be collected through this narrow filter. YFP can be collected with a 530–575 nm filter, and 514 nm excitation does not appreciably excite mTurquoise2. With alternating excitation and narrow emission filters, this combination, imaged in this way, is free from cross talk (*see* **Note 2**).

   To perform SFCCS:

   (a) ***Steps 1–4***. These steps are identical to above, with one exception—maintaining a pixel dwell time of 6.4 μs with two channels and alternating excitation results in a total line time of 15.3 ms (7.65 ms for each channel). The detrending and subsequent steps are performed with the plug-in, Plugins>Stowers>Jay_Unruh>Trajectory Tools>analysis cross corr v2.

   (b) *Calculation of the cross-correlation.* Correlation curves of the individual channels are calculated exactly as described above (Eq. 1). The cross-correlation is then calculated based on fluctuations between the channels (Fig. 2E).

$$G(\tau)_{cc} = \frac{I_B(t)\cdot I_Y(t+\tau)}{I_B(t)\cdot I_Y(t)}. \tag{4}$$

   (c) *Calculation of N_{bound}.* Once $G(\tau)$ values are known for individual channels, and thus the number of blue (N_B) and number of yellow (N_Y) are known from individualcurves, calculate $G(\tau_{CC})$ and then "N_{cc}." $N_{CC} = \frac{1}{G(\tau_{CC})}$.

   (d) Finally, the number of bound particles is as follows:

$$N_{bound} = \frac{N_B(N_Y)}{N_{CC}}. \tag{5}$$

3.3 SPA-SIM in Budding Yeast

There are several methods of determining ultrastructure of protein complexes like the yeast SPB in situ. Transmission electron microscopy (EM) has been of considerable importance in understanding the overall structure and duplication intermediates of yeast SPBs in wild-type and mutant cells, including the contribution of SUN and KASH components to this structure (reviewed in refs. 36, 37). Using immunoEM, both Mps3 and Sad1 were shown to localize to specific subregions of the SPB; as shown in Fig. 3A, Mps3 is found on the INM region of the bridge [1, 2]. FRET in combination with modeling has also played an important role in understanding the organization of proteins within the budding yeast SPB core [38]. While SPA-SIM provides lower spatial resolution than EM or FRET (reviewed elsewhere [39]), it is relatively high throughput, allowing for rapid characterization of position and dynamics of proteins within a large complex like the SPB or centrosome [20, 21, 40]. A tutorial on the alignment and averaging procedures with sample data is available at http://research.stowers.org/imagejplugins/spasim.html.

1. *Yeast fixation and mounting.*
   - (a) After cultures have reached the desired OD_{600}, pellet 1–5 mL of cells by spinning for 3 min at 3000 ×*g* in a table top centrifuge.
   - (b) Pour off media, and resuspend cell pellets in 1 mL of 4% paraformaldehyde solution.
   - (c) Transfer to an Eppendorf tube.
   - (d) Fix for 15 min at room temperature in a rotating tube rack.
   - (e) Pellet cells by spinning for 1 min at 12,000 ×*g* in Eppendorf centrifuge.
   - (f) Aspirate off the fixative and wash two times with 1 mL of PBS.
   - (g) After last wash, aspirate off all remaining PBS.
   - (h) Add 10–20 μL Dako fluorescence mounting media to each, and vortex for 1 min at highest setting (*see* **Note 8**).
   - (i) Pipet 5 μL onto a premeasured 22 × 22 coverslip, using pipet tip to spread cells over surface.
   - (j) Overlay with a clean glass slide, pressing extremely hard to ensure that the cells form a single monolayer between the two glass surfaces.
   - (k) Place with the coverslip facing down in a dark humidified chamber (*see* **Note 9**).
   - (l) Incubate for at least 12–18 h at room temperature.
2. *Image acquisition.* Refractive index matching can be crucial to good structured illumination reconstruction. Some microscopes utilize an objective correction collar for such adjustments.

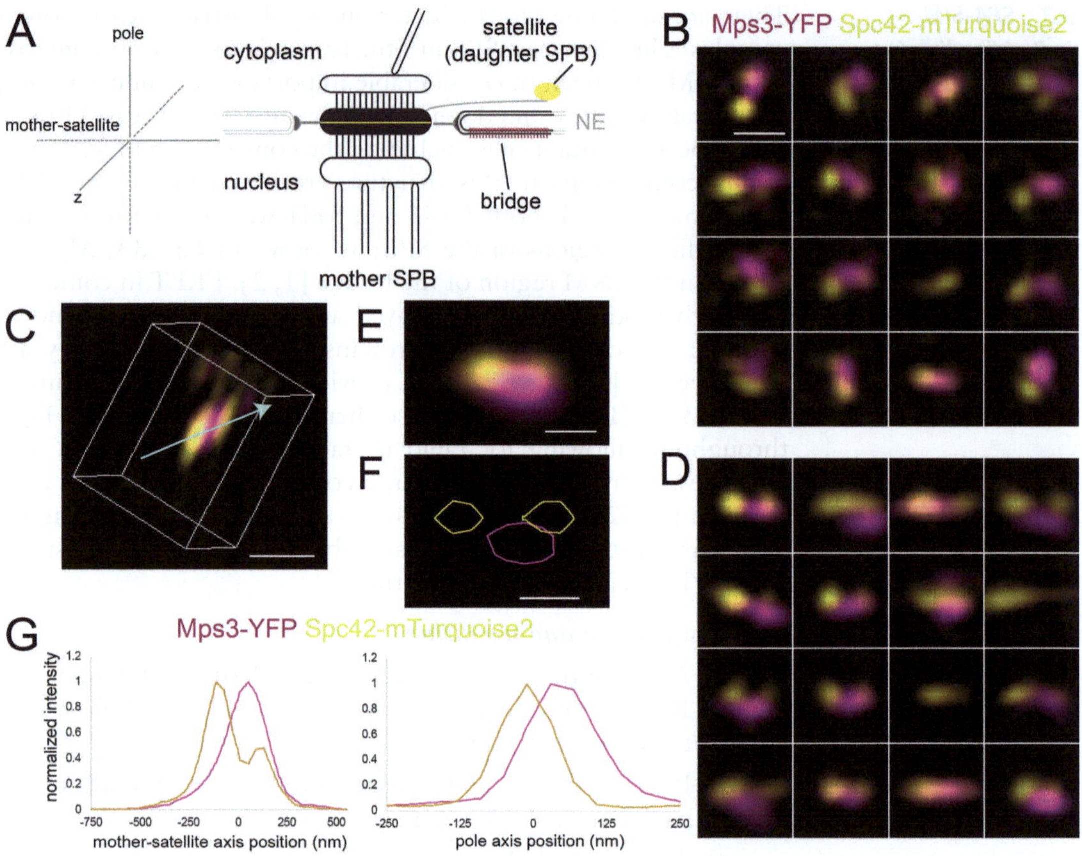

Fig. 3 SPA-SIM of Mps3. (**A**) Schematic of *S. cerevisiae* SPB undergoing duplication. Based on immunoEM, Spc42 (yellow) is found within the core of the mother SPB and at the distal cytoplasmic tip of the newly forming SPB, known as the satellite [60]. Mps3 localizes to the inner nuclear membrane region of the bridge that extends between the mother SPB and the satellite [2]. (**B**) Cells containing Mps3-YFP (magenta) and Spc42-mTurquoise2 (yellow) were arrested in G1 to enrich for the duplication intermediate depicted in (**A**) and were imaged by SIM. Shown are 16 randomly oriented SPBs. (**C**, **D**) Realignment of the 16 SPBs from (**B**) allowed for reorientation of images such that the mother SPB is positioned to the left and the daughter SPB to the right. The SPBs were also reoriented with Mps3-YFP shifted toward the nucleoplasm, which was previously determined relative to Spc42 by immunoEM. (**E**) Projection of realigned images. (**F**) Contour map showing the distribution of fluorescent intensity. (**G**) Normalized fluorescence intensity of both proteins along the mother satellite and the pole axis is plotted. Bars, 100 nm

Our OMX microscope utilizes a series of increasing refractive index immersion oils. Every experiment is a trade-off between exposure time, laser power, and image stack depth. Intensity levels above 5000 are crucial for good reconstruction, but saturation will cause the reconstruction to fail. The order of image acquisition is important to avoid photobleaching; because red fluorophores will bleach when excited in blue wavelengths, we acquire mCherry, then GFP images or YFP, and then mTurquoise2 images. We often use an image stack depth of eight slices using the default setting of 125 nm per *z* slice. This covers most of the nucleus in both *S. cerevisiae* and *S. pombe* and results in a good reconstruction. We use the 60× 1.42 NA oil objective for our SIM measurements, resulting in a 40 nm pixel size.

3. *Reconstruction.* We essentially follow the recommended reconstruction procedures from GE Healthcare for reconstruction. We do however occasionally change the Wiener filter from its typical value of 0.001 in the attempt to achieve higher resolution for low-noise data sets or lower resolution for high-noise data sets. Lateral (*x–y*) color alignment is performed using the reference slide provided by GE Healthcare and is calibrated within a week of data acquisition. Axial (*z*) alignment is performed monthly with TetraSpeck microspheres from Thermo Fisher (Madison, WI). Realignment is particularly important when switching between fluorophore pairs.
4. *Projection and selection and duplication of centrosomes.* Image realignment is performed on reconstructed 3D images, but it is easier to select individual SPBs from a sum projected image created in ImageJ. Individual SPBs are then selected with square regions of interest in the ImageJ RoiManager. These are then duplicated out of the original image. SPBs with their maximum intensity in the first or last slice or without an obvious satellite structure in the fiducial reference channel (Spc42-YFP for *S. cerevisiae* and Ppc89-mCherry for *S. pombe*) are eliminated. This process can be automated somewhat by using a peak finder plug-in. It is useful to save the list of chosen selections for future reference, especially if cell cycle analysis is desired in an asynchronous population. Sixteen examples from cells containing Mps3-YFP (magenta) and Spc42-mTurquoise2 (yellow) are shown in Fig. 3B.
5. *Selection of mother and daughter centrosomes.* While it may be possible to automate the discovery and duplication of SPB images, it is quite difficult to identify the mother and daughter (known as the satellite) spots reliably. We typically perform this operation by hand selecting the brighter SPB first and the dimmer 1 s with point selections in the RoiManager. We pre-verified that the brighter spot is the mother SPB by colocalization with a late duplicating SPB component (Spc110 for *S. cerevisiae* and Kms2 for *S. pombe*) [20, 21].
6. *Multi-Gaussian fitting and realignment.* Once mother and daughter spots are identified, these positions are used to fit the two spots to three-dimensional Gaussian functions with nonlinear least squares (Plugins>Stowers>Jay_Unruh>Image_Tools>fit 3D multi gaussian jru v1). The resulting center positions are used to reslice the image in three dimensions so

that the midpoint between the mother and satellite is in the center of the resulting image and both centers are in the resulting image plane (*see* Fig. 3C, D) (Plugins>Stowers>Jay_Unruh>Segmentation_Tools>thick 3D polyline profile jru v1).

7. *Image merging, flipping, averaging, and scaling.* Next, all realigned images are opened together and merged into a false time stack. Realigned profiles can vary in length, but the plug-in enlarges the merged stack and maintains the center of each image at the center of the final stack (Plugins>Stowers>Jay_Unruh>Image_Tools>merge all stacks jru v1). Note that the Gaussian fitting does not assign a right or left side to the image. As a result, some of the images will have to be flipped manually, taking into account positional information obtained from immunoEM [20] or imaging relative to a reference protein [21]. This can be done by hand before merging or afterward by selecting one side of the SPB (not in the fiducial reference channel) and running Plugins>Stowers>Jay_Unruh>Misc Tools>flip roi mirror right jru v1. This tool compares the intensity between a ROI and its mirror image on the right-hand side. If the right-hand intensity is higher, the image is flipped. This is a good time to inspect the quality of your realignments by creating a montage with Stowers Fiji as shown in Fig. 3D. Next, the images are averaged by creating a time projection of the entire image stack (Plugins>Stowers>Jay_Unruh>Image Tools>bin image jru v1). Finally, averaged profiles are scaled using bilinear interpolation with the built-in ImageJ scaling tool (Image>Scale) to generate the merged image in Fig. 3E.

8. *Contour creation.* While the averaged images represent a raw form of the data (fitting was only utilized for image alignment), they can be hard to compare. For that, we can either create contours or represent the data as intensity peaks. We typically report contours at 75% of the maximum intensity for each visible peak. Because mother and daughter intensities can differ dramatically in intensity, we contour them separately. We also typically scale the image (with bilinear interpolation) four-fold more than for direct viewing, resulting in an image that is scaled 32-fold compared to the original averaged image. Contouring is performed by selecting the desired peak and running Plugins>Stowers>Jay_Unruh>Misc Tools>contour selection jru v1. The contour is added to the RoiManager. The displayed image along with RoiManager rois can be exported to vector graphics using Plugins>Stowers>Jay_Unruh>File Tools>export emf image rois jru v1. These images can be edited for publication in vector graphics and layout editing programs. An example of Mps3-mTurquoise2 (magenta) relative to Spc42-YFP (yellow) is shown in Fig. 3F.

9. *Profile creation and fitting.* While contours offer a way to compare the qualitative shapes of SPA-SIM peaks, they do not provide a quantitative method for comparison of peak positions and sizes. For this we measure average intensity profiles (Plugins>Stowers>Jay_Unruh>Image_Tools>avg profile jru v1) and fit them to single or double Gaussian functions (Plugins>Stowers>Jay_Unruh>Trajectory_Tools>fit traj double gaus jru v1). This plug-in also provides Monte Carlo error analysis on the fits for downstream hypothesis testing. The position of Mps3-mTurquoise2 (magenta) relative to Spc42-YFP (yellow) along the axis from mother to satellite and along the pole axis is shown in Fig. 3G.

4 Notes

1. The choice of yeast strain is often historical, selected based on the lab or the field. However, for imaging experiments it is important to consider that mutations of the *ADE1* or *ADE2* genes found in many commonly used laboratory yeast (e.g., W303, A364a) result in the formation of red pigmented colonies due to the accumulation of a biosynthetic intermediate during purine biosynthesis. This is also linked to high levels of autofluorescence. Addition of extra adenine to the media is not sufficient to reverse the genetic defect (*see* Fig. 1), which also impinges on the production and utilization of other amino acids [41]. Thus, for optimal imaging, strains should be made prototrophic for adenine, or an upstream mutation such as *ade3Δ* or *ade8Δ* should be introduced to block accumulation of the red pigment.
2. The performance of most FPs in yeast is often difficult to evaluate based on published literature since new FPs are typically characterized in vitro or in cultured cells. Yeast codon-optimized versions of FPs generally perform best in yeast imaging experiments and have been cloned into commonly used plasmids for PCR tagging (e.g., [23, 42–45]). The most popular are deposited in plasmid distribution repositories such as Addgene or Euroscarf. We have found that yeast codon-optimized mTurquoise2 and YFP are particularly well-suited to both SFCCS and SPA-SIM since their folding/maturation time is similar and their brightness is adequate for repeated imaging [18, 20, 34]. Although GFP and mCherry are also bright and able to withstand multiple imaging cycles, the altered folding kinetics of mCherry (or superfolder RFP) relative to GFP make experiments difficult to interpret since a population of the red protein is "dark" [46]. This problem is particularly acute in *S. cerevisiae* but does not occur to the

same extent in *S. pombe*. To evaluate if this is an issue, we swap the FP tags as well as analyze the behavior of a tandemly tagged protein. Although it is feasible to create strains containing GFP and YFP, for example, this combination of fluorophores is not suitable for SFCCS or SPA-SIM because of overlap in both excitation and emission spectra.

3. Imaging media results in superior growth of *S. cerevisiae* compared to synthetic complete (SC) media, so we use it when we do not need to maintain selection for plasmids. We typically do not culture cells in YPD liquid media since it produces high background levels of fluorescence. While it is possible to transfer cells from YPD to imaging media/SC media, this results in a transient lag in cell growth as cells acclimate to the new growth media.
4. In adapting these protocols for fission yeast, imaging is best if cells are grown overnight in YES and then transferred to EMM5S for at least 4 h prior to imaging.
5. For SFCS and SFCCS, we adjust the correction collar on our objective to maximum intensity using a solution of dye on a premeasured coverslip. Typically, we use a 40× water and 1.2 NA Plan Apochromat objective for SFCS or SFCCS experiments. Once this adjustment has been made, only coverslips of the same thickness are used throughout experiments. It is also important to align the pinhole of the system using this same dye solution. Improper alignment of the pinhole will greatly reduce the observed intensity. Commercial confocal microscopes have a built-in automated alignment program.
6. Because FCCS and SPA-SIM do not involve long-term imaging (longer than ~15 min), we typically do not prepare agar or gelatin pads for imaging. However, some aspects of chromosome organization are altered if cells are placed directly on glass surfaces, so agar pads should be used in these experiments [47]. A description of the construction of pads suitable for imaging budding and fission yeast can be found in [48, 49].
7. The long-chain polymers present in powdered paraformaldehyde will not adequately fix samples, so it is typically purchased in a soluble form as 37% formaldehyde (which contains methanol) or a 16% paraformaldehyde (EM grade, in ampules from Ted Pella, Redding, CA). Because methanol leads to deterioration of FP fluorophores, we use EM grade paraformaldehyde to fix our yeast. Over time, the paraformaldehyde in the fixation solution will break down due to exposure to air. Prepare this solution fresh every 1–2 months for optimal results.

8. Mounting media for SIM performs two functions: it ensures that cells lie in a single monolayer adjacent to the coverslip, and it protects against photobleaching during SIM image acquisition. The Dako mounting media used in our protocol is specifically protective for YFP [50]. For samples with other fluorophores, this step can be omitted, or other mounting media can be used.
9. We create a humidified chamber by moistening a few paper towels and placing them at the bottom of a sealable plastic container that will hold the slides. The entire container can be placed in a drawer or cabinet overnight.
10. Background effects are one of the largest sources of variability in analysis of endogenously expressed FPs in yeast. In addition to analyzing isogenic strains that have been grown under identical conditions, preparation of slides immediately before imaging helps eliminate artifacts.
11. In the case of NE proteins that are also localized to the SPB or other punctate structures such as the nuclear-vacuolar junction or clustered NPCs, care should be taken to draw the line in such a way as to avoid the puncta. If, during data acquisition, the SPB moves into the path of the continuous line scan, that data set should be discontinued and a second attempt made with the line in a different location. To study protein movement from the NE into or out of these large puncta, techniques such as fluorescence recovery after photobleaching (FRAP) can be used.
12. The correlation equation (Eq. 2) generally includes a factor known as γ to account for the shape of the focal volume. We often omit this since we do not measure the absolute concentration of molecules, but rather compare the relative amounts of NE components. A discussion of γ can be found in [51].
13. Split GFP, also known as bimolecular fluorescence complementation or BiFC, is a method that takes advantage of the propensity of GFP to fold together into a characteristic β-barrel structure, with the protected fluorophore folded inside. GFP, and other FPs, can be split into segments in various ways. The segments are nonfluorescent on their own yet will fold together and fluoresce upon coming into proximity (reviewed in refs. 52, 53). One application is a FRET-type assay, where appearance of fluorescence signal is taken as evidence that the two proteins carrying the split halves of GFP interacted (e.g., [54, 55]). A second application takes advantage of the affinity between the two GFP halves. If split in a certain way, the affinity between the segments can be quite strong [56, 57]. While this is not advantageous when searching for protein-protein interactions, it is a sensitive method to detect the presence of

proteins within the same subcellular compartment since the mere presence of both halves will lead to self-assembly and fluorescence.

We recently completed a large-scale study with this methodology to find novel INM proteins in budding yeast [34]. We expressed over 1000 putative membrane proteins with one half of split GFP (GFP_{1-10}), while the other half of split GFP (GFP_{11}) was fused to a nucleoplasmic protein. If the protein containing GFP_{1-10} localized to the INM and the tag was on the end of the protein exposed to the nucleoplasm, we observed GFP fluorescence at the NE surface. Thus, split GFP can specifically detect proteins such as Mps3 at the INM and determine protein topology. The appearance of GFP signal in a split GFP assay is qualitative. However, a main output of the SFCS fit is concentration. Though it has reduced molecular brightness relative to GFP, the reconstituted split GFP is sufficiently bright for SFCS [34]. In the case of split GFP, the *N* from SFCS is the average number of INM particles in the focal volume. Comparison of this *N* to the NE concentration of the same protein tagged with GFP gives the percentage of the protein localized to the INM:

$$\%\text{INM} = \frac{N_{\text{SFCS}}\text{GFP}_{1-10}}{N_{\text{SFCS}}\text{GFP}}. \tag{6}$$

As a proof of principle, we compared split GFP and GFP concentrations of Asi1, a protein believed to be localized exclusively to the INM [58]. As expected, amplitudes of the correlation curves were statistically identical. On the other hand, Sec62, which was found to have a small INM pool, has a dramatically increased amplitude of its correlation curve (reduced N) with split GFP relative to GFP. This demonstrates that only a small percentage of Sec62 is at the INM, perhaps explaining why it has not been appreciated as an INM protein previously [34, 59]. Data comparing the concentration of Mps3-GFP and Mps3-GFP_{1-10} confirms that, as expected, most of diffusing species of Mps3 are at the INM (Fig. 2D).

Acknowledgments

We are grateful to members of the Microscopy Center and the Jaspersen lab for stimulating discussions and to Briana Holt, Jennifer Gardner, Christine Smoyer, Ann Cavanaugh, Andrew Bestul, and Jeff Lange for their comments on the manuscript.

References

1. Hagan I, Yanagida M (1995) The product of the spindle formation gene sad1+ associates with the fission yeast spindle pole body and is essential for viability. J Cell Biol 129:1033–1047
2. Jaspersen SL, Giddings TH Jr, Winey M (2002) Mps3p is a novel component of the yeast spindle pole body that interacts with the yeast centrin homologue Cdc31p. J Cell Biol 159:945–956
3. Nishikawa S, Terazawa Y, Nakayama T et al (2003) Nep98p is a component of the yeast spindle pole body and essential for nuclear division and fusion. J Biol Chem 278:9938–9943
4. Jaspersen SL, Martin AE, Glazko G et al (2006) The Sad1-UNC-84 homology domain in Mps3 interacts with Mps2 to connect the spindle pole body with the nuclear envelope. J Cell Biol 174:665–675
5. Walde S, King MC (2014) The KASH protein Kms2 coordinates mitotic remodeling of the spindle pole body. J Cell Sci 127:3625–3640
6. Oza P, Jaspersen SL, Miele A et al (2009) Mechanisms that regulate localization of a DNA double-strand break to the nuclear periphery. Genes Dev 23:912–927
7. Swartz RK, Rodriguez EC, King MC (2014) A role for nuclear envelope-bridging complexes in homology-directed repair. Mol Biol Cell 25:2461–2471
8. Bupp JM, Martin AE, Stensrud ES et al (2007) Telomere anchoring at the nuclear periphery requires the budding yeast Sad1-UNC-84 domain protein Mps3. J Cell Biol 179:845–854
9. Horigome C, Okada T, Shimazu K et al (2011) Ribosome biogenesis factors bind a nuclear envelope SUN domain protein to cluster yeast telomeres. EMBO J 30:3799–3811
10. Hiraga S, Botsios S, Donze D et al (2012) TFIIIC localizes budding yeast ETC sites to the nuclear periphery. Mol Biol Cell 23:2741–2754
11. Fernandez-Alvarez A, Bez C, O'Toole ET et al (2016) Mitotic nuclear envelope breakdown and spindle nucleation are controlled by interphase contacts between centromeres and the nuclear envelope. Dev Cell 39:544–559
12. Chang W, Worman HJ, Gundersen GG (2015) Accessorizing and anchoring the LINC complex for multifunctionality. J Cell Biol 208:11–22
13. Chikashige Y, Tsutsumi C, Yamane M et al (2006) Meiotic proteins bqt1 and bqt2 tether telomeres to form the bouquet arrangement of chromosomes. Cell 125:59–69
14. Koszul R, Kim KP, Prentiss M et al (2008) Meiotic chromosomes move by linkage to dynamic actin cables with transduction of force through the nuclear envelope. Cell 133:1188–1201
15. Conrad MN, Lee CY, Chao G et al (2008) Rapid telomere movement in meiotic prophase is promoted by NDJ1, MPS3, and CSM4 and is modulated by recombination. Cell 133:1175–1187
16. Conrad MN, Lee CY, Wilkerson JL et al (2007) MPS3 mediates meiotic bouquet formation in Saccharomyces cerevisiae. Proc Natl Acad Sci U S A 104:8863–8868
17. Lee CY, Conrad MN, Dresser ME (2012) Meiotic chromosome pairing is promoted by telomere-led chromosome movements independent of bouquet formation. PLoS Genet 8:e1002730
18. Chen J, Smoyer CJ, Slaughter BD et al (2014) The SUN protein Mps3 controls Ndc1 distribution and function on the nuclear membrane. J Cell Biol 204:523–539
19. Chial HJ, Rout MP, Giddings TH et al (1998) Saccharomyces cerevisiae Ndc1p is a shared component of nuclear pore complexes and spindle pole bodies. J Cell Biol 143:1789–1800
20. Burns S, Avena JS, Unruh JR et al (2015) Structured illumination with particle averaging reveals novel roles for yeast centrosome components during duplication. eLife 4:e0858
21. Bestul AJ, Yu Z, Unruh JR et al (2017) Molecular model of fission yeast centrosome assembly determined by superresolution imaging. J Cell Biol 216:2409–2424
22. Gardner JM, Jaspersen SL (2014) Manipulating the yeast genome: deletion, mutation, and tagging by PCR. Methods Mol Biol 1205:45–78
23. Bahler J, Wu JQ, Longtine MS et al (1998) Heterologous modules for efficient and versatile PCR-based gene targeting in Schizosaccharomyces pombe. Yeast 14: 943–951
24. Hentges P, Van Driessche B, Tafforeau L et al (2005) Three novel antibiotic marker cassettes for gene disruption and marker switching in Schizosaccharomyces pombe. Yeast 22:1013–1019

25. Sato M, Dhut S, Toda T (2005) New drug-resistant cassettes for gene disruption and epitope tagging in Schizosaccharomyces pombe. Yeast 22:583–591
26. Kner P, Chhun BB, Griffis ER et al (2009) Super-resolution video microscopy of live cells by structured illumination. Nat Methods 6:339–342
27. Chen BC, Legant WR, Wang K et al (2014) Lattice light-sheet microscopy: imaging molecules to embryos at high spatiotemporal resolution. Science 346:1257998
28. TLambert TJ, Waters JC (2017) Navigating challenges in the application of superresolution microscopy. J Cell Biol 216:53–63
29. Magde D, Elson EL, Webb WW (1974) Fluorescence correlation spectroscopy. II. An experimental realization. Biopolymers 13:29–61
30. Kim SA, Heinze KG, Schwille P (2007) Fluorescence correlation spectroscopy in living cells. Nat Methods 4:963–973
31. Slaughter BD, Li R (2010) Toward quantitative "in vivo biochemistry" with fluorescence fluctuation spectroscopy. Mol Biol Cell 21(2010):4306–4311
32. Ries J, Schwille P (2006) Studying slow membrane dynamics with continuous wave scanning fluorescence correlation spectroscopy. Biophys J 91:1915–1924
33. Hess ST, Webb WW (2002) Focal volume optics and experimental artifacts in confocal fluorescence correlation spectroscopy. Biophys J 83:2300–2317
34. Smoyer CJ, Katta SS, Gardner JM et al (2016) Analysis of membrane proteins localizing to the inner nuclear envelope in living cells. J Cell Biol 215:575–590
35. Goedhart J, von Stetten D, Noirclerc-Savoye M et al (2012) Structure-guided evolution of cyan fluorescent proteins towards a quantum yield of 93%. Nat Commun 3:751
36. Winey M, Bloom K (2012) Mitotic spindle form and function. Genetics 190:1197–1224
37. Cavanaugh AM, Jaspersen SL (2017) Big lessons from little yeast: budding and fission yeast centrosome structure, duplication, and function. Annu Rev Genet 51:361–383
38. Muller EG, Snydsman BE, Novik I et al (2005) The organization of the core proteins of the yeast spindle pole body. Mol Biol Cell 16:3341–3352
39. Winey M, Meehl JB, O'Toole ET et al (2014) Conventional transmission electron microscopy. Mol Biol Cell 25:319–323
40. Mennella V, Keszthelyi B, McDonald KL et al (2012) Subdiffraction-resolution fluorescence microscopy reveals a domain of the centrosome critical for pericentriolar material organization. Nat Cell Biol 14:1159–1168
41. Fraenkel DG (2011) Yeast intermediary metabolism. Cold Spring Harbor Laboratory Press, Cold Spring Harbor, NY
42. Longtine MS, McKenzie A 3rd, Demarini DJ et al (1998) Additional modules for versatile and economical PCR-based gene deletion and modification in Saccharomyces cerevisiae. Yeast 14:953–961
43. Sheff MA, Thorn KS (2004) Optimized cassettes for fluorescent protein tagging in Saccharomyces cerevisiae. Yeast 21:661–670
44. Janke C, Magiera MM, Rathfelder N et al (2004) A versatile toolbox for PCR-based tagging of yeast genes: new fluorescent proteins, more markers and promoter substitution cassettes. Yeast 21(2004):947–962
45. Thorn K (2017) Genetically encoded fluorescent tags. Mol Biol Cell 28:848–857
46. Wu B, Chen Y, Muller JD (2009) Fluorescence fluctuation spectroscopy of mCherry in living cells. Biophys J 96:2391–2404
47. Hediger F, Taddei A, Neumann FR et al (2004) Methods for visualizing chromatin dynamics in living yeast. Methods Enzymol 375:345–365
48. Tran PT, Paoletti A, Chang F (2004) Imaging green fluorescent protein fusions in living fission yeast cells. Methods 33:220–225
49. Pemberton LF (2014) Preparation of yeast cells for live-cell imaging and indirect immunofluorescence. Methods Mol Biol 1205:79–90
50. Malkani N, Schmid JA (2011) Some secrets of fluorescent proteins: distinct bleaching in various mounting fluids and photoactivation of cyan fluorescent proteins at YFP-excitation. PLoS One 6:e18586
51. Slaughter BD, Unruh JR, Li R (2011) Fluorescence fluctuation spectroscopy and imaging methods for examination of dynamic protein interactions in yeast. Methods Mol Biol 759:283–306
52. Kodama Y, Hu CD (2012) BioTechniques 53:285–298
53. Miller KE, Kim Y, Huh WK, Park HO (2015) J Mol Biol 427:2039–2055
54. Sung MK, Lim G, Yi DG, Chang YJ, Yang EB, Lee K, Huh WK (2013) Genome Res 23:736–746
55. Khmelinskii A, Blaszczak E, Pantazopoulou M et al (2014) Protein quality control at the inner nuclear membrane. Nature 516:410–413
56. Cabantous S, Terwilliger TC, Waldo GS (2005) Protein tagging and detection with engineered

self-assembling fragments of green fluorescent protein. Nat Biotechnol 23:102–107
57. Cabantous S, Waldo GS (2005) In vivo and in vitro protein solubility assays using split GFP. Nat Methods 3:845–854
58. Boban M, Zargari A, Andreasson C et al (2006) Asi1 is an inner nuclear membrane protein that restricts promoter access of two latent transcription factors. J Cell Biol 173:695–707
59. Deshaies RJ, Schekman R (1990) Structural and functional dissection of Sec62p, a membrane-bound component of the yeast endoplasmic reticulum protein import machinery. Mol Cell Biol 10:6024–6035
60. Adams IR, Kilmartin JV (1999) Localization of core spindle pole body (SPB) components during SPB duplication in Saccharomyces cerevisiae. J Cell Biol 145:809–823

Chapter 13

Genetic Analysis of Nuclear Migration and Anchorage to Study LINC Complexes During Development of *Caenorhabditis elegans*

Heidi N. Fridolfsson, Leslie A. Herrera, James N. Brandt, Natalie E. Cain, Greg J. Hermann, and Daniel A. Starr

Abstract

Studying nuclear positioning in developing tissues of the model nematode *Caenorhabditis elegans* greatly contributed to the discovery of SUN and KASH proteins and the formation of the LINC model. Such studies continue to make important contributions into both how LINC complexes are regulated and how defects in LINC components disrupt normal development. The methods described explain how to observe and quantify the following: nuclear migration in embryonic dorsal hypodermal cells, nuclear migration through constricted spaces in larval P cells, nuclear positioning in the embryonic intestinal primordia, and nuclear anchorage in syncytial hypodermal cells. These methods will allow others to employ nuclear positioning in *C. elegans* as a model to further explore LINC complex regulation and function.

Key words LINC, KASH, SUN, *C. elegans*, Nuclear migration, Nuclear anchorage, Nuclear envelope

1 Introduction

The normal development of most eukaryotes depends on actively moving and anchoring the nucleus to a specific location within the cell. Failures in nuclear positioning lead to a wide variety of defects and diseases [1, 2]. The connection between the cytoskeleton and the nuclear envelope is essential for these processes and is often mediated by a nuclear envelope bridge of Sad1p/UNC-84 (SUN) proteins at the inner nuclear membrane and Klarsicht/ANC-1/Syne homology (KASH) proteins in the outer nuclear membrane. SUN and KASH proteins interact with each other in the perinuclear space, and together, they form LINC complexes to connect the nucleoskeleton to the cytoskeleton [3–5]. LINC complexes are found throughout eukaryotes and have been shown in a variety of systems to mediate nuclear migration and anchorage [1, 2].

Gregg G. Gundersen and Howard J. Worman (eds.), *The LINC Complex: Methods and Protocols*, Methods in Molecular Biology, vol. 1840, https://doi.org/10.1007/978-1-4939-8691-0_13,

Caenorhabditis elegans played a central role in the discovery of SUN and KASH proteins and the development of the LINC complex model [3]. *C. elegans* is a particularly well-suited system for studying LINC components and nuclear positioning because it combines powerful genetics with the ability to study nuclear positioning events by real-time imaging in the context of a developing organism. In the early 1980s, mutations in *unc-83* and *unc-84* that disrupted nuclear migration were isolated by Horvitz and Sulston [6], while mutations in *anc-1* that disrupted nuclear anchorage were found by Hedgecock and Thomson [7]. Malone, Starr, and Han molecularly characterized UNC-84 (SUN), UNC-83 (KASH), and ANC-1 (KASH), which led to the discovery of SUN and KASH domains at the nuclear envelope and the creation of the LINC model [8–11].

Here we discuss three tissues in *C. elegans* that serve as excellent models for studying nuclear migration events and focus on the adult hypodermis to study nuclear anchorage. Together, these invariant and developmentally regulated nuclear positioning events have played huge roles in characterizing the mechanisms of LINC complexes. The goal of this methods paper is to make these experimental models accessible to others interested in studying LINC complexes in *C. elegans.*

The first nuclear migration event described here occurs during embryogenesis, when left and right groups of dorsal epithelial cells intercalate, and their nuclei migrate contralaterally across the length of hyp7 precursor cells (Fig. 1A). These cells subsequently fuse, forming the dorsal hypodermal syncytium with laterally positioned nuclei [12–14]. Mutations in *unc-83* or *unc-84* completely disrupt hyp7 cell nuclear migration (Fig. 1B), resulting in nuclei that are mispositioned to the dorsal cord (Fig. 1C, D) [8, 10, 15]. Subheading 3.1 describes how nuclear migration is studied in hyp7 precursors by scoring finalized nuclear positioning defects in larvae, and Subheading 3.2 describes how to view the migration defect by live imaging in the embryo.

In the second discussed tissue, larval hypodermal cells, P-cell nuclei migrate from a lateral to a ventral position (Fig. 2A). These nuclear migration events are particularly interesting because nuclei must flatten to ~5% of their width to squeeze through a constricted space between muscles and the cuticle [16] (Fig. 2B, C). Thus, P cells are an excellent model for how nuclei squeeze through constricted spaces in humans, such as leukocyte extravasation and some cancer metastases [17, 18]. After the completion of nuclear migration to the ventral cord, P cells divide and give rise to the vulva, hypodermal cells, and motor neurons. Failure in P-cell nuclear migration results in P-cell death and, in turn, Egl (egg-laying deficient) and Unc (uncoordinated) animals due to the lack of vulval cells and motor neurons, respectively. These phenotypes in *unc-83* and *unc-84* null animals are temperature sensitive; at the

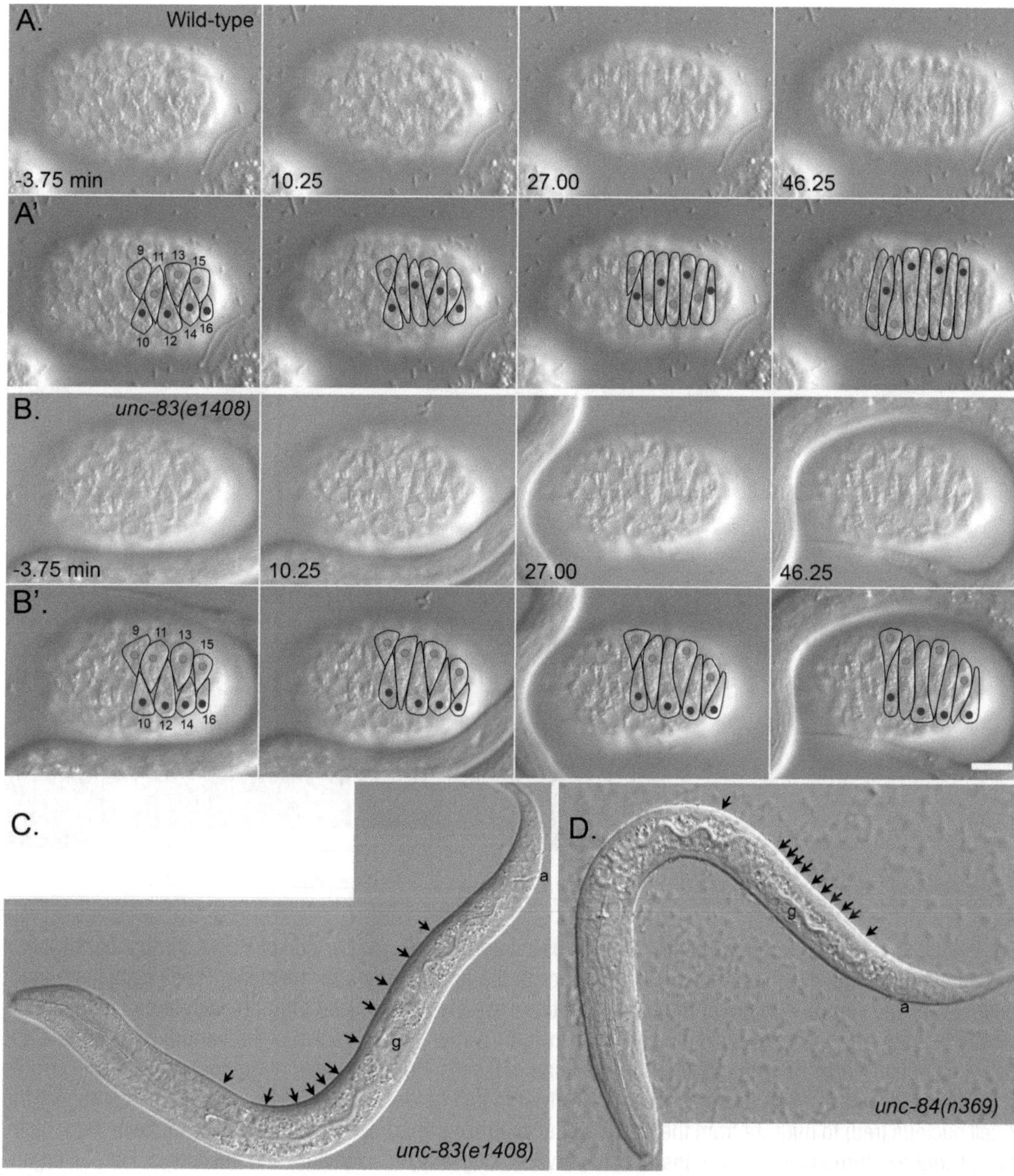

Fig. 1 Nuclear migration in embryonic hyp7 precursors. (**A**, **B**) DIC images from a time-lapse series of images of nuclear migration in dorsal hyp7 precursors in wild-type (**A**) and *unc-83(null)* (**B**) embryos. The time that cell 12 completed intercalation was defined as $t = 0$. Dorsal view, anterior is to the left. (**A'**, **B'**) Cell borders are outlined in black, nuclei migrating left to right are purple, and nuclei migrating right to left are green. Bar, 10 μm. Reproduced from [15] with permission from *The Journal of Cell Biology*. (**C**, **D**) L1 larva showing hyp7 nuclei mislocalized in the dorsal cord (arrowheads) in *unc-83(null)* (**C**) and *unc-84(null)* (**D**). In wild type, there would be no nuclei in the dorsal cord. Not all nuclei are seen in this focal plane. Lateral view; anterior is to the left, and dorsal is up. The four-cell germ line (g) and the anus (a) are marked to help with dorsal-ventral orientation. Note that the anchorage phenotype shown in (**D**) is the most severe phenotype seen in *unc-84(null)* larvae

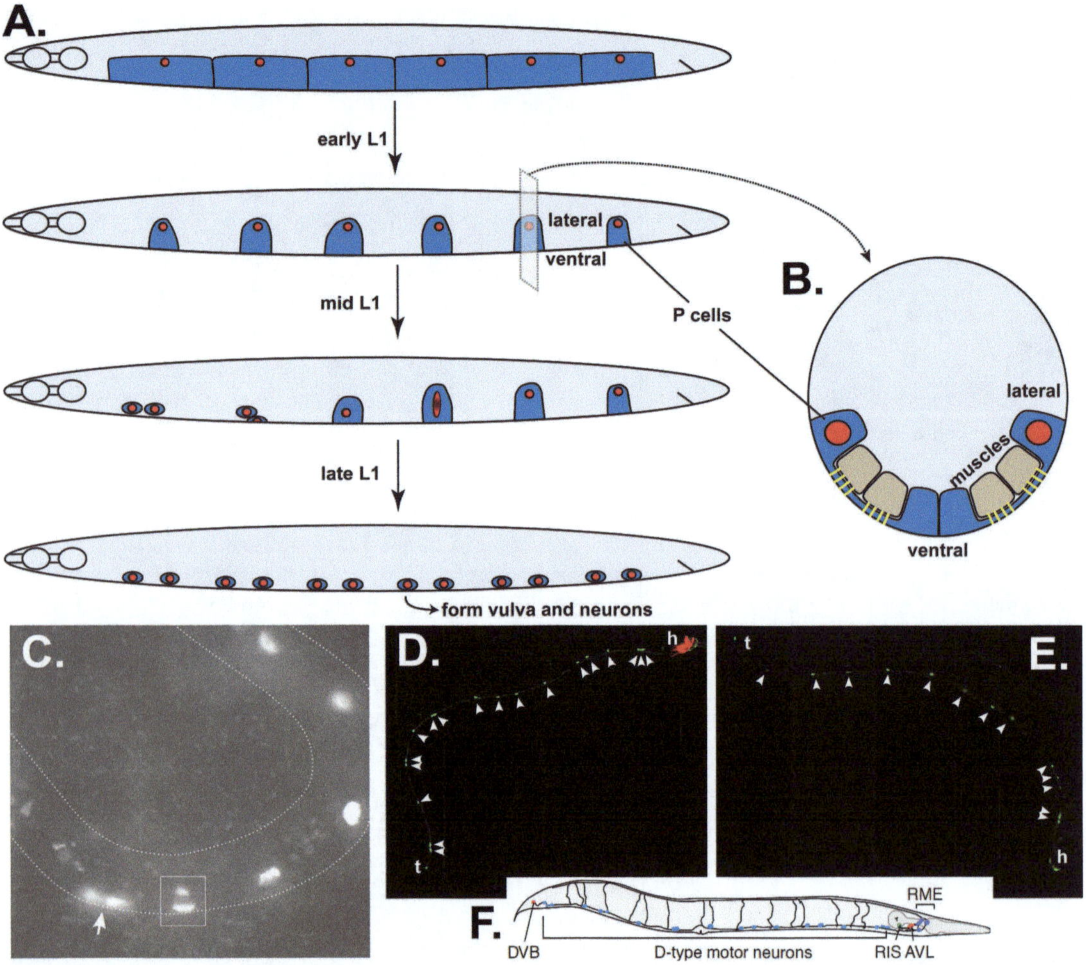

Fig. 2 Nuclear migration through constricted spaces in larval P cells. (**A**) Cartoon of P cells throughout L1 larval development; lateral view, ventral is down, anterior to the left. Shortly after hatching, P-cell cytoplasm (dark blue) covers the ventral surface of the larva, and there are six P-cell nuclei (red) on each lateral side. P cells narrow in early L1 and migrate during mid L1 to form a row of 12 P cells in the ventral cord by late L1. Migration usually initiates with the anterior-most pair of P cells. During nuclear migration, P-cell nuclei stretch from lateral to ventral. (**B**) A cross section of an L1 larva just before P-cell nuclear migration. In order for the P-cell nucleus (red) to migrate from the lateral to the ventral compartments of the P-cell cytoplasm (dark blue), it must squeeze through a narrow constriction between body wall muscles (tan) and the cuticle. Fibrous organelles (yellow) form posts in this constricted space to attach muscles to the cuticle. (**C**) A mid-L1 larva expressing an RFP P-cell nuclear marker in an otherwise wild-type animal. Anterior is left, and ventral is down and on the outside of the curve. The anterior-most pair of P cells has completed migration to the ventral cord (arrow). The next P-cell nucleus is stretched between lateral and ventral compartments; the dark space within the nucleus is nucleoplasm within the constricted space under body wall muscles. P-cell nuclei more posterior have yet to enter the constricted space. Image kindly provided by Courtney Bone (BioMarin). (**D**–**F**) Adult animals expressing *unc-47::gfp* to mark GABA neurons. Tails (t) are left, and heads (h) are right. Lateral view; ventral is down. (**D**) An *unc-84(null)* animal expressing an *unc-84(+)* transgene (transgenic animals express a red marker in the head) with 18 GABA-positive neurons (arrowheads). (**E**) An *unc-84(null)* animal with only 13 GABA-positive neurons (arrowheads) between the head and tail. (**F**) A cartoon showing all the GFP-marked GABA neurons in wild type. Reproduced from [34] with permission from Cell Press. We only count the 19 D-type motor neurons in the ventral cord (light blue), 12 of which are derived from P cells. The DVB in the tail and neurons in the head is derived from other lineages

restrictive temperature of 25 °C, about 50% of P-cell nuclei fail to migrate to the ventral cord, resulting in Egl and Unc phenotypes. However, at the permissive temperature of 15 °C, 90% of P-cell nuclei migrate normally to the ventral cord, and the animals have no obvious phenotype [6, 8, 10, 19]. In Subheading 3.3, we describe how to score P-cell nuclear migration by counting missing P-cell progeny using GABA neuron (Fig. 2D–F) or P-cell nuclear markers (Fig. 2C).

The third set of nuclear migration events occur in the embryonic intestinal primordium. These nuclei move toward the future apical surface where the intestinal lumen forms [20] (Fig. 3). *unc-83* and *unc-84* null animals have strong nuclear migration defects in the developing embryonic intestine as indicated by nuclei failing to position at the midline of the primordium at both 25 °C and 15 °C [10] (Fig. 3). However, mutant embryos raised at 25 °C have slightly less severe defects than embryos raised at 15 °C, opposite the effect of temperature on P-cell nuclear migrations. In Subheading 3.4 we describe how to assay nuclear positioning in the intestinal primordium.

There are at least two other *C. elegans* tissues with interesting nuclear migration events mediated by LINC complexes. Pronuclear migration after fertilization, which is mediated by the SUN protein SUN-1 and the KASH protein ZYG-12, can be easily filmed in the one-cell embryo [21, 22]. The syncytial germ line of the *C. elegans* hermaphrodite also relies on SUN-1 and ZYG-12 to organize nuclei [23]. As they are both well-established systems used by many different labs, we will not discuss the methods for studying them here.

After nuclei migrate to a specific location, they are anchored in place. Defects in nuclear anchorage can be seen in many of the syncytia in *C. elegans* including binucleated intestinal cells, pharynx isthmus muscle cells, seam cells, and the adult hyp7 syncytium [7, 9]. The phenotype is most dramatic in the adult hyp7 syncytium, which contains 139 nuclei that are normally evenly spaced apart [24]. Mutations in either *unc-84* or *anc-1* have a nuclear anchorage defect that can be observed with DIC optics (Fig. 1D) but is best quantified when the hyp7 nuclei are marked with a nuclear GFP (Fig. 4). Subheading 3.5 describes how we quantify nuclear anchorage in the adult hyp7.

In order to enhance the study of nuclear positioning, we have developed fluorescent fusion proteins for live imaging in transgenic animals. We have taken advantage of three tissue-specific markers: the promoter of *lbp-1* for expression of fluorescent proteins in embryonic hyp7 precursors [15, 25], the *hlh-3* promoter for expression in larval P cells [19, 26], and the *col-19* promoter for expression in the adult hyp7 syncytium [27]. Expression of fluorescent fusion proteins only in the desired tissue improves imaging quality. Refer to the referenced papers to see how these

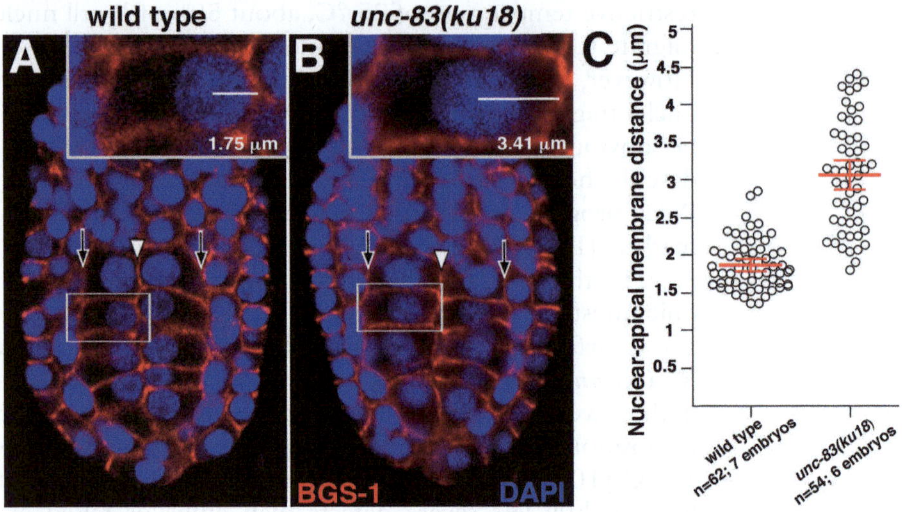

Fig. 3 Nuclear migration in the *C. elegans* intestinal primordium. Wild-type and *unc-83(ku18)* E16 stage embryos were stained with a cortical marker (anti-BGS-1, red) and a nuclear dye (DAPI, blue). (**A**, **B**) Representative single optical sections from a ventral view (anterior up) are shown with white arrowheads denoting the apical surfaces and black arrows denoting the basal surfaces of intestinal cells. The nuclear apical membrane distance was determined by measuring the distance between the center of the nucleus and the apical membrane (insets). (**C**) Each circle represents the nuclear apical membrane distance of a single intestinal cell. The mean distance is shown in red, and the error bars denote the 95% confidence interval

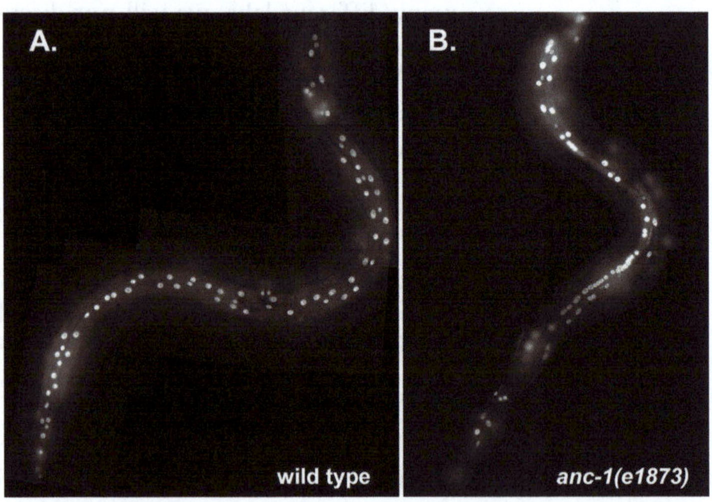

Fig. 4 Nuclear anchorage in the adult hyp7 and seam cell syncytia. Adult animals expressing nuclear-localized GFP in hypodermal cells under control of the *col-19* promoter are shown. (**A**) Wild-type animal with well spaced-out hypodermal nuclei. (**B**) *anc-1(e1873)* null animal with many clustered hypodermal nuclei. Heads are up

plasmids were cloned and how they could be of use for you to clone your gene of interest to be expressed at the time of nuclear migration or anchorage.

2 Materials

2.1 Microscopy

1. Dissecting microscope with epifluorescence.
2. Wide field fluorescence (or confocal) microscope with 10× and 63× or 100× objectives with DIC optics and epifluorescence (filters for GFP and RFP).
3. Agarose pad made of 2% agarose in H_2O on a glass microscope slide (*see* **Note 1**).
4. Worm pick (*see* **Note 2**).
5. 25 × 75 × 1 mm glass microscope slides.
6. M9 buffer: 3 g KH_2PO_4, 6 g Na_2HPO_4, 5 g NaCl, 1 mL 1 M $MgSO_4$, and H_2O up to 1 L. Autoclave to sterilize [28].
7. 1 mM tetramisole, 25 mM sodium azide, or other paralytic in M9 buffer.
8. 25 × 25 × 1 mm coverslips.

2.2 Staining Nuclei and the Cell Cortex of *C. elegans* Embryos

1. 0.1% (w/v) poly-L-lysine.
2. Custom epoxy-coated slides with three square 14 mm wells (30-2066A-Brown. Cell-Line/Thermo Fisher Scientific).
3. 25 × 25 × 1 mm and 25 × 40 × 1 mm coverslips.
4. Methanol or 4% buffered paraformaldehyde: 60 mM Pipes pH 6.8, 25 mM Hepes pH 6.8, 10 mM EGTA pH 6.8, 2 mM $MgCl_2$, 0.1 mg/mL L-α-lysolecithin.
5. 1× PBS-T: 1× PBS with 0.1% Tween-20.
6. PBS-BSA: 1× PBS-T, 1% bovine serum albumin, 0.1% sodium azide.
7. Rabbit anti-BGS-1 (diluted 1:900 in PBS-BSA) [29] and Donkey anti-Rabbit IgG-Rhodamine Red (Jackson ImmunoResearch; diluted 1:400 in PBS-BSA).
8. 0.33 μM Alexa Fluor 488 phalloidin (Thermo Fisher Scientific; diluted in 1× PBS + 0.2% TX-100).
9. PBS-DAPI: 1× PBS, 0.1 μg/mL DAPI.
10. Mounting media: 80% glycerol, 0.223 M DABCO.
11. Clear nail polish.
12. ZEN software (Zeiss) or equivalent.

3 Methods

3.1 Counting Mispositioned hyp7 Nuclei in the Dorsal Cord

1. Culture *C. elegans* on NGM agar plates spotted with OP50 *E. coli*.
2. Prepare microscope slides with 2% agarose pad (*see* **Note 1**).
3. Pipette ~5 μL of M9 buffer to the agarose pad. 1 mM tetramisole or other paralytic can be optionally added to the M9.
4. Using a dissecting microscope and worm pick (*see* **Note 2**), transfer a large number (at least 20) of L1 and L2 worms to the M9 buffer.
5. Cover the agarose pad with a 25 × 25 × 1 mm coverslip.
6. Use DIC optics and a 10× objective on a compound microscope to locate an L1 or L2 worm.
7. Switch to a higher objective (63× or 100×) using DIC optics to zoom in on the selected worm (*see* **Note 3**).
8. Ensure that the larva is on its lateral side by identifying the four-cell germ line and anus on the ventral side (Fig. 1C, D). Identify the dorsal cord, which is opposite the developing germ line (*see* **Note 4**).
9. Score "fried egg"-shaped nuclei in the dorsal cord posterior of the pharynx to the anus.
10. Repeat **steps 5–8** on additional L1 and L2 worms. Wild-type animals should have no hypodermal nuclei in the dorsal cord, whereas *unc-83* or *unc-84* null mutants have about 14 nuclei in the dorsal cord per L1 or L2 animal.

3.2 Filming and Analysis of hyp7 Nuclear Migration

1. Culture *C. elegans* on NGM agar plates spotted with OP50 *E. coli*.
2. Prepare microscope slides with 2% agarose pad (*see* **Note 1**).
3. Pipette ~5 μL of M9 buffer to the agarose pad.
4. Using a dissecting microscope and worm pick (*see* **Note 2**), transfer a large number of gravid adults and/or embryos to the M9 buffer. It is best to use a plate that has just starved, so the adults hold their embryos. These embryos will be at a later stage than normally found in the adult worm and will be close to the pre-comma stage needed for imaging. A newly starved plate will also have a lot of laid embryos close to the correct stage.
5. Cover the agarose pad with a 25 × 25 × 1 mm coverslip. When the coverslip is placed on top of the gravid adults, apply gentle pressure, and embryos will squeeze out of the vulva.
6. Use DIC optics and a 10× objective on a compound microscope to locate a pre-comma stage embryo (*see* **Note 5**).

7. Switch to a higher objective (63× or 100×) using DIC optics to zoom in on the selected embryo (*see* **Note 3**).
8. Capture images at one frame every 15 s, for at least 40 min (*see* **Note 6**).
9. Export the images as an AVI at 15 frames per second without compression.
10. Import the video into ImageJ [30], and crop embryo to 640 × 480 pixels for analysis.
11. Analyze nuclear movement in each cell individually by designating the center of the nucleus and the forward border of the hyp7 precursor using the Manual Tracking plugin for ImageJ (*see* **Note 7**).
12. Track the center of a single nucleus at each time point of the movie and then repeat with the forward border of the same hyp7 precursor (*see* **Note 8**).
13. Import the *x*/*y* coordinates for both the center of the nucleus and the forward border of the cell into Microsoft Excel to calculate the distance and time of migration (*see* **Note 9**).
14. Migration is completed when the nucleus reaches the opposite side of the cell (or when the nucleus/opposite border cannot be seen anymore due to the embryo rolling).
15. Make measurement for distance traveled from center of nucleus to the hyp7 boundary that the nucleus migrates toward (*see* **Note 10**).
16. Repeat **steps 12–15** to analyze additional hyp7 nuclear migrations in the embryo. Only analyze nuclei 11–16 as designated in [14].

3.3 Counting Missing P-Cell Progeny Using GABA Neuron or P-Cell Nuclear Markers

1. Culture *C. elegans* on NGM agar plates spotted with OP50 *E. coli* at the required temperature (*see* **Note 11** about temperature and **Note 12** about the strain with the GFP marker for the assay).
2. Using a dissecting microscope with an RFP filter (*see* **Note 12**) and a worm pick (*see* **Note 2**), transfer an equal number of young adult animals with and without the red rescuing construct onto a fresh NGM plate.
3. Prepare microscope slides with 2% agarose pad (*see* **Note 1**).
4. Pipette ~5 μL of M9 with 1 mM tetramisole or other paralytic on the center of the agarose pad.
5. With a worm pick, transfer all of the young adult animals from **step 2** into the drop of paralytic and gently swirl pick tip without damaging the agarose pad.
6. Cover the agarose pad with a 25 × 25 × 1 mm coverslip.

7. Place slide on the stage of an epifluorescence compound microscope equipped with GFP and RFP fluorescence filters.
8. Use DIC optics and a 10× objective on a compound microscope to locate a young adult animal.
9. Switch to a higher objective (63× or 100×) to zoom in on the selected young adult (*see* **Note 3**). Once an animal is in focus, make sure the anus and pharynx are in the same focal plane. If not, the specimen should not be counted. If so, switch to epifluorescence in the GFP channel.
10. Score the number of GABA neurons. Count green fluorescent cell bodies along the ventral side of the worm. Take care not to count autofluorescent gut granules, which are less bright and a slightly more yellow shade of green and in a different plane of focus. Record the number of GABA neuron cell bodies identified in the ventral cord of the animal (*see* **Note 13**).
11. Switch to the RFP fluorescence, and check for the presence of the *ycEx60* rescuing array as marked by bright red neurons in the head of the animal. Record the presence or absence of the *ycEx60* array with the respective GABA neuron cell body count.
12. Alternatively, score P-cell nuclear migration in late L1 larvae using a P-cell-specific nuclear red fluorescent marker (*see* **Note 14**).

3.4 Scoring Nuclear Migration in the Embryonic Intestinal Primordium

1. Culture *C. elegans* on NGM agar plates spotted with OP50 *E. coli* at the required temperature (*see* **Note 11**).
2. Wash hatched animals off of NGM plates by gently adding and removing successive rounds of H_2O being careful not to detach embryos from the agar surface.
3. Aspirate embryos by briskly pipetting H_2O until the force of the water releases the embryos from the agar surface. Pipette the dislodged embryos into a microfuge tube.
4. Pellet the embryos by centrifuging for 10 s at maximum speed in a benchtop microfuge.
5. Remove the bacteria containing supernatant, and resuspend the embryos in 1 mL H_2O.
6. Repeat **steps 4** and **5** until the supernatant is clear and free of bacterial contamination.
7. Remove the supernatant, leaving approximately 100 μL above the embryo pellet.
8. Resuspend the embryos in the remaining supernatant, and pipette onto the center well of a poly-L-lysine-coated slide (*see* **Note 15**).

9. After 1–2 min, remove excess H_2O by dabbing the corner of the embryo-containing well with a Kimwipe. Move briskly to **step 10** to avoid embryos drying out.
10. To stain the intestinal cell cortex with (1) antibodies, such as anti-BGS-1, permeabilize, and fix with −20 °C MeOH or (2) fluorescent phalloidin, permeabilize and fix with 3–4% paraformaldehyde as described [20].
11. After blocking and staining, rinse one time with PBS-DAPI and one time with PBS, 10 min each.
12. Remove excess PBS from the slide as described in **step 9**. Then place 5 μL of mounting media and a coverslip on the embryos. Seal the slide with nail polish. Place slides at 4 °C or view immediately.
13. Use wide-field or confocal microscopy to capture two-channel fluorescence Z stacks through E16 bean stage embryos of the correct orientation (*see* **Note 16**).
14. Open the stack with ZEN or other image analysis software. Identify a focal plane where the nucleus, apical, and basal surfaces of a single intestinal cell are in focus (Fig. 3A, B).
15. For each nucleus in the intestinal primordium, measure the distance from the center of the nucleus to the apical membrane.

3.5 Quantifying Nuclear Anchorage Defects in the Adult hyp7 Syncytium

1. Create a line of *C. elegans* with the genetic mutation of interest in a background expressing a GFP nuclear marker in the hypodermis (*see* **Note 17**).
2. Culture *C. elegans* to young adults on NGM agar plates spotted with OP50 *E. coli*.
3. Prepare microscope slides with 2% agarose pad (*see* **Note 1**).
4. Pipette ~5 μL of M9 with 1 mM tetramisole or other paralytic on the center of the agarose pad.
5. Using a dissecting microscope and worm pick (*see* **Note 2**), transfer a large number (~20) of young adults to the M9 buffer.
6. Cover the agarose pad with a 25 × 25 × 1 mm coverslip.
7. Place slide on the stage of an epifluorescence compound microscope equipped with GFP fluorescence filters.
8. Use DIC optics and a 10× objective on a compound microscope to locate an adult animal.
9. Switch to a higher objective (63× or 100×) to zoom in on the selected adult (*see* **Note 3**). Once an animal is in focus, make sure the anus and pharynx are in close to the same focal plane. If not, the specimen should not be counted. If so, switch to epifluorescence in the GFP channel.
10. Assay nuclear anchorage by counting the number of GFP-positive hypodermal nuclei that are clustered (*see* **Note 18**).

4 Notes

1. Preparing 2% agarose pads. Layer two pieces of lab tape along the length of a 25 × 75 × 1 mm slide. Repeat this on an additional slide. Flank each long side of a 25 × 75 × 1 mm slide with a taped slide (Fig. 5A). The layered tape on the flanking slides ensures proper thickness of the agarose pad. Melt 2% agarose in water in the microwave. Drop a spot of molten agarose (~100 μL) in the center of the middle slide (Fig. 5B). Immediately drop another slide, face down, on the molten drop perpendicular to the middle slide. Gently press the top slide where it is overlapping the tape, and apply gentle pressure until the agarose solidifies (Fig. 5C). Finally, gently separate the top slide to expose the agarose pad to be used for microscopy. Try to limit the number of bubbles in the agarose pad, as this will disrupt the DIC visualization of the animals. Move quickly to prevent desiccation of your pad and sample.
2. Worm picks. We typically use 30 gauge platinum wire to pick up and transfer worms between plates and to an agarose pad for imaging. About an inch of wire is attached by melting to the end of a glass pipette. The worm-picking end of the wire is flattened with a blunt instrument like the back of a scissor blade. Alternatively (recommended) we buy premade worm picks from www.wormstuff.com that are made of a stiffer alloy of 90% platinum/10% iridium and nicely flattened and feature an ergonomic design that is more comfortable than a glass pipette. Scrape a small amount of bacteria from a plate onto the tip of the worm pick to move worms/embryos to a new

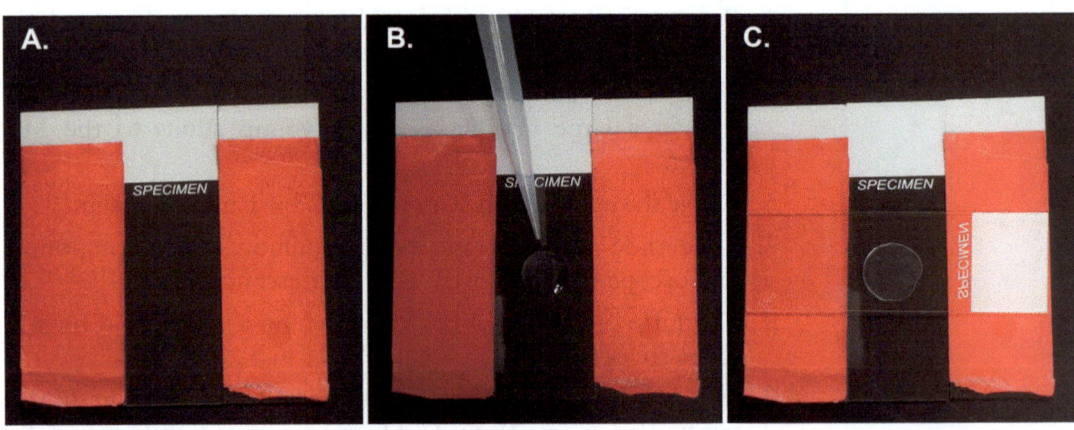

Fig. 5 Making agarose pads. (**A**) The base slide is placed between two other slides with two layers of tape (red) to act as spacers for the pad. (**B**) A drop of about 100 μL of molten 2% agarose is placed on the slide. (**C**) Immediately, another slide is placed on top of the molten agarose to create a pad of uniform thickness. *See* **Note 1** for details

plate or agarose pad. Try to limit the amount of bacteria transferred to an agarose pad because it will hinder imaging.

3. We use a Leica DM6000 compound microscope with DIC optics and a 63× Plan Apo 1.40 NA objective for imaging, but any high-quality setup with a 63× or 100× objective and DIC optics should suffice.
4. Identifying dorsal vs. ventral in an L1 larva. Worms of all stages usually lay or crawl on either their left or right side. The easiest way to identify the ventral side is to spot the anus or the developing germ line (four large cells in an elliptical organ in L1), both of which reside on the ventral side (Fig. 1C, D). One trick to see nuclei in the dorsal cord is to get nuclei on the ventral side in focus. Hypodermal nuclei abnormally in the dorsal cord are more circular and slightly larger than body wall muscle nuclei, which are right next to the dorsal cord and sometimes mistaken for hypodermal nuclei.
5. To image nuclear migration in hyp7 precursors, embryos must be properly staged and oriented with the dorsal surface on top. A pre-comma stage embryo is rounded and fills up the eggshell. The intercalation of hyp7 precursors occurs between 250 and 390 min after first cleavage and has a well-established order of intercalation [14]. As a general rule, hyp7 precursors intercalate in a posterior to anterior fashion [12]. The exception to this pattern is a pair of hyp7 precursors called pointer cells, which do not intercalate until the ventral hypodermal cells have migrated to the nascent ventral midline at 370–385 min post-cleavage [12, 13]. At the posterior end, hyp7 precursors appear as wedge-shaped cells with the cellular boundaries creating depressions in the hypodermis. The nuclei are large and can be seen by the exclusion of cytoplasmic granules (for example of pointer cells, *see* nuclei 9 and 10 in Fig. 1A, B).
6. We capture images with a Leica DC350 camera and Leica LAS AF software. Any imaging software that is able to take time-lapse images and export the images in the AVI format should suffice. The embryo is continuously moving during filming, and it is necessary to check the focal plane before each image is captured to ensure that the hyp7 nuclei are in focus. Small adjustments to the focus are required before almost every image. To image another embryo, a new slide will need to be prepared due to desiccation of the agarose pad and the lack of embryos at pre-comma stage.
7. For the Manual Tracking plugin, set the *x/y* calibration to 0.102 μm (for 640 × 480 image) and then 10 pixels = 1 μm for a Leica DC350 camera. These numbers will need adjustments for different cameras.
8. When tracking a nucleus, align the marker with the center of the nucleus at each time point. To make measurements of distance

traveled, also track the hyp7 cell boundary opposite of where the nucleus being analyzed begins (*not* the edge of the cell itself) as a reference point. This will be approximately where the tip of the hyp7 cell touches when intercalation is complete.

9. Intercalation is considered complete in the first frame that the tip of the cell can be seen touching the opposite hyp7 cell border. Define the completion of intercalation as time = 0 for each cell individually; this marks the beginning of migration for that nucleus. The distance between the nucleus and the forward border is calculated using Pythagorean theorem and the *x*/*y* coordinates determined at each time point.
10. Two quantitative traits have proven most useful for analyzing nuclear migration in hyp7 embryonic precursors [15, 31]. First is the distance a nucleus travels in the 10 min after its cell completes intercalation, and second is the time it takes for a nucleus to cross the dorsal midline of the embryo. A wild-type nucleus will migrate an average of 3.3 μm in 10 min and cross the midline at an average of 11.2 min after intercalation. An *unc-83* or *unc-84* null mutant nucleus will move an average of less than 0.1 μm in 10 min. More than half the mutant nuclei will never cross the midline, and those that do so, cross at an average of 36.4 min after intercalations are complete [15].
11. Null mutations in *unc-83* or *unc-84* are temperature sensitive. About 50% of P-cell nuclei fail to migrate at 25 °C, while almost all P-cell nuclei complete their migrations when raised at 15 °C [8, 10]. The temperature-sensitive period is only during P-cell nuclear migration when the animal is in the mid-L1 larval stage. For embryonic intestinal cells, the temperature sensitivity is opposite. Nuclear localization in *unc-83* or *unc-84* mutant backgrounds is slightly more severe at 15 °C than at 25 °C, and the temperature-sensitive period is in the pre-comma stage embryo [10].
12. We score P-cell nuclear migration defects blindly. We score a mixture of *unc-84(null)* and *unc-84(null)*; *ycEx60[odr-1::rfp; unc-84 (+)]* rescued animals. The rescued animals express a red fluorescent marker in a few chemosensory neurons in the head of the animal off of an extrachromosomal array [19]. It is important to count the GFP GABA neurons before determining if the animal you are counting is a null mutant or a rescued animal in order to reduce experimental bias.
13. UD87 (*unc-84(n369)*, *oxIs[$p_{unc\text{-}47}$::gfp]*) is an excellent control strain for counting GABA neurons; these transgenic animals are homozygous for the GABA neuron marker that is integrated on the X chromosome at genetic position 2.8 [32, 33]. The *oxIs12[$p_{unc\text{-}47}$::gfp]* marker is expressed in 19 D-type motor GABA neurons in the ventral cord (Fig. 2D–F), a single neuron in the tail (DVB) and 6 neurons in the head (RMEL, RMED,

RMER, RMEV, AVL, RIS) of the adult [32]. Make sure not to count larvae, as it takes until early adulthood for all 19 GABA neurons to express the GFP marker. Twelve of these 19 D-type GABA neurons, named VD2 to VD13 [24], are derived from P cells. If there are no P-cell nuclear migration defects, 19 GABA neurons are present in the ventral cord. Animals with P-cell nuclear migration defects have less than 19 GABA neurons in the ventral cord. The single most posterior neuron and the six neurons around the head are excluded when counting because they are not derived from P cells and are also in the interior of the animal, out of the ventral cord [34].

14. The transgene $p_{hlh\text{-}3}$*::nls::tdTomato* specifically labels P-cell nucleoplasm in L1 animals. This marker is integrated in strains UD381 (*ycIs11[*$p_{hlh\text{-}3}$*::nls::tdTomato]*) and UD59 (*unc-84(n369)*; *ycIs11[*$p_{hlh\text{-}3}$*::nls::tdTomato]*) [19]. This marker is used to score P-cell nuclear migration in L1 larvae [19]. Transgenic animals can be synchronized by bleaching as described [35] and hatched into M9 without food. Arrested L1 larvae are then fed standard OP50 to release the arrest for 16, 13, or 10 h at 15 °C, 20 °C, or 25 °C, respectively [19]. This will enrich for L1 larvae around the time of P-cell nuclear migration. Migrating P-cell nuclei flatten to move through the constricted space formed between the muscle and cuticle (Fig. 2). It takes 15–25 min for one nucleus to completely move through this constriction [16]. These migrations can be filmed live using protocols described in [16, 36]. tdTomato-labeled nuclei unable to complete the migration through this constriction remain on the lateral side of the late L1 animal and are classified as failed migration events [16, 19].

15. Prepare poly-L-lysine slides. Hold a microscope slide so that the epoxy-coated side is facing up, and place a small drop of 0.1% poly-L-lysine to the glass surface in the center well. Use the barrel of a long Pasteur pipette to quickly spread a thin layer of the solution over the entire well. Quickly flame the bottom of the slide until the liquid evaporates. Embryos can be placed on the slide once it has cooled.

16. To measure the distance between the nucleus and intestinal cell apical surface, the embryos must be properly staged and oriented in a precise dorsal- or ventral-up position. The location of the dorsally positioned hyp7 precursor cells can be used to easily determine if the embryo is properly oriented. At the early E16 stage, intestinal nuclei are centrally positioned and are not polarized along the apical-basal axis [20]. We therefore score nuclear position at the mid/late E16 stage when the nuclei have migrated and are apically polarized. To identify this stage, we observe hyp7 precursor intercalation as described in **Note 5**. The intestinal primordium is at the E16

(16 E descendants) stage between 262 and 408 min after first cleavage [13, 20]. Therefore, an embryo displaying intercalation of all hyp7 precursors except for the pointer cells indicates that the embryo is at the mid/late E16 stage of intestinal development and properly staged to score the location of intestinal nuclei.

17. A fluorescent marker to follow hypodermal nuclei was constructed by insertion of a 1 kb fragment upstream of the *col-19* gene, which is expressed in the hypodermal cells (including the dorsal and ventral hyp7 and lateral seam cells) of late larvae and adults [27], into pPD96.04 (from Andrew Fire, Addgene plasmid #1502) to create $p_{col\text{-}19}$*::gfp-nls::lacZ* (pSL779). N2 animals were injected with pSL779 at 40 ng/μL + 100 ng/μL of *odr-1::rfp* to make strain UD522 [*ycEx249*]. This marker can be crossed into other strains using standard protocols [37].

18. Nuclei were scored as clustered if within 10% nuclear diameter proximity to another nucleus in the same focal plane along the longitudinal axis of the worm, as determined by DIC microscopy. Contacts between nuclei on the perpendicular axis were not counted, as the marker could not distinguish seam cell nuclei in proximity to hyp7 nuclei from clusters of hyp7 nuclei [24]. Only nuclei situated between the pharynx and the anus were counted, as nuclei near the mouth and at the very end of the tail were observed to cluster in wild-type animals. For most strains, resolution of nuclei on the opposite lateral side of the worm is too poor to allow accurate counting, and therefore, clustering should only be counted on the top-facing lateral side. In wild type, an average of 0.8 ± 0.4 (average ± 95% CI) GFP-positive nuclei per animal were clustered, while in an *anc-1(e1873)* null background, 52.5 ± 3.7 nuclei per animal were clustered. Thus, this assay is much more quantitative than previously published nuclear anchorage assays [8, 38].

Acknowledgments

We thank past and present members of the Starr and Hermann labs for the continual development of these assays over the past 15 years. We thank Courtney Bone for the image in Fig. 2C and Venecia Valdez, Hongyan Hao, and Linda Ma for commenting on the manuscript. Studies in the Starr Lab are supported by the National Institutes of Health grant R01 GM073874. Studies in the Hermann lab are supported by the National Science Foundation grant MCB1613804 and the National Institutes of Health grant R15 GM120639.

References

1. Bone CR, Starr DA (2016) Nuclear migration events throughout development. J Cell Sci 129:1951–1961. https://doi.org/10.1242/jcs.179788
2. Gundersen GG, Worman HJ (2013) Nuclear positioning. Cell 152:1376–1389. https://doi.org/10.1016/j.cell.2013.02.031
3. Starr DA, Fridolfsson HN (2010) Interactions between nuclei and the cytoskeleton are mediated by SUN-KASH nuclear-envelope bridges. Annu Rev Cell Dev Biol 26:421–444. https://doi.org/10.1146/annurev-cellbio-100109-104037
4. Tapley EC, Starr DA (2013) Connecting the nucleus to the cytoskeleton by SUN-KASH bridges across the nuclear envelope. Curr Opin Cell Biol 25:57–62. https://doi.org/10.1016/j.ceb.2012.10.014
5. Luxton GG, Starr DA (2014) KASHing up with the nucleus: novel functional roles of KASH proteins at the cytoplasmic surface of the nucleus. Curr Opin Cell Biol 28:69–75. https://doi.org/10.1016/j.ceb.2014.03.002
6. Horvitz HR, Sulston J (1980) Isolation and genetic characterization of cell-lineage mutants of the nematode Caenorhabditis elegans. Genetics 96:435–454
7. Hedgecock EM, Thomson JN (1982) A gene required for nuclear and mitochondrial attachment in the nematode Caenorhabditis elegans. Cell 30:321–330
8. Malone CJ, Fixsen WD, Horvitz HR et al (1999) UNC-84 localizes to the nuclear envelope and is required for nuclear migration and anchoring during C. elegans development. Development 126:3171–3181
9. Starr DA, Han M (2002) Role of ANC-1 in tethering nuclei to the actin cytoskeleton. Science 298:406–409. https://doi.org/10.1126/science.1075119
10. Starr DA, Hermann GJ, Malone CJ et al (2001) unc-83 encodes a novel component of the nuclear envelope and is essential for proper nuclear migration. Development 128:5039–5050
11. McGee MD, Rillo R, Anderson AS et al (2006) UNC-83 IS a KASH protein required for nuclear migration and is recruited to the outer nuclear membrane by a physical interaction with the SUN protein UNC-84. Mol Biol Cell 17:1790–1801. https://doi.org/10.1091/mbc.E05-09-0894
12. Williams-Masson EM, Heid PJ, Lavin CA et al (1998) The cellular mechanism of epithelial rearrangement during morphogenesis of the Caenorhabditis elegans dorsal hypodermis. Dev Biol 204:263–276. https://doi.org/10.1006/dbio.1998.9048
13. Chisholm AD, Hsiao TI (2012) The Caenorhabditis eleganseepidermis as a model skin. I: development, patterning, and growth. Wiley Interdiscip Rev Dev Biol 1:861–878. https://doi.org/10.1002/wdev.79
14. Sulston JE, Schierenberg E, White JG et al (1983) The embryonic cell lineage of the nematode Caenorhabditis elegans. Dev Biol 100:64–119
15. Fridolfsson HN, Starr DA (2010) Kinesin-1 and dynein at the nuclear envelope mediate the bidirectional migrations of nuclei. J Cell Biol 191:115–128. https://doi.org/10.1083/jcb.201004118
16. Bone CR, Chang Y-T, Cain NE et al (2016) Nuclei migrate through constricted spaces using microtubule motors and actin networks in C. elegans hypodermal cells. Development 143:4193–4202. https://doi.org/10.1242/dev.141192
17. Shah P, Wolf K, Lammerding J (2017) Bursting the Bubble—Nuclear Envelope Rupture as a Path to Genomic Instability? Trends Cell Biol. https://doi.org/10.1016/j.tcb.2017.02.008
18. Ungricht R, Kutay U (2017) Mechanisms and functions of nuclear envelope remodelling. Nat Rev Mol Cell Biol 18:229–245. https://doi.org/10.1038/nrm.2016.153
19. Chang Y-T, Dranow D, Kuhn J et al (2013) toca-1 is in a novel pathway that functions in parallel with a SUN-KASH nuclear envelope bridge to move nuclei in Caenorhabditis elegans. Genetics 193:187–200. https://doi.org/10.1534/genetics.112.146589
20. Leung B, Hermann GJ, Priess JR (1999) Organogenesis of the Caenorhabditis elegans intestine. Dev Biol 216:114–134. https://doi.org/10.1006/dbio.1999.9471
21. Malone CJ, Misner L, Le Bot N et al (2003) The C. elegans hook protein, ZYG-12, mediates the essential attachment between the centrosome and nucleus. Cell 115:825–836
22. Meyerzon M, Gao Z, Liu J et al (2009) Centrosome attachment to the C. elegans male pronucleus is dependent on the surface area of the nuclear envelope. Dev Biol 327:433–446. https://doi.org/10.1016/j.ydbio.2008.12.030
23. Zhou K, Rolls MM, Hall DH et al (2009) A ZYG-12-dynein interaction at the nuclear envelope defines cytoskeletal architecture in the C. elegans gonad. J Cell Biol 186:229–241. https://doi.org/10.1083/jcb.2009 02101
24. Hall DH, Altun ZF (2008) C. Elegans Atlas, 1st edn. Cold Spring Harbor Laboratory Press, Cold Spring Harbor, NY

25. Plenefisch J, Xiao H, Mei B et al (2000) Secretion of a novel class of iFABPs in nematodes: coordinate use of the Ascaris/Caenorhabditis model systems. Mol Biochem Parasitol 105:223–236. https://doi.org/10.1016/S0166-6851(99)00179-6
26. Doonan R, Hatzold J, Raut S et al (2008) HLH-3 is a C. elegans Achaete/Scute protein required for differentiation of the hermaphrodite-specific motor neurons. Mech Dev 125:883–893. https://doi.org/10.1016/j.mod.2008.06.002
27. Liu Z, Kirch S, Ambros V (1995) The Caenorhabditis elegans heterochronic gene pathway controls stage-specific transcription of collagen genes. Development 121:2471–2478. https://doi.org/10.1126/science.6494891
28. Stiernagle T (2006) Maintenance of C. elegans (February 11, 2006), WormBook, ed. The C. elegans Research Community, WormBook. https://doi.org/10.1895/wormbook.1.101.1
29. Moorthy S, Chen L, Bennett V (2000) Caenorhabditis elegans β-G spectrin is dispensable for establishment of epithelial polarity, but essential for muscular and neuronal function. J Cell Biol 149:915–930. https://doi.org/10.1083/jcb.149.4.915
30. Schindelin J, Rueden CT, Hiner MC et al (2015) The ImageJ ecosystem: an open platform for biomedical image analysis. Mol Reprod Dev 82:518–529. https://doi.org/10.1002/mrd.22489
31. Bone CR, Tapley EC, Gorjanacz M et al (2014) The Caenorhabditis elegans SUN protein UNC-84 interacts with lamin to transfer forces from the cytoplasm to the nucleoskeleton during nuclear migration. Mol Biol Cell 25:2853–2865. https://doi.org/10.1091/mbc.E14-05-0971
32. McIntire SL, Reimer RJ, Schuske K et al (1997) Identification and characterization of the vesicular GABA transporter. Nature 389:870–876. https://doi.org/10.1038/39908
33. Gysi S, Rhiner C, Flibotte S et al (2013) A network of HSPG core proteins and HS modifying enzymes regulates netrin-dependent guidance of D-type motor neurons in Caenorhabditis elegans. PLoS One 8:e74908. https://doi.org/10.1371/journal.pone.0074908
34. Schuske K, Beg AA, Jorgensen EM (2004) The GABA nervous system in C. elegans. Trends Neurosci 27:407–414
35. Porta-de-la-Riva M, Fontrodona L, Villanueva A et al (2012) Basic Caenorhabditis elegans methods: synchronization and observation. J Vis Exp. https://doi.org/10.3791/4019
36. Chai Y, Li W, Feng G, Yang Y et al (2012) Live imaging of cellular dynamics during Caenorhabditis elegans postembryonic development. Nat Protoc 7:2090–2102. https://doi.org/10.1038/nprot.2012.128
37. Fay DS (2013) Classical genetic methods. WormBook 30:1–58. https://doi.org/10.1895/wormbook.1.165.1
38. D'Alessandro M, Hnia K, Gache V et al (2015) Amphiphysin 2 orchestrates nucleus positioning and shape by linking the nuclear envelope to the actin and microtubule cytoskeleton. Dev Cell 35:186–198. https://doi.org/10.1016/j.devcel.2015.09.018

Chapter 14

High-Resolution Imaging Methods to Analyze LINC Complex Function During Drosophila Muscle Development

Alexander L. Auld, Mary Ann Collins, Torrey R. Mandigo, and Eric S. Folker

Abstract

Using *Drosophila* muscle development as a model system makes possible the identification of genetic pathways, temporal regulation of development, mechanisms of cellular development, and physiological impacts in a single system. Here we describe the basic techniques for the evaluation of the cellular development of muscle in *Drosophila* in both embryos and in larvae. These techniques are discussed within the context of how the LINC complex contributes to muscle development.

Key words LINC complex, *Drosophila* muscle, Myonuclear position, Muscle development, Myogenesis

1 Introduction

The LINC complex which is composed of SUN proteins that span the inner nuclear envelope and nesprin proteins that span the outer nuclear envelope has been implicated in a number of cellular functions ranging from cytoskeletal organization to genome organization [1–3]. These data have been acquired in disparate cell types including yeast, fibroblasts, neurons, and muscles [4–8]. Studying LINC complex function in muscle is particularly compelling because mutations in genes that encode for the LINC complex and many associated proteins have been linked to the disease Emery-Dreifuss muscular dystrophy [2, 8–12].

There are several systems available for studying muscle cell biology and muscle development. In vitro cell culture systems [13, 14] are optically clear and are therefore amenable to high-resolution microscopy that is necessary to identify molecular mechanisms that underlie muscle cell biology [13, 14]. However, these systems are artificial and therefore cannot be used to evaluate muscle function. Conversely, mouse models present an ideal system for the evaluation of muscle function [8, 15]. Furthermore, modern

Gregg G. Gundersen and Howard J. Worman (eds.), *The LINC Complex: Methods and Protocols*, Methods in Molecular Biology, vol. 1840, https://doi.org/10.1007/978-1-4939-8691-0_14, © Springer Science+Business Media, LLC, part of Springer Nature 2018

technology has made it possible to investigate the subcellular structure of mammalian muscle [16]. But, the temporal resolution of this system is limited, and it is difficult to examine the changes that occur over developmental time.

Drosophila melanogaster combines genetic tractability, in vivo muscle function assays, and optical tractability making it an ideal model to investigate the role of LINC complex proteins in muscle function [7, 17–20]. Using *Drosophila*, we have identified roles for the LINC complex in the movement and the positioning of nuclei during the embryonic and larval stages of muscle development. Furthermore, the LINC complex is critical for the interaction between the nucleus and the sarcomere and contributes to the assembly and stability of the myofibril network [20]. All of these conclusions relied on high-resolution imaging of muscles at distinct developmental stages and time-lapse microscopy of developing embryos. Therefore, *Drosophila* was the ideal model system in which to complete these experiments.

Here we describe the basic techniques (Fig. 1) that we have used to evaluate and demonstrate LINC complex functions in *Drosophila* during specific developmental stages.

2 Materials

2.1 Embryo Sample Collection and Fixation

1. Embryo collection basket.
2. Bleach: 50% solution in water.
3. Paintbrush (size 2).
4. Agar plate covered with thin layer of wet yeast.
5. 1.7 mL microcentrifuge tubes.
6. Heptane.
7. 10% Formalin buffered solution.
8. Methanol.

2.2 Embryo Immunohistochemistry

1. 600 μL microcentrifuge tubes.
2. PBT-BSA (bovine serum albumin): phosphate-buffered saline (PBS) 1×, 0.1% BSA, 0.3% Triton X-100.
3. PBT: PBS 1×, 0.3% Triton X-100.
4. Rabbit anti-DsRed primary antibody.
5. Rat anti-tropomyosin primary antibody.
6. Mouse anti-GFP primary antibody.
7. Alexa Fluor 555 donkey anti-rabbit IgG secondary antibody.
8. Alexa Fluor 488 donkey anti-rat IgG secondary antibody.
9. Alexa Fluor 647 donkey anti-mouse IgG secondary antibody.

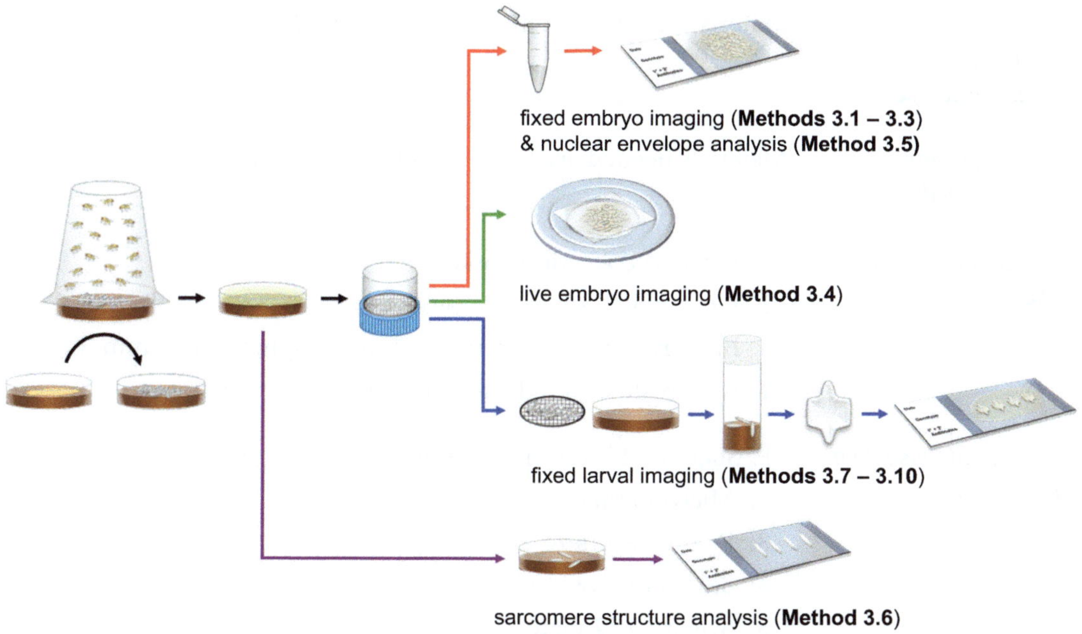

Fig. 1 Workflow of techniques used to prepare *Drosophila* embryos and larvae for imaging. Overview of the different methods used to collect, prepare, and image *Drosophila* embryos and larvae to analyze the role of the LINC complex in nuclear positioning. *Drosophila* embryos are collected from agar plates, dechorionated, and fixed in formalin (red arrow). Once the vitelline membrane is removed by heptane and methanol, embryos are ready for standard immunohistochemistry and mounting. This fixation procedure is similar for nuclear-dependent sarcomere analysis. However, the heptane and methanol fixation devitellinization step is not required due to the analysis being carried out on the fluorescence of GFP and DsRed. *Drosophila* embryos can also be prepared for in vivo time-lapse imaging to observe nuclear positioning during muscle development (green arrow). *Drosophila* embryos can also be collected and staged to the third instar larval stage (blue arrow). At this stage, larvae are dissected and fixed in formalin. After fixation, larvae are stained and mounted for imaging. Additionally, first instar larvae can be selected directly from the agar plate for sarcomere structure analysis (purple arrow)

2.3 Embryo Mounting

1. Microscope slides (3″ × 1″ × 1 mm).
2. Glass coverslips (#1.5 thickness).
3. Small coins.
4. Scotch Magic tape (#810).
5. Kimwipes.
6. ProLong Gold Antifade reagent.

2.4 Live Embryo Imaging

1. Halocarbon oil.
2. Wooden dowels.
3. Microscope imaging disk.
4. Glass coverslips (#1.5 thickness).
5. Scotch Magic tape (#810).

2.5 ZASP Recruitment to the Nuclear Envelope

Materials needed are listed in Subheadings 2.1–2.3.

2.6 Sarcomere Structure Analysis

Materials needed are listed in Subheadings 2.1–2.3.

2.7 Larval Sample Collection

1. Embryo collection basket.
2. Bleach: 50% solution in water.
3. Biology grade tweezers: Steel, nonmagnetic, 110 mm, #5.
4. Agar plate covered with thin layer of wet yeast.

2.8 Larval Dissection and Fixation

1. Sucrose: 15% (weight/volume) solution in water.
2. Microspatula.
3. Paintbrush (size 2).
4. Stereo microscope.
5. Sylgard 182 plate.
6. 0.1 mm minuten pins.
7. McPherson-Vannas dissecting spring scissors: 8 cm, straight 3 mm blades, 0.1 mm tip.
8. 2 Biology grade tweezers: Steel, nonmagnetic, 110 mm, #5.
9. PIPES dissection buffer: 100 mM PIPES-HCl pH 6.8, 115 mM D-sucrose, 5 mM D-trehalose dehydrate, 10 mM sodium bicarbonate, 75 mM potassium chloride, 4 mm magnesium chloride, 1 mM EGTA pH 7.5.
10. 10% Formalin-buffered solution.
11. PBS.
12. 1.7 mL microcentrifuge tubes.
13. PBT-BSA: PBS 1×, 0.1% BSA, 0.3% Triton X-100.

2.9 Larval Immunohistochemistry

1. 600 μL microcentrifuge tubes.
2. PBT-BSA: PBS 1×, 0.1% BSA, 0.3% Triton X-100.
3. PBT: PBS 1×, 0.3% Triton X-100.
4. Mouse anti-α-tubulin primary antibody.
5. Alexa Fluor 488 donkey anti-mouse IgG secondary antibody.

2.10 Larval Mounting

1. Microscope slides (3″ × 1″ × 1 mm).
2. Glass coverslips (#1.5 thickness).
3. Small coins.
4. Kimwipes.
5. ProLong Gold Antifade reagent.

3 Methods

3.1 Embryo Collection and Fixation

1. To set up an experiment, add males and virgin females of desired genotypes (Table 1) together in a laying pot. Cover an agar plate with wet yeast, and add it to the bottom of the laying pot. Allow embryos to lay for 20 h at 25 °C.
2. Remove the old agar plate from the laying pot, and replace with a new plate covered with wet yeast.
3. Assemble an embryo collection basket with a mesh filter for each genotype.
4. Add 50% bleach to the agar plates to cover embryos for 3–5 min (*see* **Note 1**).
5. Gently brush embryos off the plate using a paintbrush and pour into the designated collection basket (*see* **Note 2**).
6. Rinse embryos in ~1 mL of dH_2O three times.
7. Add 650 μL heptane to a 1.7 mL microcentrifuge tube for each genotype.
8. Remove the mesh filter from collection baskets. Brush up embryos using a paintbrush, and place embryos in the correctly labeled microcentrifuge tube filled with heptane (*see* **Note 3**).
9. Add 650 μL of the 10% formalin-buffered solution to each microcentrifuge tube (*see* **Note 4**).
10. Tape the tubes to a shaker and shake for 20 min at 300 rpm.
11. After 20 min, remove tubes from the shaker and remove the fixing liquid.
12. Add a 1:1 mixture of heptane/methanol (650 μL of each).
13. Vortex tubes for 1 min (*see* **Note 5**).
14. Remove the heptane/methanol mixture (*see* **Note 6**).
15. Wash embryos with ~1.5 mL of methanol.
16. Add fresh methanol to embryos and store tubes at −20 °C (*see* **Note 7**).

3.2 Embryo Immunohistochemistry and Mounting

1. For each genotype, transfer the collected embryos from the 1.7 mL microcentrifuge tube into a 600 μL microcentrifuge tube (*see* **Note 8**). Remove the methanol and add PBT-BSA.
2. Rinse each tube three times with ~500 μL of PBT-BSA.
3. Wash in PBT-BSA on rotator for 15 min to 1 h as time permits.
4. During the washing step, prepare 400 μL aliquots of PBT-BSA for the primary antibody mixture. Dilute the rabbit anti-DsRed antibody 1:400 and the rat anti-tropomyosin antibody 1:200 (*see* **Note 9**).
5. After washing, remove remaining liquid and dispense 400 μL of the primary antibody mixture to each tube.

Table 1
Transgenic Drosophila lines used for LINC complex analysis

Genotype	Description	Bloomington Stock #
1. w*; *twist-GAL4,apRed*	Expression of the GAL4 protein under *twist* enhancer control	[22]
2. *IF/CyO; Dmef2-GAL4,apRed*	Expression of the GAL4 protein GAL under *Dmef2* enhancer control	[22]
3. w*; *MHC-GAL4*	Expression of the GAL4 protein GAL under *MHC* enhancer control	
4. *twist-GAL4,apRed; Zasp-GFP*	GFP protein trap insertion for Z band alternatively spliced PDZ-motif protein 66 (Zasp66)	B6824
5. y^{1},sc^{*},v^{1};; *UAS-klar-TRiP*	Expresses dsRNA for RNAi of klarsicht under UAS control	B36721
6. y^{1},sc^{*},v^{1};; *UAS-Msp300-TRiP*	Expresses dsRNA for RNAi of muscle-specific protein-300 under UAS control	B32848
7. y^{1},sc^{*},v^{1}; *UAS-koi-TRiP*	Expresses dsRNA for RNAi of klaroid under UAS control	B40924
8. y^{1},v^{1}; *UAS-bocks-TRiP*	Expresses dsRNA for RNAi of bocksbeutel under UAS control	B38349
9. y^{1},v^{1}; *UAS-Ote-TRiP*	Expresses dsRNA for RNAi of Otefin under UAS control	B39009
10. y^{1},v^{1};; *UAS-LamC-TRiP*	Expresses dsRNA for RNAi of Lamin C under UAS control	B31621
11. *TG4,apRed*; $ru^{1}klar^{1}$/*TM6DGY*	Null allele of klarsicht; lacks the C-terminal 286 amino acids due to a point mutation causing premature termination	B3256
12. $Msp300^{compl}$; *MG4,apRed*	Null allele of muscle-specific protein-300; deletion of the complete *Msp300* locus through FRT-bearing insertions	[7]
13. $koi^{HRko80.w}$/*CyODGY; MG4,apRed*	Null allele of klaroid; replacement of the entire *koi* locus with w^{+} marker	B25105
14. *TG4,apRed*; $bocks^{DP01391}$/*TM6DGY*	A transgenic insertion derived by TE mobilization using a P-element construct within the 5′UTR	B21846
15. Ote^{B279}/*CyODGY; MG4,apRed*	Null allele of Otefin; a transgenic insertion derived by TE mobilization using a P-element construct within the first exon	[23]
16. $LamC^{K11904}$/CyODGY: MG4,apRed	Null allele of Lamin C; a transgenic insertion derived by TE mobilization using a P-element construct	B11050

List of alleles and RNAi *Drosophila* lines used to disrupt LINC Complex expression and function

6. Stain embryos on rotator overnight at 4 °C.
7. Remove the primary antibody mixture, 200 μL at a time as to not disturb embryos, and store the antibody mixture at 4 °C (*see* **Note 10**).
8. Rinse three times in ~500 μL of PBT-BSA.
9. Wash three times on rotator in ~500 μL of PBT-BSA for 5–10 min each.
10. During the last wash, prepare 400 μL aliquots of PBT-BSA for the secondary antibody mixture. Dilute both the anti-rabbit and anti-rat secondary antibodies 1:200 (*see* **Note 11**).
11. After washing, remove remaining liquid, and dispense 400 μL of the secondary antibody mixture to each tube.
12. Stain embryos on rotator for 1–2 h at room temperature, with limited exposure to light.
13. Remove the secondary antibody mixture, 200 μL at a time as to not disturb embryos, and discard.
14. Rinse three times in ~500 μL of PBT-BSA.
15. Wash three times on a rotator in ~500 μL of PBT-BSA for 5–10 min each (*see* **Note 12**).
16. During the last wash, prepare 400 μL aliquots of PBT-BSA for the additional primary antibody mixture. Dilute the mouse anti-GFP antibody 1:50 (*see* **Note 13**).
17. After washing, remove the remaining liquid, and dispense 400 μL of the primary antibody mixture to each tube.
18. Stain embryos on rotator for 1 h at room temperature or overnight at 4 °C (*see* **Note 14**).
19. Remove the primary antibody mixture, 200 μL at a time as to not disturb embryos, and store at 4 °C (*see* **Note 10**).
20. Rinse three times in ~500 μL of PBT-BSA.
21. Wash three times on rotator in ~500 μL of PBT-BSA for 5–10 min each.
22. During the last wash, prepare 400 μL aliquots of PBT-BSA for the additional secondary antibody mixture. Dilute the anti-mouse secondary antibody 1:200 (*see* **Note 15**).
23. After washing, remove the remaining liquid, and dispense 400 μL of the secondary antibody mixture to each tube.
24. Stain embryos on rotator for 1–2 h at room temperature, with limited exposure to light.
25. Remove the secondary antibody mixture, 200 μL at a time as to not disturb embryos, and discard.
26. Rinse three times in ~500 μL of PBT-BSA.
27. Wash three times on rotator in ~500 μL of PBT for 5–10 min each.

28. During each washing step, prepare slides and coverslips for mounting (Fig. 2A).
29. Label a slide denoting the appropriate genotype and antibodies used. Place a coin in the middle of each slide.
30. Make the tape bridge. Tape a coverslip to the workbench with two strips of tape on each end of the coverslip, leaving the middle section free of tape. Cut the tape surrounding the coverslip with a razor blade to release it from the bench top (*see* **Note 16**).
31. Remove the PBT from embryos so that little liquid remains.
32. Resuspend embryos and transfer them to the coverslip, 20 μL at a time (*see* **Note 17**).
33. Once all embryos have been transferred, use a Kimwipe to remove excess liquid. When most of the liquid is removed, tilt the coverslip to get any remaining liquid.
34. Add 65 μL of the mounting agent ProLong Gold and evenly spread over the embryos using a pipette tip (*see* **Note 17**).
35. Add a second coverslip gradually on top of the embryos.
36. Place the mounted coverslips on top of the coin on the labeled slide and store in slidebook overnight, with limited exposure (*see* **Note 18**).

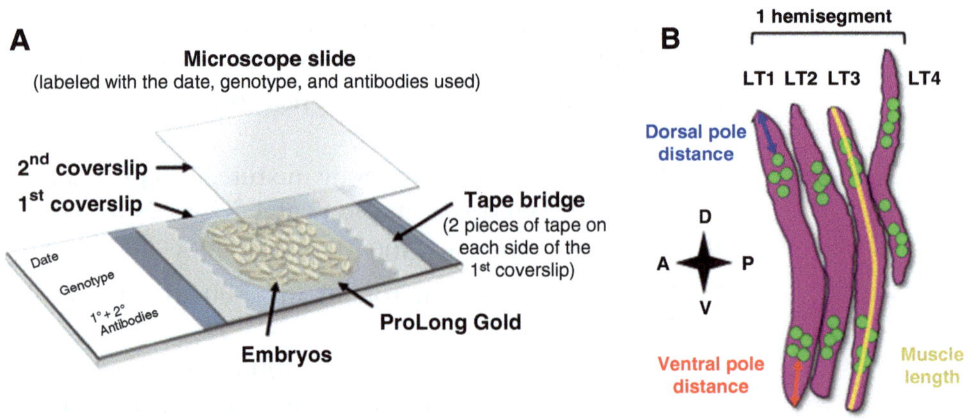

Fig. 2 Slide mounting and analysis for fixed *Drosophila* embryo imaging. (**A**) Cartoon depicting the standard setup used to mount fixed *Drosophila* embryos on a microscope slide. Embryos are placed onto a coverslip that has been prepared with a tape bridge and then the ProLong Gold mounting reagent is added to the embryos. Once the ProLong Gold has been spread evenly across the embryos, a second coverslip is placed on top. After the ProLong Gold has fully cured, the coverslip with the mounted embryos can be taped directly to the microscope slide and imaged. (**B**) Analysis of nuclear position in fixed *Drosophila* embryos, with the muscles in magenta and the nuclei in green. Measurements are taken in the four lateral transverse (LT) muscles within each hemisegment. Embryos should be orientated such that the LT1 on the left and LT4 on the right are shifted upward. The three measurements analyzed are the dorsal pole distance (blue arrow), the ventral pole distance (red arrow), and the length of the muscle (yellow arrow). A total of four different hemisegments should be measured per image

37. Remove the coin and tape coverslips directly to the microscope slide before imaging.

3.3 Imaging and Analyzing Fixed Drosophila Embryos

1. Fixed embryo imaging should be performed on a standard laser scanning confocal microscope.
2. Select a stage 16 embryo to image based on the morphology of the gut (*see* **Note 19**).
3. Check to make sure the embryo is in the correct orientation before imaging with the lateral transverse (LT) muscles positioned within the middle of the embryo (*see* **Note 20**).
4. When imaging an embryo, ensure the zoom is adequate in order to visualize the LT muscles in four hemisegment (*see* **Note 21**).
5. Set the *z*-stack while in the 488 nm channel so that the entire length of the LT muscles in all four hemisegments are visible (*see* **Note 22**).
6. Check the 555 nm channel to see that the nuclei fall within the *z*-stack selected before imaging.
7. All embryo analysis should be done in ImageJ or FIJI. Import image files into ImageJ and make each image a maximum projection (*see* **Note 23**).
8. Rotate each image to the correct orientation (Fig. 2B) such that each hemisegments is positioned with the first lateral transverse muscle (LT1) on the left and LT4 on the right, shifted upward (*see* **Note 24**).
9. Measure the length of each LT muscle using the segmented line tool and follow along the length while staying within the middle of the muscle (*see* **Note 25**).
10. To measure the dorsal end distances for each LT muscle, use segmented line tool, and measure the distance between the dorsal end of the muscle and the nearest nucleus.
11. Similarly, use the segmented line tool, and measure the distance between the ventral end of the muscle and the nearest nucleus to obtain ventral end distances for each LT muscle.
12. For statistical analysis, first average the muscle length measurements for all LT muscles present in each individual hemisegment. Then average the average length values for each hemisegment together to obtain an average length value for the entire embryo (*see* **Note 26**).
13. Repeat **step 12** for dorsal distance measurements and again for ventral distance measurements.
14. Plot the average of the average values obtained in **steps 12** and **13**.

3.4 Live Imaging of Drosophila Embryos

1. Collect embryos as described in Subheading 3.1, **steps 1–6**.
2. Float embryos by placing the collection basket in a petri dish top filled with water.
3. Dip a wooden dowel in halocarbon oil, and gently swirl it around the collection basket to pick up embryos (*see* **Note 27**).
4. Transfer the halocarbon oil/embryos onto an imaging disk (*see* **Note 28**).
5. Place a coverslip over the embryos, and tape down the sides of the coverslip to the imaging disk. Gently press the plastic barrier of the imaging disk to the underside of the coverslip (Fig. 3A).
6. Image embryos using a standard laser scanning confocal microscope.
7. Select stage 15 embryos to image based on the morphology of the gut and the position of nuclei at this stage (*see* **Note 19**).
8. Set the *z*-stack and zoom to adequately image the LT muscles in four hemisegment as described in Subheading 3.3, **steps 3–6** (*see* **Note 29**).
9. Set the time-lapse to image every 2 min for at least 2 h.
10. For live imaging analysis, import the movie into ImageJ or FIJI and orientate the image as described in Subheading 3.3, **steps** 7 and **8**.
11. To measure the separation speed between nuclear clusters, measure the shortest distance between the dorsal and ventral nuclear clusters at 0 min and at 60 min (Fig. 3B).

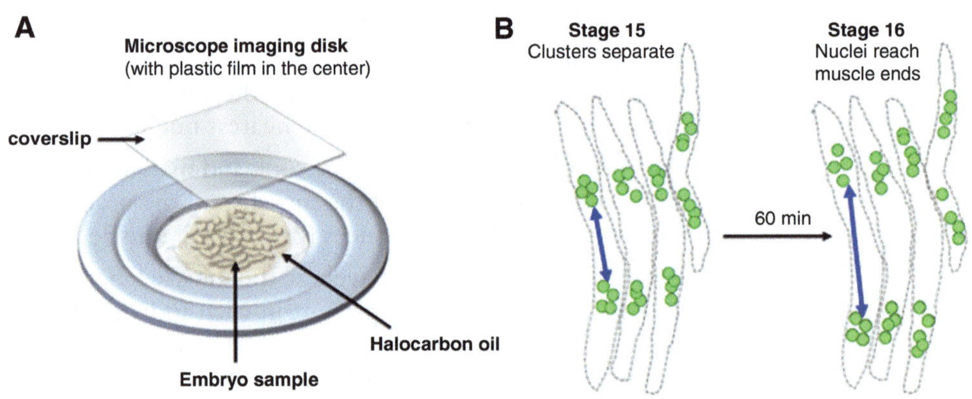

Fig. 3 Slide mounting and analysis for live *Drosophila* embryo imaging. (**A**) Cartoon depicting the standard setup used to mount *Drosophila* embryos for live-embryo time-lapse microscopy. Halocarbon is used to transfer embryos onto the plastic film in the center of the imaging disk. A coverslip is then placed on top of the embryos and taped to the imaging disk. (**B**) Analysis of nuclear position in live *Drosophila* embryos, with the nuclei in green and muscle border outline in gray. Separation speed between nuclear clusters is determined by measuring the shortest distance between the dorsal and ventral clusters at 0 min and again at 60 min

3.5 Analysis of ZASP Recruitment to the Nuclear Envelope

1. Add virgin flies that have *Twist-GAL4,apRed* on the second chromosome and *ZASP-GFP* on the third chromosome to a laying pot along with males of the desired control or LINC complex UAS-RNAi (*see* **Note 30**), and allow embryos to lay for 18–20 h at 25 °C.
2. To fix the embryos, follow **steps 2–11** in Subheading 3.1.
3. Add ~75–100 μL of water to the embryos (*see* **Note 31**).
4. To mount, follow **steps 32–35** in Subheading 3.2.
5. Before imaging select appropriately aged embryos (*see* **Note 32**).
6. Check to make sure the embryo is in the correct orientation before imaging with the lateral transverse (LT) muscles positioned within the middle of the embryo (*see* **Note 20**).
7. When imaging the embryos, ensure the zoom is adequate in order to visualize all LT muscles in a single hemisegment, and ensure that all the myonuclei are fully captured (*see* **Note 33**).
8. Import images to ImageJ or FIJI, and crop individual nuclei using the rectangle tool. Then make a maximum projection (*see* **Notes 23** and **34**).
9. In the cropped image, outline the nucleus, using the apRed signal as a guide, with the polygon tool.
10. Proceed to take one mean fluorescent reading in the apRed channel and another reading in the ZASP-GFP channel across the exact same area of the single nucleus (Fig. 4A, *see* **Note 35**).

3.6 Sarcomere Structure Analysis

1. Add virgin flies that have *Twist-GAL4,apRed* on the second chromosome and *ZASP-GFP* on the third chromosome to a laying pot along with males of the desired control or LINC complex UAS-RNAi (*see* **Note 30**), and allow embryos to lay for 18–20 h at 25 °C.
2. Remove the old agar plate from the laying pot and replace with a new plate covered with wet yeast.
3. Using a stereo microscope, pick first instar larvae (L1) from the agar plate with a fine-tipped paintbrush, and add the larvae directly onto a coverslip (*see* **Note 36**).
4. Add another coverslip on top of the L1 larvae (*see* **Note 37**), and tape this to a microscope slide.
5. Image the entire first ventral longitudinal (VL1) muscle (from top to bottom in the *Z*-plane) using the ZASP-GFP signal (*see* **Note 38**) at a step size of 0.5 μm.
6. Import images to ImageJ or FIJI. To standardize the depth of each muscle for analysis, use the rectangle tool to draw a box around the VL1 muscle and crop the image to a depth of

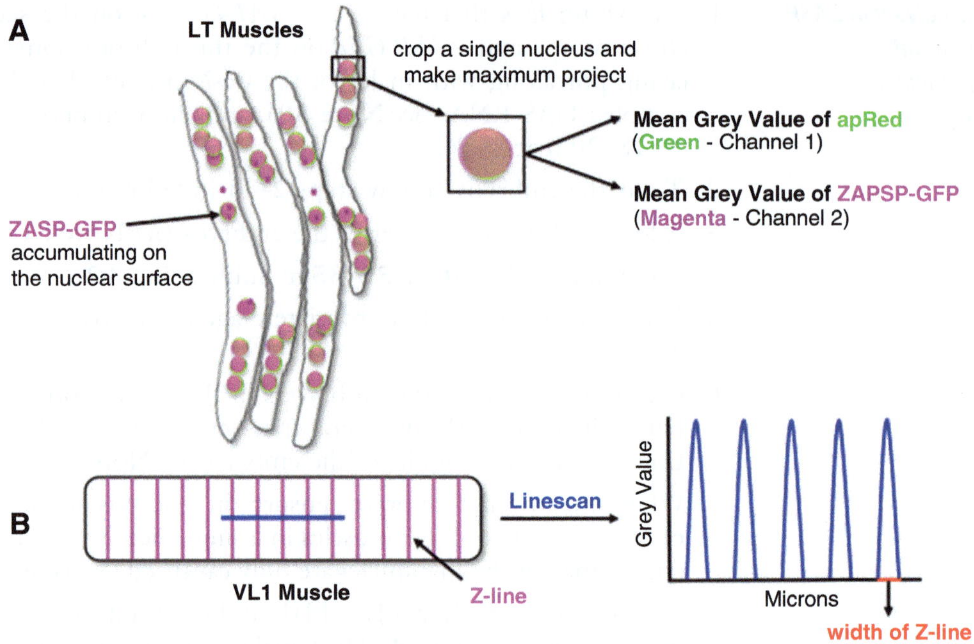

Fig. 4 Analysis of nuclear dependent sarcomere assembly and sarcomere structure. (**A**) ZASP-GFP (magenta) accumulation on the nuclei (green) in LT muscles. As the nuclei move back into the middle of the muscle, the ZASP-GFP will begin to form puncta on the nuclear surface and deposit these ZASP-GFP accumulations in the muscle. To analyze ZASP-GFP recruitment to the nucleus, each nucleus is cropped, one at a time, from all four LT muscles within one hemisegment and made into a maximum projection. The mean gray values for the apRed channel and ZASP-GFP channel can then be taken. (**B**) *Z*-lines formed by ZASP-GFP (magenta) in the VL1 muscle of a first instar larva. After a straight line has been drawn over the desired *Z*-lines, an intensity profile reading of the ZASP-GFP channel is taken. This will produce a graph showing the intensity profile over the distance of the line. From this graph, the raw values can be extracted to determine the width of each *Z*-line

2.5 μm from the top of the muscle, and then make a maximum projection (*see* **Notes 23**, **34**, and **39**).

7. To analyze the structure of the sarcomere take a line scan across five *Z*-lines in the center of the muscle using the straight line tool (Fig. 4B). This will then plot the fluorescent intensity profile over the distance of five *Z*-lines (*see* **Note 40**).
8. Press the list button on the intensity profile plot to list the position on the line in microns and the corresponding ZASP-GFP fluorescent intensity gray value for that position (*see* **Note 41**).

3.7 Larval Sample Collection

1. Set up an experiment with males and virgin females of interest together in a laying pot, and collect embryos as described in Subheading 3.1, **steps 1–6**.
2. Cover half of an agar plate with a thin layer of wet yeast (*see* **Note 42**).

3. Remove the mesh filter from the collection basket and place under a stereo microscope.
4. Use biology grade tweezers to pick stage 17 embryos and place in a line in the middle of the agar plate (*see* **Note 43**).
5. Leave the plate with the picked embryos at room temperature overnight.
6. The following day collect the hatched larvae with a paintbrush and transfer to a vial containing standard fly food.
7. Place the vial of larvae at 25 °C for 4 days allowing them to age to the third instar larvae (*see* **Note 44**).

3.8 Larval Dissection and Fixation

1. Add room temperature 15% sucrose solution to the vial of larvae, and allow larvae to float to the surface (*see* **Note 45**).
2. Pick larvae from the sucrose solution using a paintbrush and place on an agar plate.
3. Place six minuten pins in a Sylgard 182 plate per dissection being conducted.
4. Add a couple drops of ice-cold PIPES dissection buffer near the six minuten pins on the Sylgard plate.
5. Using a paintbrush, remove a larva from the agar plate and place within the ice-cold PIPES dissection buffer (*see* **Note 46**).
6. Position the larva dorsal side up, and place a pin between the trachea just posteriorly to the anterior spiracles (Fig. 5A).
7. Place a second pin between the trachea just anteriorly to the posterior spiracles.
8. Pin down the posterior of the larva such that the larva is stretched (*see* **Note 47**).
9. Using the McPherson-Vannas Scissors, make a cut perpendicular to the larval length posteriorly to the anterior pin, and make a second cut anteriorly to the posterior pin (Fig. 5A, *see* **Note 48**).
10. Next make a cut along the length of the larva between the trachea extending from one perpendicular cut to the other (*see* **Note 49**).
11. Using the tweezers, remove the intestines and adipose tissue without touching the musculature (*see* **Note 50**).
12. Once most of the intestines and adipose tissue have been removed, place a pin within each of the corners created by the perpendicular and lengthwise cuts making sure to stretch the epidermis to ensure the musculature is flat (Fig. 5A, *see* **Note 51**).
13. Remove the PIPES dissection buffer, and cover the larva with 10% formalin buffered solution

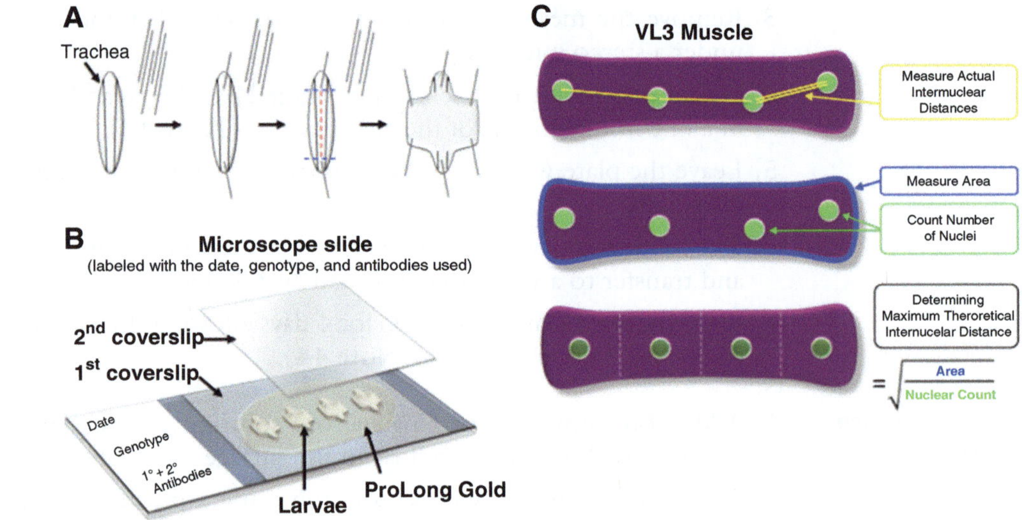

Fig. 5 Dissection, slide mounting, and analysis for *Drosophila* larvae. (**A**) Cartoon depicting the dissection process for *Drosophila* larvae. A larva is placed dorsal side up within ice-cold PIPES dissection buffer. Two pins are placed between the trachea as depicted. Two perpendicular cuts are made near the anterior and posterior pins (blue dotted lines). Then a lengthwise cut is made extending from one perpendicular cut to the other (red dotted line). Additional pins are placed in each of the corners created by the cuts, and the larva is stretched to ensure the larval musculature is flat. (**B**) Cartoon depicting the standard setup used to mount fixed *Drosophila* larvae on a microscope slide. Larvae are placed onto a coverslip dorsal side up and unfolded. Any excess liquid is removed and then the ProLong Gold mounting agent is added on top of the larvae. A second coverslip is placed on top, and the mounting agent is allowed to cure overnight before being taped directly to the microscope slide and imaged. (**C**) Analysis of nuclear position in *Drosophila* larvae as previously published [21] (used with permission of the publisher, The American Society for Cell Biology). The muscle is in magenta and the nuclei are green. First, the distance between the center of each nucleus and the center of its nearest neighbor are measured. Then the area of the muscle is measured and the number of nuclei are counted. To determine the theoretical maximum internuclear distance for each muscle, the square root of the area is divided by the number of nuclei present

14. Allow the dissected larva to fix for 20 min.
15. After 20 min remove the formalin and rinse the larva with PBS.
16. Remove the pins and transfer the larva to a 1.7 mL microcentrifuge tube containing ice-cold PBT-BSA using biology grade tweezers (*see* **Note 52**).
17. Repeat for a total of four larvae per genotype.
18. Store at 4 °C (*see* **Note 53**).

3.9 Larval Immunohistochemistry and Mounting

1. In a 600 μL microcentrifuge tube, prepare 300 μL of PBT-BSA for the primary antibody mixture. Dilute the mouse anti-α-tubulin antibody 1:200 (*see* **Note 54**).
2. Transfer dissected larvae using biology grade tweezers to the 600 μL microcentrifuge tube containing the primary antibody solution (*see* **Note 53**).

3. Place on a rotator to gently mix for 1 h.
4. After 1 h, remove the primary antibody solution, and store the antibody mixture at 4 °C (*see* **Note 55**).
5. Rinse the larvae three times with ~500 μL of PBT-BSA allowing the larvae to settle at the bottom of the tube before removing the PBT-BSA.
6. After three rinses, once again add ~500 μL of PBT-BSA and place on a rotator for 5 min. Repeat three times.
7. During the last wash, prepare 300 μL aliquots of PBT-BSA for the secondary antibody mixture. Dilute the anti-mouse secondary antibody 1:200, the Acti-stain 555 fluorescent phalloidin 1:400, and the Hoechst stain 1:1000 (*see* **Note 56**).
8. After washing, remove the remaining liquid, and dispense 300 μL of the secondary antibody mixture to each tube.
9. Place on a rotator to gently mix for 1 h making sure to limit exposure to light.
10. After 1 h, remove the secondary antibody mixture and discard.
11. Rinse the larvae three times with ~500 μL of PBT-BSA allowing the larvae to settle at the bottom of the tube before removing the PBT-BSA.
12. After three rinses, once again add ~500 μL of PBT and place on a rotator for 5 min. Repeat three times.
13. Under a stereo microscope, place a coverslip and microscope slide separated by a small coin (*see* **Note 18**).
14. Remove a dissected and stained larva using biology grade tweezers and place on the coverslip (Fig. 5B).
15. Unfold the dissected larva and place it dorsal side up (*see* **Note 57**).
16. Repeat for the remaining three dissected larvae placing them on the same coverslip.
17. Once all the dissected larvae are added to the coverslip, use a Kimwipe to remove the excess PBT while avoiding introducing bubbles under the larvae.
18. Add 75 μL of ProLong Gold to the coverslip, and gently place a second coverslip on top of the larvae.
19. Allow to cure overnight with limited light exposure before removing the small coin and taping the coverslips to the microscope slide.

3.10 Imaging and Analyzing Fixed Drosophila Larvae

1. Fixed larval imaging should be performed on a standard laser scanning confocal microscope.
2. Identify a third ventral longitudinal (VL3) muscle to image (*see* **Note 58**).
3. When imaging a VL3 muscle, ensure the entire muscle is being imaged (*see* **Note 59**).
4. Set the *z*-stack while in the 488 nm channel so that the mesh-like upper population of microtubules is fully imaged throughout the muscle (*see* **Note 60**).
5. Check in the 405 nm channel to ensure that all the nuclei in the muscle are fully imaged within the set *z*-stack.
6. All larval analysis should be done in ImageJ or FIJI. Import image files into ImageJ, and make each image a maximum projection.
7. Measure the area of the muscle by outlining the phalloidin staining with the polygon selection tool (Fig. 5C).
8. Using the straight line tool, measure the distance from the center of each nucleus to the center of the nearest nuclear neighbor. Additionally record the number of nuclei present.
9. Calculate the maximal internuclear distance by taking the square root of the muscle area divided by the nuclear count (*see* **Note 61**).
10. Determine the internuclear distance for the imaged muscle by averaging the internuclear distance measured in **step 8**.
11. Divide the internuclear distance by the calculated maximal internuclear distance to determine how evenly nuclei are positioned within the muscle.
12. Plot the ratio of actual to maximal internuclear distance obtained in **step 11**.

4 Notes

1. Bleach removes the chorion of the embryo.
2. Make sure to wet the mesh filter to allow bleach to pass through the basket. Otherwise, the basket will not drain and overflow.
3. Brush along the side of the collection basket in addition to the mesh filter, as many embryos will stick to the edge.
4. Heptane is necessary to make the embryos permeable to formalin for fixation.
5. Vortexing in a 1:1 heptane/methanol solution removes the vitelline membrane from the embryos.

6. Remove any embryos that remain floating in solution, as these embryos have not been properly devitellinized. Be careful to not disturb or remove embryos that have settled to the bottom. If embryos are accidently collected in the pipette during liquid removal, wait for embryos to settle in the bottom of the pipette, and add them back to tube. Let embryos settle back down to the bottom of the tube before removing remaining liquid again.
7. Fixed embryos can be stored in methanol at −20 °C up to 1 month.
8. If the number of total embryos collected in the 1.7 mL microcentrifuge tube is over the 0.1 mL mark, divide the fixed embryos up into two 600 microcentrifuge μL tubes to allow for an even distribution of antibody during staining. These embryos will be recombined onto the same slide during mounting in **step 32**.
9. The rabbit anti-DsRed antibody will label a subset of nuclei in the *Drosophila* embryo through the use of the apRed transgene. Flies that carry apRed will express a nuclear localization signal fused to the fluorescent protein DsRed downstream of the *apterous* mesodermal enhancer. This results in the specific labeling of the nuclei within the lateral transverse muscles of the *Drosophila* embryo. Rat anti-tropomyosin is used to label the muscles of *Drosophila* embryos since the actin epitope is destroyed during the fixation process. Although any antibody should be suitable, we find that the rabbit anti-DsRed from Clontech (632496) and then mouse anti-tropomyosin from Abcam (ab50567) are the most consistent when used for staining *Drosophila* embryos.
10. Primary antibodies may be saved and used again one additional time. After second use, discard antibody mixture in waste.
11. Although any secondary antibody should be suitable, we find that the Alexa Fluor 555 donkey anti-rabbit and Alexa Fluor 488 donkey anti-rat both from Life Technologies are the most consistent when used for staining *Drosophila* embryos.
12. If no additional antibodies will be used (such as mouse anti-GFP), proceed to **step 27** and wash in PBT.
13. Mutant alleles that are carried over a fluorescent balancer (DGY) are identified using an additional mouse anti-GFP antibody. However, to avoid cross reactively with the rat anti-tropomyosin antibody, the mouse anti-GFP antibody has to be added separately. Although any GFP antibody should be suitable, we find that the mouse anti-GFP from Developmental Studies Hybridoma Bank (GFP-G1) is the most consistent when used for staining *Drosophila* embryos.

14. If embryos are on the rotator for 1 h at room temperature, staining and mounting have to be completed the same day. Let embryos stain overnight at 4 °C if staining and mounting will be complete the next day.
15. Although any secondary antibody should be suitable, we find that the Alexa Fluor 647 donkey anti-mouse IgG from Life Technologies is the most consistent when used for staining *Drosophila* embryos.
16. The tape bridge ensures that the embryos are not crushed when another coverslip is added to the top of the embryos, as shown in Fig. 2A.
17. To transfer embryos/mounting media more easily, cut the end of the pipette tip off. Remember to transfer both sets of embryos for each genotype to the appropriate slide.
18. The coin prevents the coverslips from adhering to the slide and allows for even distribution of the mounting agent as it cures. Wait at least a day before imaging the slide to ensure the mounting agent has fully cured.
19. In *Drosophila* embryos, the gut autofluoresces at 488 nm making it a useful marker to select for embryos that are properly staged. At stage 16 the gut becomes segmented and is divided into distinct sections. Prior to stage 16 (stage 15 and earlier), the gut appears as a single lobe with no segmentation.
20. If the embryo is rolled over, the ends of the LT muscles may not be fully visible. Therefore, length and distance measurements will not be accurate. Only image embryos where the LT muscles are laying flat within the center body.
21. Using a 40× objective, a zoom of 1× is sufficient. Select four hemisegments that are positioned within the center of the embryo. Avoid imaging hemisegments at the very ends of the embryo, as the LT muscles in these hemisegments are often smaller, causing nuclei to crowd. It is important to keep microscope settings the same between embryos and genotypes.
22. The selected *z*-stack should not exceed 15 μm total (30 *z*-slices at 0.5 μm step intervals). If it does, the LT muscles may be too close to the edge.
23. To make a max projection, select the following commands: Image > Stacks > Z Project… > Max Intensity.
24. This orientation ensures that the dorsal side of the embryo is up, ventral side is down, anterior side is on the left, and posterior side is one the right, as shown in Fig. 2B. To achieve proper orientation, the image may have to be flipped horizontally and/or vertically. To flip the image horizontally, select the commands: Image > Transform > Flip Horizontally. To flip the image vertically, select the commands: Image > Transform > Flip Vertically.

25. Do not take any measurements from hemisegments that have missing or extra LT muscles.
26. The average of the average values obtained will be the values that are plotted. Thus, each data point indicates the average distance measurement within a single embryo.
27. Halocarbon oil allows for the diffusion of oxygen and thus prevents dehydration and hypoxia over the course of a few hours. Additionally, halocarbon oil has good optical properties due to its refractive index, which is similar to the refractive index of most oils used for confocal objectives, making it ideal for imaging embryos over time.
28. More halocarbon oil may be needed to collect any remaining embryos in the collection basket. Clean the stick before getting more halocarbon oil.
29. For live imaging, no muscle marker is present, and only the DsRed within the nuclei is expressed. Additionally, since the embryo is developing, it will occasionally move and drift over the course of the time-lapse. Therefore, set the *z*-stack a bit wider to ensure all nuclei within the LT muscles to be imaged remain within the field of view.
30. The nuclei of the LT muscles are labeled with the apRed marker. This marker is a DsRed tag fused to a nuclear localization signal that is downstream of the mesodermal enhancer *apterous,* which is a transcription factor, only expressed in the LT muscles. The *Twist-GAL4* is required to drive the expression of the RNAi construct specifically in the mesoderm. *ZASP-GFP* is a *Z*-line protein that localizes to myonuclei at late stage 15.
31. The amount of water added depends on the number of embryos. Add more water if the embryos cannot be aspirated properly using a micropipette.
32. ZASP-GFP accumulates on nuclei at late stage 15 with nuclei in the LT muscles being visible in the ZASP-GFP channel until early/mid stage 17, just prior to sarcomere formation. It is possible to analyze these nuclei between these stages. At stage 16 the gut becomes segmented and is divided into distinct sections. Prior to stage 16, the gut appears as a single lobe with no segmentation. At early stage 17, nuclei are starting to become dispersed throughout the muscle. Also note that imaging is performed on the fluorescence of GFP and DsRed and is not indirect immunofluorescence.
33. Using a 40× objective, a zoom of 2× is sufficient. It is also important to keep microscope settings the same between different embryos and genotypes so intensities can be compared accurately.

34. Use the apRed channel as a guide to crop a single nucleus as ZASP-GFP can extend beyond the nuclear rim, as shown in Fig. 4A. To crop an image after a rectangle has been drawn around a single nucleus, go to Image > Duplicate..., and then select the number of slices in the *z*-stack that is necessary to crop the entire nucleus from top to bottom.
35. Mean gray values can be taken by using the following commands in ImageJ: Analyze > Set Measurements... > and select mean gray value. The apRed signal is used as an internal control in order to normalize the ZASP-GFP signal. To normalize the values, divide the ZASP-GFP mean fluorescent value by the apRed mean fluorescent value.
36. L1 larvae will be the only larvae visibly crawling in the agar after a 20 h incubation.
37. The additional top coverslip halts larval crawling.
38. The VL muscle is located at the base of the LT muscles. Although the coverslip on top of the larva will halt crawling, larval muscle contraction can still occasionally occur. Therefore, it is important to only image VL muscles that are not contracting.
39. The VL muscle size can vary in the *Z*-plane. This step ensures that the same region of each VL muscle is analyzed.
40. Once the line has been drawn over the desired five *Z*-lines a line scan can be performed in ImageJ by going to Analyze > Plot Profile.
41. Measuring the width of the base of each peak on the intensity profile in microns correlates to the width of the corresponding *Z*-line as shown in Fig. 4B.
42. Adding too much yeast can make it difficult to find the hatched larvae the next day. In order to create a thin layer of yeast, add a small amount of yeast, and use a Kimwipe to spread the yeast over half of the agar plate.
43. Stage 17 embryos can be identified by the presence of trachea in the embryo. Placing embryos in a line within the middle of the agar plate will make it easier to count the number of embryos that hatched into larvae.
44. If larvae have started to crawl up the sides of the vials after 4 days, the larvae are too old to be dissected, and incubation time should be adjusted for further experiments.
45. The 15% sucrose solution is used to allow the larvae to float to the surface so they can be easily collected. If larvae remain stuck within the fly food, a microspatula can be used to break up the food.
46. Once the larva is placed in the PIPES dissection buffer, **step 13** should be reached within 5 min in order to ensure that the muscle tissue does not begin to break down.

47. The larva should be stretched so that it is unable to move laterally. If the epidermis is ripped by the minuten pin, the larva is stretched too far.
48. Do not completely cut through the larva when making the perpendicular cuts as a small amount of uncut epidermis will be needed to keep the larva stretched during fixation.
49. When making the lengthwise cut, make sure not to touch the bottom of the larva where the muscles are located as scissors will damage the muscles.
50. If not all the intestines and adipose tissue can be removed during this step, small pieces can be more easily removed after **step 12**. Additionally, if the drop of PIPES dissection buffer becomes filled with intestines and adipose tissue, the PIPES dissection buffer can be removed and replaced with fresh ice-cold PIPES dissection buffer.
51. Any additional intestines and adipose tissue can be removed more easily once the larva is fully pinned down.
52. When transferring the fixed larva, make sure to only touch the non-dissected head or tail of the larva to avoid damaging the muscle tissue.
53. Fixed larvae can be stored at 4 °C overnight.
54. Although any tubulin antibody should be suitable to label the microtubules, we find that the mouse anti-α-tubulin from Sigma-Aldrich (T6199) is the most consistent when used for staining in *Drosophila* larva.
55. Primary antibodies may be saved and used again two additional times. After third use, discard antibody mixture in waste.
56. Although any secondary antibody should be suitable, we find that the Alexa Fluor 488 donkey anti-mouse IgG from Life Technologies is the most consistent when used for staining *Drosophila* larvae. For labeling the nuclei and actin network in *Drosophila* larvae, we find that Hoechst 33342 from Life Technologies and Acti-stain 555 fluorescent phalloidin from Cytoskeleton (PHDH1-A) are the most consistent.
57. In order to avoid damaging the muscles of interest, only grab the larva by the locations where the larva was pinned when unfolding. To facilitate unfolding a small amount of PBT can be added to the larvae.
58. The VL3 muscle is positioned near the center of the ventral region of the larvae. The ventral oblique muscles 4–6 will form a V-like pattern along the center of the dissected larva. The ventral longitudinal muscles are a set of four longitudinal muscles numbered such that VL1 is furthest from the ventral oblique muscles and VL4 is closest to the ventral oblique muscles.

59. If the VL3 muscle is unable to fit within a single imaging frame, multiple tiles can be imaged and stitched together.
60. Since nuclei are one of the major microtubule organizing centers in muscle, ensuring this population of microtubules is full imaged should ensure that all nuclei from the muscle are within the *z*-stack. Also, rips in this microtubule meshwork can be a sign of mechanical damage during dissection. Therefore, any muscles with rips in this microtubule population should not be imaged.
61. The maximal internuclear distant represents how far apart nuclei should be if they are fully maximizing their distance between one another.

References

1. Chang W, Worman HJ, Gundersen GG (2015) Accessorizing and anchoring the LINC complex for multifunctionality. J Cell Biol 208:11–22. https://doi.org/10.1083/jcb.201409047
2. Meinke P, Nguyen TD, Wehnert MS (2011) The LINC complex and human disease. Biochem Soc Trans 39:1693–1697. https://doi.org/10.1042/BST20110658
3. Meinke P, Schirmer EC (2015) LINC'ing form and function at the nuclear envelope. FEBS Lett 589:2514–2521. https://doi.org/10.1016/j.febslet.2015.06.011
4. Luxton GWG, Gomes ER, Folker ES et al (2010) Linear arrays of nuclear envelope proteins harness retrograde actin flow for nuclear movement. Science 329:956–959. https://doi.org/10.1126/science.1189072
5. King MC, Drivas TG, Blobel G (2008) A network of nuclear envelope membrane proteins linking centromeres to microtubules. Cell 134:427–438
6. Zhang X, Lei K, Yuan X et al (2009) SUN1/2 and Syne/Nesprin-1/2 complexes connect centrosome to the nucleus during neurogenesis and neuronal migration in mice. Neuron 64:173–187. https://doi.org/10.1016/j.neuron.2009.08.018
7. Elhanany-Tamir H, Yu YV, Shnayder M et al (2012) Organelle positioning in muscles requires cooperation between two KASH proteins and microtubules. J Cell Biol 198:833–846. https://doi.org/10.1083/jcb.201204102
8. Zhang Q, Bethmann C, Worth NF et al (2007) Nesprin-1 and -2 are involved in the pathogenesis of Emery Dreifuss muscular dystrophy and are critical for nuclear envelope integrity. Hum Mol Genet 16:2816–2833. https://doi.org/10.1093/hmg/ddm238
9. Nagano A, Koga R, Ogawa M et al (1996) Emerin deficiency at the nuclear membrane in patients with Emery-Dreifuss muscular dystrophy. Nat Genet 12:254–259. https://doi.org/10.1038/ng0396-254
10. Bonne G, Di Barletta MR, Varnous S et al (1999) Mutations in the gene encoding lamin A/C cause autosomal dominant Emery-Dreifuss muscular dystrophy. Nat Genet 21:285–288. https://doi.org/10.1038/6799
11. Puckelwartz MJ, Kessler E, Zhang Y et al (2009) Disruption of nesprin-1 produces an Emery Dreifuss muscular dystrophy-like phenotype in mice. Hum Mol Genet 18:607–620. https://doi.org/10.1093/hmg/ddn386
12. Bione S, Maestrini E, Rivella S et al (1994) Identification of a novel X-linked gene responsible for Emery-Dreifuss muscular dystrophy. Nat Genet 8:323–327. https://doi.org/10.1038/ng1294-323
13. Wilson MH, Holzbaur ELF (2012) Opposing microtubule motors drive robust nuclear dynamics in developing muscle cells. J Cell Sci 125:4158–4169. https://doi.org/10.1242/jcs.108688
14. Cadot B, Gache V, Vasyutina E et al (2012) Nuclear movement during myotube formation is microtubule and dynein dependent and is regulated by Cdc42, Par6 and Par3. Nat Publ Group 13:741–749. https://doi.org/10.1038/embor.2012.89
15. Iyer SR, Shah SB, Valencia AP et al (2016) Altered nuclear dynamics in MDX myofibers. J Appl Physiol 122:470–481. https://doi.org/10.1152/japplphysiol.00857.2016

16. Oddoux S, Zaal KJ, Tate V et al (2013) Microtubules that form the stationary lattice of muscle fibers are dynamic and nucleated at Golgi elements. J Cell Biol 203:205–213. https://doi.org/10.1083/jcb.201304063
17. Metzger T, Gache V, Xu M et al (2012) MAP and kinesin-dependent nuclear positioning is required for skeletal muscle function. Nature 484:120–124. https://doi.org/10.1038/nature10914
18. Folker ES, Schulman VK, Baylies MK (2012) Muscle length and myonuclear position are independently regulated by distinct Dynein pathways. Development 139:3827–3837. https://doi.org/10.1242/dev.079178
19. Folker ES, Schulman VK, Baylies MK (2014) Translocating myonuclei have distinct leading and lagging edges that require kinesin and dynein. Development 141:355–366. https://doi.org/10.1242/dev.095612
20. Auld AL, Folker ES (2016) Nucleus-dependent sarcomere assembly is mediated by the LINC complex. Mol Biol Cell 27:2351–2359. https://doi.org/10.1091/mbc.E16-01-0021
21. Collins MA, Mandigo TR, Camuglia JM et al (2017) Emery-Dreifuss muscular dystrophy-linked genes and centronuclear myopathy-linked genes regulate myonuclear movement by distinct mechanisms. Mol Biol Cell. https://doi.org/10.1091/mbc.E16-10-0721
22. Richardson BE, Becket K, Nowak SJ, Baylies MK (2007) SCAR/WAVE and Arp2/3 are crucial for cytoskeletal remodeling at the site of myoblast fusion. Deveolpment 134: 4357-67. https://doi.org/10.1242/dev.010678
23. Barton LJ, Pinto BS, Wallrath LL, and Geyer PK (2016) The Drosophila nuclear lamina protein otefin is required fro germline stem cell survival. Dev Cell 25: 645-54. https://doi.org/10.1016/j.devcel.2013.05.023

Chapter 15

Computational Methods for Studying the Plant Nucleus

Axel Poulet, Xiao Zhou, Kentaro Tamura, Iris Meier, Christophe Tatout, Katja Graumann, and David E. Evans

Abstract

The analysis of nuclear envelope components and their function has recently been progressed by the use of computational methods of analysis. The methods in this chapter provided by members of the International Plant Nucleus Consortium address the identification of novel nuclear envelope proteins and the study of structure and mobility of the nucleus. DORY2 is an upgrade of the KASH-finder DORY, and NucleusJ is used to characterize the three-dimensional structure of the nucleus in light microscope images. Finally, a method is provided for analysis of the migration of the nucleus, a key technique for exploring the function of plant nuclear proteins.

Key words Higher plant, Nucleus, ImageJ, Nucleus, Nuclear migration, Nuclear structure

1 Introduction

The International Plant Nucleus Consortium (http://bms.brookes.ac.uk/ipnc; [1]) is an international group of researchers with interests in developing knowledge and techniques to describe the nature and function of the proteins of the nuclear envelope and their interaction partners. The collection of computational techniques in this chapter has been contributed by members of the consortium.

1.1 DORY2

DORY2 is an upgrade of the DORY program used by Zhou et al. [2] to identify plant KASH proteins. In comparison to DORY, which requires manual confirmation using BLASTP to confirm the predicted KASH proteins, DORY2 utilizes the local-installed BLAST+ program to fully automate the process. DORY2 also implemented multithreading and better search algorithms to improve the search efficiency. Similar to DORY (described in Zhou et al. [3]), DORY2 contains two functional units—the KASHFilter and the HomologyFilter. The KASHFilter collects protein

Gregg G. Gundersen and Howard J. Worman (eds.), *The LINC Complex: Methods and Protocols*, Methods in Molecular Biology, vol. 1840, https://doi.org/10.1007/978-1-4939-8691-0_15,

sequences that contain a putative KASH domain, and the HomologyFilter divides these protein sequences into homologous groups (Fig. 1). DORY2 can then automatically obtain homologs of each homologous group from a local nr database and generate a final report file (Fig. 1).

1.2 NucleusJ

The 3D nucleus is a good model for the development of methods such as the spatial analysis of 3D images. It is a spatial structure often described as a sphere, and alteration of its size and shape has been shown in specific tissue (placental trophoblast in human, root hair cells in plants; [4]), affected by ploidy levels (guard cells and pavement cells in plant epidermis) and to be linked to human diseases [5] or increased sensitivity to DNA damage [6]. NucleusJ

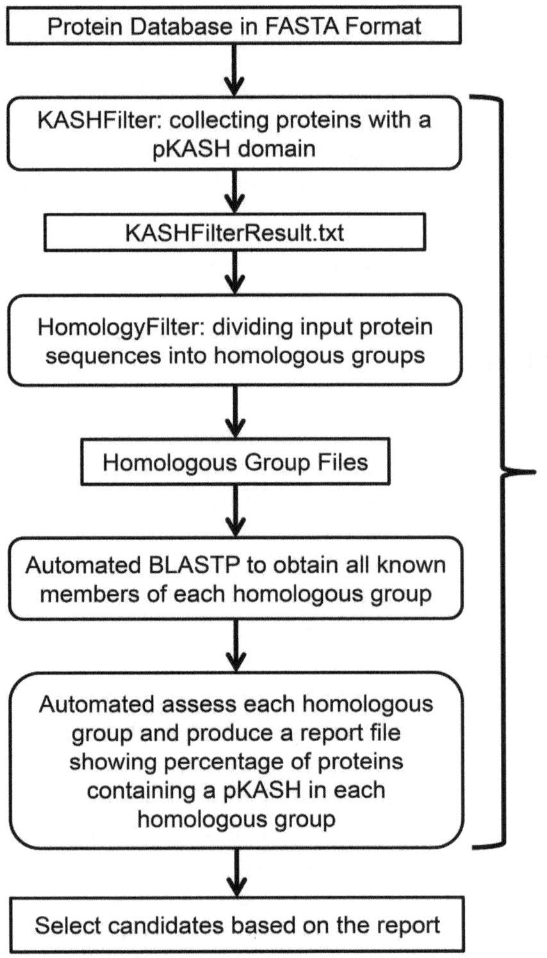

Fig. 1 Flowchart describing the operation of the DORY2 program. DORY2 contains two functional units, the KASHFilter and the HomologyFilter. The KASHFilter collects protein sequences that contain a putative KASH domain, and the HomologyFilter divides these protein sequences into homologous groups

is a simple and user-friendly ImageJ plugin dedicated to the characterization of nuclear morphology and chromatin organization in 3D [7]. Starting from image stacks, the nuclear boundary is delimited by the segmentation method developed. Chromocenters are segmented by partitioning the nucleus using a 3D watershed algorithm and by manual thresholding a contrast measure over the resulting regions. As output, NucleusJ quantifies parameters including shape and size of nuclei as well as intranuclear objects and their position in the nucleus.

Computational image analysis provides precise, objective, and reproducible quantitative data from images. The NucleusJ plugin has been developed to provide 3D quantitative measurements from single images or large datasets, without requiring expertise in image analysis. It is available on ImageJ and Fiji (an image processing package) web sites with documentation for the user (http://imagejdocu.tudor.lu/doku.php?id=plugin:stacks:nuclear_analysis_plugin:start).

1.3 Nuclear Migration Assay

Plant nuclei rapidly move over a large distance along the actin-myosin network [4, 8]. This nuclear movement is supported by the LINC complex, which consists of multi-protein complexes at the nuclear envelope. To better understand the function of each component of the nucleocytoplasmic linker and characterize the nuclear movement, a nuclear migration assay can be used. The *Arabidopsis* root epidermis proves an ideal model system for tracking nuclei in living cells. We describe a method to quantify nuclear movement in these cells. After obtaining the time-lapse images, a Fiji software equipped with a simple plugin allows tracking of each nucleus semi-manually and quantification of the distance that each nucleus of interest has travelled.

2 Materials

2.1 DORY2

2.1.1 Software

1. Updated Java runtime environment (http://www.oracle.com/technetwork/java/javase/downloads/jre8-downloads-2133155.html) and development kit (http://www.oracle.com/technetwork/java/javase/downloads/jdk8-downloads-2133151.html).
2. DORY2 (http://sourceforge.net/projects/doryforkash/).
3. Stand-alone BLAST+ program (https://blast.ncbi.nlm.nih.gov/Blast.cgi?PAGE_TYPE=BlastDocs&DOC_TYPE=Download). Choose the 64-bit version suitable for your operation system.
4. Transmembrane domain prediction web service: Phobius (http://phobius.sbc.su.se/).

5. Multiple sequence alignment web service: MAFFT (http://mafft.cbrc.jp/alignment/server/).
6. Alignment view tool: Jalview (http://www.jalview.org/Web_Installers/install.htm).

2.1.2 Database Files

1. FASTA format protein database to search for KASH proteins. The FASTA format of the non-redundant protein database (nr) can be downloaded at ftp://ftp.ncbi.nlm.nih.gov/blast/db/FASTA/nr.gz.
2. nr database for BLAST+ (not the FASTA format one). This database can be downloaded at ftp://ftp.ncbi.nlm.nih.gov/blast/db/. The database file has a name pattern: "nr.dd.tar.gz," and dd stands for two digits. It can also be installed using the "update_blastdb.pl" provided in the BLAST+ installation folder. More information can be found in the BLAST+ help file: https://www.ncbi.nlm.nih.gov/books/NBK1762/.

2.1.3 Hardware

1. Due to the heavy-duty task and large data volume of the nr database, a 64-bit computer equipped with a multi-core CPU, 16GB memory, and a fast solid state hard drive is required.

2.2 NucleusJ

A computer with ImageJ or Fiji installed and updated and with at least 2GB RAM available for the program. Download NucleusJ and dependencies (jama, MorpholibJ_ and imagescience; https://github.com/PouletAxel/NucleusJ_/releases/tag/v1.0.2 jama.jar: http://math.nist.gov/javanumerics/jama/Jama-1.0.3.jar : https://github.com/ijpb/MorphoLibJ/releases imagescience.jar: http://www.imagescience.org/meijering/software/download/imagescience.jar) in your ImageJ plugins folder, and then restart ImageJ or simply apply the command Help>Refresh Menus.

2.3 Nuclear Migration Assay

1. Murashige and Skoog and gellan gum medium (for Arabidopsis seedlings): 4.6 g/L Murashige and Skoog, 1% sucrose, in H_2O, pH 5.7 (adjust with KOH), 0.5% gellan gum.
2. 35 mm glass-based dishes with a cover glass (0.16–0.19 mm thickness) affixed on the bottom surface to enable high-magnification observations with a fluorescence microscope.
3. One- to 2-week-old transgenic *Arabidopsis* seedlings stably expressing a nucleus-targeted fluorescent protein (e.g., histone "B-GFP, Nup50a-GFP; [8]).
4. Confocal laser scanning microscope.
5. FIJI software (https://fiji.sc/) with Manual Tracking plugin (https://imagej.nih.gov/ij/plugins/track/track.html) and Microsoft Excel for data analysis.

3 Methods

3.1 DORY2

3.1.1 Install Software

1. Java, BLAST+, and Jalview can be downloaded and installed according to the instructions on the corresponding web sites.
2. DORY2 can run directly once Java is installed.

3.1.2 Configure DORY2

The parameters of DORY2 can be configured in the friendly graphic user interface. The parameters are self-explanatory, and the default values work very well. The explanation below will help to configure DORY2 for your specific usage.

1. *TMD Frame Length*, the amino acid length of a transmembrane domain.
2. *TMD Hydrophobic Threshold*, the hydrophobic threshold of a transmembrane domain. DORY2 slides a frame of *TMD Frame Length* along the protein sequence; if the hydrophobic value within this frame is no less than the *TMD Hydrophobic Threshold*, then it is considered a transmembrane domain (*see* **Note 1**).
3. *Maximum KASH Tail Length*, the maximum length of a KASH tail domain.
4. *Minimum KASH Tail Length*, the minimum length of a KASH tail domain.
5. *Protein Length Cutoff from to* is used to restrict the search to a certain protein length, i.e., proteins beyond the set length will be ignored during the search.
6. *Regex for KASH Tail*, the pattern (expressed in regular expression) that the KASH tail should follow. The C-terminal four-amino acid pattern can be summarized from known KASH proteins, and some presets can be chosen from the drop-down menu. To customize "Regex for KASH Tail," knowledge of regular expression is needed, and the details can be found at http://www.regular-expressions.info. Some basics of regular expression are explained below.

Symbol	Match
\S	Any non-space characters
\S+	One or multiple non-space characters
[...]	Any character listed inside the square brackets
\Z	The end of a sequence
^	Not containing the following character

7. *During KASHFilter search, keep the proteins whose protein names contain* checkbox is used to search for proteins that belong to specific species. Click "*Choose Species Name File (one line one name)*," and choose the text file that contains your desired species names. In this text file, each line should contain only one species name.
8. *Output potential KASH tail in a file during the KASHFilter search* checkbox is used to output identified KASH tails for further analysis. If this is checked, identified KASH tails will be saved in a file named "KASHTail.txt." If "*In the output file, left pad KASH tail to the Maximum KASH Tail Length*" is also checked, the sequences will be right aligned for easy view of the last four amino acids.
9. *Query NCBI Taxonomy Browser to filter non-eukaryotic proteins out*: if this is checked, DORY2 will read the species name from the protein (*see* **Note 2**) and send a request of checking this species name to the NCBI Taxonomy server. If the response text contains the text set in "*Being positive, server return text should contain*" textbox, then this protein will be kept for further analysis; otherwise, it will be ignored.
10. *E-value Cutoff*, the E-value cutoff used by BLASTP program to consider whether two proteins are homologs.
11. *Run full search*, *Run KASHFilter only*, and *Run HomologyFilter only*. If *Run full search* is chosen, then both KASHFilter and HomologyFilter will run. If you already have the previous "KASHTailResult.txt," you can choose *Run HomologyFilter only*.
12. *Perform a final BLASTP for identified KASH candidates against the nr database*. If this is checked after HomologyFilter, the first protein sequence from each homology group will be used to BLAST against the nr database to obtain all homologs that exist in the nr. These "complete homolog" groups will be saved in a new folder. DORY2 will then run KASHFilter again through all the groups in this folder to count total proteins and proteins with positive KASH tails. A final report will be generated. From this report, it is very easy to see whether the KASH tail is well conserved in each homology group (*see* **Note 3**).

3.1.3 Run DORY2

1. Set up proper parameters and then click *Run*. If the database files are not chosen, the Run button will not be enabled (*see* **Note 4**).
2. *See* result in a folder named "DORY-search tag," and a log file will be generated too which contains all the settings and error messages.
3. Click *Cancel* to cancel this run.

3.1.4 Interpret the Report File

After the final BLASTP against the nr database, a report file will be generated. In this file you can easily see the total number of proteins and the number of positive KASH proteins in each homology group. If majority of the proteins in a homology group contains a predicted KASH tail, then this group is likely a real KASH protein family (*see* **Note 5**).

3.1.5 Manual Confirmation of the Candidates

If you choose not to use the "*Perform a final BLASTP for identified KASH candidates against nr database*" function, you need to confirm the candidates manually after the HomologyFilter.

1. Choose one or more proteins from each homology group, and perform BLASTP against the non-redundant protein sequences using BLASTP web service (https://blast.ncbi.nlm.nih.gov/Blast.cgi). The "Organism" parameter can be set to "Eukaryota," because KASH proteins are specific to eukaryotes. "Expect threshold" should be set at 1e-4 or lower. However, if the threshold is too low, real homologs will be lost. "Max target sequences" can be started at 500 and increased to a higher number if the maximum target sequence number is reached at the first round of BLAST. Click the "BLAST" button to run.
2. In the result webpage, click "All" in the "Select" section. Uncheck unwanted sequences if necessary (*see* **Note 5**). In the "Download" drop-down menu, choose "FASTA (compete sequence)" to obtain the protein sequences of the selected homologs.
3. The presence of a C-terminal TMD can be tested using the online service Phobius [9]. Transmembrane domain predictions of Phobius may not be accurate, but it will provide an overview of whether a C-terminal TMD is conserved in most of the input homologs. If a C-terminal transmembrane domain is conserved, then check whether the C-terminal four amino acids of the majority of this homologous group follow the pattern set in the *Regex for KASH Tail*. If yes, then this homologous group is positively predicted to be a KASH protein family (*see* **Note 6**).

3.1.6 Improve the "Regex for KASH Tail"

A new pattern of the C-terminal four amino acids may be summarized from the proteins believed to be KASH proteins, especially when a member from this homology group is experimentally confirmed to be a KASH protein. This new pattern should be used as an improved "*Regex for KASH Tail*" to perform a new round of search (*see* **Note 6**).

3.1.7 Use DORY for Other Purpose

Similar to the identification of KASH proteins, DORY can also be used to identify proteins that contain TMDs followed by a short conserved C-terminal sequence. The source code can be modified to perform searches of interests.

3.2 NucleusJ

3.2.1 Images Used by NucleusJ

Three-dimensional light microscope image data is used in NucleusJ as Input. During image acquisition, aim for a high quality of the image (12, 16, or 32 bits), with optimal voxel calibration, to obtain the resolution required to characterize the nucleus, chromocenter, or other intranuclear objects. For example, images acquired with a structured illumination microscopy (Leica MAAF) using a x63 oil objective possess a voxel calibration equal to xy = 0:103 μm and z = 0:2 μm. These parameters are computed according to the maximal theoretical resolution of the microscope and the sampling theory of Shannon and Nyquist [10, 11]. To be used in NucleusJ, the images must be 8-bit and have only one nucleus and one channel per image. If they are not formatted in this way, some preprocessing is required (*see* below).

3.2.2 Image Preprocessing

1. If the image has several nuclei, crop each nucleus as one individual image, and if the image has several channels, select the channel of interest during the crop. The nucleus needs to be complete in the image, and it is better to have areas above and below the nucleus without fluorescence.
   (a) To crop in ImageJ, select the region of interest, and apply the command Image>FDuplicate (ctrl+shift+D); then in the pop-up windows, choose the z-coordinate of the whole nucleus.
2. Adjusting sub-image:
   (a) Extend histogram dynamic to avoid the saturation of some voxels during the image conversion (Image>Adjust> Brightness and Contrast or ctrl+shift+C).
   (b) Select 3D Gaussian blur (Process>Filters>Gaussian blur 3D). If the image calibration is 0.1 × 0.1 × 0.2 μm or with the same ratio between x and x-axis, the 3D Gaussian blur should be x = 1, y = 1, and z = 0.5
   (c) Convert the image to 8-bit and save it as a .tif file.

3.2.3 NucleusJ Plugins

1. **Step 1: Nuclear Segmentation**
   (a) Four different processes are required to analyze a single or multiple images (batch analysis):
   - **Nucleus Segmentation**: this process uses as input an opened image, and the image result is displayed on the screen.
   - **Nucleus Segmentation (Batch)**: before running the plugin, a WorkDirectory dedicated to a given analysis should be created by the user. Raw images are then to be saved in a new sub-directory created by the user and named RawDataNucleus. The result of the segmentation process is saved automatically in a new sub-directory created by the plugin and called SegmentedDataNucleus.

- **Nucleus Segmentation and Analysis**: this process uses as input an opened image. The image results are displayed on the screen, and results of the analysis are shown in the ImageJ log window.
- **Nucleus Segmentation and Analysis (Batch)**: before running the plugin, a WorkDirectory dedicated to a given analysis should be created. Raw images are then to be saved in a new sub-directory created by the user and named hereafter RawDataNucleus. The image results of the segmented nuclei are automatically saved in the SegmentedDataNucleus sub-directory in the main WorkDirectory. The results of the analysis are saved in two tabulated files named 3DNucleiParameters.tab and 2DNucleiParameters.tab.

(b) When using **Nucleus Segmentation**, a pop-up window appears, and the following parameters need to be filled out:

- **Work directory and raw data choice**
  - *Raw Data*: choose the WorkDirectory which contains the raw images saved in a single sub-directory. In this documentation, this sub-directory is called RawDataNucleus.
  - *Output Directory*: choose the WorkDirectory in which the results are to be stored. This directory must contain the RawDataNucleus sub-directory containing the raw images.
- **Voxel Calibration** corresponds to the voxel calibration used during the image acquisition:
  - *x*: width of voxel—*default value = 1.*
  - *y*: height of voxel—*default value = 1.*
  - *z*: depth of voxel—*default value = 1.*
  - *unit*: length unit (µm, ...)—*default value = pixel.*
- **Choose the minimum and maximum volume of the nucleus**, only objects with a volume between the minimum and the maximum allowed volume will be segmented:
  - *Minimum volume of the segmented nucleus: default value = 15.*
  - *Maximum volume of the segmented nucleus: default value = 2000.*
- **How many CPU** number of CPU (central processing unit) used for image segmentation.

Once the START button is pressed, the program will create a new sub-directory called SegmentedDataNucleus which contains the image of the segmented nuclei.

(c) The parameters of Nucleus Segmentation and Analysis are the same as for Nucleus Segmentation:

- **2D and 3D:** Two output files are created in the work directory 2DNucleiParameters.tab and 3DNucleiParameters.tab.
- **3D:** 3DNucleiParameters.tab is created in the work directory.
- **2D:** 2DNucleiParameters.tab is created in the work directory.

The sub-directory SegmentedDataNucleus, result file, and log file are created in the main WorkDirectory (*see* also the example section of this documentation).

2. **Step 2: Chromocenter detection**

(a) **Chromocenter Segmentation Method**

This step is based on the watershed algorithm [12–14] adapted in 3D (ijpb plugins). First the algorithm automatically computes the intensity contrast of the regions detected by the 3D watershed [15].

- **Chromocenter Segmentation:** The process takes as input the opened image, and the image results are displayed on the screen.
- **Chromocenter Segmentation (Batch):** Before running the plugin, a WorkDirectory dedicated to a given analysis should be created. Raw images are then to be saved in a new sub-directory created by the user and named hereafter RawDataNucleus. The result is saved in the ConstrastDataNucleus sub-directory in the WorkDirectory, with the same name as the raw images.

(b) When using **Chromocenter Segmentation,** a pop-up window appears, and the user needs to fill in the following parameters:

- **Work directory and raw data choice:**
  - *Raw Data*: The WorkDirectory should contain two sub-directories:
  - *RawDataNucleus*: containing the raw images of the nuclei
  - *SegmentedDataNucleus*: containing the segmented images of the nuclei
  - *Output Directory*: Choose the WorkDirectory in which the results are to be stored. This directory must contain the RawDataNucleus and

SegmentedDataNucleus sub-directories. Hereafter, this new sub-directory is called ConstrastDataNucleus.

- **Voxel Calibration** corresponds to the voxel calibration used during the image acquisition (same as Nucleus Segmentation).

3. **Step 3: Manual Chromocenter Segmentation**

First, create the SegmentedDataCc sub-directory in the WorkDirectory.

Then, to undertake the segmentation of the image of chromocenters, open three images in ImageJ:

1. The raw image of nucleus
2. The segmented image of nucleus
3. The contrast image of the nucleus

You can synchronize images with the ImageJ tool Synchronize Windows (Analyze>Tools>Synchronize Windows).

To define chromocenters, use the threshold tool (ImageJ menu: Image>Adjust>Threshold). Check the boxes "Dark Background" and "Stack Histogram," and choose the Over/Under option in the second drop-down list. Once you have chosen your threshold value, push the button Apply.

Save the segmented chromocenters (Ctrl+S or ImageJ menu: File>Save or File>Save as) with the same name as the raw image of the nucleus in the directory SegmentedDataCc.

If the nuclear segmentation results or the chromocenter segmentation is not representative of the nuclear image or nuclear object, remove the image of the dataset to avoid bias of the nuclear analysis.

4. **Step 4: Chromocenter Analysis**

This step allows computing of nuclear morphology and chromatin organization parameters (*see* Usage). The plugin can generate two output files, one for the nuclear characterization (NucAndCcParameters.tab) and one for the chromocenter organization (CcParameters.tab).

(a) **Chromocenter Analysis**

The process uses three opened images as input (i.e., raw image of the nucleus, segmented image of the nucleus, and segmented image of the chromocenter(s)). The results of the analysis are displayed in the ImageJ log window.

(b) **Chromocenter Analysis (Batch)**

- **Work directory and raw data choice**
  - *Raw Data*: The main WorkDirectory must contain three sub-directories (a given image keeps the same name in all three sub-directories):

RawDataNucleus containing the raw images of the nuclei

SegmentedDataNucleus containing the segmented images of the nuclei

SegmentedDataCc containing the segmented images of the chromocenters

- *Output Directory*: Choose the WorkDirectory the results are to be stored in.

- **Voxel calibration** which corresponds to the voxel calibration used during the image acquisition.
- **Type of relative heterochromatin fraction RHF [16].** This parameter determines the ratio of heterochromatin within the nucleus. This ratio can be computed with the volume (total chromocenter volume/ nuclear volume) or the intensity (total chromocenter intensity/nuclear intensity).
- **Result files of interest**
  - *Nucleus and chromocenter*: Two output files are created in the WorkDirectory NucAndCcParameters.tab and CcParameters.tab.
  - *Chromocenter*: CcParameters.tab is created in the WorkDirectory.
  - *Nucleus*: NucAndCcParameters is created in the WorkDirectory.

Once the START button is pressed, the program will create the result file(s) in the WorkDirectory.

3.3 Nuclear Migration Analysis

1. Grow *Arabidopsis* on MS plates in glass-based dishes for 1–2 weeks until roots reach the bottom of the dishes.
2. Set up a microscope for a nuclear migration assay (*see* **Note 7**). Use a 10× or 20× dry lens. Keep laser transmission low, typically at 1–10% to avoid photo damage during image acquisition.
3. Place a dish on a sample holder of the stage.
4. Image the mature root tissue to find an appropriate region (*see* **Note 8**).
5. Take time-lapse images as a set of z-stacks over time with eight series of optical sections every 6 μm (*see* **Note 9**). Collect the individual *z*-stacks every 30 s over a period of 45–60 min.
6. Reduce 3D optical sections to 2D maximal projection images at every time point, and transform to a 2D time-lapse movie.
7. Launch FIJI and open the 2D time-lapse movie.
8. If necessary, transform the movie to 8-bit images (Image>Type>8-bit).

9. Run Manual Tracking plugin (Plugins>Tracking>Manual Tracking).
10. Set the *x/y* calibration value (μm per pixel) in the pop-up window.
11. Click on "Add track" to start a new track.
12. Click on a nucleus of interest in the image window. By clicking on the nucleus, the image automatically proceeds to successive image, and you can track the nucleus.
13. The results table will pop up, showing the distance travelled by the nucleus between two successive images.
14. Once the current nucleus tracking is over, click on "End track."
15. Repeat from Step 11 to record a new nuclear track.
16. Export the results table as an Excel file (File>Save as…).
17. Launch Microsoft Excel and open the file.
18. Calculate total distance of each tracked nucleus by simply summing the distance values (*see* **Note 10**).

4 Notes

1. DORY2 will drop proteins with multiple transmembrane domains.
2. Most protein names in the nr database contain species names which are enclosed in brackets. However, not all the protein names contain species name. If you use other FASTA files, make sure the species names are within the "[ ]."
3. BLASTP against nr is very time-consuming on personal computers. Depending on your computer speed, each sequence may take 20 min to finish. If there are 100 homologous groups, it will take 34 h to finish. During this, make sure your computer does not go to sleep or automatically updates its operating system and restarts.
4. Make sure the input database file (the file to search for KASH proteins) is in FASTA format. DORY2 does not check this. Make sure you choose the nr database correctly if you choose to "*Perform a final BLASTP for identified KASH candidates against the nr database.*"
5. The homologs identified by BLASTP depend on the E-value threshold. BLAST is a local alignment algorithm, which means that proteins partially homologous to the query protein can pass the E-value threshold. A good example is nesprin-1. It contains actin-binding domains and spectrin repeats. In this case, proteins containing any of these two domains may be classified as "homologs." However, they may not belong to the same protein family. Therefore, a large homologous group may need to

be further analyzed, especially when it contains large proteins having multiple domains. The "Distribution of Blast Hits on the Query Sequence" section in the BLAST result webpage needs to be consulted. Only the protein sequences that have a good whole-sequence alignment should be chosen to download. Another way is to download all the protein sequences from a BLASTP result and manually check whether sub-homologous groups exist following the steps below:

(a) Use MAFFT to align the protein sequences of a homologous group. Set "Output order" to "Aligned" before starting the alignment.

(b) Download the alignment in "Clustal format," and open the alignment in Jalview.

(c) In Jalview, choose "ClustalX" in the "Colour" menu, uncheck "Wrap" in the "Format" menu, and adjust font in the "Format" menu to obtain an overview of the alignment.

(d) Scroll to check whether the aligned sequences can be divided into sub-homologous groups based on the alignment.

If proteins were predicted from genome or cDNA sequences and the C-terminal domain is not predicted correctly, some homologs might not end with an expected KASH tail. Such examples have been reported by Zhou et al. [2] Therefore, if the majority of a protein family possesses a predicted KASH tail, then this protein family should be considered as a candidate group.

6. It is noteworthy that not every protein in a homologous group will terminate in four amino acids exactly following the pattern but will terminate in amino acids very similar to the pattern. For example, when using pattern "[PATHQL]PP[QTVFILM]" to identify animal KASH proteins, in the homologous groups obtained by BLAST, proteins terminating in PLPV and PSPT can also be found. These outliers might however be KASH proteins, and their C-terminal four amino acids can be used to improve the pattern used in "*Regex for KASH Tail*".
7. An inverted microscope is recommended for performing the time-lapse imaging with a glass-based dish. In the case of using a conventional microscope, the glass-based dishes should be placed upside down and placed on a glass slide on the stage.
8. Root meristem and elongation zone should be avoided as they grow rapidly during the time-lapse imaging.
9. Focal distance is dependent on each experiment to ensure that a large number of nuclei in epidermal cells are imaged.
10. For the first time point of each track, distance cannot be calculated; instead −1 appears in the results table. This −1 value should be ignored for calculating the total distance of tracked nucleus.

Acknowledgments

X.Z. is supported by the Gordon and Betty Moore Foundation through Grant GBMF2550.06. Work in I.M.'s lab is funded by the National Science Foundation. I.M. would like to thank Norman Groves and Anna Newman-Griffis, OSU for "test running" DORY2. K.T. is supported by the Grant-in-Aid for Scientific Research from JSPS (No. 26711017 and 15K14545). The authors are participants in COST Action CA 16212 'In Depth'.

References

1. Graumann K, Bass HW, Parry G (2013) SUNrises on the International Plant Nucleus Consortium: SEB Salzburg 2012. Nucleus 4:3–7
2. Zhou X, Graumann K, Wirthmueller L et al (2014) Identification of unique SUN-interacting nuclear envelope proteins with diverse functions in plants. J Cell Biol 205:677–692
3. Zhou X, Tamura K, Graumann K et al (2016) Exploring the protein composition of the plant nuclear envelope. Methods Mol Biol 1411:45–65
4. Chytilova E, Macas J, Sliwinska E et al (2000) Nuclear dynamics in Arabidopsis thaliana. Mol Biol Cell 11:2733–2741
5. Bickmore WA, van Steensel B (2013) Genome architecture: domain organization of interphase chromosomes. Cell 152:1270–1284
6. Makova KD, Hardison RC (2015) The effects of chromatin organization on variation in mutation rates in the genome. Nat Rev Genet 16:213–223
7. Poulet A, Arganda-Carreras I, Legland D et al (2015) NucleusJ: an ImageJ plugin for quantifying 3D images of interphase nuclei. Bioinformatics 31:1144–1146
8. Tamura K, Iwabuchi K, Fukao Y et al (2013) Myosin XI-i links the nuclear membrane to the cytoskeleton to control nuclear movement and shape in Arabidopsis. Curr Biol 23:1776–1781
9. Kall L, Krogh A, Sonnhammer EL (2007) Advantages of combined transmembrane topology and signal peptide prediction--the Phobius web server. Nucleic Acids Res 35(Web Server issue):W429–W432
10. Nyquist H (1928) Certain topics in telegraph transmission theory. Trans Am Inst Electr Eng 47:617–644
11. Shannon CE (1949) Communication in the presence of noise. Proc IRE 37:10–21
12. Beucher S, Lantuéjoul C (1979) Use of watersheds in contour detection. International Workshop on Image Processing: Real-Time Edge and Motion Detection/Estimation, Rennes, France, 17–21 September 1979
13. Beucher S, Meyer F (1992) The morphological approach to segmentation: the watershed transformation. In: Dougherty ER (ed) Mathematical morphology in image processing. Marcel Decker, Inc., New York, pp 433–481
14. Vincent L, Soille P (1991) Watersheds in digital spaces: an efficient algorithm based on immersion simulations. IEEE Trans Pattern Anal Mach Intell 13:583–598
15. Andrey P, Kiêu K, Kress C et al (2010) Statistical analysis of 3D images detects regular spatial distributions of centromeres and chromocenters in animal and plant nuclei. PLoS Comput Biol 6:e1000853
16. Fransz P, De Jong JH, Lysak M et al (2002) Interphase chromosomes in Arabidopsis are organized as well defined chromocenters from which euchromatin loops emanate. Proc Natl Acad Sci U S A 99:14584–14589

Chapter 16

Investigation of Nuclear Periphery Protein Interactions in Plants Using the Membrane Yeast Two-Hybrid (MbY2H) System

Maxime Voisin, Emmanuel Vanrobays, and Christophe Tatout

Abstract

Identification of membrane protein interactomes is a key issue to better understand how these molecules carry out their functions. However, protein-protein interactions using conventional interaction assays are particularly challenging for integral membrane proteins, because of their hydrophobic nature. Here we describe the membrane yeast two-hybrid (MbY2H) system, a powerful tool for identifying the interactors of membrane and membrane-associated proteins.

Key words *Arabidopsis thaliana*, Yeast two-hybrid, Interactome, LINC complex, Membrane protein

1 Introduction

Methods to decipher physical interactions between proteins are essential to uncover the complex protein network anchored at the nuclear envelope or located at the nuclear periphery. In many cases, protein-protein interactions (PPI) have been spotlighted by a technique developed in yeast called the yeast two-hybrid, in which the protein of interest (the bait) is tested for its interaction with a protein partner (the prey). Yeast two-hybrid was used in pioneer studies [1], and then successfully applied to discover the interaction between the components of the nuclear periphery such as the Lamin B receptor (LBR) and the heterochromatin protein 1 (HP1) from *Drosophila melanogaster* [2]. However, because transmembrane domains may sequester the bait in the endoplasmic reticulum or plasma membrane impairing proper PPI, a specialized split-ubiquitin-based yeast two-hybrid system called the membrane yeast-two hybrid (MbY2H) was developed [3, 4]. The MbY2H system exploits the capacity of the yeast ubiquitin (ub) to be split into N (Nub)- and C (Cub)-terminal halves. When co-expressed, Nub and Cub are able to spontaneously reassociate into a

Gregg G. Gundersen and Howard J. Worman (eds.), *The LINC Complex: Methods and Protocols*, Methods in Molecular Biology, vol. 1840, https://doi.org/10.1007/978-1-4939-8691-0_16, © Springer Science+Business Media, LLC, part of Springer Nature 2018

functionally active ubiquitin. This spontaneous reassociation can be prevented by introducing in Nub a single mutation (Isoleucine -> Glycine; NubG); however if in close proximity, NubG and Cub can lead to the reassembly of the functional split-ubiquitin [5]. In this system, the bait anchored in the membrane is fused to Cub and associated with a transcription factor, while the prey is fused to NubG. PPI between bait and prey brings together Cub and NubG and induces the recruitment of specific proteases, which release the transcription factor resulting in the expression of reporter genes (Fig. 1A) [3, 4]. MbY2H has been successfully applied to study the linker of cytoskeleton and nucleoskeleton complex (LINC), an evolutionarily well-conserved protein complex between SUN and KASH proteins anchored in the outer and inner nuclear membrane [6–8]. The following protocol describes how to generate by gap repair vectors expressing baits and preys expressing cDNA generated from plant RNA, how to validate these constructs, and how to use them in the MbY2H system to test and quantify the strength of PPI.

2 Materials

2.1 Vectors and Strains

Plasmid vectors and yeast strain described below were purchased from Dualsystem Biotech (http://www.dualsystems.com) (*see* **Note 1**).

1. pBT3-N is a bait vector allowing N-terminal tagging to Cub which is fused to the artificial transcription factor LexA-VP16 under the control of Cyclin1 (Cyc1) promoter. This vector contains the yeast leucine (Leu2) marker, the *Escherichia coli* (*E. coli*) kanamycin resistance gene for bacterial selection, and a CEN/ARS origin of replication allowing the vector to be maintained at a low copy number (*see* **Note 2**).
2. pPR3-N is a prey vector allowing N-terminal tagging to NubG. It also expresses the hemagglutinin A (HA) tag under the control of Cyc1 promoter and contains the yeast tryptophan (Trp1) marker, the *E. coli* ampicillin resistance gene for bacterial selection, and a *2 μ* multicopy origin of replication.
3. pOst1-NubI used as a positive control is a prey vector expressing an endoplasmic reticulum protein Ost1 [9] fused to the wild-type Nub moiety of yeast ubiquitin (NubI) under the control of the Adh1 promoter. This vector contains the yeast tryptophan (Trp1) marker, the *E. coli* ampicillin resistance gene for bacterial selection, and a *2 μ* multicopy origin of replication.
4. pNubG-Fe65 used as noninteracting control is a prey vector expressing the cytosolic protein Fe65 [10] fused to NubG under the control of Adh1 promoter. This vector contains the yeast tryptophan (Trp1) marker, the *E. coli* ampicillin

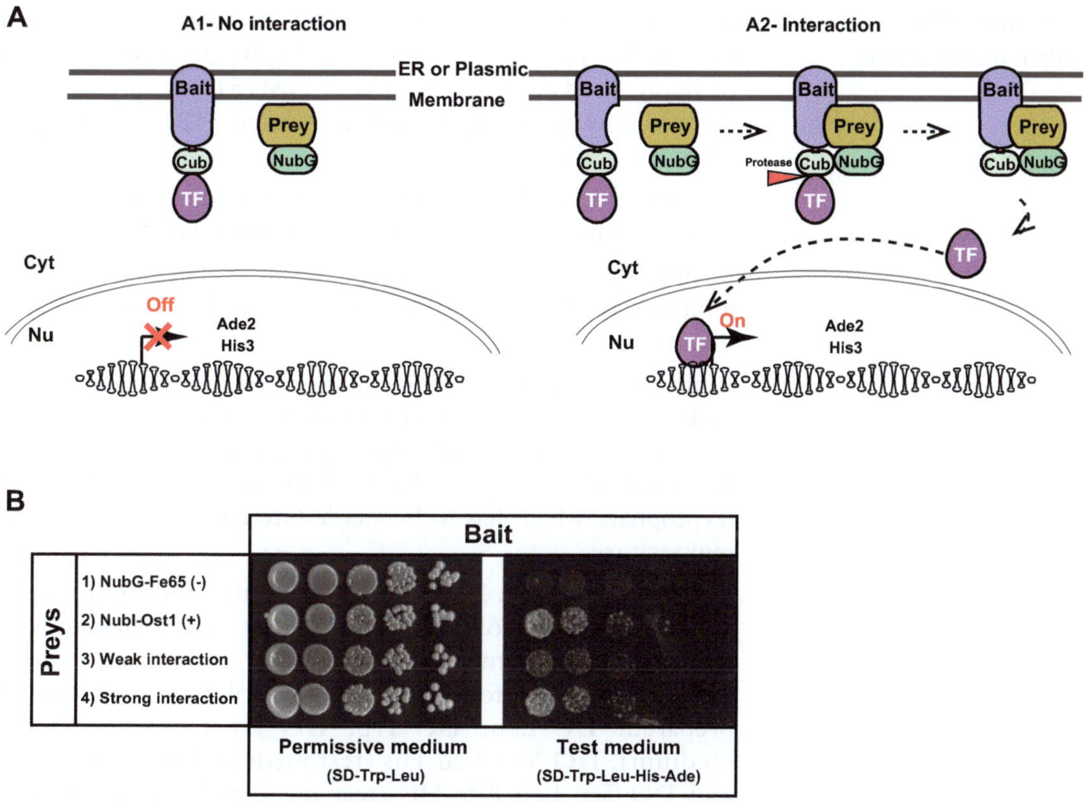

Fig. 1 Principle and typical results from the membrane yeast two-hybrid method. (**A**1) No interaction between bait and prey. MbY2H is a two component system based on a split-ubiquitin. The bait protein anchored into the endoplasmic reticulum (ER) or plasma membrane is fused to the C-terminal half of ubiquitin (Cub) and the LexA-VP16 transcription factor (TF). The prey is fused to a mutated form of the N-terminal end of ubiquitin (NubG) unable to associate with Cub. In that condition, the TF is not released by proteases, and the promoter (arrow) of the selectable markers Adenine2 (Ade2) and Histidine3 (His3) is transcriptionally inactive (OFF). (**A**2) Interaction between bait and prey. If bait and prey interact, the split-ubiquitin is reformed and allows the proteolytic cleavage of LexA-VP16, which is subsequently translocated into the nucleus where it activates the selectable markers needed for the growth on Test medium (Cyt, cytoplasm; Nu, nucleus). (**B**) Efficiency of bait and prey interaction. Bait and prey are selected on SD-Trp-Leu (Selective medium) and then tested for interaction on SD-Trp-Leu-Ade-His (Test medium) using serial dilution. (1) The absence of bait self-activation is tested by using Fe65 as a noninteracting prey control. (2) Validation of the bait construct is performed using NubI which spontaneously interacts with Cub fused to Ost1 a yeast ER protein. Example of (3) weak and (4) strong interactions between bait and prey are shown

resistance gene for bacterial selection, and a *2 μ* multicopy origin of replication.

5. The yeast strain NMY51 (*MATa*, *his3Δ200*, *trp1-901*, *leu2-3,112*, *ade2*, *LYS2::(lexAop)$_4$-HIS3*, *ura3::(lexAop)$_8$-lacZ*, *ade2::(lexAop)$_8$-ADE2*, *GAL4*) is used to express bait and prey vectors.
6. High-efficiency electrocompetent *E. coli* DH5α is used for plasmid propagation in bacteria.

2.2 Yeast Media and Transformation

1. Yeast extract-peptone-dextrose (YPD)-rich medium: 1% yeast extract, 2% peptone, 2% glucose, 2% Bacto Agar for solid medium, used for propagation of NMY51, are prepared in deionized water (dH_2O) and autoclaved (121 °C, 15 psi, 15 min).
2. Glucose (20%) 10× solution is prepared by dissolving glucose in dH_2O, autoclaved and stored at room temperature.
3. 3-Amino-1,2,4-Triazole (3-AT) 1 M, a His3-competitive inhibitor, is prepared in dH_2O, and sterilized by filtration, and then stored at 4 °C.
4. Amino acids and bases 100× solutions are prepared in dH_2O at the following concentrations: histidine (His) 2 g/L, leucine (Leu) 6 g/L, tryptophan (Trp) 2 g/L, adenine (Ade) 2 g/L, and lysine (Lys) 3 g/L. All solutions are autoclaved except tryptophan, which has to be sterile filtered and kept at room temperature.
5. Dropout (DO) mix 10× is a combination of the amino-acids and bases 100× solutions diluted ten times lacking the appropriate supplement in sterile dH_2O and stored at room temperature. The following dropout solutions need to be prepared: DO-Leu, DO-Trp, DO-Trp-Leu (permissive medium), DO-Trp-Leu-His (test medium low stringency), and DO-Trp-Leu-Ade-His (test medium high stringency) (*see* **Note 3**).
6. Yeast nitrogen base (YNB) 10× solution: 1.7% yeast nitrogen base and 5% ammonium sulfate without amino acids or bases dissolved in dH_2O and autoclaved.
7. Synthetic defined medium (SD medium): dilute ten times the YNB 10× solution, the glucose 10×, and the Dropout 10× mix in dH_2O with or without 2% Bacto Agar for solid medium. After mixing, the SD medium is autoclaved. The following SD medium solutions are prepared (*see* **Note 3**): SD-Leu for vector bait selection, SD-Trp for vector prey selection, SD-Trp-Leu for prey and bait selection, and SD-Trp-Leu-Ade-His to test interactions between baits and preys. 3-AT can be added to reduce the background growth (*see* **Note 4**).
8. Salmon sperm DNA (ssDNA) at 10 mg/mL used as carrier for yeast transformations is dissolved in sterile dH_2O, then boiled for 5 min at 95 °C, and stored at −20 °C.
9. Tris-HCl (1 M), pH 7.5 and EDTA (0.5 M), and pH 8 are prepared in dH_2O; the solutions are autoclaved after adjustment of pH.
10. Tris EDTA (TE) 10×: Tris 100 mM, EDTA 50 mM, is prepared by diluting ten times Tris 1 M pH 7.5 and EDTA 0.5 M pH 8 in dH_2O. The final solution is autoclaved.

11. Lithium acetate (LiOAc) 10× solution (1 M) is prepared by dissolving Lithium acetate in dH_2O. The final solution is autoclaved.
12. TE/LiOAc 1× (LiOAC 0.1 M, Tris 10 mM, EDTA 5 mM) is prepared by diluting ten times TE 10× solution and LiOAc 10× in dH_2O. The final solution is autoclaved.
13. 50% polyethylene glycol (PEG) solution (PEG-3350) is dissolved in dH_2O and sterilized by filtration.
14. PEG/TE/LiOAc (PEG 40%, LiOAc 1× TE 1×) is produced by diluting ten times TE 10× solution and LiOAc 10× in dH_2O in 50% PEG solution and sterilized by filtration.
15. 0.9% NaCl solution is prepared by dissolving NaCl in dH_2O. The final solution is autoclaved.
16. 100% dimethyl sulfoxide (DMSO).

2.3 *E. coli* Media, Transformation, and Plasmid Extraction

1. Kanamycin (100 mg/mL, 1000×) and ampicillin (100 mg/mL, 1000×) stocks are prepared in sterile dH_2O, and stored at −20 °C.
2. Luria-Bertani (LB) medium: 1% bacto-tryptone, 0.5% yeast extract, 0.5% NaCl, and 2% agar; for solid medium, used for propagation of *E. coli*, DH5α is dissolved in dH_2O and autoclaved. Appropriate antibiotic (kanamycin or ampicillin) is added after cooling when the solution is hand warm.
3. Miniprep DNA plasmid isolation kit (Macheray-Nagel).
4. 50 mL Falcon tubes.
5. 0.5 mm metal beads.

2.4 Arabidopsis cDNA Preparation

1. Tissue lyser (QIAGEN).
2. Liquid nitrogen.
3. 10–15-day-old *Arabidopsis thaliana* plants.
4. Safe-Lock and microfuge tubes.
5. Soda-lime glass beads.
6. Trizol® (Invitrogen).
7. 1 U/μL DNAse I (Promega).
8. Chloroform and isopropyl alcohol.
9. 70% ethanol in RNase-free water.
10. RNase-free dH_2O.
11. RNAsin RNAse inhibitor (Promega).
12. Random hexamer primers, oligo-dT, and deoxyribonucleotides (Promega).
13. Murine Leukemia Virus Reverse Transcriptase (M-MLV RT) (Promega).
14. Nanodrop-1000 (Thermo Fisher).

2.5 PCR and Enzymatic Digestion

1. 0.5 mL PCR tubes (Eppendorf).
2. SfiI restriction enzyme.
3. 10× NEBuffer 2.1 (New England Biolabs).
4. 5 U/μL GoTaq and 5× buffer (Promega).
5. 2 U/μL Phusion and 5× buffer HF (Thermo Fisher).
6. 10 mM dNTPs, 25 mM $MgCl_2$ and 1 M EDTA pH 8.
7. 1% agarose gel with 0.5 μg/L ethidium bromide (EtBr).
8. 10× DNA loading buffer (Thermo Fisher).
9. 1 kb DNA ladder (Promega).

2.6 Primers

2.6.1 Bait Primers

1. Bait Forward: 5′-TCGATAAGCTTGATATCGAATTCCTGCAGGGCCATTACGGCC-3′ followed by the 18–22 first nucleotides to your bait cDNA, starting by ATG.
2. Bait Reverse: 5′-CGCTCCGCGGTTAGCTACTTACCATGGGGCCGAGGCGGCC-3′ followed by the 18–22 reverse complement nucleotides to your bait including the STOP codon.

2.6.2 Prey Primers

1. Prey Forward: 5′-GGATCCAAGCAGTGGTATCAACGCAGAGTGGCCATTACGGCC-3′ followed by the 18–22 first nucleotides to your prey cDNA, starting by ATG.
2. Prey Reverse: 5′-GATAAGCTTGATATCGAATTCTCGAGAGGCCGAGGCGGCC-3′ followed by the 18–22 reverse complement nucleotides to your prey including the STOP codon.

2.6.3 Other Primers

1. pBT3-N Forward: 5′-CAGAAGGAGTCCACCTTAC-3′.
2. pPR3-N Forward: 5′-GTCGAAAATTCAAGACAAGG-3′.
3. pPR3-N and pBT3-N Reverse: 5′-AAGCGTGACATAACTAATTAC-3′.

3 Methods

3.1 cDNA Preparation from Plants

In order to prepare bait and/or prey cDNA, here we provide a short protocol to extract plant RNA and perform cDNA synthesis of any given RNA. The cDNA can then be used for the interaction assay using the MbY2H system.

3.1.1 Plant RNA Extraction

1. Harvest a dozen of 10–15-day-old *Arabidopsis thaliana* plants (50–100 mg) in 2 mL Safe-Lock tubes containing one 0.5 mM metal bead. Freeze in liquid nitrogen and grind tissue (tissue lyser set up at twice 30 s, 30 Hz).
2. Add 1 mL of Trizol® to the powder, vortex briefly, and repeat pipetting to lyse the cells if necessary.

3. Leave all the samples on ice until your last sample is ready, and then incubate for 5 min at room temperature.
4. Add 200 μL of chloroform, shake vigorously by hand 15 s, and then incubate 5 min at room temperature.
5. Centrifuge at 13,000 × *g* for 15 min at 4 °C, and then transfer 600 μL of the colorless upper aqueous phase to a fresh 1.5 mL microfuge tube.
6. Add 600 μL of isopropanol, mix and incubate 30 min at −20 °C.
7. Centrifuge at 13,000 × *g* for 15 min at 4 °C, carefully remove the supernatant, and wash the RNA pellet with 1 mL of 70% EtOH.
8. Centrifuge at 13,000 × *g* for 5 min at 4 °C, remove all the supernatant, and briefly air-dry the RNA pellet for 5–10 min.
9. Dissolve the pellet in 50 μL RNase-free dH_2O by pipetting up and down (*see* **Note 5**). Store at −20 °C or −80 °C for long-term storage. We typically get ~30 μg RNA/sample.

3.1.2 DNAse I Treatment

1. To the 50 μL of RNA sample, add 6 μL of 10× DNAse I buffer and 4 μL of Dnase I enzyme.
2. Mix and briefly spin down samples, then incubate 1 h at 37 °C.
3. Add 140 μL of dH_2O and 200 μL of phenol, and then shake vigorously by hand.
4. Centrifuge 10 min à 13,000 × *g* at 4 °C.
5. Transfer the supernatant to a 1.5 mL microfuge tube containing 200 μL of chloroform/isoamyl alcohol (24:1).
6. Centrifuge 5 min at 13,000 × *g* at 4 °C.
7. Transfer the supernatant (200 μL) into a 1.5 mL microfuge tube containing 440 μL of EtOH 100% and 20 μL NaOAc 3 M pH 5.2.
8. Precipitate overnight at −20 °C.
9. Centrifuge 30 min at 13,000 × *g* at 4 °C.
10. Wash the pellet with 1 mL of EtOH 70%.
11. Centrifuge 5 min at 13,000 × *g* at 4 °C.

3.1.3 Reverse Transcription (RT)

1. Resuspend the pellet in 30 μL of RNase-free dH_2O.
2. Use 1 μL to determine the RNA concentration with a NanoDrop.
3. Pipet 1 μg of RNA into a PCR tube, add 1 μL of oligo-dT or hexamers (0.5 μg/μL), and complete to 15 μL with RNase-free dH_2O (*see* **Note 6**).
4. Denature for 5 min at 70 °C, and place immediately on ice.

5. Then add 5 μL M-MLV 5× buffer, 1.5 μL dNTP 10 mM, 0.25 μL RNAsin (40 U/μL), and 1 μL M-MLV RT enzyme (200 U/μL), and complete to 20 μL final with RNase-free dH_2O.
6. Incubate 1 h at 42 °C.
7. Store at −20 °C.

3.2 Bait and Prey Vectors Construction and Verification

The cDNA sequence encoding the bait and prey proteins of interest are cloned, respectively, into the vector pBT3-N and pPR3-N by double-strand break repair also known as gap repair. The gap repair is based on homologous recombination in yeast and allows the constitution of the bait plasmid. Recombination requires 35–40 nucleotides identity between the cDNA amplified by PCR and the linearized vector [11, 12]. Alternative strategies such as conventional restriction digest or Gateway (Invitrogen) could also be used to generate these vectors.

When designing the bait vector, it is important to check if the membrane topology of the candidate protein is known to determine the location of its N and C terminus ends with respect to the cytosol, as the cleavage of the Cub-LexA-VP16 requires its location in the cytoplasmic compartment (*see* **Note 7**).

Empty pBT3-N bait and pPR3-N prey vectors are linearized by the SfiI restriction enzyme as follows:

1. Digest 1 μg of the vector with SfiI according to manufacturer's protocol in 50 μL final volume. The quantity of DNA vector digested here is sufficient for ten gap repair reactions.
2. Incubate overnight at 50 °C.
3. Add EDTA to 1 mM final concentration, and store at −20 °C.
4. To check for the digestion efficiency, run a 1% agarose gel stained with EtBr, load 6 μL of undigested vector (100 ng) in loading buffer 1×, 6 μL of the digested vector (100 ng) in loading buffer 1× and the recommended amount of 1 kb DNA ladder. Compare the profiles.

3.2.1 Primers Design and cDNA Amplification

After preparation of total cDNA from plants (Subheading 3.1), cDNAs are amplified by PCR.

1. The forward and reverse bait primers (Subheading 2.6.1) or prey primers (Subheading 2.6.2) required for the cDNA amplification must contain 35–45 nucleotides homologous to the insertion site of the bait or prey vector, respectively, followed by 18 nucleotides of the cDNA. Amplified cDNA fragments have to start with an ATG codon in frame with Cub (bait vector) or with NubG (prey vector) to ensure a continuous translation and should end with a Stop codon.

2. Perform PCR reactions using 0.2 μL 2 U/μL PhusionTaq or any appropriate proof reading polymerase, 0.5 μM final forward and reverse primers (*see* Subheading 2.6.3), 2 μL of the cDNA generated from plant material as template (Subheading 3.1) 4 μL 5× buffer HF, and 0.4 μL 10 mM dNTPs, and complete to 20 μL final with dH_2O.
3. Use the following PCR cycling conditions: 4 min at 95 °C, 35 cycles [30 s at 95 °C, 30 s at 55 °C, 40 s/kb at 72 °C], and finally 10 min at 72 °C.
4. The PCR amplification is checked by electrophoresis by loading 2 μL of the PCR product in a 1% agarose gel with EtBr and the recommended amount of 1 kb DNA ladder as control.

3.2.2 Yeast Transformation for Gap Repair Cloning

The protocol of yeast transformation for gap repair cloning [12] to generate the bait and prey vectors is very similar, the only differences will be the target yeast lines (**step 1**) and the medium used to plate and select the transformants. Bait vector will be built in the initial line NMY51, while the prey vectors will be assembled in the NMY51 already containing the bait vector validated beforehand.

1. For the bait vector construction, grow a culture of fresh yeast NMY51 in 5 mL of YPDA overnight at 30 °C. For the prey vector construction, grow the line NMY51 expressing a validated bait vector in 5 mL of SD-Leu overnight at 30 °C.
2. Dilute the yeast into 50 mL of YPDA (bait) or SD-Leu (prey) to an OD_{600} between 0.4 and 0.8.
3. Centrifuge at 11,000 × *g* for 5 min, and resuspend the pellet in 1 mL of TE/LiOAc 1×.
4. Transfer in 1.5 mL microfuge tube.
5. Pellet the yeast cells by centrifugation for 1 min at 11,000 × *g*.
6. Discard the supernatant, and resuspend cells into 1 mL of TE/LiOAc 1× to wash the cells.
7. Pellet the yeast cells, and repeat the wash steps with TE/LiOAc twice.
8. Discard the supernatant, and resuspend cells into 500 μL of TE/LiOAc 1× (sufficient for at least ten transformations).
9. Denature the ssDNA at 10 mg/mL at 95 °C for 5 min and leave in ice.
10. Mix 40 μL NMY51/TE/LiOAc 1×, 300 μL PEG/TE/LiOAc, and 5 μL boiled ssDNA, and add 5 μL of cDNA (from 3.1) and 50 ng of linearized empty bait or prey vector (from Subheading 3.2.1, **step 1**) (*see* **Note 8**).
11. Briefly vortex, incubate at 30 °C for 30 min, add 12 μL DMSO, and heat shock at 42 °C for 15 min.

12. Centrifuge at 11,000 × *g* at room temperature for 1 min, and resuspend in 300 μL of 0.9% NaCl.
13. Plate the totality of the yeast cells onto a SD-Leu selective medium for the bait transformation or SD-Leu-Trp for the prey, and incubate at 30 °C for 2–3 days.
14. After 3–5 days, the number of transformants should be at least ten times higher in gap repair samples compared to the linearized pBT3-N alone used as control.

3.2.3 Construct Verification

To check the vector, clones are first screened by a PCR on yeast colonies and for positive colonies; the vectors generated by gap repair are extracted and verified by sequencing.

1. Select six yeast colonies using a sterile toothpick or a plastic loop, and resuspend each of them in a PCR tube containing 50 μL dH_2O (*see* **Note 9**).
2. Microwave the yeast/dH_2O mix for 2 min at 900 W, and put immediately in ice (*see* **Note 10**).
3. Set up in a PCR tube a PCR reactions using 2 μL of the above yeast preparation as template, 0.1 μL GoTaq 5 U/μL, 0.5 μM final pBT3-N (for the bait) or pPR3-N (for the prey) forward and reverse primers (Subheading 2.6.3), 4 μL 5× GoTaq buffer, 0.4 μL 10 mM dNTPs, and 1.2 μL 25 mM MgCl2, and complete to 20 μL final with dH_2O.
4. We routinely use the following PCR cycling conditions: 4 min at 95 °C, 35 cycles [30 s at 95 °C, 30 s at 55 °C, 1 kb/min at 72 °C], and finally 10 min at 72 °C.
5. Load 6 μL of each sample in a 1% agarose gel with EtBr to determine the amplification of your prey or bait cDNA of interest.
5. Select two PCR positive colonies, and inoculate each colony into 25 mL SD-Leu (bait constructs) or SD-Leu-Trp (prey constructs), and grow at 30 °C overnight.
6. Pellet cells by centrifugation into 50 mL Falcon tube, and resuspend into the lysis solution of any miniprep DNA plasmid isolation kit.
7. Add a small volume of 0.5 mM glass beads and vortex vigorously for 3 min to ensure sufficient lyse of yeast cells.
8. Transfer the liquid into a new microfuge tube, and perform DNA minipreps according to the manufacturer's protocol.
9. Transform the yeast DNA plasmid isolated into electrocompetent *E. coli* strain suitable for plasmid propagation (*DH5α*) with a transformation efficiency of at least 1.10^7 cells per μg DNA.

10. Plate on LB containing kanamycin (to select the bait vector) or LB + ampicillin (to select the prey vector), and incubate overnight at 37 °C.
11. Isolate the vectors from transformed *E. coli* by performing DNA minipreps as above.
12. Verify the bait constructs by sequencing using pBT3-N (for the bait vector) or pPR3-N (for the prey vector) using the appropriate forward and reverse primers (Subheading 2.6.3).

After verification and sequencing, the extracted prey vectors can be directly transformed into NMY51 carrying any bait vector.

3.3 Bait Vector Validation

3.3.1 Bait Expression Validation

Correct expression of the bait is tested using the control prey construct pOst1-NubI [4] that expresses the yeast endoplasmic reticulum protein Ost1 [9] fused to NubI (the ubiquitin wild-type Nub half). If the bait is localized properly, the co-expression of the bait and Ost1-NubI results in the reconstitution of split-ubiquitin, cleavage of LexA-VP16 transcription factor and activation of the reporter genes His3 and Ade2 allowing yeast growth on SD-Leu-Trp-His-Ade test medium.

1. Grow yeast NMY51 containing the bait in 50 mL of SD-Leu overnight at 30 °C in a shaking incubator.
2. When the cells have reached an OD_{600} between 0.4 and 0.6, transfer the cells in a 50 mL Falcon tube, and pellet the cells by a centrifugation at 11,000 × *g* for 5 min.
3. Resuspend the pellet in 1 mL of TE/LiOAc 1×, and transfer in a 1.5 mL microfuge tube.
4. Pellet the yeast cells by centrifugation for 1 min at 11,000 × *g*.
5. Discard the supernatant, and resuspend cells into 1 mL of TE/LiOAc 1× to wash the cells.
6. Pellet the yeast cells and repeat the wash steps twice, and resuspend cells into 500 μL of TE/LiOAc 1×.
7. Denature the ssDNA at 95 °C for 5 min and put it in ice.
8. Mix 40 μL NMY51/TE/LiOAc 1×, 300 μL PEG/TE/LiOAc, 5 μL boiled ssDNA 10 mg/mL, and 1 μL pPR3N-NubI vector (100 ng/μL).
9. After a brief vortex, incubate mix at 30 °C for 30 min, add 12 μL DMSO and heat shock at 42 °C for 15 min.
10. Plate half of the yeast cells on SD-Leu-Trp selective medium and the other half on SD-Leu-Trp-His-Ade test medium.
11. After 3–5 days at 30 °C, estimate the percentage of growth on test medium versus selective medium. If the bait is functional, we typically observe more than 20% growth on the test medium SD-Leu-Trp-His-Ade.

3.3.2 Bait Self-Activation

In two-hybrid screens, background can be caused by self-activation of bait proteins that provokes the gene reporter expression and consequently the growth on selective medium without any interaction. The aim of the bait self-activation test is to estimate this background when the bait is co-expressed with a noninteracting prey control to adjust the selection conditions used for the screen by adding an His3-competitive inhibitor, the 3-amino-1,2,4-triazole (3-AT) on test medium plates.

1. Grow yeast NMY51 containing the bait in 50 mL of YPDA overnight shaking incubator at 30 °C.
2. Yeast cells are transformed by a prey vector pNubG-Fe65 expressing the cytosolic protein Fe65 [10] fused to the mutated portion of the yeast ubiquitin; NubG is used as noninteracting (negative) control. Transformation is performed as described in Subheading 3.2.2.
3. Plate transformed yeast cells on SD-Leu-Trp selective medium, SD-Leu-Trp-His-Ade test medium, and SD-Leu-Trp-His-Ade containing increasing concentration of 3-AT (*see* **Note 4**).
4. After 3–5 days at 30 °C, count the number of colonies on all plates to estimate the percentage of growth between SD-Leu-Trp selective medium and test medium containing or not 3-AT.

The lowest concentration of 3-AT inhibiting the bait background growth on test medium is then used for all the interaction tests (*see* **Note 4**).

3.4 MbY2H Interaction

When all the strains carrying the prey and bait combinations are ready, the evaluation of interaction can be pursued. The condition to analyze the interaction on test medium must be adjusted according to the 3-AT concentration determined for each bait protein (Subheading 3.3.2).

3.4.1 Interaction Analysis

1. Transfer NMY51 yeast cells carrying the baits and preys (including positive and negative control) to be tested to a fresh test medium SD-Leu-Trp (*see* **Note 11**).
2. After 2–3 days, the colonies can be tested on test medium. For each bait, adjust 3-AT optimal condition as determined in Subheading 3.3.2.
3. Streak the cells containing all the combinations between particular bait and the different preys onto high stringency test medium SD-Leu-Trp-Ade-His containing or not 3-AT and onto permissive SD-Leu-Trp to control for yeast growth.
4. On each plate, include the yeast cells containing the bait and pOst1-NubI as positive control (Subheading 2.1.3) and the yeast cells containing the bait and pNubG-Fe65 as negative control (Subheading 2.1.4).
5. After 3–5 days read the plates.

3.4.2 Evaluation of the Interaction Strengths by Drop Test

The following protocol can be used to compare the cell growth rate of yeast under different growth conditions to evaluate the interaction strength. This is achieved by serial dilutions and spotting of yeast on permissive and test media. This assay is referred to as drop test.

1. From fresh SD-Leu-Trp plates, pick a yeast colony expressing the bait and prey to be tested (*see* **Note 12**), the negative control and (pNubG-Fe65) and the positive prey (pOst1-NubI).
2. Resuspend the yeast cells into 1 mL SD-Leu-Trp medium, and adjust OD_{600} to 2.
3. Prepare 1:10, 1:100, 1:1000, 1:10,000 and 1:100,000 dilutions in 0.9% NaCl.
4. Spot 8 μL drops on SD-Leu-Trp growth control plate, SD-Leu-Trp-Ade-His test medium plate (*see* **Note 12**) containing 3-AT (according to the optimal condition of the bait determined in Subheading 3.3.2).
5. Let the spots to dry at room temperature until the liquid is completely absorbed by the medium.
6. Return the plate, and incubate at 30 °C. After 3–5 days examine the plates (Fig. 1B). The difference in growth is a measure for the strength of protein-protein interaction.

4 Notes

1. Dualsystem does not distribute anymore the MbY2H system, but vectors and yeast strain are available from several companies such as Bioquote Limited or MoBiTec.
2. pBT3-N does not contain any specific tag fused to the bait; however antibodies against VP16 are available several manufacturers and can be used to monitor the bait expression using Western blot analysis.
3. Basically we distinguish two types of media: the "transformation selection medium" (permissive medium) to select bait (DO-Leu), prey (DO-Trp), or bait + prey (DO-Trp-Leu) vectors containing yeast cells, and "interaction selection media" (test medium) to test interaction between baits and preys (DO-Trp-Leu-His (low stringency) or DO-Trp-Leu-Ade-His (high stringency)).
4. The background of the bait on test medium can easily be removed by the addition of 3-AT (usually used below 10 mM). All the baits do not require the same amount of 3-AT, the concentration has to be adjusted depending on the bait used. To find the optimal 3-AT concentration, SD-Leu-Trp-His-Ade test medium are supplemented with 0, 1, 2.5, 5, 10, 50,

and 100 mM 3-AT. For weak interaction, the stringency of the test medium plate can be adjusted either by decreasing the level of 3-AT in the SD-Leu-Trp-Ade-His test medium or by using SD-Leu-Trp-His plates.

5. If RNA pellets after precipitation are difficult to resuspend, incubate the samples 2–5 min to 65 °C.
6. Oligo-dT and hexamers can be used to produce the cDNA fragment. Oligo-dT is usually the best choice to amplify a cDNA full size, because it initiates reverse transcription at the 3′ end of the transcript however occasionally for some cDNAs hexamers gives a better result.
7. For most of the baits, the topology in respect to the membrane will be unknown. It is therefore difficult to choose witch terminal end will be fused to Cub-LexA-VP16. To predict which terminal end(s) of the protein will be outside the ER, we usually predict the position and number of putative transmembrane domains using online tools such as TMHMM (http://www.cbs.dtu.dk/services/TMHMM/) [13].
8. We routinely do not purify the PCR products to produce the cDNA fragments nor the linearized vectors after digestion to avoid loss of DNA during the gap repair procedure.
9. It is very important to take a very small quantity of yeast (the tips of a sterile toothpick); too much material will inhibit the PCR reaction.
10. Yeast cells are encapsulated by a rigid cell wall structure and its disruption will allow a better amplification. Here we use microwaves to quickly and efficiently break down the yeast cell wall, but alternatively you can use lytic enzymes such as lyticase or zymolyase to digest the cell wall.
11. We recommend using at least two clones for each combination of baits and preys.
12. We use cells directly from fresh plates to realize the drop test; alternatively you can use overnight liquid cultures. Dry the plates in sterile condition by opening the cover for 30 min. Be careful that the cover do not contains droplets of water. It is also essential to spot slowly the 8 μL of medium to avoid projections.

Acknowledgment

M.V. is supported by a fellowship from the Ministère de l'Enseignement Supérieur et de la Recherche (MESR). Work in C.T.'s lab is funded by the Université Clermont Auvergne, CNRS and INSERM. C.T. would like to thank Aline Probst for critical reading of the manuscript.

References

1. Fields S, Song O (1989) A novel genetic system to detect protein–protein interactions. Nature 340:245–246
2. Ye Q, Callebaut I, Pezhman A et al (1997) Domain-specific interactions of human HP1-type chromodomain proteins and inner nuclear membrane protein LBR. J Biol Chem 272:14983–14989
3. Snider J, Kittanakom S, Damjanovic D et al (2010) Detecting interactions with membrane proteins using a membrane two-hybrid assay in yeast. Nat Protoc 5:1281–1293
4. Snider J, Kittanakom S, Curak J et al (2010) Split-ubiquitin based membrane yeast two-hybrid (MYTH) system: a powerful tool for identifying protein-protein interactions. J Vis Exp 36:1698
5. Johnsson N, Varshavsky A (1994) Split ubiquitin as a sensor of protein interactions in vivo. Proc Natl Acad Sci U S A 91:10340–10344
6. Chen J, Smoyer CJ, Slaughter BD et al (2014) The SUN protein Mps3 controls Ndc1 distribution and function on the nuclear membrane. J Cell Biol 204:523–539
7. Graumann K, Vanrobays E, Tutois T et al (2014) Characterization of two distinct subfamilies of SUN-domain proteins in Arabidopsis and their interactions with the novel KASH-domain protein AtTIK. J Exp Bot 65:6499–6512
8. McGee MD, Rillo R, Anderson AS et al (2006) UNC-83 is a KASH protein required for nuclear migration and is recruited to the outer nuclear membrane by a physical interaction with the sun protein UNC-84. Mol Biol Cell 17:1790–1801
9. Pathak R, Parker CS, Imperiali B (1995) The essential yeast NLT1 gene encodes the 64 kDa glycoprotein subunit of the oligosaccharyl transferase. FEBS Lett 362:229–234
10. McLoughlin DM, Miller CC (1996) The intracellular cytoplasmic domain of the Alzheimer's disease amyloid precursor protein interacts with phosphotyrosine-binding domain proteins in the yeast two-hybrid system. FEBS Lett 397:197–200
11. Szostak JW, Orr-Weaver TL, Rothstein RJ et al (1983) The double-strand-break repair model for recombination. Cell 33:25–35
12. Oldenburg KR, Vo KT, Michaelis S et al (1997) Recombination-mediated PCR-directed plasmid construction in vivo in yeast. Nucleic Acids Res 25:451–452
13. Krogh A, Larsson B, von Heijne G et al (2001) Predicting transmembrane protein topology with a hidden Markov model: application to complete genomes. J Mol Biol 305:567–580

Chapter 17

Immunolabeling Protocols for Studying Meiosis in Plant Mutants Defective for Nuclear Envelope Components

Javier Varas and Mónica Pradillo

Abstract

The nuclear envelope (NE) is a dynamic boundary that allows the communication between nuclear and cytoplasmic components. It has essential roles in a variety of physiological processes including cell division. The linker of nucleoskeleton and cytoskeleton (LINC) complexes span the NE and are important during meiosis, the specialized cell division needed for sexual reproduction. During this division, the LINC complex proteins AtSUN1 and AtSUN2, located in the inner nuclear membrane (INM), are involved in tethering telomeres to the NE. This attachment promotes chromosome movements by the forces that are generated in the cytoplasmic face. In *Arabidopsis*, the double mutant *Atsun1 Atsun2* exhibits a delayed prophase I meiotic progression, partial synapsis, and recombination defects that lead to the formation of unbalanced gametes and sterility. In meiocytes from these mutants, immunolabeling can be applied to analyze possible changes in the dynamics of different meiotic proteins. In addition, if the specific antibodies are available, this technique is an easy and useful tool to determine the spatial distribution of NE proteins.

Key words Homologous recombination, Meiosis, Nuclear envelope, Pollen mother cell, Synapsis

1 Introduction

SUN domain proteins belong to the conserved eukaryotic linker of nucleoskeleton and cytoskeleton (LINC) complexes that connect chromatin to the cytoskeleton network. These proteins are required for accurate pairing and recombination between homologous chromosomes during meiosis. In *Arabidopsis*, the C-terminal SUN domain proteins AtSUN1 and AtSUN2 are located at the nuclear envelope (NE) in prophase I pollen mother cells (PMCs). Both proteins seem to be functionally redundant, at least during meiosis, as revealed by the wild-type (WT) phenotype of the corresponding single mutants. However, meiosis is disturbed in the double knockout *Atsun1 Atsun2* [1]. The meiotic defects include incomplete synapsis and frequent interlocks (entanglements among chromosomes) at prophase I, presence of univalents at metaphase I, and unbalanced segregations at second division. In this chapter, we

Gregg G. Gundersen and Howard J. Worman (eds.), *The LINC Complex: Methods and Protocols*, Methods in Molecular Biology, vol. 1840, https://doi.org/10.1007/978-1-4939-8691-0_17, © Springer Science+Business Media, LLC, part of Springer Nature 2018

describe the immunofluorescence techniques applied to gain insight into the ability of *Atsun1 Atsun2* plants to perform synaptonemal complex (SC) formation and meiotic recombination and in order to determine the localization of both SUN proteins in this double mutant.

In recent years, many different immunolabeling protocols have been designed to study the distribution of proteins on plant meiotic chromosomes. Choosing a proper method is crucial to get the desired results. The protocols can be classified into two types: the squash- and the spreading-based methods. Both procedures have different advantages and disadvantages and should be selected depending on the aim of the analysis. Squash preparations are required to preserve the volume of the nuclei, but slides contain a high background due to cytoplasm residues, and chromosomes are not properly separated. On the other hand, the spreading method does not retain the spatial organization of the meiocyte, but allows to obtain high-quality slides with a reduced background noise. Therefore, we apply the squash technique to analyze the spatial distribution of NE-associated proteins, whereas the spreading protocol is more useful to examine the dynamics of the SC formation and the kinetics of proteins involved in meiotic homologous recombination (HR). The description of both methodologies currently used in our lab is detailed in this chapter.

2 Materials

2.1 Plants

Arabidopsis thaliana plants are grown in a growth room under a 16 h light/8 h dark photoperiod, at 18–20 °C with adequate watering. They are sown in pots containing a soil mixture (rich soil/vermiculite, 3:1). Cover with a plastic film (not tightly sealed) for at least 1 week. Plants produce their first flowers within 6–7 weeks (*see* **Note 1**).

2.2 Equipment and Tools

1. Filter paper.
2. Petri dishes.
3. Pasteur pipettes.
4. Poly-L-lysine-coated and standard slides.
5. Coverslips (24 × 24 and 24 × 32).
6. Parafilm.
7. Forceps with fine and long tips.
8. Dissection needles.
9. Scalpel.
10. Aluminum bar.
11. Diamond pen.

12. Watch glasses.
13. Alcohol lamp.
14. Moist box (a plastic box containing a wet paper) (*see* **Note 2**).
15. Coplin jars.
16. Hot plate.
17. Incubator at 37 °C.
18. Humidifier.
19. Stereomicroscope with a cold light source.
20. Microscope with phase contrast.
21. Fluorescence microscope with appropriate filters, camera, and image acquisition software.

2.3 Reagents

1. Acetocarmine solution: add 5 g carmine to 45% of glacial acetic acid, and boil for 3 h on a hot plate inside of a fume hood. A magnetic stirrer can be used to produce turbulence in the solution. After cooling, pass the solution through a paper filter, and store at room temperature.
2. 10× phosphate-buffered saline (PBS): solution A (16.02 g $Na_2HPO_4{\cdot}2H_2O$ + 73.84 g NaCl in 900 mL of distilled water) + solution B (2.76 g $NaH_2PO_4H_2O$ + 16.56 g NaCl in 100 mL of distilled water). Adjust pH to 7 and autoclave. Dilute at 1:10 with distilled water before use. Store at room temperature.
3. 10× citrate buffer: mix 40 mL of 0.1 M sodium citrate and 60 mL of 0.1 M citric acid, pH 4.6. Store at 4 °C.
4. Washing buffer (PBS-T): PBS + 0.1% Triton X-100.
5. Blocking solution: 3% Bovine Serum Albumin (BSA) in PBS-T. Dispense 500 μL aliquots and store at −20 °C.
6. 6-Diaminido-2-phenlyinidole (DAPI) (1 μg/mL) in Vectashield antifade mounting medium. Store at 4 °C.

2.4 Specific Solutions for the Squash Technique

1. Liquid nitrogen.
2. 4% paraformaldehyde—0.15% Triton X-100: dissolve 4 g paraformaldehyde in 45 mL of PBS and heat while stirring to 60–70 °C on a hot plate. Be aware that the solution does not boil. Add some drops of 1 N NaOH to slowly raise the pH until the solution clears. Complete the volume to 50 mL and adjust the pH to 8 with HCl. Aliquots of 500 μL should be dispensed in Eppendorf tubes and can be frozen. After thawing one aliquot, add 500 μL of PBS and 1.5 μL of Triton X-100 (*see* **Note 3**).
3. Enzyme digestion medium: dissolve cellulose, pectolyase, and cytohelicase in citrate buffer for final concentration of 1% (w/v) each. In addition prepare a solution with cytohelicase

(0.4% w/v), sucrose (1.5%), and polyvinylpyrrolidone (PVP) (1%) in sterile deionized water. Dispense 500 μL aliquots, and store at −20 °C in both cases. Prepare a 1:1 mix of the two solutions.

2.5 Specific Solutions for the Spreading Technique

1. 4% paraformaldehyde: dissolve 4 g paraformaldehyde in 95 mL of distilled water and heat while stirring to 60–70 °C on a hot plate. Be aware that the solution does not boil. Add some drops of 1 N NaOH to slowly raise the pH until the solution clears. Complete the volume to 50 mL and adjust the pH to 8 with HCl. Store at 4 °C.
2. Enzyme digestion medium: dissolve cytohelicase (0.4% w/v), sucrose (1.5%), and polyvinylpyrrolidone (PVP) (1%) in sterile deionized water. Store in 500 μL aliquots at −20 °C.
3. Detergent medium: 1% (v/v) lipsol detergent and 0.1% (v/v) Triton X-100 in distilled water (freshly prepared) (*see* **Note 4**).

2.6 Antibodies

2.6.1 Primary Antibodies

The convenient dilutions for antibodies depend on their quality and the genetic background of the plants analyzed, among other factors. The use of an inappropriate dilution is the reason for a high background. Table 1 summarizes the working dilutions for antibodies we have used to successfully detect SUN, SC, and HR proteins in *Arabidopsis* male meiocytes. The references include information about the source of the antibodies (*see* **Note 5**).

Table 1
Primary antibodies

Antibody	Raised in	Working dilution	Reference
Anti-AtSUN1	Rabbit	1:100	[2]
Anti-AtSUN2	Rabbit	1:100	[2]
Anti-AtASY1	Rat Rabbit	1:1000	[3]
Anti-AtZYP1	Rat Rabbit	1:500	[4]
Anti-AtSYN1	Rabbit	1:500	[5]
Anti-AtSMC3	Rat	1:500	[6]
Anti-γH2AX (Ser139)	Rabbit	1:100	Upstate Biotechnology Catalog no. 07-164
Anti-AtRAD51	Rabbit	1:300	[6]
Anti-AtDMC1	Rabbit	1:300	[7]
Anti-AtMSH4	Rabbit	1:300	[8]
Anti-AtMLH1	Rabbit	1:300	[9]

2.6.2 Secondary Antibodies

There is a wide choice of secondary antibodies conjugated to Cy3 (red signal, 1:100), Alexa 555 (red signal, 1:500), or FITC (green signal, 1:50) that can be used (*see* **Notes 6–8**).

3 Methods

3.1 Bud Selection

It is necessary to select anthers which contain meiocytes at prophase I. To do this material is stained with acetocarmine (*see* **Note 9**).

1. Collect the inflorescences in a Petri dish with a moist piece of filter paper.
2. Divide up the inflorescence into individual buds and discard buds with yellow anthers containing pollen.
3. Place one bud on a slide and cover it with a single drop of acetocarmine.
4. Heat slightly the slide over a flame of an alcohol lamp.
5. Put on the coverslip, place the slide inside a folded filter paper, blot the excess liquid, and press down to release PMCs (*see* **Note 10**).
6. Check the meiotic stage under a phase contrast microscope.
7. Collect floral buds according to their size and gently remove the sepals.

3.2 Squash Protocol

1. Transfer the buds into a watch glass, and fix the material in 4% paraformaldehyde—0.15% Triton X-100 during 30 min.
2. Wash for 10 min at least three times in 1× PBS.
3. Digest the buds with the enzyme digestion medium at 37 °C for 45 min (*see* **Notes 11** and **12**).
4. Wash again in PBS three times for 5 min.
5. Put eight to ten buds on a poly-L-lysine-coated slide using a Pasteur pipette (*see* **Note 13**). Dissect the buds and try to release the pollen sacs. Tap out the anthers using an aluminum bar until a fine suspension is formed (*see* **Note 14**). Check the process under a stereomicroscope.
6. Cover with a coverslip and place the slide in a folded paper filter, blot the excess liquid with filter paper to dry the slide, and carefully disperse the material by tapping out the coverslip gently with the back of a wood handle needle. Squash down on the coverslip with a strong vertical pressure, using the thumb (*see* **Note 15**).
7. Examine the preparation under a phase contrast microscope. If the material is not sufficiently extended, squash again.
8. Soak the preparation in liquid nitrogen, and remove the coverslip with a scalpel quickly (*see* **Note 16**). Make lines on the

microscope slide with a diamond pen to indicate where the coverslip is.

9. Wait for 10 min and rinse in PBS in a Coplin jar (*see* **Note 17**).

3.3 Spreading Protocol

1. Select around eight buds, and dissect them to extract the pollen sac using a needle and fine forceps in 2–4 μL citrate buffer on a poly-L-lysine-coated slide (*see* **Note 18**). Avoid drying by adding buffer citrate as needed.
2. Add 10 μL of enzyme digestion medium and incubate in a moist box on a hot plate for 8 min at 37 °C (*see* **Note 19**).
3. Under the stereomicroscope cut the anthers in half with a needle and release the PMCs. Finally tap out the material by using an aluminum bar. Add a small drop of digestion medium if it is necessary. It is critical that the slide does not dry.
4. Add again 10 μL of enzyme digestion medium, and incubate for additional 8 min at 37 °C in the pre-warmed moist box on the hot plate.
5. Add 20 μL of detergent medium to the slide (*see* **Note 20**). Mix and gently spread by using the aluminum bar without touching the slide during 2 min (*see* **Note 21**).
6. Let the slide stand for 8 min at room temperature.
7. The meiocytes are fixed on the slide by adding 40 μL of cooled 4% paraformaldehyde. Extend the liquid over the slide surface with the help of a pipette tip without touching the slide. Let dry slowly over the hot plate at around 25 °C at least for 2 h (*see* **Note 22**).
8. After drying mark the chromosome spreading area on slides with a diamond pen (*see* **Note 23**).
9. Rinse quickly in distilled water and then in PBS in a Coplin jar.

3.4 Immunolocalization

1. Block the preparations in 50 μL of blocking solution (3% BSA). Add the solution to coverslips made with parafilm. Incubate the slides with cells facing down for 30 min at room temperature inside a moist box (*see* **Note 24**).
2. Remove the pieces of parafilm with forceps.
3. Prepare the mixture of primary antibodies diluted in blocking buffer (1% BSA) and add 75 μL to parafilm coverslips as mentioned before. Incubate overnight at 4 °C (*see* **Note 25**).
4. Remove carefully the parafilm. Wash the slides in PBS-T for 3 min five times (*see* **Note 26**).
5. Prepare the mixture of secondary antibodies diluted in blocking buffer (1% BSA), and add 75 μL to parafilm coverslips as mentioned before. Incubate for 90 min at room temperature. Slides should be protected from light from this moment on.

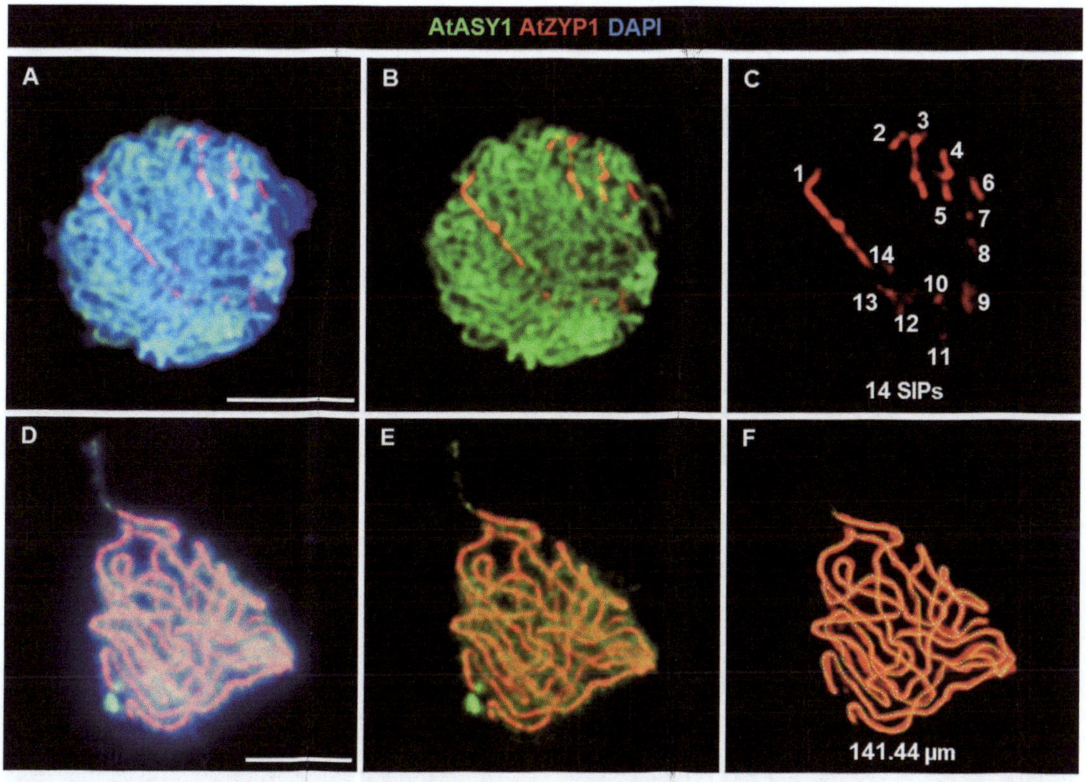

Fig. 1 Examples of images obtained for analyzing the synaptic process. Dual immunolocalization of AtASY1 (green) and AtZYP1 (red) on chromosome spread preparations from WT PMCs at (**A–C**) zygotene and (**D–F**) pachytene. (**A–C**) This example illustrates the quantification of SIPs in a zygotene in which the percentage of synapsis (AtZYP1 signal) is less than 10%. (**D–F**) The SC length is determined in pachytene meiocytes with full synapsis. Bars represent 5 μm

6. Wash slides in washing buffer for 3 min five times.
7. Stand the slides on filter paper to remove the excess wash buffer, and stain them with DAPI in Vectashield antifade mounting medium. Use a 24 × 32 coverslip, sandwich the preparations in a folded paper, and apply gentle pressure to remove excess PBS-T.
8. Examine the samples with a fluorescence microscope equipped with appropriate filters and image acquisition system.

3.5 Image Analysis

1. Synaptic initiation points (SIPs) can be scored in meiocytes with AtZYP1 signal covering at most 10% of the total of the chromosome axis (Fig. 1).
2. SC length measurements can be achieved using the Image J software. The tool set scale allows to define the spatial scale of the image, and the tool segmented line creates a segmented line selection by repeatedly clicking with the mouse. The SC length is displayed in the status bar during drawing the segmented line

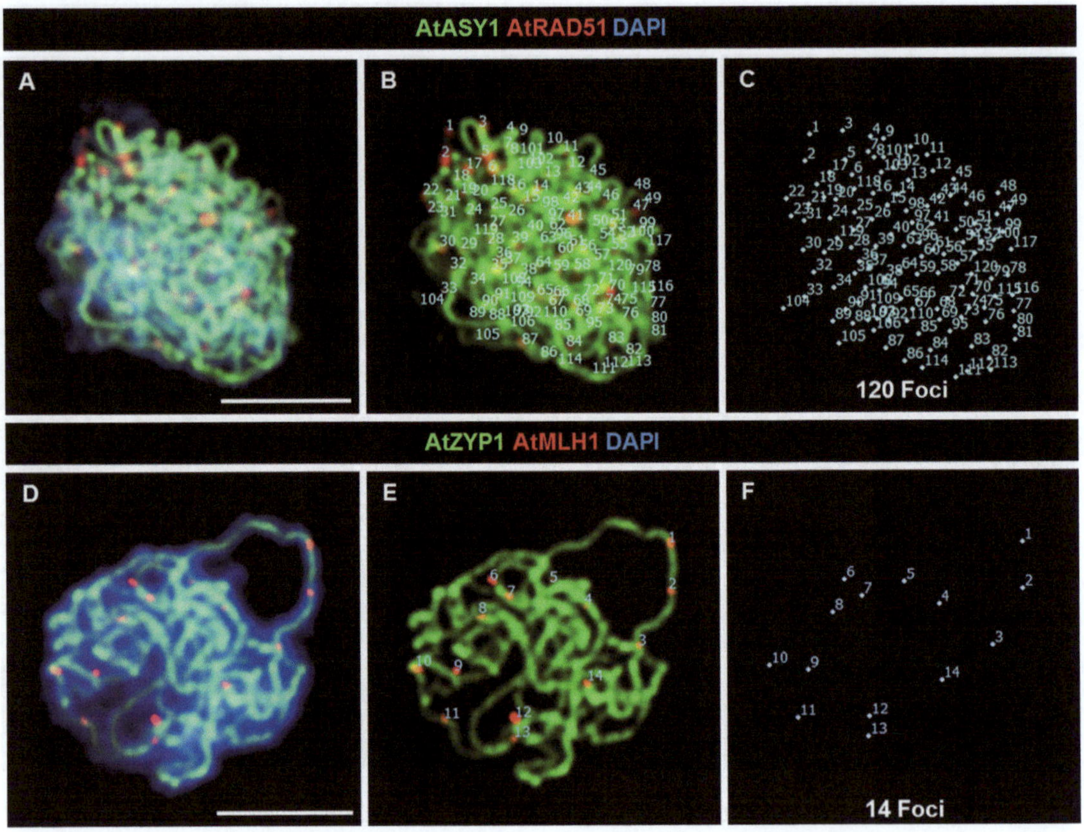

Fig. 2 Examples of images obtained for analyzing the meiotic recombination process. (**A–C**) WT zygotene displaying dual immunolocalization of AtASY1 (green) and AtRAD51 (red). (**D–F**) WT pachytene showing AtZYP1 (green) and AtMLH1 (red). Immunodetection of AtASY1 and AtZYP1 allows to establish the chronology of prophase I. AtASY1, an axial/lateral element protein, is detected from leptotene and persists until late pachytene. The localization of this protein is convenient for the quantification of foci corresponding to the recombinases AtRAD51 and AtDMC1 and also to the ZMM protein AtMSH4. However, the signal corresponding to AtZYP1, a central element protein, is more useful for the quantification of proteins involved in later stages of meiotic recombination, as is the case with AtMLH1. Only AtASY1 or AtZYP1-associated foci are scored. Bars represent 5 μm

over the AtZYP1 signal. Only cells with an AtZYP1 signal covering all the chromosome axes can be analyzed (Fig. 1).

3. The quantification of foci corresponding to HR proteins can be determined by manual counting. For doing this, it is essential to obtain enough image magnification and proper image processing. Although this method is criticized for being time-consuming, it is often more accurate than automatic foci counting. Elevated background signals and a low-defined focal plane can easily lead to incorrect numbers by automatic methods. We only score axis-associated foci applying the count tool of Adobe Photoshop software (Fig. 2) (*see* **Note 27**).

4 Notes

1. To have high-quality images, it is essential to have healthy plants. Soil can be autoclaved to eliminate pests. It is also important to use clean pots to avoid pest contamination. Low densities increase the number of flowers per plant.
2. Pasteur pipettes could be used for holding the slides inside the moist box. Incubate the moist box at 37 °C until it is needed.
3. If there is a precipitate after thawing, warm the solution until it clears.
4. For the detection of SC-associated proteins, bring distilled water to pH 9 using borate buffer.
5. Double immunodetections require the primary antibodies to be raised in different species.
6. For double immunolocalizations the secondary antibodies must recognize one of the species exclusively. To avoid cross-reactivity choose secondary antibodies which have been obtained from different species.
7. The proper working solution usually is more diluted than suggested by the manufacturer. It should be determined experimentally.
8. The fluorophore must be selected according to the available filters in the microscope and the quality of the antibodies. Generally, the brightest signals correspond to red fluorophores, and they are recommended for punctate foci, but they usually cause a higher background noise.
9. Primary shoots usually produce a few buds per inflorescence. Flowers from the lateral shoots are the best to use because they generate large inflorescences with many buds. Cutting off the primary shoot triggers most lateral shoots that emerge rapidly from the basal rosette.
10. Be aware of any movement between the slide and the coverslip.
11. The appropriate duration of the digestion should be determined experimentally for each accession. 45 min is suitable for Col-0.
12. It is important to dip completely the buds into the solution to get a good digestion.
13. The poly-L-lysine-coated slides provide a sticky surface which enhances the adhesion of the cells. However, it also increases the non-specific binding of the antibodies and the corresponding background noise. Optionally, you can treat the slides with a quick wash in PBS and air dry afterward.
14. Add more PBS as needed to assure that the material does not dry out.

15. Make sure the slide is on a flat surface. In order to get a monolayer of cells, it is very important to avoid excess (the cells will move to the edge of the coverslip) and too little liquid (air bubbles will appear).
16. Blowing breath over the coverslip helps to remove the coverslip.
17. Slides can be maintained for several months in 100% glycerol at 4 °C. If this is the case, it is necessary to wash the preparations four times in PBS (10 min) with gentle agitation.
18. To get better results, it is important to work in a humidified atmosphere (switch on a humidifier) and at 20 °C. We usually perform the protocol in the plant growth room.
19. Pre-warm the moist box which facilitates the digestion.
20. Cut the end of a yellow tip to pipette the lipsol because it is very viscous.
21. The movement is similar to the one you apply to beat an egg. The best results are obtained when bubbles appear on surface.
22. Try to not stir the fixative around a surface higher than the coverslip (up to 40 mm).
23. After drying out a white precipitate can be observed. Preserve the slides in a humidity atmosphere to avoid this problem.
24. This step is usually not necessary for the immunolocalization of the SC proteins.
25. The incubation time varies according to the antibodies used and the nature of the signals. We commonly incubate the preparations for 24 h to detect SC or chromosomal axis-associated proteins, whereas longer times are needed to successfully detect HR proteins (punctate foci).
26. Slow agitation can contribute to reduce the background noise.
27. Since this manual method is operator-biased, it is recommendable that the same person performs all the analyses. It should be useful to blind the images by randomizing them in a single group (without any distinction between WT and mutant cells) before the analysis. Results corresponding to WT cells deviated from reference less than 10% will confirm the reliability of quantifications.

Acknowledgment

We are grateful to Prof. Chris Franklin and Dr. Susan Armstrong (University of Birmingham, UK) and Prof. David Evans and Dr. Katja Graumann (Oxford Brookes University) for providing the antibodies mentioned in this study.

The work in MP laboratory is supported by grants from the Ministerio de Economía y Competitividad of Spain (AGL2015-67349-P) and from the European Union by the advanced grant Marie Curie Initial Training Network (ITN) "COMREC" (Grant agreement number: 606956).

References

1. Varas J, Graumann K, Osman K et al (2015) Absence of SUN1 and SUN2 proteins in Arabidopsis thaliana leads to a delay in meiotic progression and defects in synapsis and recombination. Plant J 81:329–346
2. Graumann K, Runions J, Evans DE (2010) Characterization of SUN-domain proteins at the higher plant nuclear envelope. Plant J 61:134–144
3. Armstrong SJ, Caryl AP, Jones GH et al (2002) Asy1, a protein required for meiotic chromosome synapsis, localizes to axis-associated chromatin in *Arabidopsis* and *Brassica*. J Cell Sci 115:3645–3655
4. Higgins JD, Sanchez-Moran E, Armstrong SJ et al (2005) The Arabidopsis synaptonemal complex protein ZYP1 is required for chromosome synapsis and normal fidelity of crossing over. Genes Dev 19:2488–2500
5. Zamariola L, De Storme N, Tiang CL et al (2013) SGO1 but not SGO2 is required for maintenance of centromere cohesion in *Arabidopsis thaliana* meiosis. Plant Reprod 26:197–208
6. Mercier R, Armstrong SJ, Horlow C et al (2003) The meiotic protein SWI1 is required for axial element formation and recombination initiation in Arabidopsis. Development 130:3309–3318
7. Sánchez-Morán E, Santos JL et al (2007) ASY1 mediates AtDMC1-dependent interhomolog recombination during meiosis in Arabidopsis. Genes Dev 21:2220–2233
8. Higgins JD, Armstrong SJ, Franklin FCH et al (2004) The Arabidopsis MutS homolog AtMSH4 functions at an early step in recombination: evidence for two classes of recombination in Arabidopsis. Genes Dev 18:2557–2570
9. Jackson N, Sanchez-Moran E, Buckling E et al (2006) Reduced meiotic crossovers and delayed prophase I progression in AtMLH3-deficient Arabidopsis. EMBO J 25:1315–1323

Part IV

LINC Complex in Mammalian Tissue, Organs and Disease

Chapter 18

Generation and Analysis of Striated Muscle Selective LINC Complex Protein Mutant Mice

Matthew J. Stroud, Xi Fang, Jennifer Veevers, and Ju Chen

Abstract

The linker of nucleoskeleton and cytoskeleton (LINC) complex mediates intracellular cross talk between the nucleus and the cytoplasm. In striated muscle, the LINC complex provides structural support to the myocyte nucleus and plays an essential role in regulating gene expression and mechanotransduction. A wide range of cardiac and skeletal myopathies have been linked to mutations in LINC complex proteins. Studies utilizing tissue-specific knockout and mutant mouse models have revealed important insights into the roles of the LINC complex in striated muscle. In this chapter, we describe several feasible approaches for generating striated muscle-specific gene knockout and mutant mouse models to study LINC complex protein function in cardiac and skeletal muscle. The experimental procedures used for phenotyping and analysis of LINC complex knockout mice are also described.

Key words LINC complex, Nuclear envelope, Knockout mouse, Knock-in mouse, Cre/loxP, CRISPR/Cas9, Striated muscle, Cardiac muscle, Skeletal muscle

1 Introduction

The application of gene targeting technology to create modifications in a tissue-specific manner and at a precise stage in development is a powerful tool to elucidate the functions of *l*inker of *n*ucleoskeleton and *c*ytoskeleton (LINC) complex proteins in vivo. Targeting specific genes for modification by homologous recombination in embryonic stem (ES) cells has become a routine procedure [1, 2]. Over the past decade, we have routinely reported the use ES cell gene targeting technology to generate null or point mutation alleles in the study of LINC complex protein function [3–6]. More recently, a growing body of studies has utilized clustered regulatory interspaced short palindromic repeats (CRISPR)/ CRISPR-associated protein 9 (Cas9) technology to generate genetically modified mice with extraordinary simplicity and speed

Matthew J. Stroud, Xi Fang and Jennifer Veevers contributed equally to this chapter.

Gregg G. Gundersen and Howard J. Worman (eds.), *The LINC Complex: Methods and Protocols*, Methods in Molecular Biology, vol. 1840, https://doi.org/10.1007/978-1-4939-8691-0_18, © Springer Science+Business Media, LLC, part of Springer Nature 2018

[7–11]. Here, we briefly describe the methodology of traditional gene targeting (Fig. 1) and report in detail a cloning-free method to successfully generate floxed as well as single-amino-acid-substituted mice that can be used for striated muscle-specific gene targeting of LINC complex proteins. The general approaches and techniques described here can also be applied to establish other gene knockout and mutant mouse models.

1.1 Generation of Floxed Mice

Conditional knockout of a target gene in mice is based on generating a floxed allele by inserting two 34 base pair (bp) sequences of DNA referred to as loxP sites that flank either side of a critical exon in a gene of interest. The loxP sites are recognized by a Cre recombinase, which mediates recombination to excise the floxed exon and achieve gene inactivation [3, 12–14]. Many of the genes encoding the LINC complex are available as floxed alleles on the International Mouse Phenotype Consortium (IMPC) website (https://www.mousephenotype.org/data/search), which greatly expedites the process of generating conditional null alleles. Generally, the IMPC mouse lines contain the desired floxed allele with LacZ and neomycin cassettes still present (Fig. 1A) (*see* **Note 1**), which are removed by crossing the mice with a flippase (FLP) deleter mouse [15]. Alternatively, global or conditional tissue-specific knockout mice can be generated by crossing the IMPC mice with a desired Cre deleter mouse (Fig. 1A). For mouse lines unavailable through the IMPC website, two approaches can be used to generate floxed alleles as outlined below.

In traditional gene targeting, a conditional construct is first generated (Fig. 1B) (*see* **Notes 2** and **3**), linearized with a restriction enzyme, and electroporated into ES cells. Targeted ES cells are identified by Southern blot analysis. ES cells from a homologous recombinant clone are then microinjected into mouse blastocysts. Male chimeras are bred with female breeder mice to generate germline-transmitted heterozygous mice with a neomycin cassette, which are confirmed by PCR or Southern blot analysis of mouse tail DNA [3–6, 16].

In this chapter, we focus on the clustered regulatory interspaced short palindromic repeats (CRISPR)/CRISPR-associated protein 9 (Cas9) system as a powerful tool to generate striated muscle-specific gene knockout and mutant mouse models to study LINC complex protein function in cardiac and skeletal muscle. The CRISPR/Cas9 system consists of a Cas9 nuclease and two small RNAs: *CR*ISPR RNA (crRNA), which acts as a guide for gene targeting, and *tra*ns-activating crRNA (tracrRNA), which binds to crRNA and forms a ribonucleoprotein complex with Cas9 to direct sequence-specific double-stranded breaks (DSBs) [17–19]. Recently, we developed a cloning-free method to generate floxed alleles by pronuclear injection of a commercial Cas9 protein:crRNA:tracrRNA:single-strand oligodeoxynucleotide (ssODN) complex into mouse zygotes [9] (Fig. 2A). In this method, two crRNAs are designed to direct the Cas9 to target the upstream and downstream introns of the target

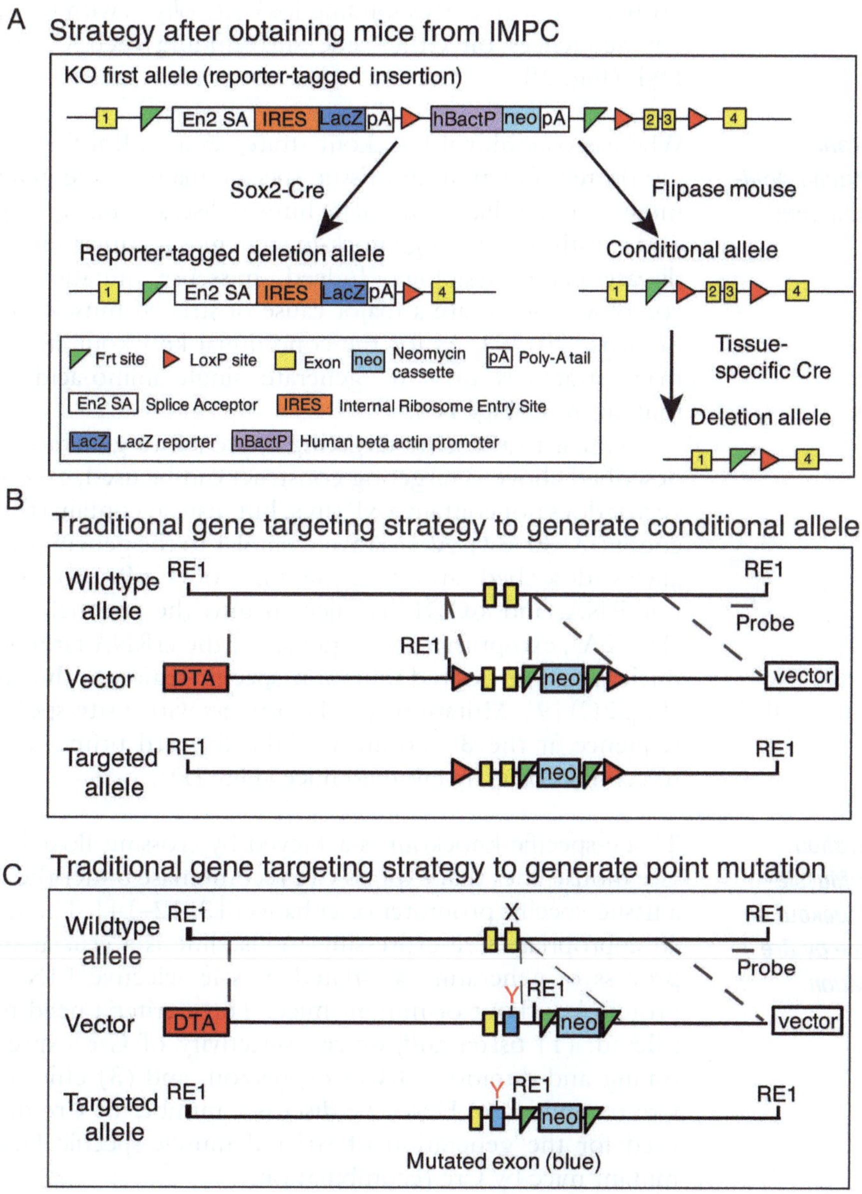

Fig. 1 Mouse generation strategies. (**A**) Mice obtained from the International Mouse Phenotype Consortium (IMPC) can be crossed with Sox2-Cre mice to generate global knockout mice. Alternatively, they can be crossed with flippase deleter mice to generate the conditional allele followed by a tissue-specific Cre deletion in the relevant tissue or cell type. (**B**) The traditional gene targeting approach inserts two 34 base pair sequences referred to as loxP sites (red triangles) that flank either side of a critical exon in a gene of interest and a neomycin cassette for embryonic stem (ES) cell selection. Diptheria toxin A gene is incorporated outside of the homology arms to select against ES colonies that have not taken up the insert by homologous recombination. A new restriction enzyme site is introduced in the mutant construct to aid screening with Southern blotting. (**C**) Traditional gene targeting approach to generate the substitution mutation from nucleotide X to Y in an exon (blue)

exon, along with corresponding loxP site oligos with 60 bp homology sequences on either side surrounding each Cas9-mediated DSB (Fig. 2B).

1.2 Generation of Single-Amino-Acid-Substituted Mutant Mice

Whereas conditional knockout strategies are ideal for understanding protein function in a tissue-specific manner, the generation of mouse models that recapitulate human disease-causing single amino acid substitutions using knock-in strategies is critical in the study of disease pathophysiology. Indeed, missense mutations in LINC complex proteins are a major cause of striated muscle disorders in humans [20–29]. As for the conditional knockout strategies, two main strategies exist to generate single-amino-acid-substituted mutant mice (Fig. 1C).

Similar to the gene targeting approach to generate knockouts described above, a targeting construct can be used, except that the vector does not contain loxP sites, but instead contains the mutated codon(s) [30–32]. Alternatively, similar to the generation of floxed alleles described above, a mixture of Cas9 protein, crRNA, tracrRNA, and ssODN is injected into the pronuclei of zygotes (Fig. 2A), except that the sequence of the crRNA targets the Cas9 nuclease to the desired mutation-specific region of the target gene (Fig. 2C) [9]. Mutation-specific primers with a site-specific variant sequence at the 3′ terminus of the forward primer are used to screen for correctly targeted mice (Fig. 2D).

1.3 Generation of Striated Muscle-Specific Knockout or Mutant Mice by Cre Recombination

Tissue-specific knockout is achieved by crossing floxed mice with Cre mouse lines that express Cre recombinase under the control of a tissue-specific promoter or enhancer [3, 12–14]. The selection of an appropriate Cre-expressing mouse line is a crucial step in the process of generating a striated muscle selective LINC complex protein knockout or mutant mice. Three criteria need to be considered: (1) tissue and/or cell selectivity of Cre expression, (2) timing and duration of Cre expression, and (3) efficiency of the Cre recombinase. Here, we discuss a number of Cre mouse lines used for the generation of striated muscle-specific knockout or mutant mice by Cre recombination.

There are many cardiac-specific Cre lines available for ablating gene expression in cardiomyocytes (*see* Table 1 and references therein) [33]. Here we describe those we have used in our studies of cardiac function. The most widely used Cre lines for inducing gene ablation in cardiomyocytes are those utilizing promoter regions from the alpha (α)-myosin heavy chain (MHC) gene [34–36]. However, a potential caveat of using certain αMHC lines is that those with high levels of myocardial expression of Cre recombinase lead to dilated cardiomyopathy [36]. In contrast and in our hands, the αMHC-Cre line developed by Dale Abel's laboratory has shown no evidence of Cre toxicity [32].

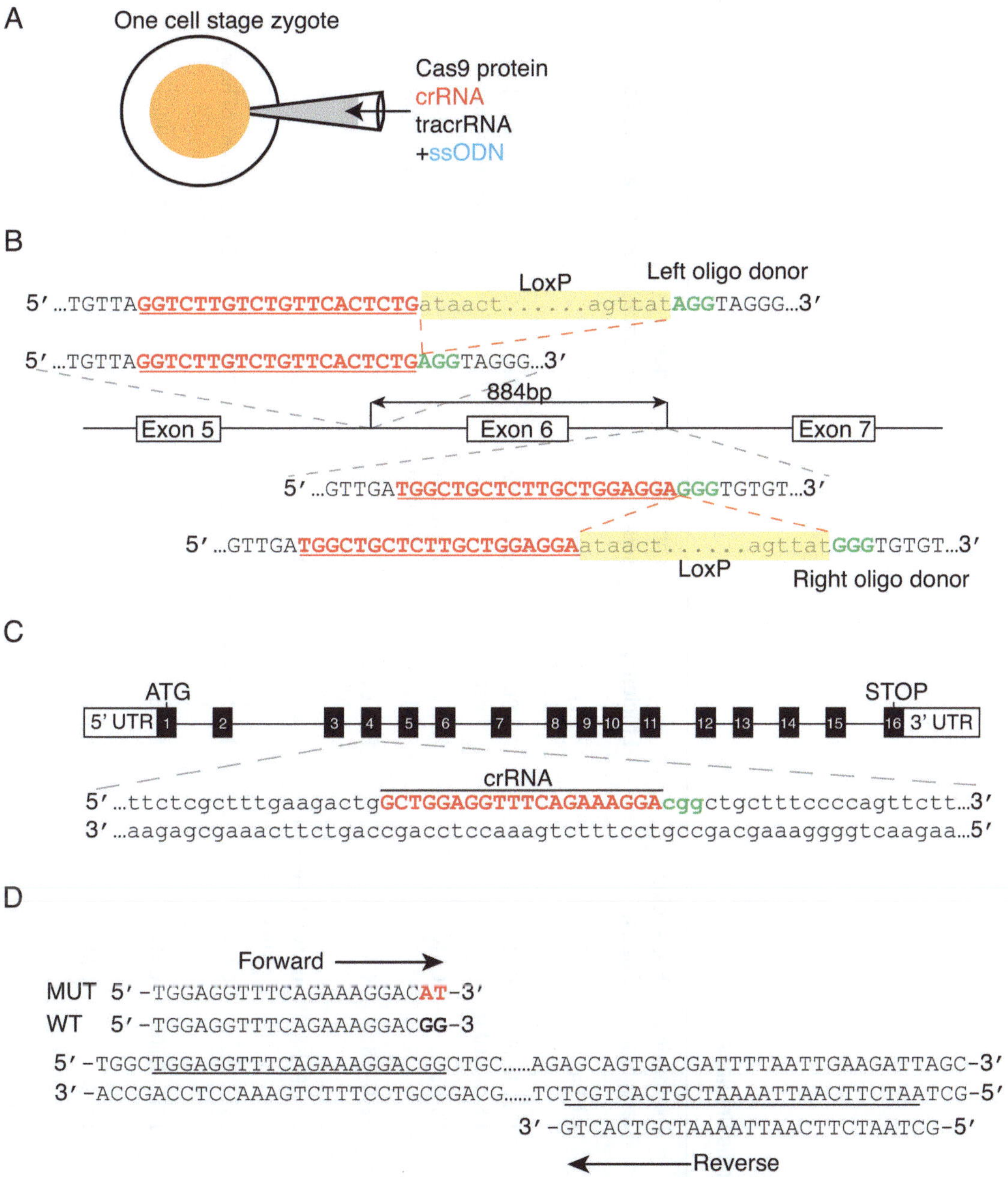

Fig. 2 CRISPR/Cas9-mediated gene manipulation to generate floxed and point mutation alleles. (**A**) A cloning-free CRISPR/Cas9 system by pronuclear injection into a one-cell stage zygote of commercial Cas9 protein combined with chemically synthesized CRISPR RNA (crRNA), trans-activating crRNA (tracrRNA), and single-strand oligodeoxynucleotides (ssODNs). (**B**) Schematic illustration to generate a conditional allele by insertion of two loxP sites. In this method, two crRNAs are designed to direct the Cas9 to target the upstream and downstream introns of the target exon, along with corresponding loxP site oligos with 60 bp homology sequences on either side surrounding each Cas9-mediated double-strand break. The protospacer adjacent motif (PAM) is shown in green. In the oligo donor sequence, the loxP site is highlighted in yellow. (**C**) To generate point mutation mice, the sequence of the crRNA (red) targets the Cas9 nuclease to the desired mutation-specific region (in this example, CGG to CAT) of the target gene. The PAM is shown in green. (**D**) Design of the mutation-specific primers with a site-specific variant sequence at the 3′ terminus of the forward primer to screen for correctly targeted mice. Figure modified from [9]

Table 1
Cardiac and skeletal muscle-specific Cre lines

Tissue specificity	Cre line	Expression Temporal	Expression Spatial	Efficiency	Laboratory source	Original reference	Potential caveats
Cardiac constitutive	Alpha MHC	E10.5	CMs	70% excision in adult CMs	Abel	[34]	–
	Alpha MHC	Low levels E8.0–E10.0	CMs	70–80% excision by 3 weeks	Schneider	[35]	High expression level leads to DCM
	Beta MHC	E12.5	CMs	70% excision by E17.5	Yutzey and Molkentin	[66]	Expressed in somites and soleus muscle
	Nkx2.5	E8.0–E8.5	CMs	88% excision in adult CMs	Olson	[37]	–
	Nkx2.5	E7.75–E8.0	CMs, pharyngeal endoderm	High	Harvey	[38]	–
	Nkx2.5	E7.5	CMs	100% excision by P1	Schwarz	[39]	Development of heart abnormalities
	cTnT	E7.5–E10.5	CMs	High	Hogun	[42]	Breeding problems in older mice
	XMLC2	E7.5–adulthood	CMs	High	Mohun	[44]	–
	MLC2v	E8.5	Ventricular CMs	Low	Chen	[45]	–

Inducible cardiac	MerCreMer-Alpha MHC	N/A	CMs	>80% excision after 5 days	Molkentin	[48]	Cre toxicity observed at high levels
	MerCreMer-cTnT	N/A	CMs	High	Cai	[55]	–
	Tet-TnT	N/A	CMs	Not reported	Zhou	[56]	–
Constitutive skeletal	HSA	E9.5	Skeletal muscle, some evidence in heart	High	Melki	[57]	Low level expression in the heart
	Myo	E8.5	Skeletal muscle lineage	>90% excision	Olson	[60]	–
Inducible skeletal	HSA-MerCreMer	–	Skeletal muscle, some evidence in heart	High	Esser	[58]	Low level expression in the heart
	HSA-Tet	–	Skeletal muscle	Not reported	Monks	[59]	–
	CreRT2 lines	–	Various		Lepper	[61]	–

CMs: cardiomyocytes, *E:* embryonic day, *MHC:* myosin heavy chain, *Nkx2.5:* Nk2 homeobox 5, *cTnT:* cardiac troponin T, *XMLC2:* Xenopus myosin light chain 2, *MLC2V:* myosin light chain 2 ventricular, *MerCreMer:* mutated estrogen receptor, *Dox:* doxycycline, *HSA:* human alpha skeletal actin, *DCM:* dilated cardiomyopathy

As one of the earliest markers of heart progenitor cells, the regulatory regions of Nkx2.5 have been used to generate several Cre lines for inducing gene ablation during early cardiogenesis [37–39]. In our experience, the Nkx2.5-Cre transgenic line developed by the Olson laboratory reduced the RNA level of nesprin 1 by 88% in adult cardiomyocytes [3, 37]. The Nkx2.5-Cre developed by the Harvey laboratory is also highly efficient; however, it is expressed in derivatives of the pharyngeal endoderm, which include the stomach, spleen, pancreas, and liver [38]. The Schwarz laboratory developed an Nkx2.5 knock-in Cre line, which is heterozygous null for Nkx2.5 [39]. However, owing to the importance of the Nkx2.5 gene, these mice develop cardiac abnormalities [40, 41].

The cardiac troponin T (cTnT)-Cre line generated by Brigid Hogan's laboratory, in which Cre is driven by the rat cTnT promoter, induces early recombination at E7.5 with high efficiency in cardiomyocytes [42]. In our experience of using this line, we achieved ~80% reduction at the protein level of the two floxed alleles, Numb and NumbL, from whole hearts at E10.5 [43]. For this line, it is important to maintain male breeders of a young age (<6 months of age) as older breeders develop a shorter stature and don't reproduce as well (unpublished observations).

The Mohun laboratory has developed a *Xenopus myosin light chain 2* promoter-driven Cre recombinase (XMLC2-Cre) that is functional in all myocardial cells throughout embryonic development and adulthood [44]. Importantly, the onset of recombinase activity occurs very early in cardiogenesis at the cardiac crescent stage. Given that the XMLC2-Cre mouse line has high efficiency, precise tissue specificity, and is early onset, it is a powerful tool to achieve gene ablation in a myocardial-specific manner in both embryos and adults.

Our laboratory generated a ventricle-restricted Cre line by knocking-in Cre into the myosin light chain 2v (MLC2v) locus [45]. This MLC2v-Cre mouse is heterozygous for MLC2v, but the mice are normal, display no morphogenic defects, and express normal levels of MLC2v protein [46]. The Cre recombinase expresses at the earliest stages of ventricular chamber specification; however, the efficiency of the recombinase is relatively low in embryonic ventricular cardiomyocytes [45, 47].

Temporal regulation of gene excision is highly desirable for studying gene function in the adult heart, especially if ablation during development results in embryonic lethality. This can be achieved through using inducible tamoxifen- or tetracycline-driven Cre lines (Table 1). The αMHC-MerCreMer transgenic line, in which the cardiac-specific αMHC promoter directs expression of tamoxifen-inducible Cre recombinase (MerCreMer) [48], has been used extensively. We and others have shown that Cre activity is tightly regulated and that Cre-mediated recombination occurs only in response to the injection of tamoxifen [49, 50]. There have been reports of tamoxifen treatment induced phenotypes using this line; therefore, we recommend using αMHC-MerCreMer mice alone and/or αMHC-MerCreMer mice with the heterozygous floxed allele as controls [51–54].

Recently, a tamoxifen-inducible cTnT-Cre was developed by introducing the MerCreMer cassette upstream of the first exon of the cTnT gene [55]. Minimal Cre "leakiness" prior to tamoxifen injection and robust recombination in cardiomyocytes was reported 24 h after administration. Tamoxifen administration alone may cause behavioral alterations; therefore, dosage optimization is pertinent when using tamoxifen-inducible mouse lines [54]. As an alternative to tamoxifen, for cardiac-specific ablation we have used the tetracycline-inducible system developed by the Zhou laboratory. This approach uses the rat troponin T promoter to express the reverse tetracycline-controlled transactivator, which in turn drives Cre expression via a tetracycline-responsive promoter [43, 56].

A number of constitutive and inducible skeletal muscle-specific Cre mice have also been developed (Table 1). The human α-skeletal actin (HSA) promoter has been used to drive constitutive [57], tamoxifen-inducible [58], and tetracycline-inducible [59] Cre expression. HSA-Cre expression is largely restricted to skeletal muscle; however, there is evidence of mosaicism between different skeletal muscle types and evidence of low-level expression in the heart. Myogenin-Cre transgenic mice were developed by the Olson laboratory [60] in which Cre expression is under the control of the mouse myogenin promoter and mouse myocyte enhancer factor 2C (MEF2C) enhancer region. There are an extensive number of tamoxifen-inducible skeletal muscle knock-in Cre lines developed by the Lepper laboratory, which include Pax3, MyoD, Myog, Myf6, and Myl1, and have been well characterized [61].

If global knock-in homozygous mice are not viable or unable to reproduce, tissue-specific mutant mice strategies can be employed to circumvent this. The Kontaridis laboratory generated Cre-dependent conditional knock-in mice harboring a Y279C mutant of the Ptpn11 gene [62]. In this case, the Y279C mutant is only expressed after Cre-mediated excision. Therefore, a tissue-specific Cre could be used to drive expression only in the tissue of interest. As a simpler alternative, we crossed conditional knockout mice (f/f; Cre+) with heterozygous mutants (m/+) to generate conditional mutant mice (f/m; Cre+) mice [32]. Conditional mutant mice were subsequently crossed with floxed (f/f) mice to generate relevant control littermates (Fig. 3) (*see* **Note 4**).

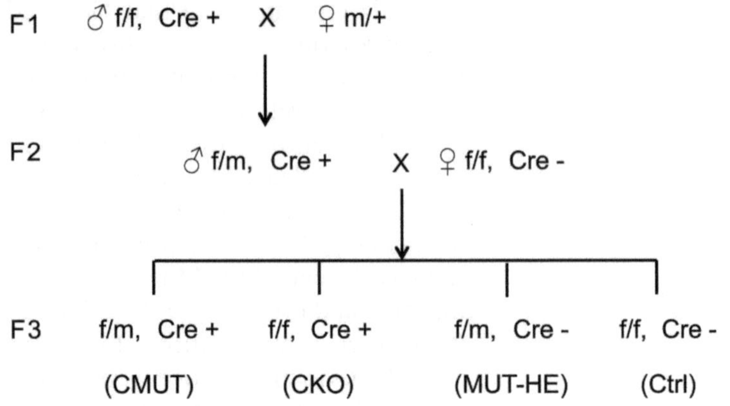

Fig. 3 Breeding strategy of conditional knock-in mice. *See* text for description

2 Materials

2.1 CRISPR/Cas9-Mediated Gene Manipulation to Create Floxed and Point Mutation Mice with a Cloning-Free Method

2.1.1 CRISPR/Cas9 Injection

1. Mice. C57BL/6 J female mice at age 3–4 weeks and male mice at 9–24 weeks for zygote collection. ICR (CD-1) mice for pseudopregnant mother and vasectomized males.
2. Recombinant Cas9 proteins.
3. CrRNA and tracrRNA: chemically synthesized and RNase-free HPLC purified.
4. Single-strand ODN: chemically synthesized and standard desalted.
5. 1× Tris-EDTA (TE) solution.
6. Pregnant mare serum gonadotropin (PMSG).
7. Human chorionic gonadotropin (hCG).
8. KSOM medium.
9. Reagents and equipment used for routine pronuclear microinjection are provided in the protocol on the University of California, San Diego Transgenic Mouse Core website (https://healthsciences.ucsd.edu/research/moores/shared-resources/transgenic-core/services/Pages/pronuclear-injection.aspx).

2.1.2 Genotyping and Screening of Targeted Alleles

1. Primers.
2. 50 mM NaOH.
3. 1 M Tris–HCl (pH 8.0).
4. *Taq* PCR Kit.
5. Agarose.
6. Tris-acetate-EDTA (TAE) buffer.
7. Zero Blunt TOPO PCR cloning kit.
8. LB plates containing 50 μg/mL kanamycin.

2.2 Maintenance of Mouse Stains and Generation of Conditional Knockout and Mutant Mice

1. Mice. C57BL/6J.
2. Flippase deleter mice [15].
3. Appropriate striated muscle-specific Cre-expressing mouse line (*see* Table 1).

2.3 Induction of Cre Expression

2.3.1 Tamoxifen

1. Chow containing 0.4–1 g/kg tamoxifen.
2. Water containing 0.5–1 mg/mL tamoxifen.
3. Sesame oil containing 10 mg/mL tamoxifen.

2.3.2 Doxycycline

1. Water containing at 1–2 mg/mL doxycycline.

2.4 Analysis of Hearts

2.4.1 Echocardiographic Analysis of Cardiac Function

1. Electric razor.
2. Hair removal cream (Nair).
3. Echocardiography machine (FujiFilm Visualsonics 2.0) with heating pad and electrodes.
4. Echo software.
5. Echo gel.
6. Isofluorane.

2.4.2 Sample Preparation

1. Heparin (1000 U/mL).
2. Ketamine (100 mg/kg).
3. Xylazine (10 mg/kg).
4. Insulin needles.
5. Fine scissors.
6. Extra fine Graefe forceps.
7. Ice cold phosphate-buffered saline (PBS).

2.4.3 Measurement of Heart Weight/Tibia Length and Heart Weight/Body Weight Ratios

1. 70% ethanol.
2. Medium-size forceps.
3. Vernier calipers (Kingsmart 6 in. LCD digital/micrometer gage 150 mm).

2.4.4 Immunofluorescence Analysis

1. Cryostat.
2. Cryostat blades.
3. Wax pen.
4. Superfrost Plus slides.
5. Humidified slide-incubation chamber.
6. Thickness 1.5 coverslips.
7. Inverted confocal microscope.
8. PBS.
9. Acetone.

10. 4% paraformaldehyde (PFA) in PBS (wt/vol).
11. Optimal cutting temperature (OCT) compound.
12. Sucrose.
13. Dry ice.
14. 3% bovine serum albumin (99.5% pure) in PBS.
15. Normal donkey serum.
16. DAKO mounting medium.
17. Clear nail polish.
18. Vectashield mounting medium.
19. 0.2% Triton-X100 in PBS (PBS/T).
20. Relevant antibodies (*see* Table 2).

2.4.5 Histological Preparation for Chemical Staining

1. Tissue cassettes.
2. Wide-necked Erlenmeyer flask.
3. Glass histology staining chambers.
4. Microtome.
5. Water bath.
6. Oven.
7. Ethanol.
8. Xylene.
9. 10% neutral buffered formalin.
10. 4% PFA in PBS.
11. Paraffin.
12. Hematoxylin and eosin stain.
13. Masson's trichrome stain.
14. Brightfield microscope.

2.4.6 RNA Extraction

1. RNA extraction kit.
2. cDNA synthesis kit.
3. Quantitative real-time PCR machine.
4. Primer pairs (*see* Table 3).

2.4.7 Protein Extract Preparation

1. Pestle and mortar.
2. Sonicator.
3. Spatula.
4. Western blotting apparatus.
5. Liquid nitrogen.
6. Lysis buffer: 8 M urea, 2 M thiourea, 3% SDS (wt/vol), 75 mM DTT, 0.03% bromophenol blue (wt/vol), 50 mM Tris–HCl pH 6.8.

Table 2
Recommended antibodies for detecting LINC complex proteins and sarcomeres

Protein	Company/source	Cat. No./clone	Application	Epitope	Host species and nature	Validated with knockout tissue?	Reference
Nesprin 1	Glenn E Morris	7A12	WB and IF	Human Nesprin 1α2	Rabbit monoclonal	Yes	[67]
SUN1	Millipore	ABT285	WB	Mouse C-terminus	Rabbit polyclonal	–	[3]
SUN1	Abcam	ab103021	IF	Mouse AAs 200–300	Rabbit polyclonal	–	[5]
SUN2	Abcam	EPR6557	WB and IF	Human C-terminus	Rabbit monoclonal	–	[5]
LUMA	Abcam	EPR15378	WB and IF	Human AAs 100–200	Rabbit monoclonal	Yes	[69]
Emerin	Novocastra	NCL-EMERIN (4G5)	IF	Human N-terminus	Mouse monoclonal	–	[68]
Emerin	Santa Cruz	sc-15378 (FL-254)	WB	Human AAs 3–254	Rabbit polyclonal	–	[5]
LEM2	Sigma	HPA017340	WB and IF	Human AAs 243–361	Rabbit polyclonal	Yes	[69]
Sarcomeric α-actinin	Sigma	A7811	WB and IF	Rabbit skeletal α-actinin	Rabbit polyclonal	–	[30]

WB: Western blot, *IF:* immunofluorescence

2.5 Analysis of Skeletal Muscle

2.5.1 Preparation of Tools for TA Muscle Fiber Isolation

1. P200 tips.
2. Bunsen burner.
3. Razor blade.
4. 0.2% Triton-X100 in PBS (PBS/T).
5. Superfrost Plus slides.
6. Wax pen.

2.5.2 Extraction of TA Muscle from Leg

1. Dissection microscope.
2. Lightsource.
3. Dumont #5 fine forceps.
4. Spring scissors.
5. Dumont AA polished forceps.
6. 4% PFA in PBS.

2.5.3 Analysis of Myonuclear Positioning and Immunofluorescence Staining

1. Thermomixer (Eppendorf Thermomixer F).
2. Superfrost Plus slides.
3. Thickness 1.5 coverslips.
4. DAKO mounting medium.

7. Wax pen.

5. 10 M NaOH.
6. PBS.
7. 4′,6-diamidine-2′-phenylindole dihydrochloride (DAPI).
8. 3% bovine serum albumin (99.5% pure) in PBS.
9. Normal donkey serum.
10. 0.2% Triton-X100 in PBS (PBS/T).
11. Relevant antibodies (*see* Table 2).

3 Methods

3.1 CRISPR/Cas9-Mediated Gene Manipulation to Create Floxed and Point Mutation Mice with a Cloning-Free Method

3.1.1 CRISPR/Cas9 Targeting Design

1. For design of the guide CrRNA, follow the protocol described by the Zhang laboratory [63]. Briefly, input the target genomic DNA sequence into the online CRISPR design tool: http://crispr.mit.edu/. The CRISPR design tool takes an input sequence (e.g., a 1 kb genomic fragment from the region of interest), identifies and ranks suitable target sites, and computationally predicts off-target sites for each intended target. Alternatively, one can manually select guide sequences by identifying the 20 bp sequence directly upstream of a protospacer adjacent motif (PAM) sequence (5′-NGG) [9] (Fig. 2B).

Table 3
Sequences of qRT-PCR primers used to detect fetal, pro-fibrotic, early activator, and reference genes

	Forward	Reverse
Fetal genes		
ANP	GATAGATGAAGGCAGGAAGCCGC	AGGATTGGAGCCCAGAGTGGACTAGG
BNP	TGTTTCTGCTTTTCCTTTATCTGTC	CTCCGACTTTTCTCTTATCAGCTC
AMHC	CTGCTGGAGAGGTTATTCCTCG	GGAAGAGTGAGCGGCGCATCAAGG
BMHC	TGCAAAGGCTCCAGGTCTGAGGGC	GCCAACACCAACCTGTCCAAGTTC
Pro-fibrotic genes		
COL1A1	TCACCAAACTCAGAAGATGTAGGA	GACCAGGAGGACCAGGAAG
COL3A1	ACAGCAGTCCAACGTAGATGAAT	TCACAGATTATGTCATCGCAAAG
Early activator genes		
Egr-1	CCTATGAGCACCTGACCACA	TCGTTTGGCTGGGATAACTC
Iex-1	TTATAGGGTCGGTAAGACAGAGTTG	GACGGAGTGTTACCCCTAATCTTAT
c-fos	AGCCCCTGTGTACTCCCGTG	GCCTTGCCTTCTCTGACTGC
c-jun	TTCCTCCAGTCCGAGAGCG	TGAGAAGGTCCGAGTTCTTGG
c-myc	ATGCCCCTCAACGTGAACTTC	GTCGCAGATGAAATAGGGCTG
Reference genes		
18S	GGAAGGGCACCACCAGGAGT	TGCAGCCCCGGACATCTAAG
GAPDH	CTCAAGATTGTCAGCAATGCATCC	CCAGTGGATGCAGGGATGATGTTC

Primers are in 5′ to 3′ orientation

2. For conditional allele generation, two sgRNAs are designed to elicit DSBs that flank the sequence to be deleted. In addition, an ssDNA oligo is designed that contains the corresponding loxP site flanked by two 40- to 60-base homology arms that correspond to the sequence surrounding each sgRNA-mediated DSB (Fig. 2B).
3. To introduce a point mutation, in addition to the sgRNA targeting the site of interest, an ssDNA oligo is designed that contains the desired alteration flanked on each side by 40–60 bases that are homologous to the sequences directly upstream and downstream of the DSB.

3.1.2 Zygote Preparation and Microinjection

1. Inject 12–15 female C57BL/6 J (3–4 weeks old) mice with PMSG (5 IU) at 1:00–2:00 p.m. on day 1.
2. After 48 h, inject female mice with hCG (5 IU). After hCG injection, house female mice with C57BL/6 J male mice overnight.
3. Prepare the medium for embryo culture. Place several drops (30–50 μL for each drop) of KSOM medium on a 6-cm dish and cover with mineral oil; place the dish into a 37 °C incubator for at least 30 min before use.
4. At 20–21 h after hCG injection, euthanize the mice and collect zygote-cumulus complexes from the oviduct where it is most swollen.
5. Move the zygote-cumulus complexes into M2 + Hy medium; gently triturate three times with a P200 tip.
6. Gently aspirate complexes with a transfer pipette, wash three times in M2 medium, and place the embryos into KSOM medium at 37 °C in a 5% CO_2 incubator.
7. Prepare the injection mix (Cas9 protein, crRNA, and tracrRNA, with or without ssODN (experiment dependent)) and mix in TE buffer to a working concentration of 30 ng/μL, 0.6 pmol/μL, 0.6 pmol/μL, and 20 ng/μL, respectively. Incubate the mixture at 37 °C for 5 min (Fig. 2A).
8. The standard protocol describing pronuclear microinjection is provided on the University of California, San Diego Transgenic Mouse Core website (https://healthsciences.ucsd.edu/research/moores/shared-resources/transgenic-core/services/Pages/pronuclear-injection.aspx).

3.1.3 Primer Design, Genotyping, and Screening of Targeted Alleles

1. Primer design: To screen for correctly targeted point mutation mice, a mutation-specific forward primer is designed with the site-specific variant sequence at the 3′ end, and a generic wild-type reverse primer is used [9] (Fig. 2D). To detect floxed alleles, a loxP-specific and wild-type primer pair are designed.

2. Genomic DNA preparation:
   (a) Collect tail biopsies from 3-week-old mice and add 300 μL of 50 mM NaOH.
   (b) Incubate at 98 °C for 30 min; then add 50 μL of 1 M Tris–HCl (pH 8.0). 2 μL of genomic DNA sample is used for genotyping PCR.
3. Genotyping PCR of F0 founders:
   (a) Amplify the extracted DNA using gene-specific primers under the following conditions: 30 cycles at 94 °C for 20 s, 60 °C for 20 s, and 72 °C for 30 s.
   (b) Run the PCR product on a 2% (wt/vol) agarose gel in TAE buffer to verify that the product is unique and of the expected size.
4. PCR for sequencing validation of mutated alleles of F0 founders:
   (a) PCR amplify the extracted genomic DNA using primers flanking the targeted region.
   (b) Clone the PCR product using a Zero Blunt TOPO PCR cloning kit according to manufacturer's instructions.
   (c) To validate mutant alleles, extract plasmid DNA from six individual colonies and sequence (*see* **Note 5**).

3.2 Maintenance of Mouse Stains and Generation of Conditional Knockout and Mutant Mice

1. Breed F0 mutant mice with wild-type C57BL/6 J mice.
2. The first-generation offspring (F1) from each founder are subjected to genotyping using PCR and are sequenced to confirm germline transmission (as described in Subheading 3.1.3) (*see* **Note 6**).
3. Cross F1 mutant heterozygous mice with wild-type C57/B6J mice to maintain the mutant line.
4. Floxed (f/f) females are mated with Cre-positive (+/+; Cre+) males to generate f/+; Cre+ mice. Floxed females are then mated with f/+; Cre+ males to generate tissue-specific knockout mice (f/f; Cre+) (CKO) and their control littermates (f/f) and (f/+; Cre+) (*see* **Notes 7** and **8**).
5. Heterozygous mutant (m/+) females are mated with CKO (f/f; Cre+) males to generate cell-selective specific mutant mice (f/m; Cre+) (CMUT). CMUT males are bred with homozygous floxed (f/f) females to generate CMUT (f/m, Cre+) and littermate controls (Fig. 3) [32].

3.3 Induction of Cre Expression

Tamoxifen can be administered via food (custom-made chow containing 0.4–1 g/kg tamoxifen, Harlan), water intake (0.5–1 mg/mL), or 1–5 consecutive days of intraperitoneal injection (0.03–0.09 mg/g body weight) [48, 53]. Doxycycline is administered in drinking water at 1–2 mg/mL [56, 59].

3.4 Analysis of Hearts

3.4.1 Echocardiographic Analysis of Cardiac Function

1. Weigh mouse.
2. Shave the middle and the left-hand side of the chest on the underside of the mouse with a small electric razor (*see* Fig. 4A).
3. Using a Q-tip, place a small amount of hair removal cream on the chest around the center and to the left of the rib cage on the underside of the mouse (*see* **Note 9**). Leave for 10–20 s, and then using a surgical swab, wipe off the hair removal cream and hair. Wipe the chest with a wet surgical swab to remove the excess hair removal cream and hair.
4. Place mouse in supine position on pre-warmed echo pad at 39 °C.
5. Place one electrode on the right leg and another on the right arm to measure heart rate.
6. Tape the limbs to the pad.
7. Anaesthetize the mouse using a quick burst of 5% isofluorane in oxygen for a few seconds.
8. Once anesthetized, reduce to 0.5% isofluorane to ensure the heart rate is maintained above 500 beats per minute (*see* **Note 10**).
9. Squeeze a 4 mm cylinder of conductive echo gel on the probe to be used for echocardiography (use a 45 MHz probe).
10. Place the probe on the chest and locate the heart (*see* **Note 11**).
11. For a four-chamber (or long axis) view using B-mode, place the probe perpendicular to the chest at approximately a 45° angle to make an imaginary line between the right arm and middle of the left side of the abdomen (*see* Fig. 4A, left).
12. For a two-chamber (or short axis) view or short axis using M-mode, rotate the probe by 90° to make an imaginary line between the left arm and middle of the right side of the abdomen (*see* Fig. 4A, right).
13. Images are taken at the level of the papillary muscles, which are very prominent, echogenic regions (appear white) in mouse hearts.
14. Record three consecutive cycles of diastole (during heart relaxation) and systole (heart contraction).
15. Remove the excess echo gel and place the mouse back in cage.
16. Download the echo images and perform analysis on a computer with Vevo 2100 software installed.
17. Measure the left ventricle chamber sizes, called the left ventricle internal dimension (LVID), and interventricular septum (IVS) wall thickness and left ventricle posterior wall (LVPW) thickness in diastole and systole, and then calculate the mean values between the three beats (*see* Fig. 4B) (*see* **Note 12**).

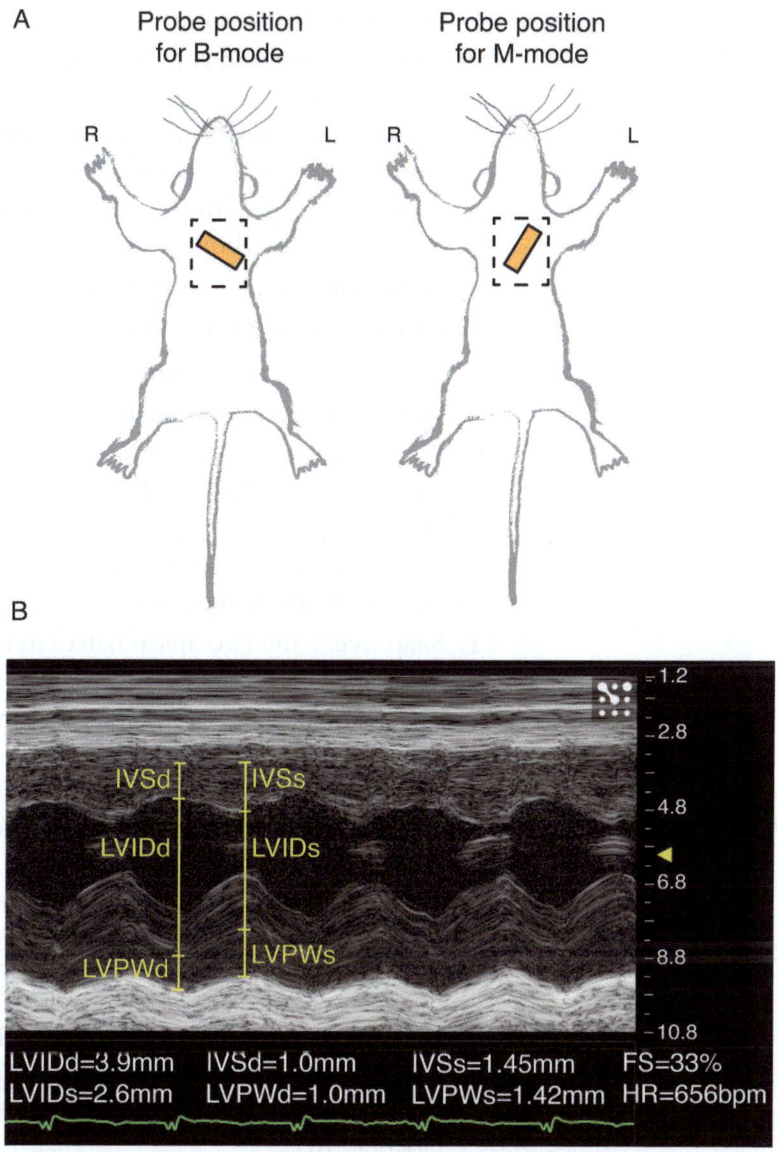

Fig. 4 Echocardiography recording and analysis. (**A**) Approximate positioning of echocardiography probe (orange rectangle) for a 2D long-axis B-mode view (left) and short-axis M-mode view (right). Dashed rectangle indicates the recommended area from which to remove hair. (**B**) Parasternal short-axis M-mode with left ventricular functional measurements. Post capture, left ventricle internal dimension in diastole/systole (LVIDd/s); left ventricle posterior wall thickness in diastole/systole (LVPWd/s); interventricular septum thickness in diatole/systole (IVSd/s); fractional shortening (FS); and heart rate (HR) are measured

18. Left ventricle systolic function is measured using the parameter called fractional shortening, which is measured as a percentage of the amount the chamber sizes change between systole and diastole divided by the chamber size in diastole (FS = [LVIDd−LVIDs]/LVIDd).

3.4.2 Sample Preparation

1. Weigh the animal.
2. Intraperitoneal inject 0.2 mL heparin (1000 U/mL) (*see* **Note 13**).
3. Open up the chest by cutting around the edge of the rib cage to avoid cutting the heart.
4. Using forceps grip the base of the heart and with sharp scissors excise the heart above the aorta to avoid cutting the atria.
5. Rinse heart in ice cold PBS to remove excess blood.
6. Remove the aorta and other non-cardiac tissue.
7. Briefly dry the heart using a surgical swab.
8. Weigh the heart on a fine balance.
9. Using a scalpel blade, dissect the heart into four pieces (*see* Fig. 5A, B) (*see* **Notes 14** and **15**).
10. Place the middle section in a glass vial containing 5 mL 10% neutral buffered formalin and fix at 4 °C overnight for histology (*see* Subheading 3.4.5).
11. Snap freeze the two apical halves in liquid nitrogen (for RNA and protein analysis, *see* Subheadings 3.4.6 and 3.4.7).
12. Take tail biopsy for genotype confirmation and snap freeze.

3.4.3 Measurement of Heart Weight/Tibia Length and Heart Weight/ Body Weight Ratios

1. Spray the lower legs with 70% ethanol to dampen the hair.
2. Using medium-size forceps or between fingers, grab one of the feet.
3. With a pair of medium-size sharp-ended forceps, pierce and grab the skin above the foot and pull skin over the front of the knee joint to expose the patella tendon (shiny, almost metallic-looking strip that is approx. 1 mm wide by 5 mm long) that runs over the front of the knee (*see* **Note 16**).
4. Using Vernier calipers, measure the distance between the middle of the exposed patella tendon and the bottom of the ankle (*see* Fig. 5B).
5. Divide heart weight by tibia length or body weight for global indices of cardiac hypertrophy (*see* **Note 17**).

3.4.4 Immunofluorescence Analysis

1. Place the base of the heart in isopentane to dehydrate (*see* **Note 18**).
2. Remove after 30–60 s and place in a small pool of OCT compound on a sectioning mold.
3. Place the mold on a block of dry ice, and let the tissue/OCT slowly freeze into place.
4. Slowly fill the mold with OCT, making sure the orientation of the heart does not change until the mold is filled with OCT.

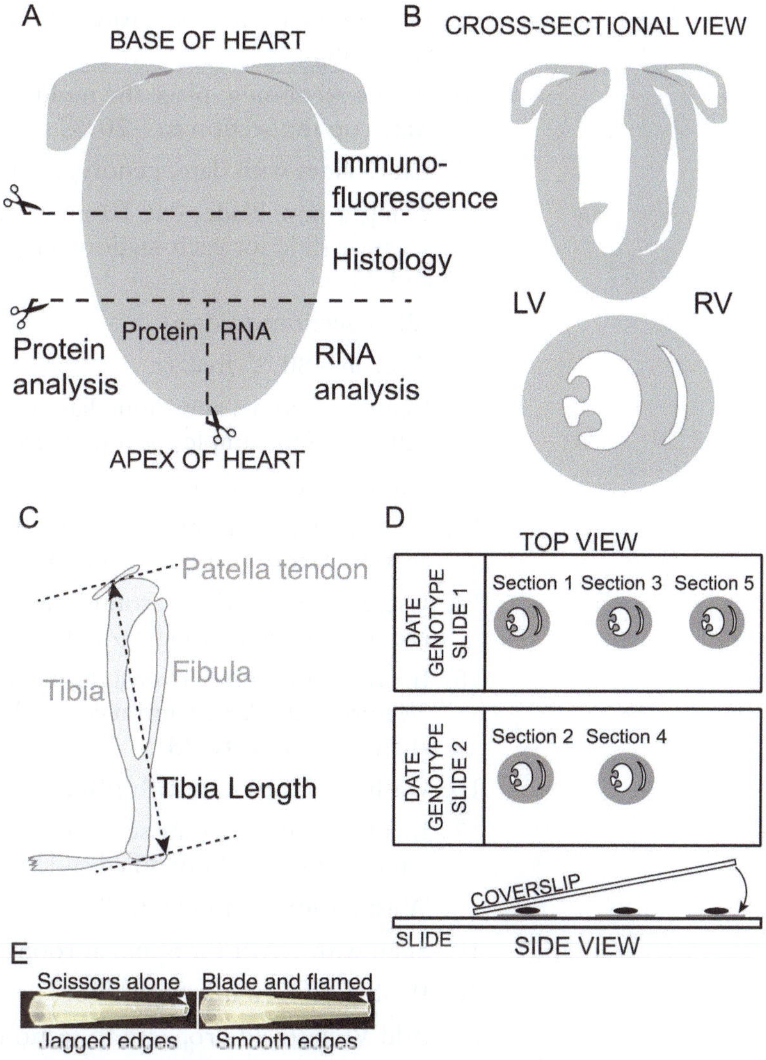

Fig. 5 Processing of heart for analysis. (**A**) Post-extraction, the heart is cut into three pieces: The base is used for immunofluorescence and the midsection for histology; the apex is divided into two parts, for RNA and protein extraction. (**B**) Cross-sectional view of the heart showing all four chambers (top) and two chambers (bottom). (**C**) Schematic of a mouse tibia with measurement points indicated between the bottom of the tibia and the patella tendon. (**D**) Top view: After sectioning the heart for immunofluorescence analysis, sequential sections are placed on consecutive slides to allow direct comparison of different antibodies at similar spatial locations. Side view: Post-antibody staining and wash steps, a small amount of mounting medium is placed directly on the section, and the coverslip is slowly lowered using forceps to avoid generating air bubbles. Depending on the mounting medium, coverslips are either left to dry overnight or sealed with clear nail polish. (**E**) Preparation of P200 tips to pipette isolated myofibers. Note the importance of using a razor blade and gentle Bunsen burner flame to generate smooth edges on the pipette tip to minimize sample loss

5. Use immediately or store at −80 °C until ready for sectioning.
6. Before sectioning, place the mold in the cryostat at −20 °C to warm up the section to −20 °C (*see* **Note 19**).
7. Label slides with date, genotype, slide number.
8. Using a new blade, cut 10 μm sections sequentially using a separate slide for each section cut (*see* Fig. 5D) (*see* **Notes 20** and **21**).
9. Allow sections to dry.
10. Place in −80 °C freezer.
11. Remove sections and immediately place in −20 °C acetone at −20 °C to fix sample for 5 min (*see* **Note 22**).
12. Remove and wash section with PBS.
13. Permeabilize sections with PBS containing 0.2% Triton-X100 at room temperature for 15 min.
14. Use wax pen to draw around individual sections to allow for use of multiple antibodies/slide.
15. Incubate antibodies overnight in PBS containing 3% BSA and 2% normal donkey serum at 4 °C in humidified slide-incubation chamber (*see* **Note 23**).
16. Wash 3× for 10 min with PBS.
17. Incubate with relevant secondary antibodies for 1 h at room temperature in a humidified slide-incubation chamber.
18. Wash 3× for 10 min with PBS.
19. Stain with DAPI for 5 min at room temperature.
20. Wash 3× for 10 min with PBS.
21. Add Vectashield dropwise to sections (*see* **Note 24**).
22. Carefully place coverslip (type 1.5 high tolerance) down at one end of the slide holding the other end up with forceps; then slowly lower the forceps (*see* Fig. 5D) (to avoid generating air bubbles).
23. Let solidify overnight at 4 °C.
24. Image on confocal microscope/deconvolution microscope.

3.4.5 Histological Preparation for Chemical Staining

1. After fixing the heart overnight at 4 °C wash 3× in PBS.
2. Label the tissue cassette using a pencil (*see* **Note 25**).
3. Prepare molten paraffin wax by placing paraffin in a bottle in an incubator set to 55–65 °C.
4. Remove tissue from glass vial and transfer to tissue cassette.
5. Place cassettes in an Erlenmeyer flask (with a neck wide enough to slot the cassettes in) or beaker containing 70% ethanol (filled as necessary to completely cover the cassettes) and wait 20 min.

6. Replace 70% ethanol with 95% ethanol and wait 20 min.
7. Remove 95% ethanol and replace with fresh 95% ethanol and wait 20 min.
8. Remove 95% ethanol and replace with 100% ethanol and wait 20 min.
9. Remove 100% ethanol and replace with fresh 100% ethanol and wait 20 min.
10. Remove 100% ethanol and replace with xylene and wait 20 min.
11. Remove xylene and replace with fresh xylene.
12. Open cassette and pour molten paraffin wax into the cassette.
13. Carefully adjust the orientation of the heart to ensure a two-chamber view.
14. Allow paraffin to set around tissue.
15. Turn on water bath at 35–40 °C.
16. Mount hardened section on microtome with new blade attached.
17. Cut 5 μm sections and place sections on the surface of the water in the bath.
18. Immerse the slide below the floating section and scoop up the section onto the slide.
19. Place slide in an oven at 65 °C for 10–15 min to melt the paraffin and mount the section to the slide.
20. Remove slide.
21. Stain slides with commercially available staining kits for hematoxylin and eosin (nuclei, dark blue/violet; tissue, red/pink) or Masson's trichrome (nuclei, dark red/purple; tissue, red/pink; collagen, blue).

3.4.6 RNA Extraction

1. Take half of the apical section of the heart (Fig. 5A).
2. Extract RNA using the Promega RNA Mini-prep kit (or equivalent).
3. Generate cDNA using a reverse transcription kit.
4. Perform quantitative real-time PCR using primers listed in Table 3.

3.4.7 Protein Extract Preparation

1. Pre-chill pestle and mortar with liquid nitrogen.
2. Take half of the apical section of the heart (Fig. 5A) out of liquid nitrogen.
3. Weigh tissue.
4. Calculate volume of lysis buffer needed to achieve 50 μL/mg tissue (*see* **Note 26**).

5. For tissue weighing >6 mg, place frozen tissue in mortar and add 300 μL lysis buffer (the lysis buffer will freeze, you want to keep everything frozen during this process, and more liquid nitrogen can be added gently if necessary).
6. For tissue weighing <6 mg, add the calculated volume of lysis buffer and continue as below.
7. Homogenize tissue and lysis buffer with pestle and mortar into a fine powder.
8. Use a spatula to scrape the powdered lysate into a 2.0 mL tube.
9. Put sample on ice.
10. Sonicate sample to shear DNA, until the lysate is easy to triturate.
11. Add remaining lysis buffer to top up to the calculated volume necessary for 50 μL/mg.
12. Triturate 3× to mix lysis buffer.
13. Spin at 17,000 G for 15 min.
14. Aspirate supernatant.
15. Aliquot into working aliquots of 300 μL and snap freeze in liquid nitrogen, snap freeze the remaining lysate.
16. Use for Western blot or other downstream analysis.

3.5 Analysis of Skeletal Muscle

3.5.1 Preparation of Tools for TA Muscle Fiber Isolation

1. Before starting, prepare some pipette tips to handle the fibers and a "dissection slide" for the dissection of individual fibers. These steps are important to avoid losing a lot of fibers in the pipette tip and also to avoid contaminating your samples with plastic contaminants.
2. Cut the end from a P200 tip using a razor blade (not scissors), and then under a gentle Bunsen burner flame, flame the edges of the cut tip (*see* Fig. 5E) (*see* **Note 27**). Rinse the P200 tip with PBS/Triton mix to prevent fibers sticking to the inside of the tip.
3. Prepare a "dissection slide" by drawing a square in the middle of a glass slide with a wax pen and leave to dry. This slide will be used to tease apart the individual fibers in a pool of PBS/T.

3.5.2 Extraction of TA Muscle from Leg

1. Cut the whole legs from an E18.5 or neonatal pup and place each in 1 mL of 4% PFA.
2. Fix the samples overnight at 4 °C.
3. Wash legs with PBS 3× to remove the PFA.
4. Remove the skin from the leg using fine forceps.
5. Using a dissection microscope, remove the fascia (thin membrane) surrounding the TA muscle by piercing it and dragging it away from the muscle with sharp, fine forceps.

6. Slide the fine forceps in between the tibia bone and the TA muscle at the front of the shin.
7. Cut/remove the tendons (shiny structures connecting the end of the muscle to bone) with fine scissors (*see* **Note 28**).

3.5.3 Analysis of Myonuclear Positioning

1. Place a 0.5 mm × 2 mm strip of the TA muscle in 200 μL of 10 M NaOH in a 1.5 mL microcentrifuge tube.
2. Place in a thermomixer set to 1000 rpm at 20 °C for 20–30 min depending on the size of the TA fragment. Alternatively, a regular vortex can be used, and pulsed until the fibers start to separate.
3. Inspect the fibers every 10 min to look at the progress of fiber separation.
4. Allow fibers to settle by gravity.
5. Wash 5× with 1 mL PBS to remove the NaOH.
6. Resuspend in 200 μL of PBS and add DAPI for 5 min.
7. Wash 3× for 5 min with PBS.
8. Using a wax pen, draw around the edges of the slide and leave to dry (so that you won't lose any fibers when you squash the coverslip down on the fibers).
9. Remove as much PBS as possible leaving ~20 μL.
10. Resuspend fibers in an equal volume of DAKO mounting medium.
11. Pipette the fibers directly onto the slide, spreading them along the length of the slide.
12. Lower the coverslip and seal with nail polish.
13. Image on a confocal or wide-field deconvolution microscope.
14. Nuclear lengths and internuclear distances were measured using ImageJ software. In brief, for internuclear distances, a line was drawn between the nuclear centroid to centroid; for nuclear lengths, a line was drawn along the axis of the myofiber between the shortest widths of the nuclei [5].

3.5.4 Immunofluorescence Staining

1. Pipette a pool of PBS/T into the square on the "dissection slide." Place the fiber in the center of the square containing PBS/T.
2. Using two very small/fine forceps, tease apart the fibers into the smallest bundles as best you can. It will take about 1 h per TA muscle, which will yield plenty of fibers for staining; therefore, only about half of the TA muscle will be required and three to four per genotype.
3. It is important to keep the fibers wet during this process, as they will dry up if left.
4. Once you have dissected a good number of fibers, using the pipette tips prepared earlier, transfer the fibers into a 1.5 mL microcentrifuge tube.

5. The amount of fibers should come to about the 200 μL marker line on the side of the tube. This will provide enough fibers for up to six antibody stainings.
6. Resuspend the fibers in an equal volume of PBS/T (~400 μL total including fibers).
7. Aliquot the fibers according to the number of antibody stainings required. You may want to store a number of dissected fibers in the fridge in case problems are encountered or it requires optimization.
8. Dilute the antibodies in 4% BSA/PBS/T at double the concentration to the recommended dilution.
9. Add the antibody mix to an equal volume of the fibers (this will lead to the recommended dilution).
10. Put on rotator overnight at 4 °C.
11. In the morning, wash with PBS/T 3× for 5 min. For the washes, *do not* spin the fibers in a centrifuge, let them settle by gravity.
12. Add secondary antibodies for 1 h and DAPI using the same procedure as primary antibodies.
13. Wash 3× for 5 min with PBS.
14. Using a wax pen, draw around the edges of the slide and leave to dry (so that you won't lose any fibers when you squash the coverslip down on the fibers).
15. Resuspend the fibers in an equal volume of DAKO mounting medium (about 10–15 μL/slide).
16. Pipette the stained fibers on to the center of the slide (about 10–15 μL/slide).
17. Separate the fibers around the slide (you want to avoid clustering in the center of the slide).
18. Lower a large, rectangular coverslip onto the fibers avoiding bubbles.
19. Seal the coverslip with nail varnish.
20. Image using a confocal microscope or wide-field deconvolution microscope (*see* **Note 29**).

4 Notes

1. It is important to check for LacZ expression on a mouse line-by-line basis, as some of the mutant alleles don't express LacZ.
2. For homologous recombination to occur, a minimum of 2 kb of sequence homology is required [64]. Five to fourteen kilo base of sequence homology is typical for targeting constructs. The availability of suitable restriction enzymes within the locus

of interest and the sizes of inserted DNA fragments are common limitations of conventional cloning strategies. The following restriction enzymes are suitable for digesting genomic DNA of ES cells for Southern blotting: BamHI, HindIII, Acc65I/KpnI, EcoRV, SpeI, StuI, BglI.

3. Upon deciding which exon to flox, ideally, the loxP site should be more than 250 bp away from splice donor and acceptor sites. We also try and avoid floxing exon 1 and exons that are divisible by 3 (so to avoid potential frameshift mutations).
4. Note that the conditional mutant mice have the genotype "−/m" in the tissue that the Cre is specifically expressed in, and "f/m" genotype in other tissues.
5. CRISPR/Cas9 injection in mouse zygotes may introduce mosaicism as the Cas9 enzyme may be active after the single-cell stage [65]. However, in our cloning-free method, there were no cases of mosaicism observed [9].
6. It is important to be aware that the CRISPR/Cas9 approach may introduce off-target mutations. Therefore, in addition to sequencing the predicted off-target loci, mutant mice should be backcrossed with wild-type mice to dilute potential off-target effects, and at least two independently generated lines should be used per genotype.
7. Depending on the mouse background, litter sizes can vary. Therefore, the number of animals set up for breeding should be adjusted according to preliminary breeding data.
8. To avoid the unexpected (non-specific or broader expression) pattern of Cre, we utilize male Cre carriers crossed with floxed females.
9. It is important to remove as much hair as possible as hair can interfere with the echo signal.
10. The cardiac function measurement will be dramatically reduced due to the artifact of a low heart rate.
11. Because results can vary between operators, we recommend using a trained sonographer who is blinded to the genotypes and the mice be randomized prior to performing the echo. This will ensure better reproducibility, which is essential for robust examination of cardiac function in a serial study.
12. For echocardiography physiological measurements, we usually observe standard deviation of 10% between mice of the same genotype. With this assumption, 6–12 mice per group will be needed to measure a 14–22% change in heart function, to ensure robust analysis of statistical significance ($P < 0.05$ using two-tailed *t*-test, with a power of α90%). A useful resource for calculating numbers required for physiological analyses: http://www.3rs-reduction.co.uk/html/6__power_and_sample_size.html.

13. Heparin is used to prevent blood clotting in the heart.
14. To make full use of the animals in line with the 3Rs policy, once the defined end point for cardiac function has been performed, hearts can be processed for Western blotting, histology, immunofluorescence, and RNA analyses.
15. It is important to always check where your molecule of interest is expressed in the heart by performing immunofluorescence analysis on a wild-type four-chamber view heart section.
16. To measure tibia lengths, it is also possible to dissolve the whole leg at room temperature over several days in 10 M NaOH. However, this involves handling caustic NaOH and produces an unpleasant odor and therefore should be performed in a fume cupboard.
17. Measuring tibia length is preferred to body weight for measuring cardiac hypertrophy owing to greater fluctuations in body weight over time.
18. It is also possible to perform sucrose gradients to dehydrate the heart. Start with 10% sucrose (wt/vol) followed by 12%, 16%, and 20% sucrose. For each step, incubate in sucrose for a sufficient amount of time until the heart sinks to the bottom of the tube. For the final step, incubate briefly with a 1:1 ratio of 20% sucrose/OCT before embedding on dry ice.
19. Place OCT/heart block in the cryostat set to −20 °C prior to sectioning to soften the section; otherwise, sections can be brittle if sectioned immediately after removal from a −80 °C freezer.
20. For sectioning, an anti-roll plate or fine paint brush can be used to prevent the section from folding back upon itself.
21. By putting sequential sections on sequential slides, it allows comparison of multiple antibodies at approximately the same region of heart (within 10 μm of each other).
22. It is also possible to fix with 4% PFA, but this is antibody dependent.
23. Incubation times can be varied as required, but we found the antibodies listed in Table 2 work best with overnight incubation at 4 °C.
24. Other mounting media can be used at this step. The benefit of using Vectashield is that it has a well-defined refractive index of 1.45, which is important to know for super-resolution imaging.
25. Use pencil to mark slides as markings from most pens will be erased by further processing steps.
26. Heart tissue is very proteinaceous and therefore requires well-diluted samples to run on SDS-PAGE gels.
27. It is important to prepare P200 tips in this manner as it will minimize the myofibers from becoming stuck inside the tip and therefore loss of precious material.

28. Removing the contaminating tendon will make teasing the muscle fibers apart much easier.
29. If no sarcomeric marker is available (such as sarcomeric alpha actinin), be sure to take a differential interference contrast or phase/contrast image.

References

1. Muller U (1999) Ten years of gene targeting: targeted mouse mutants, from vector design to phenotype analysis. Mech Dev 82(1–2):3–21
2. van der Weyden L, Adams DJ, Bradley A (2002) Tools for targeted manipulation of the mouse genome. Physiol Genomics 11:133–164
3. Banerjee I, Zhang J, Moore-Morris T et al (2014) Targeted ablation of nesprin 1 and nesprin 2 from murine myocardium results in cardiomyopathy, altered nuclear morphology and inhibition of the biomechanical gene response. PLoS Genet 10(2):e1004114
4. Chapman MA, Zhang J, Banerjee I et al (2014) Disruption of both nesprin 1 and desmin results in nuclear anchorage defects and fibrosis in skeletal muscle. Hum Mol Genet 23(22):5879–5892
5. Stroud MJ, Feng W, Zhang J et al (2017) Nesprin 1alpha2 is essential for mouse postnatal viability and nuclear positioning in skeletal muscle. J Cell Biol 216(7):1915–1924
6. Zhang J, Felder A, Liu Y et al (2010) Nesprin 1 is critical for nuclear positioning and anchorage. Hum Mol Genet 19(2):329–341
7. Aida T, Chiyo K, Usami T et al (2015) Cloning-free CRISPR/Cas system facilitates functional cassette knock-in in mice. Genome Biol 16:87
8. Hashimoto M, Yamashita Y, Takemoto T (2016) Electroporation of Cas9 protein/sgRNA into early pronuclear zygotes generates non-mosaic mutants in the mouse. Dev Biol 418(1):1–9
9. Ma X, Chen C, Veevers J et al (2017) CRISPR/Cas9-mediated gene manipulation to create single-amino-acid-substituted and floxed mice with a cloning-free method. Sci Rep 7:42244
10. Nakagawa Y, Sakuma T, Nishimichi N et al (2016) Ultra-superovulation for the CRISPR-Cas9-mediated production of gene-knockout, single-amino-acid-substituted, and floxed mice. Biol Open 5(8):1142–1148
11. Wang H, Yang H, Shivalila CS et al (2013) One-step generation of mice carrying mutations in multiple genes by CRISPR/Cas-mediated genome engineering. Cell 153(4):910–918
12. Rajewsky K, Gu H, Kuhn R et al (1996) Conditional gene targeting. J Clin Invest 98(3):600–603
13. Kuhn R, Schwenk F, Aguet M et al (1995) Inducible gene targeting in mice. Science 269(5229):1427–1429
14. Kilby NJ, Snaith MR, Murray JA (1993) Site-specific recombinases: tools for genome engineering. Trends Genet 9(12):413–421
15. Rodriguez CI, Buchholz F, Galloway J et al (2000) High-efficiency deleter mice show that FLPe is an alternative to Cre-loxP. Nat Genet 25(2):139–140
16. Liang X, Zhou Q, Li X et al (2005) PINCH1 plays an essential role in early murine embryonic development but is dispensable in ventricular cardiomyocytes. Mol Cell Biol 25(8):3056–3062
17. Jinek M, Chylinski K, Fonfara I et al (2012) A programmable dual-RNA-guided DNA endonuclease in adaptive bacterial immunity. Science 337(6096):816–821
18. Deltcheva E, Chylinski K, Sharma CM et al (2011) CRISPR RNA maturation by trans-encoded small RNA and host factor RNase III. Nature 471(7340):602–607
19. Chylinski K, Le Rhun A, Charpentier E (2013) The tracrRNA and Cas9 families of type II CRISPR-Cas immunity systems. RNA Biol 10(5):726–737
20. Bione S, Maestrini E, Rivella S et al (1994) Identification of a novel X-linked gene responsible for Emery-Dreifuss muscular dystrophy. Nat Genet 8(4):323–327
21. Bione S, Small K, Aksmanovic VM et al (1995) Identification of new mutations in the Emery-Dreifuss muscular dystrophy gene and evidence for genetic heterogeneity of the disease. Hum Mol Genet 4(10):1859–1863
22. Christensen AH, Andersen CB, Tybjaerg-Hansen A et al (2011) Mutation analysis and evaluation of the cardiac localization of TMEM43 in arrhythmogenic right ventricular cardiomyopathy. Clin Genet 80(3):256–264
23. Haque F, Mazzeo D, Patel JT et al (2010) Mammalian SUN protein interaction networks at the inner nuclear membrane and their role in laminopathy disease processes. J Biol Chem 285(5):3487–3498
24. Hodgkinson KA, Connors SP, Merner N et al (2013) The natural history of a genetic sub-

type of arrhythmogenic right ventricular cardiomyopathy caused by a p.S358L mutation in TMEM43. Clin Genet 83(4):321–331
25. Malhotra R, Mason PK (2009) Lamin A/C deficiency as a cause of familial dilated cardiomyopathy. Curr Opin Cardiol 24(3):203–208
26. Merner ND, Hodgkinson KA, Haywood AF et al (2008) Arrhythmogenic right ventricular cardiomyopathy type 5 is a fully penetrant, lethal arrhythmic disorder caused by a missense mutation in the TMEM43 gene. Am J Hum Genet 82(4):809–821
27. Puckelwartz MJ, Kessler EJ, Kim G et al (2010) Nesprin-1 mutations in human and murine cardiomyopathy. J Mol Cell Cardiol 48(4):600–608
28. Yamada T, Kobayashi T (1996) A novel emerin mutation in a Japanese patient with Emery-Dreifuss muscular dystrophy. Hum Genet 97(5):693–694
29. Zhang Q, Bethmann C, Worth NF et al (2007) Nesprin-1 and -2 are involved in the pathogenesis of Emery Dreifuss muscular dystrophy and are critical for nuclear envelope integrity. Hum Mol Genet 16(23):2816–2833
30. Zhang Z, Stroud MJ, Zhang J et al (2015) Normalization of Naxos plakoglobin levels restores cardiac function in mice. J Clin Invest 125(4):1708–1712
31. Sheikh F, Ouyang K, Campbell SG et al (2012) Mouse and computational models link Mlc2v dephosphorylation to altered myosin kinetics in early cardiac disease. J Clin Invest 122(4):1209–1221
32. Fang X, Bogomolovas J, Wu T et al (2017) Loss-of-function mutations in co-chaperone BAG3 destabilize small HSPs and cause cardiomyopathy. J Clin Invest 127(8):3189–3200
33. Davis J, Maillet M, Miano JM et al (2012) Lost in transgenesis: a user's guide for genetically manipulating the mouse in cardiac research. Circ Res 111(6):761–777
34. Abel ED, Kaulbach HC, Tian R et al (1999) Cardiac hypertrophy with preserved contractile function after selective deletion of GLUT4 from the heart. J Clin Invest 104(12):1703–1714
35. Agah R, Frenkel PA, French BA et al (1997) Gene recombination in postmitotic cells. Targeted expression of Cre recombinase provokes cardiac-restricted, site-specific rearrangement in adult ventricular muscle in vivo. J Clin Invest 100(1):169–179
36. Buerger A, Rozhitskaya O, Sherwood MC et al (2006) Dilated cardiomyopathy resulting from high-level myocardial expression of Cre-recombinase. J Card Fail 12(5):392–398
37. McFadden DG, Barbosa AC, Richardson JA et al (2005) The Hand1 and Hand2 transcription factors regulate expansion of the embryonic cardiac ventricles in a gene dosage-dependent manner. Development 132(1):189–201
38. Stanley EG, Biben C, Elefanty A et al (2002) Efficient Cre-mediated deletion in cardiac progenitor cells conferred by a 3'UTR-ires-Cre allele of the homeobox gene Nkx2-5. Int J Dev Biol 46(4):431–439
39. Moses KA, DeMayo F, Braun RM et al (2001) Embryonic expression of an Nkx2-5/Cre gene using ROSA26 reporter mice. Genesis 31(4):176–180
40. Jay PY, Harris BS, Maguire CT et al (2004) Nkx2-5 mutation causes anatomic hypoplasia of the cardiac conduction system. J Clin Invest 113(8):1130–1137
41. Biben C, Weber R, Kesteven S et al (2000) Cardiac septal and valvular dysmorphogenesis in mice heterozygous for mutations in the homeobox gene Nkx2-5. Circ Res 87(10):888–895
42. Jiao K, Kulessa H, Tompkins K et al (2003) An essential role of Bmp4 in the atrioventricular septation of the mouse heart. Genes Dev 17(19):2362–2367
43. Hirai M, Arita Y, McGlade CJ et al (2017) Adaptor proteins NUMB and NUMBL promote cell cycle withdrawal by targeting ERBB2 for degradation. J Clin Invest 127(2):569–582
44. Breckenridge R, Kotecha S, Towers N et al (2007) Pan-myocardial expression of Cre recombinase throughout mouse development. Genesis 45(3):135–144
45. Chen J, Kubalak SW, Chien KR (1998) Ventricular muscle-restricted targeting of the RXRalpha gene reveals a non-cell-autonomous requirement in cardiac chamber morphogenesis. Development 125(10):1943–1949
46. Minamisawa S, Gu Y, Ross J Jr et al (1999) A post-transcriptional compensatory pathway in heterozygous ventricular myosin light chain 2-deficient mice results in lack of gene dosage effect during normal cardiac growth or hypertrophy. J Biol Chem 274(15): 10066–10070
47. Shai SY, Harpf AE, Babbitt CJ et al (2002) Cardiac myocyte-specific excision of the beta1 integrin gene results in myocardial fibrosis and cardiac failure. Circ Res 90(4):458–464
48. Sohal DS, Nghiem M, Crackower MA et al (2001) Temporally regulated and tissue-specific gene manipulations in the adult and embryonic heart using a tamoxifen-inducible Cre protein. Circ Res 89(1):20–25
49. Zhang Z, Mu Y, Veevers J et al (2016) Postnatal loss of Kindlin-2 leads to progressive heart failure. Circ Heart Fail 9(8):e003129
50. Nakai A, Yamaguchi O, Takeda T et al (2007) The role of autophagy in cardiomyocytes in the

basal state and in response to hemodynamic stress. Nat Med 13(5):619–624

51. Lexow J, Poggioli T, Sarathchandra P et al (2013) Cardiac fibrosis in mice expressing an inducible myocardial-specific Cre driver. Dis Model Mech 6(6):1470–1476
52. Bersell K, Choudhury S, Mollova M et al (2013) Moderate and high amounts of tamoxifen in alphaMHC-MerCreMer mice induce a DNA damage response, leading to heart failure and death. Dis Model Mech 6(6):1459–1469
53. Koitabashi N, Bedja D, Zaiman AL et al (2009) Avoidance of transient cardiomyopathy in cardiomyocyte-targeted tamoxifen-induced MerCreMer gene deletion models. Circ Res 105(1):12–15
54. Vogt MA, Chourbaji S, Brandwein C et al (2008) Suitability of tamoxifen-induced mutagenesis for behavioral phenotyping. Exp Neurol 211(1):25–33
55. Yan J, Sultana N, al ZL (2015) Generation of a tamoxifen inducible Tnnt2MerCreMer knock-in mouse model for cardiac studies. Genesis 53(6):377–386
56. Wu B, Zhou B, Wang Y et al (2010) Inducible cardiomyocyte-specific gene disruption directed by the rat Tnnt2 promoter in the mouse. Genesis 48(1):63–72
57. Miniou P, Tiziano D, Frugier T et al (1999) Gene targeting restricted to mouse striated muscle lineage. Nucleic Acids Res 27(19):e27
58. McCarthy JJ, Srikuea R, Kirby TJ et al (2012) Inducible Cre transgenic mouse strain for skeletal muscle-specific gene targeting. Skelet Muscle 2(1):8
59. Rao P, Monks DA (2009) A tetracycline-inducible and skeletal muscle-specific Cre recombinase transgenic mouse. Dev Neurobiol 69(6):401–406
60. Li S, Czubryt MP, McAnally J et al (2005) Requirement for serum response factor for skeletal muscle growth and maturation revealed by tissue-specific gene deletion in mice. Proc Natl Acad Sci U S A 102(4):1082–1087
61. Southard S, Low S, Li L et al (2014) A series of Cre-ER(T2) drivers for manipulation of the skeletal muscle lineage. Genesis 52(8):759–770
62. Marin TM, Keith K, Davies B et al (2011) Rapamycin reverses hypertrophic cardiomyopathy in a mouse model of LEOPARD syndrome-associated PTPN11 mutation. J Clin Invest 121(3):1026–1043
63. Ran FA, Hsu PD, Wright J et al (2013) Genome engineering using the CRISPR-Cas9 system. Nat Protoc 8(11):2281–3308
64. Melton DW (2002) Gene-targeting strategies. Methods Mol Biol 180:151–173
65. Yen ST, Zhang M, Deng JM et al (2014) Somatic mosaicism and allele complexity induced by CRISPR/Cas9 RNA injections in mouse zygotes. Dev Biol 393(1):3–9
66. Parsons SA, Millay DP, Wilkins BJ et al (2004) Genetic loss of calcineurin blocks mechanical overload-induced skeletal muscle fiber type switching but not hypertrophy. J Biol Chem 279(25):26192–26200
67. Randles KN, Lam le T, Sewry CA et al (2010) Nesprins, but not sun proteins, switch isoforms at the nuclear envelope during muscle development. Dev Dyn 239(3):998–1009
68. Huber MD, Guan T, Gerace L (2009) Overlapping functions of nuclear envelope proteins NET25 (Lem2) and emerin in regulation of extracellular signal-regulated kinase signaling in myoblast differentiation. Mol Cell Biol 29(21):5718–5728
69. Matthew J Stroud, Xi Fang, Jianlin Zhang et al (2018) Luma is not essential for murine cardiac development and function, Cardiovascular Research 114(3):378–388. https://doi.org/10.1093/cvr/cvx205PMID:29040414

Chapter 19

An In Vitro System to Measure the Positioning, Stiffness, and Rupture of the Nucleus in Skeletal Muscle

William Roman, Mafalda R. Pimentel, and Edgar R. Gomes

Abstract

Nuclear positioning plays important roles for certain cellular functions. This is particularly relevant in skeletal muscle cells also known as myofibers in which nuclear positioning defects were shown to hinder muscle function. Myofibers are multinucleated cells with nuclei equally distributed at the periphery of the cell. However, nuclei can be found centrally located during myogenesis before anchoring at the periphery or in certain muscle disorders, either due to regenerating myofibers or defects in nuclear movement. As such, nuclear localization in myofibers (central or peripheral) can be used to assess myofiber maturity, regeneration, or health. To study how nuclei reach the periphery of myofibers during development, we devised a unique protocol to mature myofibers thereby recapitulating later stages of differentiation, including nuclear movement to the periphery. Here we describe how to use this system to study nuclear positioning and other nuclear characteristics such as nuclear stiffness or rupture.

Key words Nucleus, Skeletal muscle, Cytoskeleton, Microscopy, Optogenetics

1 Introduction

A hallmark of skeletal muscle is the position of its nuclei at the periphery of the muscle fiber, just below the plasma membrane [1]. Nuclei are first found in the center of immature myofibers, and upon differentiation, nuclei move to the periphery. The position of the nucleus in skeletal muscle is important for muscle function [2]. Furthermore, centrally located nuclei are commonly found in regenerative muscle and in multiple muscle disorders, further supporting a role for nuclear positioning in muscle function [3].

To study nuclear movement to the periphery of the myofiber, we developed an in vitro system that recapitulates the stages of muscle formation [4, 5]. Using this system, we are able to identify

---

William Roman and Mafalda R. Pimentel contributed equally to this chapter.

Gregg G. Gundersen and Howard J. Worman (eds.), *The LINC Complex: Methods and Protocols*, Methods in Molecular Biology, vol. 1840, https://doi.org/10.1007/978-1-4939-8691-0_19, © Springer Science+Business Media, LLC, part of Springer Nature 2018

a novel pathway that is disrupted in centronuclear myopathies, to identify how nuclei are positioned at the periphery of muscle myofibers, and to study the role of dynein on MuSK clustering [4, 6, 7]. This in vitro system is accessible for manipulation either by overexpression or knockdown of proteins. Overexpression can be done by transfection of expression vectors or infection with lentivirus. Knockdown of proteins can be done by infection of lentivirus encoding shRNA or transfection of siRNAs [4, 6, 7]. These manipulated myofibers can be analyzed by time-lapse microscopy, immunofluorescence, electron microscopy, and biochemistry. In this chapter, we describe the different methodologies to manipulate this in vitro myofibers and how to measure different parameters related to nuclear positioning in muscle myofibers.

2 Materials

2.1 In Vitro Differentiated Myofibers

1. Primary myoblast culture starting to spontaneously fuse (usually after 3 days of proliferation), as previously described [4, 5].
2. Growth medium without antibiotics: 20% fetal bovine serum (FBS), 1% chicken embryo extract, in Iscove's Modified Dulbecco's Medium Glutamax, sterile filtered.
3. Differentiation medium: 10% horse serum, 1% penicillin-streptomycin, sterile filtered.
4. 35 mm live imaging cell culture dishes, e.g., FluoroDishes from World Precision Instruments.
5. Matrigel Growth Factor Reduced, freshly diluted to 50% in cold differentiation medium.

2.2 Transfections, Infections, and RNAi

1. Sterile 2 mL microtubes.
2. siRNA or plasmid DNA for transfection.
3. Add gene plasmid #20939 pcDNA3.1/hChR2-EYFP for optogenetics and plasmid #39319 pmCherry-NLS.
4. Lipofectamine RNAiMAX, Lipofectamine 2000, or Lipofectamine 3000.
5. Opti-MEM Reduced Serum Medium (Invitrogen).
6. Lentiviruses expressing shRNA or lentiviral plasmid DNA [8].
7. Hexadimethrine bromide (Polybrene) solution at 8 mg/mL.

2.3 Immunofluorescence

1. Phosphate-buffered saline (PBS).
2. Fixation solution: 4% paraformaldehyde (PFA) in PBS. Store at 4 °C.
3. Permeabilization solution: 0.5% Triton X-100 in PBS. Store as a 40× solution at 4 °C.

4. Blocking solution: 10% goat serum, 5% bovine serum albumin (BSA), 0.1% saponine, in PBS. Keep 10× stocks of all components at −20 °C, and make a fresh mix each use.
5. 4′,6-Diamidino-2-phenylindole (DAPI), use at 0.5 μg/mL.
6. Alexa Fluor 555 Phalloidin (Invitrogen), use diluted 1:100.
7. Mounting medium, e.g., Fluoromount-G.

2.4 Microscopes

Equipment similar to the following is required:

1. Zeiss Cell Observer widefield inverted microscope equipped with 37 °C chamber and 5% CO_2 for live-cell microscopy.
2. sCMOS camera Hamamatsu ORCA-flash4.0 V2 for 10 ms/frame streaming acquisition.
3. 40× Plan-Apochromat Oil objective (NA = 1.4).
4. 63× Plan-Apochromat Oil objective (NA = 1.4).
5. LED light source, e.g., Colibri2 from Zeiss.
6. Quadruple band filter set (FS81HE), in addition to traditional filters.

The LED and quadruple filter set combination allows for eyepiece multicolor screening. This setup also enables simultaneous acquisition of a given fluorescence channel and brief channelrhodopsin activation through the LED controller.

2.5 Imaging Analysis Software

1. Zen lite edition free software (available at https://www.zeiss.com/microscopy/int/products/microscope-software/zen-lite).
2. Fiji free software (available at https://fiji.sc/).
3. ICY free software (available at http://icy.bioimageanalysis.org/).

3 Methods

Nuclear quantification at the periphery of myofibers and indirect measurements of nuclear stiffness are performed on matured myofibers (usually at day 7 after initial differentiation) generated by the protocol previously described [4, 5]. Nuclei at the periphery are distinguished from central nuclei by their bulging from the average width of the myofiber (Fig. 1). As such, nuclear quantification requires the visualization of the myofiber and nuclei. Moreover, in order to assess nuclear positioning defects, proteins can be knocked down or overexpressed by transfection or infection before initiating differentiation, when myoblasts are still fusing and before the addition of the layer of Matrigel. We will therefore begin by detailing the different methods to alter protein expressions and in a second step explicit the staining methods to quantify peripheral nuclei.

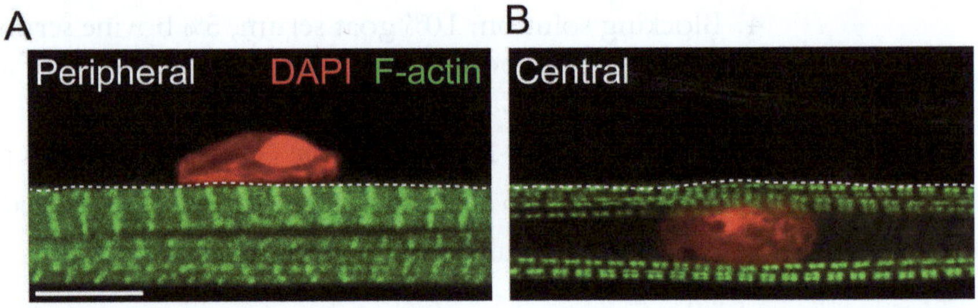

Fig. 1 Representative immunofluorescent images of fully matured myofiber stained for myofibrils (green, phalloidin) and nucleus (red, DAPI) displaying peripheral (**A**) or central (**B**) nuclei. White dashed line represents myofibril plane. Scale bar: 10 μm

3.1 Transfections (siRNA and cDNA) (See Note 1)

1. Add 1 μg of plasmid DNA or 20 pmol of siRNA in an Eppendorf tube containing 50 μL of Opti-MEM.
2. Add 1 μL of Lipofectamine reagent to a separate Eppendorf tube containing 50 μL of Opti-MEM as well.
3. Incubate separate Eppendorf tubes for 5 min.
4. Mix DNA/siRNA and Lipofectamine solutions in a single tube and, incubate at room temperature for 30 min for complex formation.
5. Add 400 μL of growth medium (at 37 °C) without antibiotics to the DNA/siRNA and Lipofectamine mixture.
6. Remove medium on myoblasts cultured in one FluoroDish, and add DNA/siRNA and Lipofectamine mixture in growth medium.
7. Incubate for 5 h.
8. Wash 35 mm live imaging cell culture dishes containing early myotubes once with differentiation medium, and resume differentiation protocol by incubating cells in differentiation medium and adding Matrigel the following day.

3.2 Infection (shRNA) (See Note 1)

1. Produce lentivirus through standard protocol [8].
2. Add 8 μg/mL of Polybrene directly to the myoblast culture and mix.
3. Add 1:10 of original lentivirus production per FluoroDish and mix.
4. Incubate for 5 h and wash once with differentiation medium. Resume differentiation protocol by incubating early myotubes in differentiation medium and adding Matrigel the following day.

3.3 Staining for Peripheral Nuclei Localization and Quantification (See Note 2)

1. Fix myofibers with 4% PFA around day 8, and incubate for 10 min at room temperature.
2. Wash myofibers twice with PBS, and add permeabilization solution for 5 min.

3. Wash twice with PBS and add blocking buffer for 1 h.
4. Remove blocking solution and add primary antibody diluted in blocking solution and incubate overnight at 4 °C.
5. Wash ten times with PBS.
6. Add secondary antibody, DAPI, and phalloidin diluted in blocking solution for 1 h.
7. Wash ten times in PBS.
8. Cover with 200 μL of Fluoromount and leave to dry overnight.
9. Transfer dish to the microscope for nuclear quantification.
10. Quantify between 10 and 100 myofibers per dish at different location in the plate, as described in the **Note 2**.

3.4 Measurement of Nuclear Stiffness, Breakage, and Shape (See Note 3)

Nuclear stiffness can be assessed by atomic force microscopy or aspiration assays [9]; however, the use of Matrigel and more importantly the presence of myofibrils within the myofiber can obstruct such measurements for centrally located nuclei. As such, we devised another method to quantify nuclear stiffness by monitoring changes in nuclear length during myofiber contraction. This method is proposed to measure stiffness of centrally located nuclei using the ChR2 optogenetic tools, very popular in the neuroscience community to control neuronal excitability [10]. Since myofiber contraction is also mediated by membrane depolarization and formation of an action potential, we transfect cells with the ChR2 optogenetic cation channel to induce myofiber contraction with light. This is performed while live stream acquisition to monitor changes in nuclear length during contraction. Length changes are then normalized on myofiber width to account for the force of contraction (Fig. 2A). The force on the nucleus during muscle contraction or squeezing to the periphery induces stress on the nuclear envelope that in certain conditions can lead to nuclear rupture, such as in nuclei depleted for lamin A/C (Fig. 2B).

1. Add 1 μg of ChR2-GFP and NLS-mCherry in an Eppendorf tube containing 50 μL of Opti-MEM.
2. Add 1 μL of Lipofectamine and 1 μL of Lipofectamine reagent to the other Eppendorf tube containing 50 μL of Opti-MEM as well.
3. Incubate separate Eppendorf tubes for 5 min.
4. Mix DNA and Lipofectamine solutions in a single tube, and incubate at room temperature for 30 min.
5. Add 400 μL of heated growth medium without antibiotics to the DNA and Lipofectamine mixture.
6. Remove medium on myoblast culture, and add DNA and Lipofectamine mixture in growth medium.

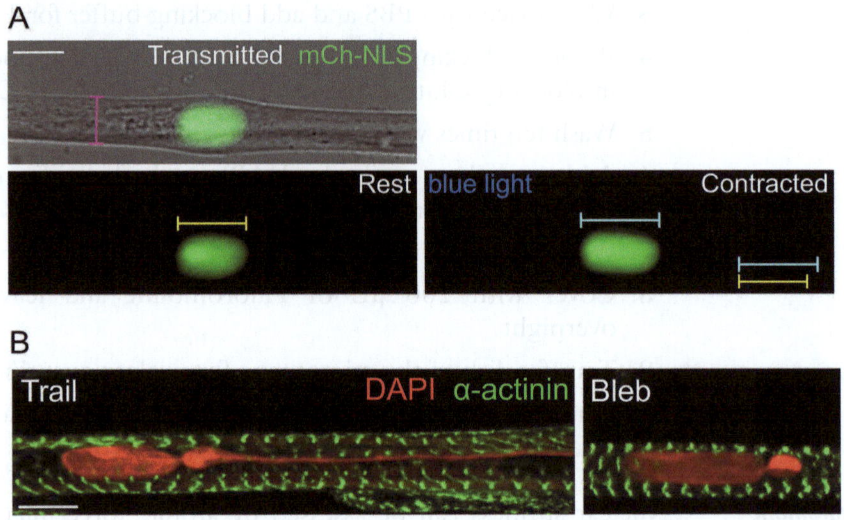

Fig. 2 (**A**) Representative immunofluorescent image of 4.5-day myofiber transfected with ChR2-EYFP (not shown) and mCherry-NLS (green) in a relaxed state (left panel) and induced to contract with blue light (right panel). Magenta bar represents myofibril width, yellow bar represents relaxed nuclear length, whereas cyan bar represents nuclear length during myofiber contraction. Scale bar: 10 μm. (**B**) Representative immunofluorescent image of fully matured myofiber knocked down for lamin A/C and stained for myofibrils (green, phalloidin) and nucleus (red, DAPI) showing nuclear rupture either by leaving a DAPI trail (left panel) or nuclear blebbing (right panel). Scale bar: 10 μm

7. Incubate for 5 h.
8. Wash myotubes once with differentiation medium, and resume differentiation protocol by incubating cells in differentiation medium and adding Matrigel the following day.
9. When nuclear stiffness is to be assessed (usually at day 4, prior to peripheral nuclear localization), heat the microscope chamber to 37 °C with 5% CO_2 for 5–10 min based on the heating capacity of your microscope.
10. Place the 35 mm dish with myofibers on the microscope, and wait until the microscope is stable (at least 90 min in our system).
11. Identify a centrally located nucleus in a myofiber that is at least 10 μm thick, but not more than 20 μm thick.
12. Acquire a transmitted light image to measure myofiber width for the normalization.
13. Acquire images using streaming mode (up to 1000 frames at 10 ms/frame) using the 587 nm excitation wavelength fluorescence in combination with a multiband filter set (containing an emission/excitation bandpass in the range of 484/519 and 553/586 nm). Use high excitation intensity to be able to visualize the nucleus during stream acquisition.

14. During stream acquisition, turn on the 480 nm LED manually to activate the optogenetic channel and induce contraction. Use low excitation intensity to avoid signal acquisition from the excited 488 nm fluorophores. ChR2 channel opens with low exposure. Turn off the 480 nm LED once you observe three contractions.
15. Use ZEN or Fiji software to measure the difference in length of the nucleus at rest and at the height of contraction and normalize on myofiber width (Fig. 2A).

4 Notes

1. Transfections are most efficient when performed in myoblast cultures after several myotubes are spontaneously formed and before inducing differentiation by starvation (usually 3 days after proliferation) (Fig. 3A). Transfections performed after inducing differentiation result in minimal or no myofiber transfection efficiency. Efficiency is maximal with Lipofectamine 3000; however, levels of cytotoxicity are also higher, when compared to Lipofectamine 2000 or LTX. As such the type of Lipofectamine should be used based on the number of cells in the dish and potential effects of cytotoxic secretions in the medium from dying cells. Co-transfection of plasmid DNA and siRNA can be done using Lipofectamine 3000. To transfect only siRNA, Lipofectamine RNAiMAX should be used as this reagent displays good efficiency without much cytotoxic effect. Expression of plasmid DNA and RNAi efficiency can be assessed either by immunofluorescence microscopy, by qPCR, or by Western blot. Since myofibers are non-dividing differentiated cells, we observed that protein knockdown is sustainable until the death of the cultures. Infections should be made in the same day as transfections. Infections performed after inducing differentiation result in minimal or none myofiber transfection efficiency. The amount of virus to use depends on the titer of each produced virus. We have better results using lentivirus expressing cDNA than shRNA.
2. The method to generate highly matured myofibers was previously published [4, 5]. We will provide here relevant details for optimal nuclear quantification.

   Timing of fixation: As the cell source for this protocol is primary myoblasts, the culture life span may vary between 7 and 11 days, depending mostly on the set of reagents that are used. Thus, once the protocol is established in a lab, the survival of each culture is quite reproducible. In order to maximize the number of peripheral nuclei, fixation should be performed before the initial stages of myofiber degradation

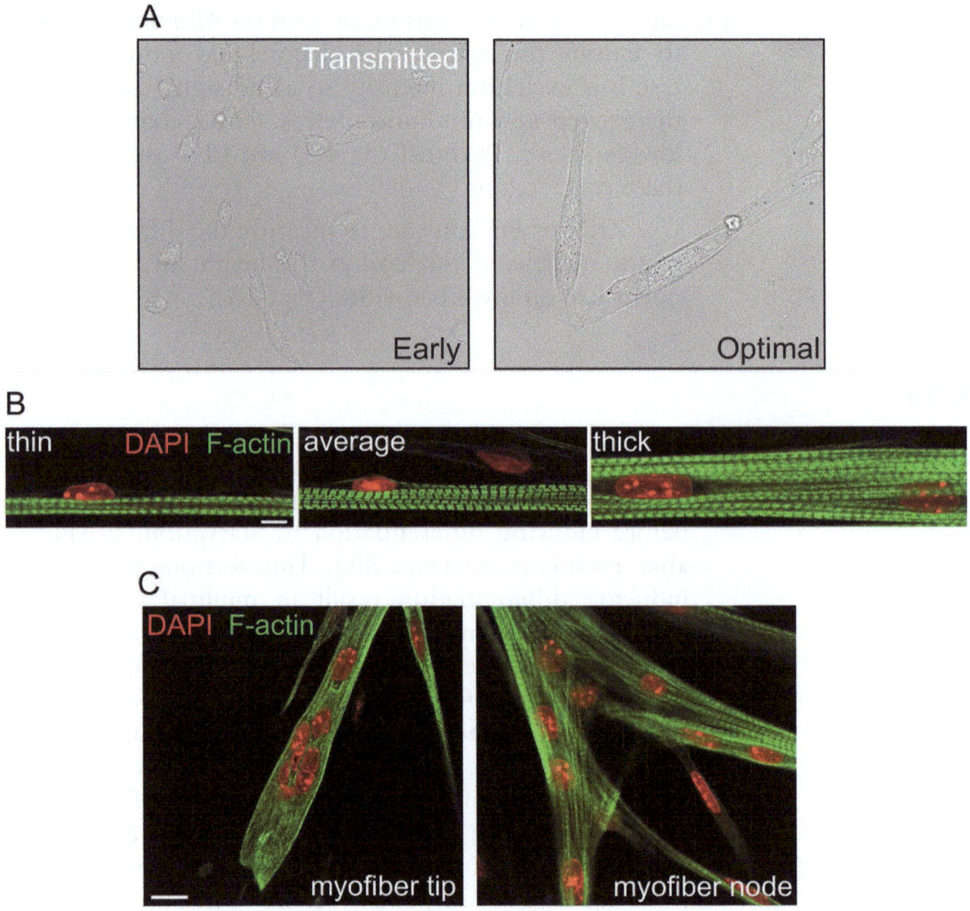

Fig. 3 (**A**) Representative transmitted light images of myoblasts and myotubes. Left panel display the aspect of cells when they are not sufficiently differentiated for adequate transfection. Right panel shows the aspect of cells when optimal transfections can be performed. (**B**) Representative immunofluorescent image of fully matured myofiber stained for myofibrils (green, phalloidin) and nucleus (red, DAPI) showing different myofiber width that are of adequate width to quantify (middle panel) or too thin or thick to quantify (left and right panel). Scale bar: 10 μm. (**C**) Representative immunofluorescent image of fully matured myofiber stained for myofibrils (green, phalloidin) and nucleus (red, DAPI) representing an accumulation of centrally located nuclei at the tip of the myofiber (left panel) or at nodes (right panel). Scale bar: 10 μm

take place. These can occur rapidly, sometimes overnight, so visual cues can be used to estimate the beginning of myofiber degradation and to set up the ideal fixation time point. Visual cues that help determining ideal fixation time point are (1) when myofibers become thinner, (2) parts of the myofibers in the dish begin to detach, and (3) higher magnification of light microscope reveals less visible sarcomeres. One should always monitor the cells before fixation as quantification of peripheral nuclear positioning could not be performed if most cells have died. It is recommended to frequently keep one untransfected culture dish to assess myofiber longevity.

Optimal visualization: Myofibers should be stained with a myofibril and a nuclear marker. DAPI (to visualize the nuclei) and phalloidin (to visualize F-actin in the myofibrils) are the easiest stainings and permit simultaneous immunofluorescence with antibodies in the remaining two fluorescence channels. Once stained, myofibers can be analyzed in a widefield epifluorescence microscope. 40× is the advised magnification although 63× can also be used. Oil immersion objectives are strongly recommended due to their higher NA and lower depth of field. As the length of the myofiber exceeds several times, the field of view and the focus plane must be changed constantly, and quantification through the eyepieces is more efficient than acquisition of images. As such a multiband filter set is highly advised to observe the DAPI and phalloidin channels simultaneously. This avoids constant shifting between channels and allows for more precise evaluation of nuclear position. Bumps on the myofiber surface may occur without nuclei being present thereby mimicking peripheral nuclei when only looking at the myofibril channel. This is why using a multiband filter is so useful, and if unavailable, shifting between channels is absolutely necessary.

Choosing which myofiber to quantify: As a primary culture, myofibers within the dish are highly variable. Some myofibers exceed 20 μm in diameter, whereas others are below 5 μm (Fig. 3B). In order to consistently assess peripheral positioning, quantifiably myofibers must be between 10 and 18 μm. Moreover, as myofibers are not aligned, they either cross each other, fuse with each other at nodes, or end in a tip. The node or myofiber tips mark the end of the myofiber (Fig. 3C), whereas quantification can be pursued when myofibers only cross each other. However, cell density may be too important when myofibers cross one another thereby limiting quantifica tion possibilities. It is therefore important to assess myofiber clarity before quantifying. Usually, quantified myofibers possess between 7 and 15 quantifiable nuclei. Note that nuclei often aggregate at the tip of the cells as seen in myotendinous junctions (MTJ). These nuclei are excluded from the quantification. Myofibers fitting the description above are randomly chosen in the center of the dish and in each quadrant. Two to ten myofibers should be quantified per center and per quadrant of the dish.

Quantifying: Since myofibers are embedded into a Matrigel matrix, they are organized in three dimensions (3D). Nuclei that are located at the side of the myofiber will be clearly observed at the periphery, but nuclei positioned on top of the myofiber (on the axis between your vision and the myofiber) will appear centrally located. It is therefore necessary to constantly shift the focus wheel of the microscope while quantifying to distinguish nuclear positioning. This can be done by acquiring the image in z and performing a 3D reconstruction although assessing nuclear positioning of top nuclei can be achieved visually with practice. We found that the most accurate quantification of peripheral nuclei is to calculate the percentage of peripheral nuclei per myofiber. With a varying number of nuclei per myofiber, calculating overall peripheral nuclei would favor myofibers with a greater number of nuclei. In the best cultures, average number of peripheral nuclei reaches up to 80% although the minimum acceptable percentage as a control is 55%.

Maturation levels of myofibers: Nuclear movement to the periphery is a step in the development of myofibers. As such an accumulation of centrally located nuclei may result from defective machinery to drive nuclei at the periphery or due to maturation arrest. To distinguish between these two causes, maturation characteristics should be assessed. This involves monitoring fusion index, frequency of myofiber contraction and width, myofibril integrity, as well as desmin and transversal triad organization. Fusion index can be calculated by counting number of nuclei per cell, whereas myofiber width can be measured [4]. Frequency of myofiber contraction can be determined by counting the number of contractions of three myofibers over 3 min each, by direct observation under a microscope [7]. Myofibers already contracting should be used to assess contraction. Myofibril integrity can be evaluated by EM and by immunostaining using certain myofibril markers such as F-actin, α-actinin, myomesin, and vinculin. Finally, desmin should be organized at the z-line in mature myofibers, whereas triads are organized as doublets on both sides of the z-line [7]. Transversal triad quantification is the best indicator of maturation as they become organized after nuclear movement to the periphery during myogenesis. Triad quantification is assessed visually by marking myofibers positive or negative based on a DHPR and triadin staining. The length of the myofiber is screened, and if more than 50% of the myofiber displays transversal doublets, the myofiber is marked as positive. In fully mature cultures, levels of transversal triad formation reach between 70 and 80% of the myofibers [4, 7].

3. Nuclear stiffness is best assessed in centrally located nuclei as they are surrounded by myofibrils. This is usually the case for myofibers with 10 μm thickness. The nucleus must be under myofibril tension to observe changes in length. Nuclei under tension can be distinguished due to an elongated shape in the axis of the myofiber (*see* Fig. 1B). To assess nuclear breakage or rupture, we have developed four assays [7]. (1) Myofibers can be stained with DAPI, and nuclear breakdown can be observed.

(2) Myofibers can be stained for CHMP4, and signal intensity in the nuclei can be measured. Both of these techniques are useful for fixed samples. (3) Myofibers can be transfected with cGAS as was previously described [11]. (4) Myofibers can be transfected with NLS-GFP, and loss of signal can be monitored live [12]. Both methods (3) and (4) are more tailored for live samples. Numerous programs such as Fiji, ICY, ZEN, or Imaris provide further options to measure multiple nuclear parameters, both 2D and 3D. Any of these software can be used, accordingly to user experience and parameter to be measured.

References

1. Cadot B, Gache V, Gomes ER (2015) Moving and positioning the nucleus in skeletal muscle—one step at a time. Nucleus 6:1–9
2. Metzger T, Gache V, Xu M et al (2012) MAP and kinesin-dependent nuclear positioning is required for skeletal muscle function. Nature 484:120–124
3. Clark KA, McElhinny AS, Beckerle MC et al (2002) Striated muscle cytoarchitecture: an intricate web of form and function. Annu Rev Cell Dev Biol 18:637–706
4. Falcone S, Roman W, Hnia K et al (2014) N-WASP is required for Amphiphysin-2/BIN1-dependent nuclear positioning and triad organization in skeletal muscle and is involved in the pathophysiology of centronuclear myopathy. EMBO Mol Med 6:1455–1475
5. Pimentel MR, Falcone S, Cadot B et al (2017) In vitro differentiation of mature myofibers for live imaging. J Vis Exp. https://doi.org/10.3791/55141
6. Vilmont V, Cadot B, Vezin E et al (2016) Dynein disruption perturbs post-synaptic components and contributes to impaired MuSK clustering at the NMJ: implication in ALS. Sci Rep 6:27804
7. Roman W, Martins JP, Carvalho FA et al (2017) Myofibril contraction and cross-linking drive nuclear movement to the periphery of skeletal muscle. Nat Cell Biol 19:1189–1201
8. Baumlin-Schmid N, Salathe M, Fregien NL (2016) Optimal lentivirus production and cell culture conditions necessary to successfully transduce primary human bronchial epithelial cells. J Vis Exp. https://doi.org/10.3791/54176
9. Lammerding J (2011) Mechanics of the nucleus. Compr Physiol 1:783–807
10. Nagel G, Ollig D, Fuhrmann M et al (2002) Channelrhodopsin-1: a light-gated proton channel in green algae. Science 296:2395–2398
11. Denais CM, Gilbert RM, Isermann P et al (2016) Nuclear envelope rupture and repair during cancer cell migration. Science 352:353–358
12. Vargas JD, Hatch EM, Anderson DJ et al (2012) Transient nuclear envelope rupturing during interphase in human cancer cells. Nucleus 3:88–100

Chapter 20

Functional Analysis of LINC Complexes in the Skin

Iakowos Karakesisoglou, Carmen Mroß, and Angelika A. Noegel

Abstract

The genome in eukaryotic cells is encased by two intricate and interconnected concentric membranes, which together with the underlying nuclear lamina form the nuclear envelope (NE). Two fundamental macromolecular structures are embedded within the nuclear envelope: the nuclear pore (NPC) and the LINC complex. The former perforates the nucleus controlling biomolecule trafficking between the nucleoplasm and the cytoplasm, while the latter integrates the nucleus via the cytoskeleton to the extracellular matrix. LINC complex structural and functional integrity is of utmost importance for various fundamental cellular functions. Mechanical forces are relayed into the nuclear interior via the LINC complex, which controls lamina organization, chromosome dynamics, and genome organization and stability. Thus, LINC constituents play pivotal roles in cellular architecture including organelle positioning, cell movement, tissue assembly, organ homeostasis, and organismal aging. The LINC complex oligomeric core contains several multi-isomeric, multifunctional, and often tissue-specific proteins. Therefore, for a proper functional analysis, genetic mouse models are an invaluable resource. Herein, we focus on the LINC complex roles in the skin and describe methods that enable the successful isolation of primary embryonic fibroblast and newborn skin cells, which can be then investigated functionally in vitro.

Key words Nesprin, SUN, Mouse embryonic fibroblasts, Primary keratinocytes, Primary fibroblasts, Directed cell migration

1 Introduction

The linker of nucleoskeleton and cytoskeleton (LINC) complex is a macromolecular protein complex of the nuclear envelope. It is conserved throughout the evolution and is composed of SUN (Sad1p, UNC-84) and KASH (Klarsicht-, ANC-, Syne homology) domain-containing proteins. SUN proteins are located in the inner nuclear membrane (INM) and interact in the perinuclear space with KASH proteins [1]. Nesprins, the prominent KASH proteins in vertebrates, are type II transmembrane proteins, which are located mostly in the outer nuclear membrane (ONM). They extend into the cytoplasm with their spectrin-repeat-rich N-termini and interact with components of the cytoskeleton, whereas SUN proteins face the nucleoplasm with their N-termini and interact

Gregg G. Gundersen and Howard J. Worman (eds.), *The LINC Complex: Methods and Protocols*, Methods in Molecular Biology, vol. 1840, https://doi.org/10.1007/978-1-4939-8691-0_20, © Springer Science+Business Media, LLC, part of Springer Nature 2018

with lamins A/C, the core intermediate filament proteins of the nuclear lamina, emerin, chromatin, and nuclear pore components [2–4]. Thus, the LINC complex mediates the connection between nucleoskeleton and cytoskeleton [5]. Furthermore, LINC complexes stabilize structurally the nuclear envelope, are responsible for positioning the nucleus during the establishment of cell polarity, and play key roles in fertilization, cell migration, and differentiation [1, 6–8]. Mutations in LINC complexes lead to severe diseases resembling those caused by mutations in the lamin A gene (*LMNA*) collectively known as laminopathies. Laminopathies are associated with a wide spectrum of phenotypes such as muscular dystrophy, Hutchinson-Gilford progeria syndrome, restrictive dermopathy, dilated cardiomyopathy, conduction system disease, and lipodystrophy [6, 9].

There are multiple nesprin- and SUN-encoding genes present in mammals, which generate a plethora of isoforms [10]. Four separate genes (i.e., *SYNE1-4*) encode nesprin-1, nesprin-2, nesprin-3, and nesprin-4, respectively, that are distinguished by the number of spectrin repeats making up their rod domain, a transmembrane domain followed by a short conserved C-terminal extension (termed KASH), which interacts with the SUN domain, an N-terminal F-actin-binding domain in the case of nesprin-1 and nesprin-2 giant [11, 12] and an intermediate filament tethering site found in nesprin-3α, which is mediated by plectin [13].

The skin is the largest organ of the body and provides a physical barrier between the organism and the surrounding environment protecting it from microbial infection, physical insults, and dehydration. Its main layers are the epidermis, dermis, and hypodermis. The most abundant cells found in the epidermis are the keratinocytes. Keratinocytes are specialized epithelial cells rich in keratin intermediate filaments. They are attached together by cell–cell junctions forming a mechanically coupled syncytium that provides protection. Fibroblasts are embedded within the dermis and play key roles in extracellular matrix (ECM) synthesis, ECM remodeling, force generation, skin wounding, and aging. The remaining cells that reside within the dermis and hypodermis nourish the epidermis, exhibit sensory roles, combat infection, and function in thermoregulation.

In the skin, the LINC complex is well conserved with nesprin-1 and nesprin-2 being the major components. Nesprin-1 and nesprin-2 have, however, different subcellular localizations in keratinocytes. Whereas nesprin-1 is present mainly at cell–cell junctions and is weakly expressed at the nucleus, nesprin-2 is restricted to the nuclear envelope. In agreement, the loss of nesprin-2 giant isoforms in mice compromises nuclear morphology and increases nuclear size in keratinocytes. Importantly, the mutant mice exhibit an increased epidermal thickness and display an impaired wound healing response [14, 15].

The accessibility of skin and the ability to isolate and culture successfully various cell types make this organ an excellent system to interrogate the functions of the various LINC complex proteins. In the following sections, we describe the experimental procedures to isolate and cultivate in vitro both primary mouse dermal fibroblasts and keratinocytes from the skin of newborn mice. We also describe the isolation of mouse embryonic fibroblasts (MEF), which might be particularly useful in case mutations in LINC complex components result in embryonic lethality. The availability of these cells offers the possibility to carry out in vitro functional tests, which otherwise would not have been possible at this level. Specifically, we describe the protocols that we implement to study directed fibroblast and keratinocyte migration. This particular experimental setup enables the documentation of the cell velocities but also allows the computation of various other valuable parameters of migration such as direction of migration, persistence, and cell shape. These experiments can further be combined with immunofluorescence experiments that indicate correct cell polarization and relocation of intracellular organelles [16]. Finally, a method for protein extraction for Western blotting from tissue and cells is also presented with a focus on the giant nesprin proteins, which have molecular weights up to 1 million Dalton [11, 12].

2 Materials

2.1 Buffers

1. RIPA lysis buffer: 50 mM Tris–HCl, pH 7.5, 150 mM NaCl, 0.1% SDS (sodium dodecyl sulfate), 1% Nonidet P-40, 0.5% sodium deoxycholate and protease inhibitors.
2. Laemmli sample buffer (5×): 0.25 M Tris–HCl, pH 6.8, 20% glycerol, 4% SDS, 1.43 M β-mercaptoethanol, 0.2% bromophenol blue.
3. SDS gel running buffer (10×): 0.25 M Tris–HCl, pH 8.3, 2 M glycine, 1% SDS.

Precast commercially availabe SDS-PAGE gradient gels ensuring high reproducibility.

2.2 Equipment for Dissection and for Cell Culture

1. Sterile scissors and tweezers.
2. Scalpel (No 21).
3. 6 cm, 10 cm Petri dishes.
4. Cell strainer, size 70 μm.
5. Collagen I-coated six-well plates.
6. Cryo tubes, 1 mL.
7. Ibidi μ-slide eight-well glass bottom dish.
8. Ibidi culture insert, two-well chamber.

9. 25 and 50 mL Falcon tubes.
10. 0.22 μm filters.
11. 10 mL syringes.

2.3 Reagents for Establishment of Primary Cells

1. Antibiotic-antimycotic solution 100× (Sigma-Aldrich).
2. Accutase (Sigma-Aldrich).
3. Betaisodona (commercial disinfectant, iodine containing).
4. Chelex 100.
5. Collagenase I (stored at -20 °C).
6. Dulbecco's Modified Eagle Medium (DMDM).
7. DMEM growth medium: DMEM containing 4.5 g glucose/L, 1 mM sodium pyruvate, 0.1 mM nonessential amino acids, 2 mM L-glutamine, 100 units/mL penicillin G, and 100 μg/mL streptomycin.
8. DMEM/Ham's F12 Medium (keratinocyte medium; Merck), low calcium (50 μM Ca^{2+}), containing 10% fetal bovine serum (FBS), which had been pretreated with Chelex 100 (*see* **Note 1**) in order to remove all calcium ions, supplemented with 0.18 mM adenine, 0.5 μg/mL hydrocortisone, 5 μg/mL insulin, 10 ng/mL EGF, 0.1 nM cholera toxin, 5 μg/mL L-glutamine, 0.05 mg/mL vitamin C, 100 units/mL penicillin G, and 100 μg/mL streptomycin.
9. Dispase II, neutral protease grade II 5 mg/mL, dissolved in phosphate-buffered saline (PBS).
10. FBS.
11. Mitomycin C (0.4 mg/mL, dissolved in PBS), stored in aliquots at −20 °C.
12. PBS, filter sterilized.
13. 70% ethanol solution.
14. Protease inhibitor cocktail.
15. Trypsin solution 10×, (Sigma-Aldrich), stored in aliquots at −20 °C.
16. Trypsin/EDTA solution: 10×: 0.5% Trypsin/0.2% EDTA in PBS. Store at −20 °C.

2.4 Other Equipment

1. Camera.
2. Temperature-controlled mixer.
3. 32 °C incubator, 5% CO_2.
4. 37 °C incubator, 5% CO_2.
5. Laminar flow hood.
6. Microscope (we use Leica DMI6000 B).
7. ULTRA-TURRAX.

3 Methods

3.1 Isolation of Mouse Skin

1. Decapitate 1–3-day-old mice, then place the bodies into a 10 cm Petri dish, and incubate on ice for 30–60 min to cool them down (*see* **Note 2**).
2. Prepare in the meantime: One 50 mL Falcon tube containing 50% Betaisodona solution in PBS; two 50 mL Falcon tubes containing 1× PBS; one 50 mL Falcon tube containing 70% ethanol; one 50 mL Falcon tube containing the 1× antibiotic-antimycotic PBS solution.
3. Disinfect and wash the carcasses (one body after the other) sequentially in Betaisodona solution (1 min), PBS (rinse briefly), 70% ethanol (1 min), and PBS (rinse briefly), and incubate the body in the 1× antibiotic-antimycotic solution (2 min). Perform all these steps in a laminar flow hood (*see* **Note 3**).
4. Transfer the body to a sterile dish for the removal of the skin. Mice that are not immediately dissected should be kept on ice.
5. Pinch the neck skin with one tissue forceps, and insert below the neck skin a sterile blunt end scissor in closed position, and then open the scissor carefully. Repeat this step several times while proceeding toward the posterior end of the body in order to separate the skin from the underlying tissue (*see* **Note 4**). Make two lengthwise head to the tail incisions on the dorsal side.
6. Cut the skin flap and rinse with ice-cold PBS.
7. Carefully remove the fat from the skin.
8. Transfer the skin into a new culture dish (35 mm in diameter), and spread the tissue with the dermis facing down. Make sure that the skin lies flat, with the correct orientation, and avoid any curling of the tissue specimen.
9. Fill the Petri dish from the side with 2 mL PBS solution containing dispase II (5 mg/mL, sterile filtered using 0.22 μm filters), and ensure that only the underside of the suspended skin tissue (dermal compartment) is exposed to the solution.
10. Incubate the Petri dish overnight at 4 °C.
11. The next day, the epidermis can be separated from the dermis using sterile tweezers.

3.1.1 Isolation of Primary Keratinocytes from Mouse Skin [15, 17]

1. Prepare on the previous day a feeder cell layer-coated six-well culture plate: Treat subconfluent feeder cells (NIH 3T3) grown on 10 cm dishes with mitomycin C (4 μg/mL in medium) for 2 h at 37 °C and 5% CO_2.
2. Wash the feeder cells twice with 10 mL PBS, then aspirate the PBS and add 3–4 mL Trypsin/EDTA, rinse the plate with

Trypsin/EDTA, and discard the Trypsin/EDTA solution (*see* **Note 5**).

3. Incubate at 37 °C and 5% CO_2 for few minutes until cells are dislodged from the dish.
4. Resuspend the trypsinized feeder cells in 12 mL DMEM/Ham's F12 medium, and rigorously pipette in order to generate a single cell suspension.
5. Wash the collagen I-precoated six-well plates twice with PBS.
6. Seed out 2 mL of feeder cells per well, and place the plate overnight in the incubator.
7. Take the skins out (**step 9** from above), and gently peel the semitransparent epidermis from the dermis using a pair of tweezers.
8. Transfer the epidermis to a fresh plastic dish (60 or 100 mm), and place it with the epidermal basal layer facing down on top of a 500 μL drop of Trypsin/EDTA solution (*see* **Note 5**). Using a pair of forceps, ensure that the tissue specimen is evenly spread out and avoid any tissue folds.
9. Incubate the covered Petri dish at room temperature for up to 20 min.
10. Add 1.5 mL DMEM/Ham's F12 Medium (low Calcium) to inactivate the Trypsin.
11. Tilt the Petri dish at an ~30° angle to collect the cells, and filter the suspension through a 70 μm porous cell strainer fitted over a 50 mL centrifugation tube.
12. Repeat the washing step vigorously shaking the epidermal fragment several times using the forceps within an additional 2 mL DMEM/Ham's F12 Medium (low calcium) in order to dislodge more basal layer cells.
13. Repeat **step 11** and collect the flow-through in the same 50 mL centrifugation tube.
14. Centrifuge the collected cells at 1000 × *g* for 5 min, at room temperature.
15. Discard the supernatant, and resuspend the epidermal cell pellet in 2 mL DMEM/Ham's F12 Medium (low calcium).
16. Aspirate and replace the medium contained within a single well of the six-well feeder layer plate (**step 6**) with the 2 mL keratinocyte-containing media obtained above (**step 15**; one well per newborn mouse).
17. Incubate cells at 32 °C, 5% CO_2, in a cell culture incubator.
18. Colonies might become visible only after 2 or 3 weeks or even later.

19. Culture propagation: Feeders, once they die away, can be replaced with fresh feeders. Change medium on Mondays, Wednesdays, and Fridays. When the dish is confluent, split 1:2 or 1:1 on collagen I-coated dishes. Wash cells 1× with PBS, incubate with PBS containing 0.02% EDTA for 5 min at RT, aspirate and incubate with Trypsin for 8 min at 37 °C in the CO_2 incubator. Cells come off by tapping the cell culture vessel. Resuspend cells in DMEM/Ham's F12 low-calcium keratinocyte medium, spin, resuspend cells in fresh medium, and seed.

3.1.2 Isolation of Primary Fibroblasts from Mouse Skin

1. Mince the dermis from **step** 7 above (Subheading 3.1.1) into very fine tissue pieces using two scalpels (No 21) in a Petri dish, and digest the tissues in a collagenase I solution (400 U/mL) for 1–2 h at 37 °C with mild shaking.
2. Stop the reaction by adding DMEM growth medium containing 10% FBS, and pellet the cells (500 × *g*, 5 min).
3. Resuspend the cells in medium, and filter them through a cell strainer with 70 μm pores.
4. Centrifuge the flow-through at 500 × *g* for 5 min, and resuspend the pelleted cells in medium again, and transfer them to a fresh Petri dish.
5. Incubate cells at 37 °C, 5% CO_2, in a cell culture incubator. All solutions were sterile-filtered for use (0.22 μm filters).

3.2 Isolation of Mouse Embryonic Fibroblasts (MEFs)

1. Check for a mating plug in the female mice in order to track the estrous stage.
2. Use cervical dislocation to sacrifice mice at day 8–12 of the pregnancy.
3. Place the mouse on its back on a dissecting board, and spray 70% ethanol on the fur.
4. Make a small midline incision on the skin using a scissor, and grab with your two hands the skin, and pull away to expose the abdominal wall.
5. Open the abdominal cavity to expose the intestines.
6. Dissect the uterus out and wash the uterus in 70% ethanol in a Falcon tube.
7. Wash the uterus twice in PBS in a Falcon tube.
8. Place the uterus into a 10 cm Petri dish that contains PBS.
9. Transfer the uterus to a sterile cell culture hood, and using a scissor cut the uterus into one embryo fragment.
10. Slice carefully through one side of the exposed uterine wall in order to isolate the embryos.

11. Transfer the embryos into a fresh Petri dish containing PBS, and swirl to remove any blood.
12. Decapitate the embryos and take a small tail biopsy for genotyping.
13. Remove all internal organs very carefully from the abdominal cavity using a pair of forceps.
14. Mince the remaining body finely using a pair of forceps in a new Petri dish.
15. Transfer the embryo into a 1.5 mL centrifuge tube containing 300 μL Trypsin/EDTA solution (0.05% Trypsin and 0.02% EDTA in PBS), and pipette up and down.
16. Incubate for 35 min at 37 °C.
17. Terminate the reaction with 600 μL pre-warmed medium.
18. Transfer the cell suspension into a 6 cm dish containing 3 mL of medium (DMEM supplemented with 10% FBS).
19. Incubate the cells at 37 °C, 5% CO_2 in a cell culture incubator for several days until the cells can be split.

3.3 Culturing of Mammalian Cells

1. Culture adherent fibroblast cells with DMEM growth medium supplemented with 10% FBS at 37 °C, 5% CO_2 and 95% humidity. Culture keratinocytes in low-calcium DMEM/Ham's F12 medium supplemented with 10% FBS (Chelex 100 pretreated, *see* **Note 1**), 0.18 mM adenine, 0.5 μg/mL hydrocortisone, 5 μg/mL insulin, 10 ng/mL EGF, 0.1 nM cholera toxin, 5 μg/mL L-glutamine, 0.05 mg/mL vitamin C, 100 units/mL penicillin G, and 100 μg/mL streptomycin at 32 °C, 5% CO_2 and 95% humidity.
2. Incubate cells for 5 min at 37 °C in 0.05% Trypsin, 0.02% EDTA for splitting.
3. Terminate the reaction by adding culture medium containing serum.
4. Seed cells at a ratio of 1:2 to 1:10 according to the experimental design.

3.4 Freezing and Thawing of Cells

1. Trypsinize cells as described above and centrifuge at 200 × *g* for 10 min at 4 °C.
2. Remove the supernatant, and resuspend the cells in ice-cold freezing medium (90% FBS, 10% DMSO) and fill into 1 mL cryo tubes.
3. Place the tubes into a freezing container (Nunc) filled with isopropanol, and store for 1 day at −80 °C. Freezing of the cells happens at 1 °C per min.
4. Thaw cells for a few minutes at room temperature.
5. Add 37 °C warm culture medium and transfer the cells to a Falcon tube.

6. Centrifuge for 5 min at 500 × *g* and remove the supernatant.
7. Resuspend the cells in fresh culture medium and plate on cell culture dishes.

3.5 Extraction of Proteins from Cells

1. Wash cells in culture plates with PBS, and scrape them off using PBS containing protease inhibitors.
2. Pellet cells by centrifugation at 500 × *g* for 5 min.
3. Dissolve cells in RIPA lysis buffer.
4. Homogenize the pellet by drawing through a 0.4 mm syringe.
5. Incubate on ice for 15 min.
6. Sonicate the solution to shear the genomic DNA (50% setting).
7. Add 5× Laemmli sample buffer, and heat the sample to 95 °C for 5 min to denature the proteins.
8. Separate Nesprin giant proteins using gradient gels and SDS-PAGE [18].
9. Transfer proteins to nitrocellulose membranes (0.22 µm pore size) for 72 h using a wet blot tank followed by probing with antibodies [18].

3.6 Extraction of Proteins from the Skin Tissue

1. Determine the weight of the skin tissue.
2. Shock freeze the skin in liquid nitrogen or on dry ice.
3. Grind the skin with a pestle and mortar with liquid nitrogen.
4. Add an equal volume of RIPA lysis buffer.
5. Homogenize the skin using an ULTRA-TURRAX.
6. Sonicate the homogenized sample.
7. Add 5× SDS sample buffer and incubate at 95 °C for 5 min.
8. The proteins are separated on gradient gels followed by Western blotting (Fig. 1).

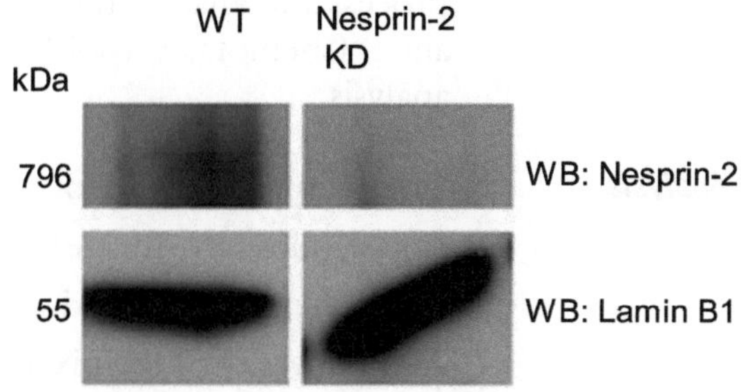

Fig. 1 Western blot analysis of mouse embryonic fibroblasts (MEFs). Cell lysates from MEFs from Nesprin-2 KO and wild-type (WT) mice were separated on a gradient gel (3–12% acrylamide) and the Western blot probed for Nesprin-2 with mAb K56-374-3 and lamin B1-specific antibodies for control

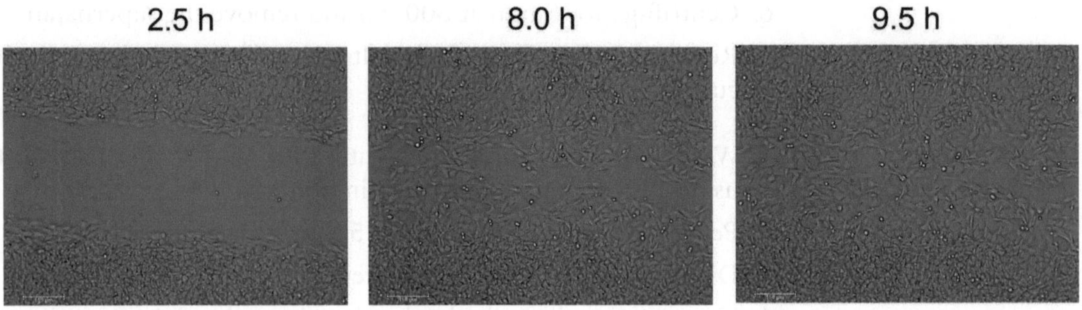

Fig. 2 Cell migration assay. MEFs from wild-type mice were deposited in an eight-well chamber, and a gap was generated by an insert which was removed after overnight incubation. Gap closure was followed by live cell microscopy. Pictures from the indicated times after start of migration are shown. Scale bars, 150 μm

3.7 Cell Migration Analysis

1. Plate 50,000 mouse primary fibroblasts each into ibidi culture inserts that are placed into wells of a microscope slide (ibidi, eight-well chamber). The culture insert provides two culture reservoirs each separated by a 500 μm wall. The cells are cultured in both reservoirs.
2. Incubate in CO_2 incubator (5% CO_2) at 37 °C.
3. Next day, wash cells once with PBS.
4. Remove the insert which results in two well-defined cell patches. The cells start immediately migrating into the gap generated by the insert.
5. Place microscope slides with fibroblasts in a humidified 5% CO_2 atmosphere. For 37 °C warm air incubation, the microscope and objectives are encased (heater and ventilation ON).
6. Capture images at 15 min intervals for 24 h with a Leica TIRF microscope (Leica DMI6000 B TIRF MC, LAS AF software version 2.0.2 build 2038) equipped with a camera (Hamamatsu) and using 10×/0.25 NA dry objective and magnification of 1.6× (Fig. 2).
7. Process and analyze the data using ImageJ "Manual Tracking" and "Chemotaxis tool." Use only the inner four wells for analysis.

3.8 Image Analysis

1. Save pictures as TIFF overlay.
2. Import files into ImageJ as image sequence.
3. Open the manual tracking window.
4. Add the value of the pixel size under "x/y calibration."
5. To track a cell, choose "show path" and "add track."
6. Track several cells (>10). The mean values will show up in an additional window.
7. Save values in an Excel file.

8. Then use the Chemotaxis plug-in in ImageJ.
9. Define the time between two pictures under "setting."
10. Add the value for the pixel size in μm² under "x/y calibration."
11. Import the Excel sheet into the Chemotaxis plug-in.
12. Press "apply settings": The migration velocity and directionality are shown.
13. Use results for evaluation.

4 Notes

1. FBS pretreatment: 2 g Chelex 100 per 50 mL FBS, overnight on a rotating wheel at 4 °C, and then sterile filtered.
2. All rules and regulations concerning animal work must be followed.
3. Whenever possible, the work should be performed in a laminar flow hood.
4. All equipment used for mouse dissection should be sterile.
5. For isolation of keratinocytes, Trypsin/EDTA is often replaced by TrypLE (Thermo Fisher Scientific), recombinant cell-dissociation enzymes, which act more gently than Trypsin and preserve cell surface epitopes.

Acknowledgments

This work was supported by Cologne Cluster of Excellence on Cellular Stress Responses in Aging-Associated Diseases (CECAD), the International Graduate School in Development Health and Disease (IGSDHD), and the BBSRC (Biotechnology and Biological Sciences Research Council).

References

1. Crisp M, Liu Q, Roux K et al (2006) Coupling of the nucleus and cytoplasm: role of the LINC complex. J Cell Biol 172:41–53
2. Xiong H, Rivero F, Euteneuer U et al (2008) Dictyostelium Sun-1 connects the centrosome to chromatin and ensures genome stability. Traffic 9:708–724
3. Haque F, Mazzeo D, Patel JT et al (2010) Mammalian SUN protein interaction networks at the inner nuclear membrane and their role in laminopathy disease processes. J Biol Chem 285:3487–3498
4. Li P, Noegel AA (2015) Inner nuclear envelope protein SUN1 plays a prominent role in mammalian mRNA export. Nucleic Acids Res 43:9874–9888
5. Luxton GW, Starr DA (2014) KASHing up with the nucleus: novel functional roles of KASH proteins at the cytoplasmic surface of the nucleus. Curr Opin Cell Biol 28:69–75
6. Worman HJ (2012) Nuclear lamins and laminopathies. J Pathol 226:316–325
7. Sosa BA, Kutay U, Schwartz TU (2013) Structural insights into LINC complexes. Curr Opin Struct Biol 23:285–291
8. Rothballer A, Kutay U (2013) The diverse functional LINCs of the nuclear envelope to

the cytoskeleton and chromatin. Chromosoma 122:415–429

9. Lattanzi G, Marmiroli S, Facchini A et al (2012) Nuclear damages and oxidative stress: new perspectives for laminopathies. Eur J Histochem e45:56
10. Rajgor D, Mellad JA, Autore F et al (2012) Multiple novel nesprin-1 and nesprin-2 variants act as versatile tissue-specific intracellular scaffolds. PLoS One 7:e40098
11. Zhen YY, Libotte T, Munck M et al (2002) NUANCE, a giant protein connecting the nucleus and actin cytoskeleton. J Cell Sci 115:3207–3222
12. Padmakumar VC, Abraham S, Braune S et al (2004) Enaptin, a giant actin-binding protein, is an element of the nuclear membrane and the actin cytoskeleton. Exp Cell Res 295:330–339
13. Wilhelmsen K, Ketema M, Truong H et al (2006) KASH-domain proteins in nuclear migration, anchorage and other processes. J Cell Sci 119:5021–5029
14. Lüke Y, Zaim H, Karakesisoglou I et al (2007) Nesprin-2 Giant (NUANCE) maintains nuclear envelope architecture and composition in skin. J Cell Sci 121:1887–1898
15. Rashmi RN, Eckes B, Glöckner G et al (2012) The nuclear envelope protein Nesprin-2 has roles in cell proliferation and differentiation during wound healing. Nucleus 3:172–186
16. Bhattacharya K, Swaminathan K, Peche VS et al (2016) Novel Coronin7 interactions with Cdc42 and N-WASP regulate actin organization and Golgi morphology. Sci Rep 6:25411
17. Zhang ZG, Bothe I, Hirche F et al (2006) Interactions of primary fibroblasts and keratinocytes with extracellular matrix proteins: contribution of alpha2beta1 integrin. J Cell Sci 119:1886–1895
18. Carthew J, Karakesisoglou I (2016) Detection of diverse and high molecular weight Nesprin-1 and Nesprin-2 isoforms using Western blotting. Methods Mol Biol 1411:221–232

Chapter 21

Detection of SUN1 Splicing Variants at the mRNA and Protein Levels in Cancer

Ayaka Matsumoto, Nariaki Matsuura, and Miki Hieda

Abstract

The linker of nucleoskeleton and cytoskeleton (LINC) complex, containing the proteins SUN and nesprin, is the fundamental structural unit of the nuclear envelope. The neoplastic-based regulation of the LINC complex in cancer tissues has become increasingly recognized in recent years, including the altered expression, somatic mutation, and methylation of genes. However, precisely how mutations and deregulated expression of the LINC complex contribute to the pathogenic mechanisms of tumorigenesis remain to be elucidated, mainly because of several technical difficulties. First, both the *SUN* and *SYNE* (encoding nesprin) genes give rise to a vast number of splicing variants. Second, immunoprecipitation experiments of endogenous SUN and nesprin proteins are difficult owing to the lack of suitable reagents as well as the limited solubility of these proteins in mild extraction conditions. Here, we describe three protocols to investigate these aspects: (1) immunohistochemistry to determine the expression levels and localization of the LINC complex in cancer tissue, (2) detection of *SUN1* splicing variants at the mRNA level, and (3) detection of SUN1 splicing variants and binding partners at the protein level.

Key words SUN1, SUN2, Nesprin, Immunohistochemistry, Splicing variants

1 Introduction

Morphological changes in the nucleus are hallmarks of cancer cells. In particular, enlarged nuclei, abnormal nuclear shape, anisokaryosis, increased chromatin staining, and altered nucleolar size are often observed in cancer cells [1–3]. These phenotypes are used for both diagnosing malignancy and to predict outcomes [4]. Further, the role of the nuclear envelope (NE) in the diagnostic and prognostic pathology of human cancers has also been described in recent years [5, 6]. A multifunctional nuclear membrane protein assembly called the linker of nucleoskeleton and cytoskeleton (LINC) complex consists of the inner nuclear membrane spanning SUN homology domain proteins (SUN proteins), and the outer nuclear membrane spanning nesprin proteins, which share a KASH homology domain [7]. Mammalian SUN proteins are encoded by

Gregg G. Gundersen and Howard J. Worman (eds.), *The LINC Complex: Methods and Protocols*, Methods in Molecular Biology, vol. 1840, https://doi.org/10.1007/978-1-4939-8691-0_21,

at least five genes: *SUN1*, *SUN2*, *SUN3*, *SUN4*, and *SUN5*. *SUN1* and *SUN2* are widely expressed in mammalian somatic cells [7, 8], whereas *SUN3, SUN4* (also known as sperm-associated antigen 4, *SPAG4*), and *SUN5* (also known as SPAG4-like, *SPAGL*) are largely, but not entirely, restricted to the germ cells [9–11]. The human genome contains six different nesprin-encoding genes, which are defined by the presence of a C-terminal KASH domain. Nesprin-1 and nesprin-2 each have multiple isoforms resulting from alternative transcriptional initiation, termination, and splicing and are expressed in a wide variety of tissues [12, 13]. The LINC complex performs diverse functions, including nuclear shaping and positioning, maintenance of the centrosome-nucleus connection, DNA repair, nuclear membrane spacing, cell migration, and moving chromosomes within the nucleus during meiosis [14–22]. In addition, lamins—intermediate filaments that interact with SUN proteins underneath the NE—also play various roles such as maintenance of nuclear integrity, cell cycle regulation, mechanotransduction, cellular signalling, and DNA repair [23]. Accordingly, deregulated LINC complex components, including both nesprin and SUN proteins, have been observed in several cancers.

The downregulation of SUN2 protein expression was first observed in atypical teratoid/rhabdoid tumor (AT/RT), a rare and highly aggressive form of embryonal tumor, and was subsequently observed in lung and breast cancer tissues [24–26]. Hsieh et al. [24] showed that SUN2 is a direct downstream target of miR-221/miR-222, which is abundantly expressed in AT/RT. Moreover, SUN2 expression inhibited cell proliferation and colony formation in embryonal tumor and lung cancer cell lines, pointing to the role of SUN2 as a tumor suppressor [24, 25]. SUN2 was also shown to exert its tumor suppressor functions by decreasing the expression levels of the glucose transporter 1 (*GLUT1*) and lactate dehydrogenase A (*LDHA*) genes to ultimately inhibit the Warburg effect in lung cancer cell lines [25]. In addition, breast cancer tissues also showed decreased SUN1 expression [26]. Depletion of SUN1 mediated by small interfering RNA in cultured mammary epithelial cells induced nucleolar hypertrophy, and a negative correlation was observed between SUN1 expression and the size of the nucleoli in human breast cancer tissue [27]. Human SUN4 (also known as SPAG4) is expressed in a limited number of normal tissues, with notable expression detected in the pancreas and testis; however, SUN4 expression has been detected in a large number of human cancer cell lines [28] and was shown to be upregulated in human renal clear cell carcinoma (RCC) [29]. The hypoxic regulation of SUN4 is mediated through hypoxia-inducible transcription factor 1 [29, 30]. SUN4 knockdown reduced the invasion capability of RCC cells in vitro, and its overexpression led to enhancement of tumor cell migration [30].

SUN4 is thus suggested as an independent prognostic factor in RCC [29, 30].

In addition to the altered expression of SUN proteins, nesprins are also deregulated in several cancer types. *SYNE1* and *SYNE2* are unusually large genes on chromosomes 6q and 14q and contain 146 and 115 exons, respectively. These genes give rise to a large majority of tissue-specific nesprin isoforms of different sizes [31]. A large genotyping study demonstrated that a missense single nucleotide polymorphism, located 19 kb downstream of the estrogen receptor alpha gene (*ESR1*) in the coding COOH-terminal region of the *SYNE1* gene, was most strongly associated with an increased risk of invasive ovarian cancer among the hormone-related genes tested [32]. In addition, somatic mutations in *SYNE1* have been observed in breast and colorectal cancer [33]. Methylation in the *SYNE1* gene has also been detected in cases of lung adenocarcinoma [34] and colorectal cancer [35]. One of the shorter transcripts of nesprin-1 that lacks the KASH domain, termed Drop1, is encoded at the 5′ end of the *SYNE1* gene, and striking downregulation of Drop1 expression has been observed in ovarian, mammary, and other carcinomas [36]. In addition to alterations of the *SYNE1* gene, array analysis of gastrointestinal stromal tumor (GIST) samples revealed recurrent copy number variations of chromosomal arm 14q, which includes the *SYNE2* locus [37]. The same study also revealed that the high expression of nesprin-2 in GIST was associated with a relatively high mitotic rate and shorter disease-free survival [37]. One of the nesprin-2 variants tethers active mitogen-activated protein kinase (MAPK)1 and MAPK2 at promyelocytic leukemia protein nuclear bodies and acts to regulate smooth muscle cell proliferation [38]. Another nesprin-2 variant was shown to interact with MAPK8IP2 [39]. Therefore, it is possible that deregulation of the *SYNE2* gene is of crucial importance in GIST progression via altered MAPK signalling [37].

Although deregulation of the LINC complex components has been observed in various cancer tissues, whether or how the altered expression of these components and the contributions of their cancer-associated mutations to tumorigenesis and tumor progression remain poorly understood. In addition, SUN1 and SUN2 are essential factors in cell migration, a characteristic feature of cancer cells; however, the expression levels of SUN1 and SUN2 are dramatically decreased in several cancer tissues. This mystery suggests that the roles of LINC complex components are likely context-dependent, and thus further study of the alterations in diverse LINC complex components will provide deeper insight into their physiological and pathological roles. Here, we describe three methods to explore the functions of the LINC complex in tumor development and progression. The first method is immunohistochemical (IHC) staining of the LINC complex. There are various methods available to determine the expression levels of the LINC complex in cancer cells, including IHC, real-time polymerase chain reaction (PCR), western blotting, in situ hybridization, array analysis, and tandem mass spectrometry. However, cancer tissues are very heterogeneous, and it is highly possible that some of these methods would point to downregulation of the LINC complex, while others would point to upregulated expression for the same component, even within an individual specimen; moreover, expression levels are also likely to vary among patients. Therefore, among the possible methods listed above, clinical studies with IHC have two key advantages: statistical power and the ability to show a direct association between malignancy and expression level in a patient. The second assay system described herein is detection of *SUN1* splicing variants at the mRNA level. Although the *SYNE1, SYNE2,* and *SUN1* genes all give rise to a vast number of different variants, investigations of nesprin variants are quite well documented; thus, we here focus on the *SUN1* variants. For detailed discussion of the nesprin variants, please refer to previous excellent papers on the topic [31, 40, 41]. The last method is the detection of SUN1 splicing variants and SUN1 binding partners at the protein level. In contrast to the *SUN2* gene, which generates only several isoforms, the *SUN1* gene generates various splicing variants that may have different functions [21, 42, 43]. Therefore, the identification and verification of SUN1 variants at the protein level is crucial to understanding the functions of SUN proteins. However, this investigation is difficult because of two main factors. First, except for the predominantly expressing variant SUN1_916, most of the SUN1 splicing variants have low protein expression levels in humans, making their expression difficult to verify. Second, the LINC complex and related proteins such as lamins and nuclear pore complex components are highly insoluble. Therefore, we used three lysis conditions to investigate SUN1 variants and binding partners.

2 Materials

2.1 IHC Staining of the LINC Complex

1. Formalin-fixed, paraffin-embedded (FFPE) specimens. Ready-to-use tissue sections are commercially available from several companies. We recommend the use of thin sections (~2 μm thick) to obtain high-quality images, although 4 μm-thick sections can also be used.
2. Primary antibodies specific to the LINC complex components. We verified the activities of the following polyclonal anti-LINC complex components antibodies: rabbit anti-SUN1 polyclonal antibody (1:200; Sigma, HPA 008346), rabbit anti-SUN2 polyclonal antibody (1:200; Millipore,

anti-UNC84B, 06-1038), rabbit anti-nesprin-1 polyclonal antibody (1:200, Sigma, HPA019113), and rabbit anti-nesprin-2 polyclonal antibody (1:200; Sigma, HPA003435). However, polyclonal antibodies may suffer from lot-to-lot variations. Thus, we recommend that the activities and specificities of antibodies be confirmed with western blotting (for size and specificity) and immunofluorescent microscopy using cultured cells before beginning the experiments.

3. Biotinylated secondary anti-rabbit IgG antibodies (DAKO).
4. Antigen retrieving buffer: DAKO Target Retrieval Solution (pH 9.0) or 0.01 M citrate buffer (pH 6.0).
5. 3,3-diaminobenzidine (DAB) solution, prepared fresh: (1) dissolve 20 mg of DAB in 100 mL of 0.05 M Tris–HCl (pH 7.6); (2) let stand for 10–15 min; (3) add 16.7 μL of H_2O_2.
6. Counter-staining solution: hematoxylin solution.
7. Xylene, Lemosol (Wako Pure Chemical Industries Ltd., Japan), or Lemosol ACE (Wako Pure Chemical Industries Ltd., Japan) (*see* **Note 1**).
8. Tris-buffered saline with Tween-20 (TBST): 0.05 M Tris–HCl (pH 7.6), 0.3 M NaCl, 0.1% Tween-20.
9. Blocking solution: 5% bovine serum albumin (BSA) in TBST.
10. Mounting medium such as Malinol (*see* **Note 2**).

2.2 Detection of Human SUN1 Variants at the mRNA Level

1. Splicing variant-specific *SUN1* primer set (sense primer, GACCACTTCTGGGGTCTTGA; antisense primer, GTTTCGAAGGCACCTGGTAA); the amplicon size is shown in Fig. 1.
2. Universal *SUN1* primer set (sense primer, GTGGACG TGCAAGTCAGAGA; antisense primer, CCCGGTCT TATCTTGGGAAT); amplicon size of 312 bp.
3. PCR reagents: most commonly used PCR reagents found in standard molecular laboratories can be used.
4. 5× Tris-borate- EDTA (TBE): dissolve the following reagents in 1000 mL H_2O: 54.0 g Tris base, 27.5 g boric acid, 3.7 g EDTA·2Na($2H_2O$); the pH of the 1× solution is 8.3 without using an acid or base to adjust the pH.
5. 1× TBE buffer: 89 mM Tris base, 89 mM boric acid, 2 mM EDTA, pH 8.3. To prepare the 1× TBE running buffer, add 200 mL 5× TBE running buffer to 800 mL of H_2O.
6. 6% TBE polyacrylamide (commercially available from Invitrogen, Novex) (*see* **Notes 3** and **4**).
7. Gel caster for the mini gel.

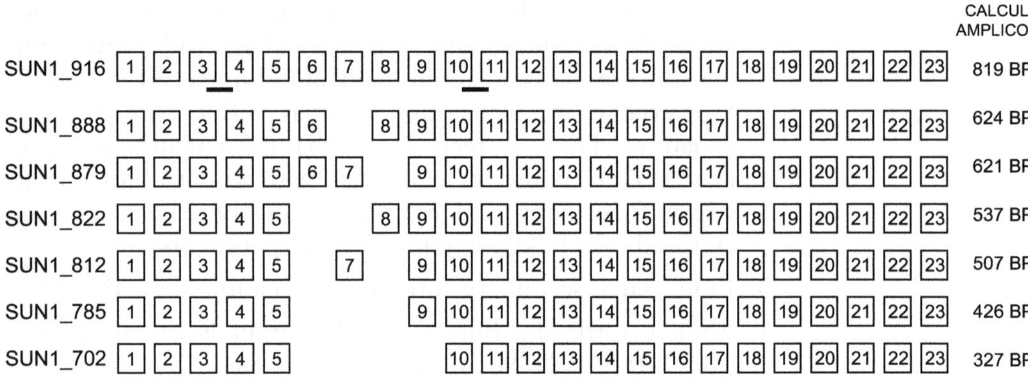

Fig. 1 The human *SUN1* gene contains 22 exons, which are indicated by white boxes. Exons 6–9 are alternatively spliced to generate numerous SUN1 variants [21]. Arrow indicates the position of primers used in the text

8. 5× sample buffer for the TBE-acrylamide gel: mix the following reagents and adjust to 10 mL with H_2O (*see* **Note 5**): 2.0 mL 5× TBE buffer, 1.5 g Ficoll type 400, 1.0 mL 1% bromophenol blue, and 1.0 mL 1% xylene cyanol.
9. Nucleic acid gel staining solution: conventionally used nucleic acid gel staining solutions can be employed, such as GelRed (Wako Pure Chemical Industries Ltd., Japan), ethidium bromide, SYBR Green, or other solutions (*see* **Note 6**).
10. Appropriate molecular weight marker such as 100-bp DNA ladder.
11. Power supply.

2.3 Detection of SUN1 Variant Proteins and Binding Partners

1. Phosphate-buffered saline (PBS).
2. RIPA buffer: 0.025 M Tris–HCl (pH 7.6), 150 mM NaCl, 1% NP-40, 1% sodium deoxycholate, 0.1% sodium dodecyl sulfate (SDS); store at 4 °C. Available from Thermo Fisher Scientific.
3. Lysis buffer A: 0.1 M Tris–HCl (pH 8.3), 0.5 M NaCl, 0.5% Triton X100; store at 4 °C
4. Lysis buffer B: 1× PBS, 0.05 M Tris–HCl (pH 7.5), 0.15 M NaCl, 1% Triton X100; store at 4 °C. Mix the following reagents and adjust to 50 mL with H_2O: 5.0 mL 10× PBS, 2.5 mL 1.0 M Tris–HCl (pH 7.5), 1.5 mL 5.0 M NaCl, 5.0 mL 10% Triton X100.
5. Cell scrapers.
6. Transfection reagents: typical reagents used for transfection in your lab can be used.
7. Protease inhibitor cocktail.
8. Rabbit anti-GFP polyclonal antibody (MBL, Japan) (*see* **Note 7**).
9. Rabbit anti-SUN1 polyclonal antibody (Sigma, HPA008346).

10. Rabbit anti-SUN2 polyclonal antibody (Millipore, anti-UNC84B, 06-1038).
11. Control rabbit IgG.
12. SDS-polyacrylamide gel.
13. 2× SDS sample buffer.
14. Dynabeads Protein G (Thermo Fisher Scientific).
15. Materials for Western blotting (e.g., transfer buffer, nitrocellulose membrane, primary and secondary antibodies, and an appropriate detection system such as enhanced chemiluminescence).

3 Methods

3.1 IHC Staining of the LINC Complex

3.1.1 Preparation of Tissue Sections on a Slide

1. To deparaffinize specimens on slides, incubate the FFPE slides in Lemosol (or Lemosol ACE or xylene) for 5 min (*see* **Notes 1** and **8**).
2. Repeat **step 1** three times using three jars.
3. Wash the specimens in ethanol to remove Lemosol (or Lemosol ACE/xylene).
4. Repeat **step 3** four times using four jars.
5. Wash with running water for 5 min (tap water can be used).
6. Process samples for antigen retrieval using an appropriate procedure. Antigen retrieval is one of the most critical steps and should be optimized for the target antigen and tissues, because many epitopes may be masked or altered by certain fixatives. In addition, some antibodies may not be able to recognize epitopes that are altered by the fixative or hidden within subcellular structures. We usually use DAKO Target Retrieval Solution (pH 9.0) or 0.01 M citrate buffer (pH 6.0) in a Pascal pressure chamber (DAKO) at 125 °C for 2 min and 90 °C for 10 s. Both buffers work effectively for anti-SUN1, anti-SUN2, anti-nesprin-1, and anti-nesprin-2 antibodies. If this equipment is not available, try a boiled water bath (place a staining jar containing the appropriate retrieval solution and slides into a 100 °C water bath for 20 min, followed by a 20-min cool-down period); also *see* **Note 9**.
7. Block endogenous peroxidase in a 3% H_2O_2 solution in methanol for 10 min.
8. Wash with running water for 5 min.
9. Block the section with 5% BSA/TBST for 30 min in a humid chamber to prevent drying.

3.1.2 Antibody Reaction

This step involves choosing antibodies and antibody-binding conditions, including the concentration, incubation time, and temperature. Even when selecting an antibody that is well-validated and/or of guaranteed quality, optimization is still required to determine the best concentration and incubation period; these parameters also depend on the tissue samples analyzed.

1. Incubate the slides with primary antibodies at 4 °C overnight in a humid chamber (*see* **Note 10**).
2. Wash the slides three times with TBST for 5 min each.
3. Incubate the slides with biotinylated secondary antibody at room temperature (RT) for 30 min in a humid chamber.
4. Wash the slides three times with TBST for 5 min each.
5. Incubate the slides with peroxidase-conjugated streptavidin at room temperature for 30 min in a humid chamber.

3.1.3 Detection and Observation (See Note 11)

1. Wash the slides three times with TBST for 5 min each.
2. Place the sections in freshly prepared DAB solution. Incubate for ~5 min to develop.
3. Place the sections in water or TBST to stop development.
4. Wash the sections with running water for 5 min.
5. Counterstain with hematoxylin for 2 min.
6. Wash the sections with running water for 10 min.
7. Dehydrate the sections three times in 100% ethanol.
8. Incubate in Lemosol (or Lemosol ACE/xylene) three times for 5 min each.
9. Mount with a coverslip using mounting medium.
10. Observation (*see* **Notes 12** and **13**). The assessment should be unbiased, and the procedure requires adequate internal negative and positive controls. To achieve unbiased assessment, at least two people should evaluate each sample independently. In addition, all independently scored samples should be assessed without knowledge of clinicopathological data.

3.2 Detection of Human SUN1 Variants at the mRNA Level

The human *SUN1* gene contains 22 exons and generates various splicing variants that are distinguished by identification of variable deletions just upstream from the transmembrane domain, between exons 6 and 9 (Fig. 1). The largest splicing variant of human *SUN1* is composed of 916 amino acids (EAW87177). In this report we refer to this variant as SUN1_916, which predominates in most human tissues and corresponds to the largest mouse SUN1 variant, composed of 913 amino acids. *SUN1* expression tends to be ubiquitous when a universal *SUN1* primer set is used (*see* Subheading 2). To examine the expression levels of individual *SUN1* splice variants, a primer set can be designed at the junctions of exon 3/4 and exon 10/11, which would amplify the variable region as different amplicon sizes (Fig. 1). Reverse transcription-PCR analysis of mRNA levels from various human tissues identified at least seven discrete *SUN1* transcripts, including a testis-specific variant [21]. Amplicon sizes can be confirmed using plasmids as positive controls.

1. Mix the following materials (another PCR mixture can also be used): 10.0 μL 2× Emerald master mix, 0.4 μL 10 μM sense primer, 0.4 μL 10 μM antisense primer, 1.0 μL cDNA, 8.2 μL H_2O.
2. Perform the PCR under the following conditions: initial denaturation at 94 °C for 5 min; 25–30 cycles of 98 °C for 10 s, 55 °C for 30 s, 72 °C for 1.5 min; final extension at 72 °C for 5 min; and hold at 16 °C. The temperature and incubation time are highly dependent on the PCR reagent used.
3. Mix 5× TBE sample buffer, PCR product, and an adequate volume of H_2O.
4. Set up the TBE gel to the gel caster and gently pull the comb out of the gel.
5. Fill the upper and lower buffer chambers of the gel caster with TBE buffer. The buffer level must exceed the level of the wells.
6. Load the molecular weight markers and samples.
7. Run the gel at a constant voltage of 200 V until the bromophenol blue tracking dye reaches the bottom of the gel; this usually takes about 50 min but depends on the gel type. The expected current is 10–18 mA at the start and 4–6 mA at the end (*see* **Note 14**).
8. Remove the gel from the plate and incubate to stain in an appropriate staining solution such as GelRed or SYBR Green for an appropriate time (*see* **Note 6**).

3.3 Detection of SUN1 Variant Proteins and Binding Partners

SUN1_916 is a predominant variant among the SUN1 variants present in most tissues. Although it is difficult to design a variant-specific antibody and no such antibody is commercially available, the size of variants differs from each other, and they are thus distinguishable. However, Western blot analysis of the total cell lysate of various cultured human cells using an anti-SUN1 antibody, which recognizes a domain common to all variants, mainly showed a single band around 100 kDa, which corresponds to SUN_916. The protein expression level of other splicing variants is usually too low to be verifiable, whereas the smaller SUN1 variant proteins are actually expressed, and a concentration procedure can reveal the smaller endogenous SUN1 variants [21]. To detect these minor SUN1 variants, the cells should be lysed with an appropriate lysis buffer, and then a concentration procedure should be performed. Here, we describe methods to lyse cells and concentrate the SUN1 variants and their binding partners. The following protocol is optimized for HeLa cells. Thus, if another cell line is used, the specific conditions should be optimized, such as the volume of the lysis buffer and incubation time. To determine the variant-specific binding partners, SUN1 variant expression can be forced in SUN1 knockout cells [21].

1. Cells are transfected with the target variants. To focus on the endogenous SUN1 variants and their binding partners, please start from **step 2**.
2. The next day, remove the media from confluent culture plates.
3. Wash the cells twice with cold PBS
4. Add 500 μL lysis buffer (A, B, or RIPA buffer) to the cells. For Nups and lamins, use lysis buffer A and lysis buffer B, respectively. To detect the SUN2-SUN1 variant interaction, use RIPA buffer. Use 0.5 mL of buffer per 75 cm^2 (10-cm dish) containing 5×10^6 HeLa cells, and swirl the plate for uniform spreading of the lysis buffer on ice.
5. Scrape the cells and transfer to a microcentrifuge tube.
6. Rotate for 30–60 min at 4 °C.
7. During step 6, wash Dynabeads protein G with lysis buffer, and mix with appropriate antibodies (anti-GFP, anti-SUN1, anti-SUN2 antibodies or control IgGs) and rotate at 4 °C.
8. Centrifuge the lysed samples (from **step 6**) at ~14,000 × *g* for 30–60 min to pellet the cell debris.
9. During **step 8**, wash the Dynabeads protein G from **step 7** with lysis buffer twice, and mix with the supernatant from **step 8**.
10. Incubate for 2 h with constant rotation at 4 °C.
11. Remove the supernatant, wash Dynabeads protein G four times with lysis buffer, and add 2× SDS sample buffer.
12. Boil the Dynabeads in SDS sample buffer and load onto the SDS-polyacrylamide gel electrophoresis (PAGE) gel.
13. Perform Western blotting using appropriate antibodies or directly perform silver staining to detect novel protein bands.

4 Notes

1. Although xylene has long been used in IHC, it is a highly volatile reagent and is hazardous to health. Thus, it can be substituted by Lemosol or Lemosol ACE (Wako Pure Chemical Industries Ltd., Japan).

2. Besides Malino, there is a wide range of commercially available resin-based mounting media.
3. An agarose gel can be used for this assay. However, TBE-PAGE is a more sensitive, accurate, and high-resolution technique. DNA fragments ranging from 10 to 3000 bp are clearly resolved into sharp, tight bands. The commercially available TBE polyacrylamide gel provides rapid results with good reproducibility.
4. The packing buffer contains sodium azide and a residual acrylamide monomer. Wear gloves when handling the gels.
5. A ready-to-use sample buffer is available from Invitrogen (Novex, Hi-density TBE sample buffer).
6. Ethidium bromide can be used for staining, but ethidium bromide staining of polyacrylamide gels requires at least 10 ng of DNA for detection due to the quenching of the fluorescence by polyacrylamide. Since ethidium bromide is a powerful mutagen, it is important to wear gloves when using this reagent.
7. Anti-GFP antibodies are available from various companies. However, their activities show striking variation. Thus, before starting the experiments, comparison of antibodies is recommended.
8. We start this protocol from paraffin-embedded specimens on glass slides, which are commercially available. The conventional methods for fixation, sectioning, and mounting on slides are not described here because various well-established methods are already available, and these steps can considerably vary depending on the tissues of interest.
9. Various procedures can also be used, including limited protease treatment, microwaving, or autoclaving. Limited protease treatment sometimes, but not always, helps to unmask hidden epitopes.
10. Incubation conditions such as temperature and time should be optimized.
11. Several detection procedures are also available. Methods involving fluorescence detection offer higher resolution, whereas enzyme-labelled chromogenic detection methods offer higher sensitivity and are compatible with most histological stains such as hematoxylin and eosin staining, which can provide detailed information about the tissue. We here describe enzyme-labelled methods using a biotin-streptavidin system for signal amplification.
12. Cancer-associated non-cancerous regions, leukocytes, endothelial cells, or fibroblasts in the same tissue section can often be used as internal positive and/or negative controls.

13. HE staining allows pathologists to distinguish between epithelial and stromal cells, but specific markers such as pan-cytokeratin of epithelial and stromal cells can be used to discriminate cancerous and non-cancerous cells.
14. The size of the DNA fragments visualized at the front edge of the dye on a 6% TBE gel is 65 bp with bromophenol blue (dark blue) and 250 bp with xylene cyanol (blue green).

Acknowledgment

We thank Ms Yu Nishioka and Ms Junko Imada for technical assistance. This work is supported by the Education and Research Grant Program of Ehime Prefectural University of Health Science to MH.

References

1. Fischer AH, Bardarov S Jr, Jiang Z (2004) Molecular aspects of diagnostic nucleolar and nuclear envelope changes in prostate cancer. J Cell Biochem 91:170–184
2. Zink D, Fischer AH, Nickerson JA (2004) Nuclear structure in cancer cells. Nat Rev Cancer 4:677–687
3. de Las Heras JI, Schirmer EC (2014) The nuclear envelope and cancer: a diagnostic perspective and historical overview. Adv Exp Med Biol 773:5–26
4. Jevtić P, Levy DL (2014) Mechanisms of nuclear size regulation in model systems and cancer. Adv Exp Med Biol 773:537–569
5. Fischer AH (2014) The diagnostic pathology of the nuclear envelope in human cancers. Adv Exp Med Biol 773:49–75
6. Bell ES, Lammerding J (2016) Causes and consequences of nuclear envelope alterations in tumour progression. Eur J Cell Biol 95:449–464
7. Crisp M, Liu Q, Roux K, Rattner JB et al (2006) Coupling of the nucleus and cytoplasm: role of the LINC complex. J Cell Biol 172:41–53
8. Hodzic DM, Yeater DB, Bengtsson L et al (2004) Sun2 is a novel mammalian inner nuclear membrane protein. J Biol Chem 279:25805–25812
9. Shao X, Tarnasky HA, Lee JP et al (1999) Spag4, a novel sperm protein, binds outer dense-fiber protein Odf1 and localizes to microtubules of manchette and axoneme. Dev Biol 211:109–123
10. Frohnert C, Schweizer S, Hoyer-Fender S (2011) SPAG4L/SPAG4L-2 are testis-specific SUN domain proteins restricted to the apical nuclear envelope of round spermatids facing the acrosome. Mol Hum Reprod 17:207–218
11. Jiang XZ, Yang MG, Huang LH et al (2011) SPAG4L, a novel nuclear envelope protein involved in the meiotic stage of spermatogenesis. DNA Cell Biol 30:875–882
12. Rajgor D, Shanahan CM (2013) Nesprins: from the nuclear envelope and beyond. Expert Rev Mol Med 15:e5
13. Roux KJ, Crisp ML, Liu Q et al (2009) Nesprin 4 is an outer nuclear membrane protein that can induce kinesin-mediated cell polarization. Proc Natl Acad Sci U S A 106:2194–2199
14. Gundersen GG, Worman HJ (2013) Nuclear positioning. Cell 152:1376–1389
15. Zhang X, Lei K, Yuan X et al (2017) SUN1/2 and Syne/Nesprin-1/2 complexes connect centrosome to the nucleus during neurogenesis and neuronal migration in mice. Neuron 2009(64):173–187
16. Lei K, Zhu X, Xu R et al (2012) Inner nuclear envelope proteins SUN1 and SUN2 play a prominent role in the DNA damage response. Curr Biol 22:1609–1615
17. Lawrence KS, Tapley EC, Cruz VE et al (2016) LINC complexes promote homologous recombination in part through inhibition of nonhomologous end joining. J Cell Biol 215:801–821
18. Luxton GW, Gomes ER, Folker ES et al (2010) Linear arrays of nuclear envelope proteins harness retrograde actin flow for nuclear movement. Science 329:956–959

19. Luxton GW, Gomes ER, Folker ES et al (2011) TAN lines: a novel nuclear envelope structure involved in nuclear positioning. Nucleus 2:173–181
20. Chang W, Antoku S, Östlund C et al (2015) Linker of nucleoskeleton and cytoskeleton (LINC) complex-mediated actin-dependent nuclear positioning orients centrosomes in migrating myoblasts. Nucleus 6:77–88
21. Nishioka Y, Imaizumi H, Imada J et al (2016) SUN1 splice variants, SUN1_888, SUN1_785, and predominant SUN1_916, variably function in directional cell migration. Nucleus 7:572–584
22. Hiraoka Y, Dernburg AF (2009) The SUN rises on meiotic chromosome dynamics. Dev Cell 17:598–605
23. Adam SA (2017) The nucleoskeleton. Cold Spring Harb Perspect Biol 9. pii a023556
24. Hsieh TH, Chien CL, Lee YH et al (2014) Downregulation of SUN2, a novel tumor suppressor, mediates miR-221/222-induced malignancy in central nervous system embryonal tumors. Carcinogenesis 35:2164–2174
25. Lv XB, Liu L, Cheng C et al (2015) SUN2 exerts tumor suppressor functions by suppressing the Warburg effect in lung cancer. Sci Rep 5:17940
26. Matsumoto A, Hieda M, Yokoyama Y et al (2015) Global loss of a nuclear lamina component, lamin A/C, and LINC complex components SUN1, SUN2, and nesprin-2 in breast cancer. Cancer Med 4:1547–1557
27. Matsumoto A, Sakamoto C, Matsumori H et al (2016) Loss of the integral nuclear envelope protein SUN1 induces alteration of nucleoli. Nucleus 7:68–83
28. Kennedy C, Sebire K, de Kretser DM et al (2004) Human sperm associated antigen 4 (SPAG4) is a potential cancer marker. Cell Tissue Res 315:279–283
29. Shoji K, Murayama T, Mimura I et al (2013) Sperm-associated antigen 4, a novel hypoxia-inducible factor 1 target, regulates cytokinesis, and its expression correlates with the prognosis of renal cell carcinoma. Am J Pathol 182:2191–2203
30. Knaup KX, Monti J, Hackenbeck T et al (2014) Hypoxia regulates the sperm associated antigen 4 (SPAG4) via HIF, which is expressed in renal clear cell carcinoma and promotes migration and invasion in vitro. Mol Carcinog 53:970–978
31. Rajgor D, Mellad JA, Autore F et al (2012) Multiple novel nesprin-1 and nesprin-2 variants act as versatile tissue-specific intracellular scaffolds. PLoS One 7:e40098
32. Doherty JA, Rossing MA, Cushing-Haugen KL et al (2010) SRI/SYNE1 polymorphism and invasive epithelial ovarian cancer risk: an Ovarian Cancer Association Consortium study. Cancer Epidemiol Biomark Prev 19:245–250
33. Sjöblom T, Jones S, Wood LD et al (2006) The consensus coding sequences of human breast and colorectal cancers. Science 314:268–274
34. Tessema M, Willink R, Do K et al (2008) Promoter methylation of genes in and around the candidate lung cancer susceptibility locus 6q23-25. Cancer Res 68:1707–1714
35. Schuebel KE, Chen W, Cope L et al (2007) Comparing the DNA hypermethylome with gene mutations in human colorectal cancer. PLoS Genet 3:1709–1723
36. Marmé A, Zimmermann HP, Moldenhauer G et al (2008) Loss of Drop1 expression already at early tumor stages in a wide range of human carcinomas. Int J Cancer 123:2048–2056
37. Schoppmann SF, Vinatzer U, Popitsch N et al (2013) Novel clinically relevant genes in gastrointestinal stromal tumors identified by exome sequencing. Clin Cancer Res 19:5329–5339
38. Warren DT, Tajsic T, Mellad JA et al (2010) Novel nuclear nesprin-2 variants tether active extracellular signal-regulated MAPK1 and MAPK2 at promyelocytic leukemia protein nuclear bodies and act to regulate smooth muscle cell proliferation. J Biol Chem 285:1311–1320
39. Vinayagam A, Stelzl U, Foulle R et al (2011) A directed protein interaction network for investigating intracellular signal transduction. Sci Signal 4(189):rs8
40. Autore F, Shanahan CM, Zhang Q (2016) Identification and validation of putative nesprin variants. Methods Mol Biol 1411:211–220
41. Holt I, Duong NT, Zhang Q et al (2016) Specific localization of nesprin-1-α2, the short isoform of nesprin-1 with a KASH domain, in developing, fetal and regenerating muscle, using a new monoclonal antibody. BMC Cell Biol 17:26
42. Göb E, Meyer-Natus E, Benavente R et al (2011) Expression of individual mammalian Sun1 isoforms depends on the cell type. Commun Integr Biol 4:440–442
43. Razafsky D, Wirtz D, Hodzic D (2014) Nuclear envelope in nuclear positioning and cell migration. Adv Exp Med Biol 773:471–490

Chapter 22

Next-Generation Sequencing and Mutational Analysis: Implications for Genes Encoding LINC Complex Proteins

Peter L. Nagy and Howard J. Worman

Abstract

Targeted panel, whole exome, or whole genome DNA sequencing using next-generation sequencing (NGS) allows for extensive high-throughput investigation of molecular machines/systems such as the LINC complex. This includes the identification of genetic variants in humans that cause disease, as is the case for some genes encoding LINC complex proteins. The relatively low cost and high speed of the sequencing process results in large datasets at various stages of analysis and interpretation. For those not intimately familiar with the process, interpretation of the data might prove challenging. This review lays out the most important and most commonly used materials and methods of NGS. It also discusses data analysis and potential pitfalls one might encounter because of peculiarities of the laboratory methodology or data analysis pipelines.

Key words DNA sequencing, DNA sequence analysis, LINC complex, Mutation, Next-generation sequencing, Polymorphism, Sequence variants

1 Introduction

Mutations in genes encoding LINC complex proteins have been linked to human disease. Mutations in *SYNE1* encoding nesprin 1 cause autosomal recessive cerebellar ataxia, either pure or with associated features such as motor neuron involvement [1–4]. *SYNE1* mutations also cause autosomal recessive arthrogryposis multiplex congenita, a disorder characterized by congenital joint contractures and reduced fetal movements [5–7]. Linkage of *SYNE1* mutations to these autosomal recessive diseases is robust, as the pathogenic alleles clearly segregate with affected individuals in several families. Similarly, homozygosity for a protein truncating mutation in *SYNE4*, which encodes nesprin-4 expressed in the hair cells of the inner ear, has been shown to segregate with progressive high-frequency hearing loss in two families of Iraqi-Jewish ancestry [8]. An autosomal dominantly inherited point mutation in *SYNE2* leading to an amino acid substation in nesprin-2β1 has also been

Gregg G. Gundersen and Howard J. Worman (eds.), *The LINC Complex: Methods and Protocols*, Methods in Molecular Biology, vol. 1840, https://doi.org/10.1007/978-1-4939-8691-0_22, © Springer Science+Business Media, LLC, part of Springer Nature 2018

shown to segregate among first-degree relatives with an Emery-Dreifuss muscular dystrophy-like phenotype [9].

There have been other reports of mutations in genes encoding LINC complex proteins leading to disease where segregation within families has not been demonstrated. Autosomal dominant sequence variations in *SYNE1* have been reported in individuals with Emery-Dreifuss muscular dystrophy-like phenotypes [9–11]. Sequence variations in *SUN1* and *SUN2* have also been reported in individuals with Emery-Dreifuss muscular dystrophy-like phenotypes [12]. Functional abnormalities in cells expressing the protein variants and in the case of *Syne1* genetically modified mice suggest that these sequence variants could be pathogenic [9–15]. Furthermore, mutations in *EMD* and *LMNA*, respectively, encoding the LINC complex-associated proteins emerin and A-type lamins that bind to SUNs, also cause Emery-Dreifuss muscular dystrophy [16, 17].

The advent of next-generation sequencing (NGS) has allowed for the analysis of large panels of genes and even whole exomes in disease gene discovery research as well as in clinical practice [18, 19]. NGS using the predominant Illumina technology, which is a highly parallelized version of Sanger sequencing generating short (up to 300 bp) reads, involves library preparation, target capture, and the sequencing process proper followed by data processing and analysis. The initial step in library generation is DNA fragmentation. Unless whole genome sequencing is performed (in which case the genomic DNA library is directly sequenced), various PCR-based or hybridization-based methodologies are used to capture the genomic regions of interest to be sequenced. Subsequently, adaptors are ligated to the DNA fragments that allow attachment of the individual library molecules to a solid surface for amplification (cluster generation) and sequencing by synthesis through annealing of sequencing primers followed by template-dependent extension. Mixing of multiple samples in a shared sequencing process is made possible by individual specific molecular tags also introduced via the adaptor molecules. If cost is an important consideration, multiple adaptor-ligated individually tagged libraries can be used for capture, although that might compromise efficiency of the process. Cluster generation is a solid phase amplification step that results in hundreds of millions of clusters each consisting of thousands of clones of the individual library molecules densely scattered on a glass slide, called the flowcell. During the sequencing process, the fluorescent signal corresponding to the incorporating nucleotides in the individual clusters is electronically converted to hundreds of millions of individual DNA sequences corresponding to the DNA molecule clones in individual clusters. Sequence data must then be aligned so variants relative to the reference can be identified and evaluated for their potential role in pathogenesis.

As more researchers utilize this technology and as more data becomes available from its use in routine clinical practice, care must be taken in concluding that sequence variants cause disease. This applies to genes encoding LINC complex proteins [3, 7]. Determining the pathogenicity of sequence variants, especially without precise phenotypic descriptions and sequences of family members, requires review of the literature and available databases, careful consideration of population allele frequency, and variant data from other individuals or other families that have the same variant. Analysis programs can also be used to determine how a variant potentially affects protein structure or expression. Complementary analyses such as repeat expansion testing, methylation testing, transcriptome analysis, and copy number assessment can provide additional information. Ultimately, bench experimentation may be required to confirm that a rare variant uncovered by NGS is functionally disruptive and potentially pathogenic. For example, when whole exome sequencing identified a missense mutation in *LMNA* reported in the literature to abolish prelamin A processing in vitro, we performed cell biological experiments on the patient's fibroblast to confirm there was accumulation of the unprocessed pathogenic protein [20].

We review the materials and methods used for Illumina NGS and the identification of disease-causing variants. The scope of this chapter does not allow for a detailed description of the entire NGS process. Rather, we provide a general overview for non-geneticists who study molecular machines/systems such as the LINC complex and how alterations in their components may cause human disease.

2 Materials

Several different kits, reagents, and devices are commercially available for library preparation, target capture, and sequencing. With regard to sequencers, Illumina has emerged as the unequivocal leader. We describe some of the instruments, reagents, and kits we use for NGS.

2.1 Library Preparation, Target Capture, and NGS Equipment

1. Sonicator for DNA fragmentation. We recommend the Covaris S2 System Sonicator from VWR or its derivatives that can handle multiple samples simultaneously. Reproducible fragment size and size distribution of the library is essential requirement for NGS sequencing. The Covaris sonicator can perform this task in a highly reproducible manner without direct contact of the instrument with the sample. The multiplexing, automated versions can handle eight samples at a time, making this tedious and time-consuming process somewhat less of a challenge.
2. Hardware for DNA quantitation and library quality assessment. We use the Qubit Fluorometer from Invitrogen (Q32857) to obtain highly accurate measurements of DNA concentration before fragmentation. This is absolutely essential for NGS sequencing. Besides concentration, the size and size distribution of the sonicated DNA fragments is also

critical. This is best assessed using the Fragment Analyzer, the Advanced Analytical Quantitation, or the Bioanalyzer from Agilent. Quantitation of the library with the successfully attached adaptors is best done using real-time PCR with CFX96 Real-Time System, BioRad, or equivalent. Precise assessment of the quality and quantity of library generated is essential for efficient clustering and representative mixing of libraries if more than one sample is sequenced at a time. Within the recommended cluster density range, sequence yield correlates directly with the cluster density obtained on the flowcell.

3. Hardware for DNA capture. Standard PCR machines with heated lids are used for hybridization-based capture (Agilent Sureselect reagents or their equivalents).
4. Sequencers. Illumina is the leader in the manufacturing of NGS instruments. Most laboratories currently use the models HiSeq 1500 or 2500 and 3000 or 4000. The numbers refer to whether the machine can run a single (1500; 3000) or two flowcells (2500; 4000) at the same time or whether cluster generation on the flowcell is randomly spaced (1500, 2500) or patterned (3000, 4000). Larger laboratories use customized and serially linked versions of these instruments (HiSeq XTen and XFive) to sequence exclusively genomes. The names of these instruments reflect their price in millions of US dollars and are out of reach of most academic research, hospital-based, or even private laboratories. New technology on the horizon is the NovaSeq machine that is predicted to drive down the cost of whole genome sequencing during its production cycle within the next few years from approximately $1000 to $300 dollars. Other NGS machines, such as Life Technologies' Proton machine, use a different chemistry and a pH-based incorporation detection system that is less accurate around homopolymer regions [21]. This limits its usefulness for discovery of novel variants on a genomic scale. The platform provided by Pacific Biosystems allows for sequencing of individual DNA molecules over 10,000 base pairs but has a high error rate and has a limited throughput. Large genome centers use it as a corollary instrument, but it is rarely seen in the clinical molecular laboratory environment [22]. We therefore focus on generation and analysis of data obtained using the Illumina instrument product line.

2.2 Kits and Custom Reagents

1. SureSelect Exome V6 Capture Library from Agilent (5190-8865); one per sample.
2. TruSeq Custom Amplicon kit for 96 samples from Illumina (FC-130-1001); one per 96 sample.
3. SureSelectXT Reagent kit for 96 samples from Agilent (G9641B); one per sample.

4. Dynabeads MyOne Streptavidin T1 from Thermo Fisher (65602); one per sample.
5. Herculase II Fusion DNA Polymerase from Agilent (600677); one per sample.
6. Library Quant Kit (Illumina Universal) from Kapa Biosystems (KK4824); three per sample.
7. AgenCourt AMPure XP from Beckman Coulter (A63882); one bottle.
8. Qubit dsDNA Broad Range Assay Kit from Life Tech (Q32853); one kit.
9. Qubit dsDNA High Sensitivity assay kit Life Tech (Q32854); one kit.

Details regarding the use of these kits and reagents are provided in the manufacturers' instructions and Illumina library preparation and sequencing protocols. In the Subheading 3, we address some important considerations relating to their use.

2.3 Computational Hardware (Recommended Minimum)

1. Network attached storage (NAS) capable of storing 20 terabytes of data.
2. Linux virtual machine: 4 processors and 32 GB RAM, running CentOS.
3. Windows Workstation: two 12 core Intel E5-2690v3 processors and 128 GB RAM.
4. Windows Server 2012: R2 Standard 64-bit.

2.4 Computational Software

1. CentOS Linux operating system.
2. NextGENe v2.4.02 from Softgenetics.
3. bcl2fastQ Conversion Software v1.8.4 from Illumina.
4. Variant annotation and filtering software: Golden Helix SNP or Variation Suite.
5. Genome MaNaGer™; current availability is limited to data reanalysis by MNG Laboratories (fee for service).

3 Methods

3.1 Library Preparation and Selection of Targeted Regions

The cost and computational complexity of whole genome sequencing makes it impractical for most laboratories. The alternative is to enrich and select genomic regions to sequence. Methodology such as targeted PCR or hybridization-based capture can be used to select relatively small targeted regions, such as specific genes, or more expansive regions, such as whole exomes. Selection of the best approach is based on the scenario, test volume, laboratory setup, and affordability.

Long-range PCR amplification is a necessity for thorough assessment of ambiguously mapping regions of the genome. Primers flanking ambiguously mapping regions should be used to avoid artifacts due to divergent variation in highly similar genomic regions. A list of such problematic regions can be found in Mandelker et al. [23]. An example is mitochondrial genome sequencing, which is performed optimally on a single amplicon of the mitochondrial genome, removing the possibility of artifacts due to sequencing of mitochondrial pseudogenes located in the nuclear genome. Long-range PCR is not easily scalable, and thus most laboratories resign to the increased false negative and false positive rates in these regions due to ambiguous mapping and unpredictable representation percentage of specific alleles. This is a serious issue, since these sequences represent about 2% of all exomic coding regions.

Multiplex PCR approaches are best suited for sequencing of relatively small (less than a megabase) noncontiguous genomic regions such as specific exons of genes. This approach allows fast, high-volume testing even with limited starting material available. Targeted screens for carriers of mutations in a specific gene are a good application for this method. We have found TruSeq Amplicon reagents by Illumina to be well suited for most applications. Limitation of this method is that it cannot be used to identify large deletions, even if the precise position of the deletion is known. Since the amplified regions from a specific target region are all the same size, the experiments should be designed to take into account the danger of duplicate reads that arise if a low number of template DNA molecules are used as starting material.

Hybridization-based selection of regions of interest is recommended if the region to be sequenced is greater than one megabase, although it also works for smaller regions. Kits containing oligonucleotide baits (RNA or DNA) synthesized using various technologies are commercially available, such as the Agilent SureSelectXT or equivalents from other manufacturers such as Illumina or IDT. Some allow or encourage capturing multiple libraries simultaneously with a single capture reagent. The smaller the region of interest, the more one can save on sequencing cost using a single capture reagent for a large number of combined libraries. Using individual capture for each sample, however, allows greater reproducibility between experiments and thus allows obtaining copy number information from the sequencing data with great reliability. This approach requires at least 100 ng DNA to perform. Agilent Sureselect reagents perform well for both custom and off-the-shelf (e.g., exome) panels. The flexibility of this platform is important if the panel of targeted genes changes over time. Since hybridization capture uses randomly fragmented DNA as an input, duplicate reads are easier to identify. The ratio of forward and reverse reads over specific nucleotides is also much better balanced than with multiplexed PCR-based methods.

Transcriptome analysis can be thought of as another approach to focus on a subset of genomic regions without the need to specifically amplify or capture by hybridization the regions of interest. The cellular transcriptional machinery essentially does the work for you. All that needs to be done is removal of the high abundance structural RNAs using a hybridization-based approach. Transcriptomes provide an integrated output of the actual living state of the cell/tissue which could be very difficult or impossible to establish from analysis of even whole genome sequencing. Transcriptome analysis is also invaluable to assess the effects of splice site variants and even regulatory mutations that are outside the scope of most capture-based targeted amplification schemes. This method is essential in cancer genomics, and in that case, generally the tumor is available for "tissue-specific" transcript evaluation.

3.2 DNA Sequencing and Data Acquisition

Illumina technology is based on synthesis of a new DNA molecule complementary to a template strand. The main difference from the Sanger method is that the sequencing reaction is massively parallelized, meaning that results can be recorded from hundreds of millions of template DNA molecules simultaneously. Another important difference is that this technology generates relatively short (100–150 base pair) paired-end reads compared to the 500–1000 base pair reads that can be generated by traditional Sanger methodology. This is a significant limitation when it comes to precise mapping of variants in non-unique sequences within the genome. In addition, identification and sizing of repeat expansions are also limited to a size of approximately 100 base pairs. However, detection of variants in a subset of the DNA molecules interrogated is more sensitive than what can be achieved using Sanger methodology. This is because each original interrogated DNA molecule generates an independent sequence and, depending on the depth of sequencing, many dozens or hundreds of molecules are investigated for each region of interest. Detection of a mutation at less than 1% representation, as can occur in patients with mitochondrial heteroplasmy, chimerism, or mosaicism, requires additional indexing [24]. Following the initial couple of cycles, the sequencer gives a quantity and quality estimate of the reads that will be obtained from the run. This is an important step that allows decision to be made whether the run should be continued or aborted potentially preventing completion of a very expensive failed experiment.

3.3 Raw Data Quality Assessment

The overall quality of the sequences obtained is largely dependent on four components: (1) the expertise of the technologist, (2) the quality and quantity of the starting material, (3) the quality of the capture, and (4) the sequencing reagents and the reliability and precision of the sequencing instrument itself. The most reliable way to assess the performance of the instrument is through the use of an internal control library generated of the phage Phi X 174 (PhiX Control v3 (catalog # FC-110-3001)). When this control library is mixed in with the sample(s), it will yield sequence and sequence quality information independent of the quality of the sample library. The machine aligns the phage-derived sequence to an internal reference, providing information about the error rate associated with the run and the overall quality of the sequences obtained. An average Phred quality score of ≥Q30 for ≥90% of the reads indicates a successful run. Setting a lower cutoff for quality depends on particular circumstances and specific limitations in sample quality and quantity. If there is a significant difference between the quality of the reads obtained from the Phi X 174 control and the sample library, there is a set of troubleshooting steps in the Illumina instruments' users' manuals to identify the source of the problem. The projected total yield (in gigabases) from a run is one metric provided by the machine that allows predictions about the number of reads that the run will generate. This is variable depending on the specific instrument used and the success of the clustering but allows generation of a very good estimate about the depth of coverage to be expected for the specific region of interest. The Illumina instruments users' manuals also provide a plethora of other metrics that allow troubleshooting of the sequence generation process.

3.4 Demultiplexing

Most NGS runs, depending on the size of the targeted regions, contain multiple samples in a single flowcell lane. Sample-specific reads are sorted from the mixed data using demultiplexing. Demultiplexing is a process whereby the sample specific indexes (short DNA sequences, usually 6-mers), introduced into the sample during the library preparation process, are read and used for sorting the reads into individual bins corresponding to the samples sequenced. The indexes are read in an independent priming reaction (Read2) and are kept in association with the forward (Read1) and reverse (Read3) reads obtained from the same cluster. This association allows assignment of the reads from specific clusters to specific samples. Selection of compatible indices for a specific run is a crucial process that has to be carefully supervised to avoid misassignment of reads to the wrong sample (or to no sample at all). In some cases, 96 or more indices can be used in the same batch. However, for most large panels and exomes, the number is more likely to be up to 20 or up to 6 samples, respectively. At the end of the process, "FastQ" files are generated which contain the individual sequences obtained from a sample with a quality score (Q1-40) assigned to each nucleotide.

3.5 Sequence Alignment and Mutation Calling

Sequence alignment and mutation calling pipelines such the Genome Analysis Toolkit (GATK; Broad Institute) have been described in detail and also have recommendations available (Best Practices) to guide the user in setting them up either in the laboratory or in the "cloud" [25–27]. Setting up these pipelines requires significant computational hardware and an experienced bioinformatics staff, which is often beyond the resources of smaller laboratories [28]. For smaller laboratories, we recommend software packages will well developed graphical user interfaces such as NextGENe produced by Softgenetics. These can be run on PCs and sometimes on MACs and provide a primary data analysis capability with some basic annotations. Knowing exactly what the software is doing or not doing, as well as being able to adapt it to the task at hand, is of huge importance. From this perspective, open-source software is preferred to software packages with inaccessible, unmodifiable code. That said, the algorithms used by most pipelines, open-source or commercial, use Burrows-Wheeler aligner and the Genome Analysis Toolkit pipeline as described in the Best Practices guidelines for mutation calling [26, 27]. The input into these pipelines is FastQ files (individual reads with quality scores for each base called) generated by the sequencer. The output of the alignment software is a Binary Alignment/Map (BAM) file that contains the genomic coordinates corresponding to the beginning and end of each read. The information relating to the variants called is summarized in a variant call format (vcf) file ordered according to specific chromosomal positions. Information in the BAM files also allows for the visualization of the aligned reads and the variants called and can be used for assessment of coverage depth throughout the region of interest. These files can be conveniently viewed using the integrated genome viewer (reference) generated by the Broad Institute.

It is important to understand the significance of mean or average coverage depth and what that means for the specificity and sensitivity of the testing. These are the metrics generally provided on clinical reports of panels/whole exome or genome sequencing and can be used to compare the products of various laboratories. Average coverage in itself provides no information about poorly covered regions and their size. We recommend the service provider gives three coverage statistics: the percent of region of interest (ROI) covered 5-, 10-, and 30-fold. Thirtyfold coverage is generally considered the desirable minimum coverage to avoid false positive and false negative calls for constitutional samples for unique genomic regions. Tenfold coverage is generally considered to be sufficient to provide a good indication for the absence of a mutation at a specific position; if the variant does not show up even once out of ten reads, the sample is almost certainly wild type at that site. Fivefold coverage is helpful to pick up homozygous or apparently homozygous variants in poorly covered regions, thus decreasing the false negative rate. In our experience, all variants that make it into a report need to be confirmed using an alternative method (such as targeted PCR). This is especially true for variants with less than 30-fold coverage. However, the most significant cause of false negative results is due to limitations in the size and content of the ROI. One has to verify the completeness of the inclusion of regions with known pathogenic variants in the capture reagent used. Not all ROIs/capture reagents are created equal. An exome capture reagent from one company can represent half of the ROI of the exome capture reagent of another company, and they might both omit a significant percentage of hard to capture regions with pathogenic variants.

3.6 Data Interpretation and Reporting

The vcf files listing the position and nature of the identified variants in a tab delimited text file format generated from most NGS platforms are too large and too poorly annotated for human interpretation. When faced with the hundreds of thousands to millions of variant calls in partial, whole exome or genome datasets, respectively, interpretation cannot be done without a database providing variant annotation and filtering. There are many publicly available databases with indications of pathogenicity of previously identified variants, such as ClinVar (National Center for Biotechnology Information), HGMD (Institute of Medical Genetics in Cardiff), OMIM (Johns Hopkins), and COSMIC (Sanger Institute). Others have data on allele frequency in the general population, such as 1000 Genomes Project (The International Genome Sample Resource), Exome Variant Server (University of Washington), ExAC (Broad Institute), and gnomAD (Broad Institute). However, having an internal dataset reference, generated with the same methodology and thus containing the same set of artifacts, is extremely helpful to provide accurate classification and reporting of variants [29]. Many off-the-shelf software packages for variant interpretation emphasize phenotype-based filtering as one of the early steps in the process. In our experience, this is not recommended, since medical and family histories for the patients are often scarce or nonexistent. Therefore, filtering should use phenotype information only as one of the last steps to avoid discarding unexpected pathogenic variants. This may be just as important for targeted panels as for interpretation of a whole exome. For this reason, most large centers have developed their own analysis pipeline/database that can be updated regularly and allow the filtering algorithm to be controlled depending on the particular dataset. The list of variants requiring expert human review can be significantly shortened if informative family members are sequenced along with the proband. Most commonly these are the parents, but if they are not available, having healthy unaffected first-degree relatives and affected distant relatives can be of great help to predict the pathogenicity of a given rare variant not previously described in public databases.

The MNG Genome MaNaGer filtering strategy for variants is summarized in Fig. 1. Briefly, it classifies variants first into four categories using reference and reportable range filters:

Category 1: variants with a clear pathogenic or likely pathogenic annotation in ClinVar that are present in the patient

Category 2: variants with a clear pathogenic or likely pathogenic annotation in ClinVar that are not covered in the patient

Category 3: all variants in disease-associated genes in ClinVar and other scientific literature (all category 1 variants are also present in this list)

Category 4: all variants in genes that have not been associated with disease yet

With the exception of the first two categories, these files are too large to be reviewed without further filtering. In a following step, we therefore use a frequency filter to generate four additional categories that contain the variants in categories 1–4 that are not present in any known population at greater than 1% allele frequency:

Category 5: rare known pathogenic variants

Category 6: rare known pathogenic variants not covered (filled in using targeted method such as PCR followed by Sanger sequencing)

Category 7: rare variants in known disease-associated genes

Category 8: rare variants in non-disease associated genes

The lists in categories 5–8 are more manageable for human review. We have developed a specific order to review them based on information obtained from individual and family histories as well as functional predictions (the most important five subcategories are listed in Fig. 1):

1. Homozygous variants
2. Disruptive variants; splice site, frameshift, stop codon variants
3. Variants unique to proband sample
4. Variants unique to case (trio)
5. De novo variants; not present in parents
6. Compound heterozygous changes
7. Variants that cause an amino acid change (missense)

Variants that after review (preferably) by three independent reviewers are judged by any one of the reviewers as of interest are further discussed in a grand rounds-like forum to determine if they should be further considered or reported based on American College of Medical Genetics and Genomics criteria.

The American College of Medical Genetics and Genomics has developed specific guidelines for the classification of variants as "pathogenic," "likely pathogenic," "unknown significance," "likely benign," and benign [30]. These guidelines are based on medical knowledge about disease frequencies, modes of inheritance, in silico prediction of the disruptive nature (nonsense, frameshift, splice site, predicted damaging/disruptive) of the

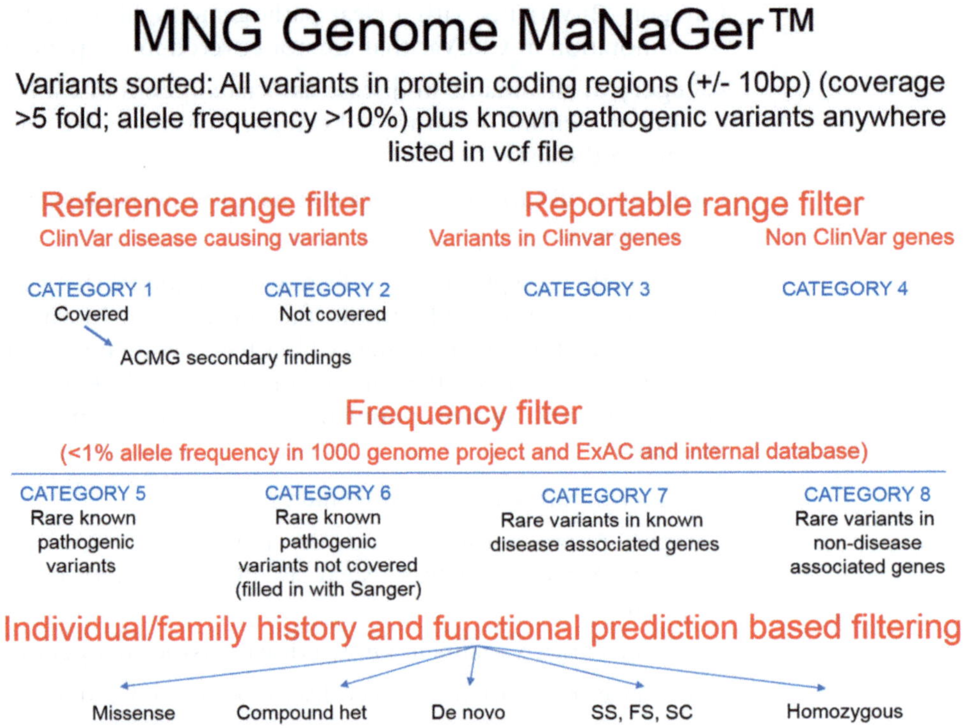

Fig. 1 The MNG Genome MaNaGer strategy for variant filtering and annotation. *See* text for details. American College of Medical Genetics and Genomics (ACMG) secondary findings are those unrelated to the indication for ordering the sequencing but of medical value for patient care. Other abbreviations used in figure: *het* heterozygous, *SS* splice site, *FS* frameshift, *SC* stop codon

variants as well as data available about specific genomic regions where known pathogenic variants cluster. The guidelines also take into account the presence or absence of the variants in healthy or affected parents (whether inherited or de novo). We use these guidelines in our daily clinical routine conservatively, recognizing that there is a danger to use many weak lines of evidence for declaring a variant pathogenic or likely pathogenic [30, 31].

Reporting variants in the clinical setting carries great responsibility. The same is true for publishing them in the biomedical literature. The public, physicians and non-physician scientists vary greatly in their ability to rationally evaluate the significance of a reported variant. What is reported to a physician and patient or in the literature, even if properly qualified as a variant of unknown significance, may be perceived as a cause of the disease. This becomes a more significant problem as the number of the genes sequenced increases, while knowledge about the genes' functions lags behind. In such instances, it is critical to obtain parental DNA to assess whether the mutation is present in unaffected parents or it arose de novo. Every clinical report should contain a sentence stating that such testing needs to be performed before the significance of the reported find can be properly established. This is even true in situations when variants carry a pathogenic or likely pathogenic label based on outdated or insufficient evidence. Notably, such information is lacking for some rare variants of genes encoding LINC complex proteins that have been claimed to cause disease [10–13].

Although estimates vary, greater than 10% of variants labeled as pathogenic in ClinVar are rare ethnicity-specific variants with no clinical significance. There have been many efforts to increase data sharing among laboratories to assist with correct interpretation but much remains to be done. [32]. Until a time when the number of sequenced genomes allows a better phenotype-genotype correlation, sequencing trios (proband and parents), rather than probands alone, should be used for clinical and research studies. Reporting should be performed cautiously and in a conservative manner, both as to the number of variants reported and assessment of their clinical significance.

Another important concept generally applied throughout medicine but somewhat neglected in clinical genetics is comprehensive rather than method-specific testing. Many successful NGS companies emphasize the low cost of their platform but fail to emphasize that NGS—even whole exome sequencing—can only provide answers to the patient's clinical problem in about 30% of cases. Combining tests for other genetic alterations or their manifestations, such as repeat expansions, methylation/imprinting disorders, copy number changes, chromosomal rearrangements, and transcript processing defects, and offering a synthesis of their results in a single report is something that needs to become the norm rather than the exception. Finally, despite comprehensive genetic testing, bench research using cultured cells or model organisms/animals is sometimes necessary to determine if a newly discovered rare variant actually affects protein function or expression that can underlie pathology.

3.7 Applying Information from Basic Research to Clinical Variant Interpretation

Unfortunately, only a fraction of variants discovered in clinical practice make it to basic scientists, and there is often an unnecessary delay in the transfer and review of information. This handicap can be overcome by forming strong collaborations between clinical molecular diagnosticians and basic scientists to develop and apply predictive screening tools and functional verification methodology for newly discovered variants. Of utmost importance is establishment of a system that facilitates or automates mapping human phenotype-associated mutations onto the functional models of not only individual genes and proteins but also multi-molecular complexes, such as the LINC complex. Clearly there are some encouraging examples of such efforts, but we believe this area requires much more attention and resources [33].

Another prerequisite for discovering new disease-gene associations is developing better structural and functional interaction networks connecting genes and encoded proteins. A systems approach focusing on disruption of molecular assemblies and pathways rather than individual proteins is a highly promising yet not fully exploited area of modern molecular diagnostic practice. Incorporating transcriptome analysis with both transcript level and processing assessment into routine genetic testing is already a reality in some laboratories. Furthermore, combining genetic data with proteomic data to identify protein levels and modifications using mass spectrometry might further enhance an understanding of the metabolic homeostasis of cells and tissues [34]. This is particularly true for molecular machines/systems such as the LINC complex, in which different protein isoforms may be expressed at different levels in various cell types or tissues.

4 Conclusions

The field of genomic scale molecular diagnosis is rapidly expanding. Genomic datasets from individuals of diverse genetic backgrounds are accumulating at an ever-increasing rate. Analysis and processing of these data and deposition of the frequency and effect prediction information for individual variants gained from it will make an increasing portion of data analysis amenable to automation. That said, there will continue to be a need for basic scientists giving "personalized" attention to newly identified variants, including those in protein components of the LINC complex.

Acknowledgments

H.J.W. is supported by National Institute of Arthritis and Musculoskeletal and Skin Diseases of the National Institutes of Health under award numbers R01AR048997 and R01AR068636. The content is solely the responsibility of the authors and does not necessarily represent the official views of the National Institutes of Health.

Disclosures

P.L.N. is an employee of MNG Laboratories and owns equity in the company. H.J.W. serves on the scientific advisory board of MNG Laboratories and receives fees for this service.

References

1. Gros-Louis F, Dupré N, Dion P et al (2007) Mutations in SYNE1 lead to a newly discovered form of autosomal recessive cerebellar ataxia. Nat Genet 39:80–85
2. Noreau A, Bourassa CV, Szuto A et al (2013) SYNE1 mutations in autosomal recessive cerebellar ataxia. JAMA Neurol 70:1296–1231
3. Synofzik M, Smets K, Mallaret M et al (2016) SYNE1 ataxia is a common recessive ataxia with major non-cerebellar features: a large multi-centre study. Brain 2016(139):1378–1393
4. Mademan I, Harmuth F, Giordano I et al (2016) Multisystemic SYNE1 ataxia: confirming the high frequency and extending the mutational and phenotypic spectrum. Brain 139:e46
5. Attali R, Warwar N, Israel A et al (2009) Mutation of SYNE-1, encoding an essential component of the nuclear lamina, is responsible for autosomal recessive arthrogryposis. Hum Mol Genet 18:3462–3469
6. Laquérriere A, Maluenda J, Camus A et al (2014) Mutations in CNTNAP1 and ADCY6 are responsible for severe arthrogryposis multiplex congenita with axoglial defects. Hum Mol Genet 23:2279–2289
7. Baumann M, Steichen-Gersdorf E, Krabichler B et al (2017) Homozygous SYNE1 mutation causes congenital onset of muscular weakness with distal arthrogryposis: a genotype-phenotype correlation. Eur J Hum Genet 25:262–266
8. Horn HF, Brownstein Z, Lenz DR et al (2013) The LINC complex is essential for hearing. J Clin Invest 123:740–750
9. Zhang Q, Bethmann C, Worth NF et al (2007) Nesprin-1 and -2 are involved in the pathogenesis of Emery Dreifuss muscular dystrophy and are critical for nuclear envelope integrity. Hum Mol Genet 16:2816–2833
10. Puckelwartz MJ, Kessler EJ, Kim G (2010) Nesprin-1 mutations in human and murine cardiomyopathy. J Mol Cell Cardiol 48:600–608
11. Zhou C, Li C, Zhou B et al (2017) Novel nesprin-1 mutations associated with dilated cardiomyopathy cause nuclear envelope disruption and defects in myogenesis. Hum Mol Genet 26:2258–2276
12. Meinke P, Mattioli E, Haque F et al (2014) Muscular dystrophy-associated SUN1 and SUN2 variants disrupt nuclear-cytoskeletal connections and myonuclear organization. PLoS Genet 10:e1004605
13. Puckelwartz MJ, Kessler E, Zhang Y et al (2009) Disruption of nesprin-1 produces an Emery Dreifuss muscular dystrophy-like phenotype in mice. Disruption of nesprin-1 produces an Emery Dreifuss muscular dystrophy-like phenotype in mice. Hum Mol Genet 18:607–620
14. Zhang J, Felder A, Liu Y et al (2010) Nesprin 1 is critical for nuclear positioning and anchorage. Hum Mol Genet 19:329–341
15. Stroud MJ, Feng W, Zhang J et al (2017) Nesprin-1α2 is essential for mouse postnatal viability and nuclear positioning in skeletal muscle. J Cell Biol 216:1915–1924
16. Bione S, Maestrini E, Rivella S et al (1994) Identification of a novel X-linked gene responsible for Emery-Dreifuss muscular dystrophy. Nat Genet 8:323–327
17. Bonne G, Di Barletta MR, Varnous S et al (1999) Mutations in the gene encoding lamin A/C cause autosomal dominant Emery-Dreifuss muscular dystrophy. Nat Genet 21:285–288
18. Bamshad MJ, Ng SB, Bigham AW et al (2011) Exome sequencing as a tool for Mendelian disease gene discovery. Nat Rev Genet 12:745–755
19. Nagy PL, Mansukhani M (2015) The role of clinical genomic testing in diagnosis and discovery of pathogenic mutations. Expert Rev Mol Diagn 15:1101–1105
20. Wang Y, Lichter-Konecki U, Anyane-Yeboa K et al (2016) A mutation abolishing the ZMPSTE24 cleavage site in prelamin A causes a progeroid disorder. J Cell Sci 129:1975–1980
21. Boland JF, Chung CC, Roberson D et al (2013) The new sequencer on the block: comparison of Life Technology's Proton sequencer to an Illumina HiSeq for whole-exome sequencing. Hum Genet 132:1153–1163
22. English AC, Salerno WJ, Hampton OA et al (2015) Assessing structural variation in a personal genome-towards a human reference diploid genome. BMC Genomics 16:286
23. Mandelker D, Amr SS, Pugh T et al (2014) Comprehensive diagnostic testing for stereocilin: an approach for analyzing medically important genes with high homology. J Mol Diagn 16:639–647

24. Kennedy SR, Schmitt MW, Fox EJ et al (2014) Detecting ultralow-frequency mutations by duplex sequencing. Nature Protoc 9:2586–5606
25. McKenna A, Hanna M, Banks E et al (2010) The Genome Analysis Toolkit: a MapReduce framework for analyzing next-generation DNA sequencing data. Genome Res 20:1297–1303
26. DePristo MA, Banks E, Poplin R et al (2011) A framework for variation discovery and genotyping using next-generation DNA sequencing data. Nat Genet 43:491–498
27. Van der Auwera GA, Carneiro MO et al (2013) From FastQ data to high confidence variant calls: the Genome Analysis Toolkit best practices pipeline. Curr Protoc Bioinformatics 43:11.10.1–11.1033
28. Tsai EA, Shakbatyan R, Evans J et al (2016) Bioinformatics workflow for clinical whole genome sequencing at Partners HealthCare Personalized Medicine. J Pers Med 6:12
29. Wang K, Li M, Hakonarson H (2010) ANNOVAR: functional annotation of genetic variants from high-throughput sequencing data. Nucleic Acids Res 38:e164
30. Richards S, Aziz N, Bale S et al (2015) Standards and guidelines for the interpretation of sequence variants: a joint consensus recommendation of the American College of Medical Genetics and Genomics and the Association for Molecular Pathology. Genet Med 17:405–423
31. Green RC, Berg JS, Grody WW et al (2013) ACMG recommendations for reporting of incidental findings in clinical exome and genome sequencing. Genet Med 15:565–574
32. Topol EJ (2015) The big medical data miss: challenges in establishing an open medical resource. Nat Rev Genet 16:253–254
33. Luu TD, Rusu AM, Walter V et al (2012) MSV3d: database of human MisSense Variants mapped to 3D protein structure. Database (Oxford) 2012:bas018
34. Alfaro JA, Sinha A, Kislinger T et al (2014) Onco-proteogenomics: cancer proteomics joins forces with genomics. Nat Methods 11:1107–1113

Index

Gregg G. Gundersen and Howard J. Worman (eds.), *The LINC Complex: Methods and Protocols*, Methods in Molecular Biology, vol. 1840, https://doi.org/10.1007/978-1-4939-8691-0, © Springer Science+Business Media, LLC, part of Springer Nature 2018

MIX
Papier aus verantwortungsvollen Quellen
Paper from responsible sources
FSC® C105338

If you have any concerns about our products,
you can contact us on
ProductSafety@springernature.com

In case Publisher is established outside the EU,
the EU authorized representative is:
Springer Nature Customer Service Center GmbH
Europaplatz 3, 69115 Heidelberg, Germany

Printed by Libri Plureos GmbH
in Hamburg, Germany